Foundations of
Computational Linguistics

Springer
Berlin
Heidelberg
New York
Barcelona
Hong Kong
London
Milan
Paris
Singapore
Tokyo

Roland Hausser

Foundations of Computational Linguistics

Man-Machine Communication in Natural Language

Springer

Roland Hausser

Professor of Computational Linguistics
Friedrich Alexander University Erlangen Nürnberg
Bismarckstr. 12, D-91054 Erlangen, Germany
rrh@linguistik.uni-erlangen.de

ISBN 3-540-66015-1 Springer-Verlag Berlin Heidelberg New York

ACM Computing Classification (1998): J.5, I.2.7, I.5.4, H.3.1, F.4.2-3, F.1.3

Library of Congress Cataloging-in-Publication Data
Hausser, Roland R.
Foundations of computational linguistics: man-machine
communication in natural language/Roland Hausser.
p. cm.
Includes bibliographical references and indexes.
ISBN 3-540-66015-1 (hardcover: alk. paper)
1. Computational linguistics. 2. Robotics. I. Title.
P98.H35 1999   99-32062
410' .285--dc21   CIP

© Springer-Verlag Berlin Heidelberg 1999
Printed in Italy

Cover design: Künkel + Lopka, Heidelberg
Cover picture: The ladder of ascent and descent, Ramon Lull, Valencia 1512
Typesetting: Computer to film from author´s data
Printing and binding: Legoprint, Italy
Printed on acid-free paper   SPIN 10730217   45/3142PS 5 4 3 2 1 0

Preface

The central task of a future-oriented computational linguistics is the development of cognitive machines which humans can freely talk with in their respective natural language. In the long run, this task will ensure the development of a functional theory of language, an objective method of verification, and a wide range of applications.

Natural communication requires not only verbal processing, but also non-verbal perception and action. Therefore the content of this textbook is organized as a theory of language for the construction of talking robots. The main topic is the *mechanism of natural language communication* in both, the speaker and the hearer.

The content is divided into the following parts:

    I.   Theory of Language
    II.   Theory of Grammar
    III. Morphology and Syntax
    IV. Semantics and Pragmatics

Each part consists of 6 chapters. Each of the 24 chapters consists of 5 sections. Altogether 772 exercises help reviewing key ideas and important problems.

Part I begins with current applications of computational linguistics. Then it describes a new theory of language the functioning of which is illustrated by the robot CURIOUS. This theory is referred to with the acronym SLIM, which stands for *Surface compositional Linear Internal Matching*. It includes a cognitive foundation of semantic primitives, a theory of signs, a structural delineation of the components syntax, semantics, and pragmatics, as well as their functional integration in the speaker's utterance and the hearer's interpretation. The presentation refers to other contemporary theories of language, especially those of Chomsky and Grice, as well as to the classic theories of Frege, Peirce, de Saussure, Bühler, and Shannon & Weaver, explaining their formal and methodological foundations as well as their historical background and motivations.

Part II presents the theory of *formal grammar* and its methodological, mathematical, and computational role in the description of natural languages. A description of categorial grammar and phrase structure grammar is combined with an introduction to the basic notions and linguistic motivation of generative grammar. Further topics are the declarative vs. procedural aspects of parsing and generation, type transparency as well as the relation between formalisms and complexity classes. It is shown that the principle of possible *substitutions* causes empirical and mathematical problems for the description of natural language. Alternatively the principle of possible *continuations* is formalized as LA-grammar. LA stands for the left-associative derivation order which models the time-linear nature of language. Applications of LA-grammar to relevant artificial languages show that its hierarchy of formal languages is orthogonal to that of phrase structure grammar. Within the LA-hierarchy, natural language is in the lowest complexity class, namely the class of C1-languages which parse in linear time.

Part III describes the *morphology* and *syntax* of natural language. A general description of the notions word, word form, morpheme, and allomorph, the morphological processes of inflection, derivation, and composition, as well as the different possible methods of automatic word form recognition is followed by the morphological analysis of English within the framework of LA-grammar. Then the syntactic principles of valency, agreement, and word order are explained within the left-associative approach. LA-grammars for English and German are developed by systematically extending a small initial system to handle more and more constructions such as the fixed vs. free word order of English and German, respectively, the structure of complex noun phrases and complex verbs, interrogatives, subordinate clauses, etc. These analyses are presented in the form of explicit grammars and derivations.

Part IV describes the *semantics* and *pragmatics* of natural language. The general description of language interpretation begins by comparing three different types of semantics, namely those of logical languages, programming languages, and natural languages. Based on Tarski's foundation of logical semantics and his reconstruction of the Epimenides paradox, the possibility of applying logical semantics to natural language is investigated. Alternative analyses of intensional contexts, propositional attitudes, and the phenomenon of vagueness illustrate that different types of semantics are based on different ontologies which greatly influence the empirical results. It is shown how a semantic interpretation may cause an increase in complexity and how this is to be avoided within the SLIM theory of language. The last two Chapters 23 and 24 analyze the interpretation by the hearer and the conceptualization by the speaker as a time-linear navigation through a database called *word bank*. A word bank allows the storage of arbitrary propositions and is implemented as an extension of a classic (i.e. record-based) network database. The autonomous navigation through a word bank is controlled by the explicit rules of suitable LA-grammars.

As supplementary reading the *Survey of the State of the Art in Human Language Technology*, Ron Cole (ed.) 1998 is recommended. It contains about 90 contributions by different specialists giving detailed snapshots of their research in language theory and technology.

Erlangen, June 1999   Roland Hausser

Table of Contents

Part II. Theory of Grammar

Introduction

I. BASIC GOAL OF COMPUTATIONAL LINGUISTICS

Transmitting information by means of a natural language like Chinese, English, or German is a real and well-structured procedure. This becomes evident when we attempt to communicate with people who speak a foreign language. Even if the information we want to convey is completely clear to us, we will not be understood by our hearers if we fail to use their language adequately.

The goal of computational linguistics is to reproduce the natural transmission of information by modeling the speaker's production and the hearer's interpretation on a suitable type of computer. This amounts to the construction of autonomous cognitive machines (robots) which can communicate freely in natural language.

The development of speaking robots is not a matter of fiction, but a real scientific task. Remarkably, however, theories of language have so far avoided a functional modeling of the natural communication mechanism, concentrating instead on peripheral aspects such as methodology (behaviorism), innate ideas (nativism), and scientific truth (model theory).

II. TURING TEST

The task of modeling the mechanism of natural communication on the computer was described in 1950 by ALAN TURING (1912–1954) in the form of an 'imitation game' known today as the Turing test. In this game, a human interrogator is asked to question a male and a female partner in another room via a teleprinter in order to determine which answer was given by the man and which by the woman. It is counted how often the interrogator classifies his communication partners correctly and how often (s)he is fooled by them.

Subsequently one of the two human partners is replaced by a computer. The computer passes the Turing test, if it simulates the man or the woman which it replaced so well that the guesses of the interrogator are just as often right and wrong as with the previous set of partners. In this way Turing wanted to replace the question "Can machines think?" by the question "Are there imaginable digital computers which would do well in the imitation game?"

III. ELIZA PROGRAM

In its original intention, the Turing test requires the construction of an artificial cognitive agent with a verbal behavior so natural that it cannot be distinguished from that of a human native speaker. This presupposes complete coverage of the language data and of the communicative functions in real time. At the same time, the test tries to avoid all aspects not directly involved in verbal behavior.[1]

However, the Turing test does not specify which cognitive structure the artificial agent should have in order to succeed in the imitation game. For this reason, it is possible to misinterpret the aim of the Turing test as fooling the interrogator rather than providing a functional model of communication on the computer. This was shown by the Eliza program of Weizenbaum 1965.

The Eliza program simulates a psychiatrist encouraging the human interrogator to talk more and more about him- or herself. The structure of Eliza is based on sentence templates into which certain words used by the interrogator, now in the role of a patient, are inserted. For example, if the interrogator mentions the word mother, Eliza uses the template Tell me more about your ___ to generate the sentence Tell me more about your mother.

Because of the way in which Eliza works we know that Eliza has no understanding of the dialog with the interrogator/patient. Thus, the construction of Eliza is not a model of communication. If we regard the dialog between Eliza and the interrogator/patient as a modified Turing test, however, the Eliza program is successful insofar as the interrogator/patient *feels* him- or herself understood and therefore does not distinguish between a human and an artificial communication partner in the role of the psychiatrist.

The purpose of computational linguistics is the real modeling of natural language communication, and not a mimicry based on exploiting particular restrictions of a specific dialog situation, as in the Eliza program. Thus, computational linguistics must (i) explain the mechanism of natural communication theoretically and (ii) verify this explanation in practice. The latter is done in terms of a complete and general implementation which must prove its functioning in everyday communication rather than the Turing test.

IV. MODELING NATURAL COMMUNICATION

Designing a talking robot provides an excellent occasion for systematically developing the basic notions as well as the philosophical, mathematical, grammatical, methodological, and programming aspects of computational linguistics. This is because modeling the mechanism of natural communication requires

---

[1] As an example of such an aspect, A. Turing 1950, p. 434, mentions the artificial recreation of human skin.

- a theory of language which explains the natural transfer of information in a way that is functionally coherent, mathematically explicit, and computationally efficient,
- a description of language data which is empirically complete for all components of this theory of language, i.e. the lexicon, the morphology, the syntax, and the semantics, as well as the pragmatics and the representation of the internal context,
- a degree of precision in the description of these components which is sufficient for computation,

Fulfilling these requirements will take hard, systematic, goal-oriented work, but it will be worth the effort.

For theory development, the construction of talking robots is of interest because an electronically implemented model of communication may be tested both, externally in terms of the verbal behavior observed, and internally via direct access to its cognitive states. The work towards realizing unrestricted man-man machine communication is facilitated by that the functional model may be developed incrementally, beginning with a simplified, but fully general system to which additional functions as well as additional natural languages are added step by step.

For practical purposes, unrestricted communication with computers and robots in natural languages will make the interaction with these machines maximally user friendly and permit new, powerful ways of information processing. Artificial programming languages may then be limited to specialists developing and servicing the machines.

V. USING PARSERS

Computational linguistics analyzes natural languages automatically in terms of software programs called parsers. The use of parsers influences the theoretical viewpoint of linguistic research, distribution of funds, and everyday research practice as follows.

- *Competition*
  Competing theories of grammar are measured with respect to the new standard of how well they are suited for efficient parsing and how well they fit into a theory of language designed to model the mechanism of natural communication.
- *Funding*
  Computationally efficient and empirically adequate parsers for different languages are needed for an unlimited range of practical applications, which has a major impact on the inflow of funds for research, development, and teaching in this particular area of the humanities.
- *Verification*
  Programming grammars as parsers allows testing their empirical adequacy automatically on arbirarily large amounts of real data in the areas of word form recognition/synthesis, syntactic analysis/generation and semantic-pragmatic interpretation in both, the speaker and the hearer mode.

The verification of theories of language and grammar by means of testing electronic models in real applications is a new approach which clearly differs from the methods of traditional linguistics, psychology, philosophy, and mathematical logic.

VI. THEORETICAL LEVELS OF ABSTRACTION

So far there are no electronic systems which model the functioning of natural communication so successfully that one can talk with them more or less freely. Furthermore, researchers do not agree on how the mechanism of natural communication really works.

One may therefore question whether achieving a functional model of natural communication is possible in principle. I would like to answer this question with an analogy[2] from the recent history of science.

Today's situation in computational linguistics resembles the development of mechanical flight before 1903.[3] For hundreds of years humans have observed sparrows and other birds in order to understand how they fly. Their goal was to become airborne in a similar manner.

It turned out, however, that flapping wings did not work for humans. This was taken by some to declare human flight impossible in principle, in concord with the pious clichè "If God had intended humans to fly, He would have given them wings."[4]

Today human air travel is common place. Furthermore, we now know that a sparrow remains air-borne in terms of the same aero-dynamic principles as a jumbo jet. Thus, there is a certain level of abstraction at which the flights of sparrows and jumbo jets function in the same way.

Similarly, the modeling of natural communication requires an abstract theory which applies to human and artificial cognitive machines alike. Thereby, one naturally runs the risk of setting the level of abstraction either too low or too high. As in the case of flying, the crucial problem is finding the correct level of abstraction.

A level of abstraction which is too low is exemplified by closed signal systems such as vending machines. Such machines are inappropriate as a theoretical model because they fail to capture the diversity of natural language use, i.e. the characteristic property that one and the same expression can be used meaningfully in different contexts.

---

[2] See also CoL, p. 317.

[3] In 1903, the brothers Orville and Wilbur Wright succeeded with the first manned motorized flight.

[4] Irrational reasons against a modeling of natural communication reside in the subconscious fear of creating artificial beings resembling humans and having superhuman powers. Such *homunculi*, which occur in the earliest of mythologies, are regarded widely as violating a tabu. The tabu of doppelganger similarity is described in Girard 1974.

Besides dark versions of homunculi, such as the cabalistically inspired Golem and the electrically initialized creature of the surgeon Dr. Frankenstein, the literature provides also more lighthearted variants. Examples are the piano-playing doll automata of the 18th century, based on the anatomical and physical knowledge of their time, and the mechanical beauty singing and dancing in *The tales of Hoffmann*. More recent is the robot C3P0 in George Lucas' film *Star Wars*, which represents a positive view of human-like robots.

A level of abstraction which is too high, on the other hand, is exemplified by naive anthropomorphic expectations. For example, a notion of 'proper understanding' which requires that the computational system be subtly amused when scanning *Finnegan's Wake* is as far off the mark as a notion of 'proper flying' which requires mating and breeding behavior from a jumbo jet.[5]

VII. ANALYZING HUMAN COGNITION

The history of mechanical flight shows how a natural process (bird flight) poses a conceptually simple and obvious problem to science. Despite great efforts it has been unsolvable for a long time. In the end, the solution turned out to be a highly abstract mathematical theory. In addition to being a successful foundation of mechanical flight, this theory is able to explain the functioning of natural flight as well.

This is why the abstract theory of aero-dynamics has led to a new appreciation of nature. Once the development of biplanes, turboprops, and jets resulted in a better theoretical and practical understanding of the principles of flight, interest was refocussed again on the natural flight of animals in order to grasp their wonderful efficiency and power. This in turn led to major improvements in artificial flight, resulting in less noisy and more fuel efficient air planes.

Applied to computational linguistics, this analogy illustrates that our highly abstract and technological approach does not imply a lacking interest in the human language capacity. On the contrary, investigating the specific properties of human language communication is theoretically meaningful only *after* the mechanism of natural language communication has been modeled computationally and proven successful in concrete applications on massive amounts of data.

VIII. INTERNAL AND EXTERNAL TRUTHS

In science we may distinguish between internal and external truths. Internal truths are conceptual models, developed and used by scientists to explain certain phenomena, and held true by relevant parts of society for limited periods of time. Examples are the Ptolemaic (geocentric) view of planetary motion or Bohr's model of the atom.

External truths are the bare facts of external reality which exist irrespective of whether or not there are cognitive agents to appreciate them. These facts may be measured more or less accurately, and explained using conceptual models.

Because conceptual models of science have been known to change radically in the course of history internal truths must be viewed as *hypotheses*. They are justified mainly by the degree to which they are useful for arriving at a systematic description of external truths, represented by sufficiently large amounts of real data.

Especially in the natural sciences, internal truths have improved dramatically over the last five centuries. This is shown by an increasingly close fit between theoretical

---

[5] Though this may seem quite reasonable from the viewpoint of sparrows.

predictions and data, as well as a theoretical consolidation exhibited in the form of greater mathematical precision and greater functional coherence of the conceptual (sub)models.

In contrast, contemporary linguistics is characterized by lack of theoretical consolidation, as shown by the many disparate theories of language[6] and the overwhelming variety of competing theories of grammar.[7] As in the natural sciences, however, there is external truth also in linguistics. It may be approximated by completeness of empirical data coverage and functional modeling.

IX. LINGUISTIC VERIFICATION

The relation between internal and external truth is established by means of a *verification method*. The verification method of the natural sciences consists in the principle that experiments must be repeatable. This means that, given the same initial conditions, the same measurements must result again and again.

On the one hand, this method is not without problems because experimental data may be interpreted in different ways and may thus support different, even conflicting, hypotheses. On the other hand, the requirements of this method are so minimal that by now no self-respecting theory of natural science can afford to reject it. Therefore the repeatability of experiments has managed to channel the competing forces in the natural sciences in a constructive manner.

Another aspect of achieving scientific truth has developed in the tradition of mathematical logic. This is the principle of formal consistency as realized in the method of axiomatization and the rule-based derivation of theorems.

Taken by itself the quasi-mechanical reconstruction of mathematical intuition in the form of axiom systems is separate from the facts of scientific measurements. As the logical foundation of natural science theories, however, the method of axiomatization has proven to be a helpful complement to the principle of repeatable experiments.

In linguistics, corresponding methods of verification have been sorely missed. To make up for this shortcoming there have been repeated attempts to remodel linguistics into either a natural science or a branch of mathematical logic. Such attempts are bound to fail, however, for the following reasons:

- The principle of repeatable experiments can only be applied under precisely defined conditions suitable for measuring. The method of experiments is not suitable for the objects of linguistic description because they are *conventions* that developed in the course of centuries and exist as the intuitions ('Sprachgefühl') of the native speaker-hearer.

---

[6] Examples are nativism, behaviorism, structuralism, speech act theory, model theory, as well Givón's iconicity, Lieb's neostructuralism, and Halliday's systemic approach.

[7] Known by acronyms such as TG (with its different manifestations ST, EST, REST, and GB), LFG, GPSG, HPSG, CG, CCG, CUG, FUG, UCG, etc. These theories of grammar concentrate mostly on an initial foundation of internal truths such as 'psychological reality,' 'innate knowledge,' 'explanatory adequacy,' 'universals,' 'principles,' etc., based on suitably selected examples. Cf. Section 9.5

– The method of axiomatization can only be applied to theories which have consolidated on a high level of abstraction, such as Newtonian mechanics, thermodynamics, or the theory of relativity. In today's linguistics, there is neither the required consolidation of theory nor completeness of data coverage. Therefore, any current attempt at axiomatization in linguistics is bound to be empirically vacuous.

Happily, there is no necessity to borrow from the neighboring sciences in order to arrive at a methodological foundation of linguistics. Instead, theories of language and grammar are to be implemented as electronic models which are tested automatically on arbitrarily large amounts of real data as well as in real applications of spontaneous man-machine communication. This method of verifying or falsifying linguistic theories objectively is specific to computational linguistics and may be viewed as its pendant to the repeatability of experiments in the natural sciences.

X. EMPIRICAL DATA AND THEIR THEORETICAL FRAMEWORK

The methodology of computational linguistics presupposes a theory of language which defines the goals of empirical analysis and provides the framework into which components are to be embedded without conflict or redundancy. The development of such a framework can be extraordinarily difficult, as witnessed again and again in the history of science.

For example, in the beginning of astronomy scientists wrestled for a long time in vain with the problem of providing a functional framework to explain the measurements that had been made of planetary motion and to make correct predictions based on such a framework. It was comparatively recent when Kepler (1571–1630) and Newton (1642–1727) first succeeded with a description which was both, empirically precise and functionally simple. This, however, required a radical revolution in the theory of astronomy.

The revolution affected the *structural hypothesis* (transition from geo- to heliocentrism), the *functional explanation* (transition from crystal spheres to gravitation in space), and the *mathematical model* (transition from a complicated system of epicycles to the form of ellipses). Furthermore, the new system of astronomy was constructed at a level of abstraction where the dropping of an apple and the trajectory of the moon are explained as instantiations of one and the same set of general principles.

In linguistics, a corresponding scientific revolution has long been overdue. Even though the empirical data and the goals of their theoretical description are no less clear in linguistics than in astronomy, linguistics has not achieved a comparable consolidation in the form of a comprehensive, verifiable, functional theory of language.[8]

---

[8] From a history of science point of view, the fragmentation of today's linguistics resembles the state of astrology and astronomy before Kepler and Newton.

XI. PRINCIPLES OF THE SLIM THEORY OF LANGUAGE

The analysis of natural communication should be structured in terms of methodological, empirical, ontological, and functional principles of the most general kind. The SLIM theory of language presented in this book is based on surface compositional, linear, internal matching. These principles are defined as follows.

1. *Surface compositional* (methodological principle)
   Syntactic-semantic composition assembles only concrete word forms, excluding the use of zero-elements, identity mappings, or transformations.
2. *Linear* (empirical principle)
   Interpretation and production of utterances is based on a strictly time-linear derivation order.
3. *Internal* (ontological principle)
   Interpretation and production of utterances is analyzed as cognitive procedures located inside the speaker-hearer.
4. *Matching* (functional principle)
   Referring with language to past, current, or future objects and events is modeled in terms of pattern matching between language meaning and context.

These principles originate in widely different areas (methodology, ontology, etc.), but within the SLIM theory of language they interact very closely. For example, the functional principle of (4) matching can only be implemented on a computer if the overall system is handled ontologically as an (3) internal procedure of the cognitive agent. Furthermore, the methodological principle of (1) surface compositionality and the empirical principle of (2) time-linearity can be realized within a functional mechanism of communication only if the overall theory is based on internal matching (3,4).

In addition to the interpretation of its letters, the acronym SLIM is motivated as a word with a meaning like *slender*. This is so because detailed mathematical and computational investigations have proven SLIM to be efficient in the areas of syntax, semantics, and pragmatics – both, relative in comparison to existing alternatives, and absolute in accordance with the formal principles of mathematical complexity theory.

XII. CHALLENGES AND SOLUTIONS

The SLIM theory of language is defined on a level of abstraction where the mechanism of natural language communication in humans and in suitably constructed cognitive machines are explained in terms of the same principles of surface compositional, linear, internal matching.[9] This is an important precondition for unrestricted

---

[9] Moreover, the *structural hypothesis* of the SLIM theory of language is a regular, strictly time-linear derivation order – in contrast to grammar systems based on constituent structure. The *functional explanation* of SLIM is designed to model the mechanism of natural communication as a speaking robot – and not some tacit language knowledge innate in the speaker-hearer which excludes language use (performance). The *mathematical model* of SLIM is the continuation-based algorithm of LA-grammar, – and not the substitution-based algorithms of the last 50 years.

man-machine communication in natural language. Its realization requires general and efficient solutions in the following areas.

First, the hearer's *understanding* of natural language must be modeled. This process is realized as the automatic reading-in of propositions into a database and – most importantly – determining their correct place for storage and retrieval (Chapter 23). The foundation of the semantic primitives is handled in terms of natural or artificial recognition and action.

Second, it must be modeled how the speaker determines the contents to be expressed in language. This process, traditionally called *conceptualization*, is realized as an autonomous navigation through the propositions of the internal database. Thereby speech production is handled as a direction reflection (internal matching) of the navigation path in line with the motto: *Speech is verbalized thought* (Chapter 24).

Third, the speaker and the hearer must be able to draw *inferences* on the basis of the contents of their respective databases. Inferences are realized as a special form of the autonomous time-linear navigation resulting in the derivation of new propositions. Inferences play an important role for the pragmatic interpretation of natural language, both in the hearer and the speaker (see Section 24.5).

The formal basis of time-linear navigation consists in concatenated propositions stored in a network database as a set of word tokens. A word token is a feature structure with the special property that it explicitly specifies the possible continuations to other word tokens, both within its proposition and from its proposition to others. This novel structure is called a *word bank* and provides the 'rail road tracks' for the navigation of a mental focus point. The navigation is powered and controlled by suitable LA-grammars (motor algorithms) which compute the possible continuations from one word token to the next (see Section 24.2).

The word bank and its motor algorithms constitute the central processing unit of an artificial cognitive agent called SLIM machine. The word bank is connected to external reality via the SLIM machine's recognition and action. The interpretation of perception, both verbal and nonverbal, results in concatenated propositions which are read into the word bank. The production of action, both verbal and nonverbal, is based on realizing some of the propositions traversed during the autonomous navigation.

Part I

Theory of Language

1. Computational language analysis

The practical development of computers began around 1940. From then on there evolved a basic distinction between numerical and nonnumerical computer science.

Numerical computer science specializes in the calculation of numbers. In the fields of physics, chemistry, economics, sociology, etc., it has led to a tremendous expansion of scientific knowledge. Also many applications like banking, air travel, stock inventory, manufacturing, etc., depend heavily on numerical computation. Without computers and their software, operations could not be maintained in these areas.

Nonnumerical computer science deals with the phenomena of perception and cognition. Despite hopeful beginnings, nonnumerical computer science soon lagged behind the numerical branch. In recent years, however, nonnumerical computer science has made a comeback as artificial intelligence and cognitive science. These new, interdisciplinary fields investigate and electronically model natural information processing.

The term computational linguistics refers to that subarea of nonnumerical computer science which deals with language production and language understanding. Like artificial intelligence and cognitive science in general, computational linguistics is a highly interdisciplinary field which comprises large sections of traditional and theoretical linguistics, lexicography, psychology of language, analytic philosophy and logic, text processing, the interaction with databases, as well as the processing of spoken and written language.

1.1 Man-machine communication

The goal of nonrestricted man-machine communication presupposes solutions to the most basic tasks of natural language analysis. Realizing this goal is therefore the ultimate standard for a successfull computational linguistics.

Today, man-machine communication is still limited to highly *restricted* forms. Consider, for example, the interaction between the user and a standard computer, such as a PC or a work station. These machines provide a key board for the input of letters and a screen for the output of letters and pictures.[1]

---

[1] For simplicity, we are disregarding additional input and output devices, such as mouse and sound, respectively.

It is conceivable to expand the notion of man-machine communication to machines which do not

Computers are comfortable for entering, editing, and retrieving natural language, at least in the medium of writing, for which reason they have replaced electric typewriters. For utilizing the computers' abilities beyond word processing, however, commands of artificial languages must be applied. These are called programming languages, and especially designed for controlling the computer's electronic operations.

In contrast to natural languages, which are flexible and rely on the seemingly obvious circumstances of the utterance situation, common background knowledge, the content of earlier conversations, etc., programming languages are inflexible and refer directly, explicitly, and exclusively to operations of the machine. For most potential users, a programming language is difficult to handle because (a) they are not familiar with the operations of the computer, (b) the expressions of the programming language differ from those of everyday language, and (c) the use of the programming language is highly regulated.

Consider, for example, a standard database[2] which stores information about the employees of a company in the form of records.

1.1.1 EXAMPLE OF A RECORD-BASED DATABASE

	last name	first name	place	...
A1	Schmidt	Peter	Bamberg	...
A2	Meyer	Susanne	Nürnberg	...
A3	Sanders	Reinhard	Schwabach	...
	⋮	⋮	⋮	

The rows, named by different attributes like first name, last name, etc., are called the fields of the record type. The lines A1, A2, etc., each constitute a record. Based on this fixed record structure, the standard operations for the retrieval and update of information in the database are defined.

To retrieve the name of the representative in, e.g., Schwabach the user must type in the following commands of the programming language (here a query language for databases) without mistake:

---

provide general input/output components for language signs. Consider, for example, the operation of a contemporary washing machine. Leaving aside the loading of laundry and the measuring of detergent, the 'communication' consists in choosing a program and a temperature and by pushing the start button. The machine 'answers' by providing freshly laundered laundry once the program has run its course.

Such an expanded notion of man-machine communication should be avoided, however, because it fosters misunderstandings. Machines without general input/output facilities for language constitute the special case of *nonverbal* man-machine communication, which may be neglected for the purposes of computational linguistics.

[2] As introductions to databases see C. Date 1990[4] and R. Elmasri & S. Navathe 1989. We will return to this topic in Chapter 22 in connection with the interpretation of natural language.

1.1.2 DATABASE QUERY

Query:

```
select A#
where city = 'Schwabach'
```

Result:

```
result: A3 Sanders Reinhard
```

The correct use of commands such as 'select' initiates quasi-mechanical procedures which correspond to filing and retrieving cards in a filing cabinet with many compartments. Compared to the nonelectronic method, the computational system has many practical advantages. The electronic version is faster, the adding and removing of information is simpler, and the possibilities of search are much more powerful because various different key words may be logically combined into a complex query.[3]

Standard computers have been regarded as general purpose machines for information processing because any kind of standard program can be developed and installed on them. From this point of view, their capabilities are restricted only by hardware factors like the available speed and memory. In another sense, the information processing of standard computers is not universal, however, because their input and output is restricted to the language channel.

A second type of computer not subject to this limitation is autonomous robots. In contradistinction to standard computers, robots are not restricted to the language channel, but designed to recognize their environment and to act in it.[4]

Corresponding to the different technologies of standard computers and robots, there have evolved two different branches of artificial intelligence. One branch, dubbed classic AI by its opponents, is based on standard computers. The other branch, which calls itself nouvelle AI,[5] requires the technology of robots.

Classic AI analyzes intelligent behavior in terms of manipulating abstract symbols. A typical example is a chess playing program.[6] It operates in isolation from the rest of the world, using a fixed set of predefined pieces and a predefined board. The search space for a dynamic strategy of winning in chess is astronomical. Yet the technology of a standard computer is sufficient because the world of chess is closed.

Nouvelle AI aims at the development of autonomous agents. In contrast to systems which respond solely to a predefined set of user commands and behave otherwise in isolation, autonomous agents are designed to interact with their real world environment. Because the environment is constantly changing in unpredictable ways they must continually keep track of it by means of sensors.

---

[3] See also Section 2.1.

[4] Today three different generations of robots are distinguished. Most relevant for computational linguistics are robots of third generation, which are designed as autonomous agents. See D.W. Wloka 1992.

[5] See for example P. Maes (ed.) 1990.

[6] A. Newell & H. Simon 1972, R. Reddy et al. 1973.

For this, nouvelle AI uses the strategy of task level decomposition. Rather than building and updating one giant global representation to serve as the basis of automatic reasoning, nouvelle AI systems aim at handling their tasks in terms of many interacting local procedures controlled by perception. Thereby low-level inferencing is defined to operate directly on the local perception data.

A third type of machine processing information – besides standard computers and robots – is systems of virtual reality (VR).[7] While a robot analyzes its environment in order to influence it in certain ways (such as moving in it), a VR system aims at creating an artificial environment for the user. Thereby the VR system reacts to the movements of the user's hand, the direction of his gaze, etc., and utilizes them in order to create as realistic an environment as possible.

The different types of man-machine communication exemplified by standard computers, robots, and VR systems may be compared schematically as follows.

1.1.3 THREE TYPES OF MAN-MACHINE COMMUNICATION

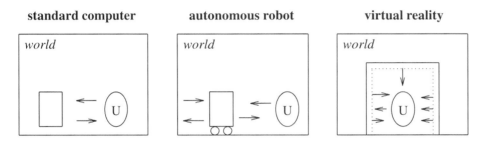

The ovals represent the users who face the respective systems in the 'world.' The arrows represent the interaction of the systems with their environment and the user.

A standard computer communicates with users who initiate the interaction. A robot communicates independently with its environment and its users. A VR system does not communicate with its environment, but rather creates an artificial environment for the user. In robots and VR systems, an interaction with the user in terms of language is optional and may be found only in advanced systems. These system must always have a language-based 'service channel,' however, for the installation and upgrading of the system software.

1.2 Language science and its components

A speaker of English knows the meaning of a word like **red**. When asked to pick the red object among a set of non-red objects, for example, a competent speaker-hearer will be able to do it. A standard computer, on the other hand, does not 'understand' what **red** means, just as a piece of paper does not understand what is written on it.[8]

---
[7] For an introduction see A. Wexelblat (ed.) 1993.

In the interaction with a standard computer, the understanding of natural language is restricted largely[9] to the user. For example, if a user searches in a database for a red object, (s)he understands the word **red** before it is put into – and after it is given out by – the standard computer. But inside the standard computer, the word **red** is manipulated as a sign which is uninterpreted with respect to the color denoted.

What is true for standard computers does not apply to man-machine communication in general, however. Consider for example a modern robot which is asked by its master to bring an object it has not previously encountered, for example the new blue and yellow book on the desk in the other room.[10] If such a robot is able to spontaneously perform an open range of different jobs like this, it has an understanding of language which at some level may be regarded as functionally equivalent to the corresponding cognitive procedures in humans.

The communication with a robot may be based on artificial or natural language. The use of natural language is much more challenging, however, and much preferable in many situations. As a first step towards achieving unrestricted man-machine communication in natural language let us consider the current state of linguistics.

In this field of research, three basic approaches to grammatical analysis may be distinguished, namely (i) traditional grammar, (ii) theoretical linguistics, and (iii) computational linguistics. They differ in their methods, goals, and applications.

1.2.1 VARIANTS OF LANGUAGE SCIENCE

– *Traditional Grammar*
   uses the method of informal classification and description based on tradition and experience,
   has the goal to collect and classify the regularities and irregularities of the natural language in question as completely as possible, and
   is applied mostly in teaching languages (originally Latin).
   While traditional grammar has long been shunted aside by theoretical linguistics, it has been of great interest to computational linguistics because of its wealth of concrete data. The notions of traditional grammar will be discussed in detail in Part III.

– *Theoretical Linguistics*
   uses the method of mathematical logic to describe natural languages by means of formal rule systems intended to derive all and only the well-formed expressions – which has the advantage of stating hypotheses explicitly,[11]

---

[8] This fact has been misunderstood to imply that a modeling of understanding in AI is impossible in principle. A prominent example is the 'Chinese room argument' in J. Searle 1992.

[9] Even a standard computer may interpret certain structural aspects of language, however, such as the categories of the word forms.

[10] Reference to objects removed in time and/or space is called *displaced reference*.

[11] As compared to traditional grammar.

has pursued the goal of describing the 'innate human language ability' (competence) whereby aspects of language use in communication (performance) have been excluded, and

has had rather limited applications because of computational inefficiency and fragmentation into different schools.

Theoretical linguistics is relevant to computational linguistics in the area of formal language analysis and mathematical complexity theory. Formal language theory will be discussed in detail in Part II.

– *Computational Linguistics*

combines the methods of traditional grammar and theoretical linguistics with the method of effectively verifying explicit hypotheses by implementing formal grammars as efficient computer programs and testing them automatically on realistic – i.e. very large – amounts of real data,

has the goal to model the mechanism of natural language communication – which requires a complete morphological, lexical, syntactic, semantic, and pragmatic analysis of a given natural language within a functional framework, and

has applications in all instances of man-machine communication far beyond letter-based 'language processing.'

Despite their different methods, goals, and applications, the three variants of language science described in 1.2.1 divide the field into the same components of grammar, namely phonology, morphology, lexicon, syntax, semantics, and the additional field of pragmatics. The components differ, however, in their respective role and scientific treatment within the three approaches:

1.2.2 THE COMPONENTS OF GRAMMAR

– *Phonology*: Science of language sounds

Phonology describes historical changes as well as synchronic alternations, such as trisyllabic laxing in English or final devoicing in German, in terms of grammatical rules.

For theoretical linguistics, phonology is important: it is used as a kind of sand table on which different schools try to demonstrate the innateness of their current universals and grammar variants.

In computational linguistics, the role of phonology is marginal at best.[12] One might conceive of using it in automatic speech recognition, but the science appropriate is in fact phonetics. Phonetics investigates the (i) articulatory, (ii) acoustic, and (iii) auditive processes of speech. In contrast to phonology, phonetics is traditionally not considered part of the grammar.

– *Morphology:* Science of word form structure

Morphology classifies the words of a language according to their part of speech

---

[12] Computational linguistics analyzes natural language at a level of abstraction which is independent of any particular medium of manifestation, e.g. sound. Cf. Section 1.4.

(category) and describes the structure of word forms in terms of inflection, derivation, and composition. To traditional grammar, morphology has long been central, as shown by the many paradigm tables in, e.g., grammars of Latin.

In theoretical linguistics, morphology has played a minor role. Squeezed between phonology and syntax, morphology has been used mostly to exemplify principles of either or both of its neighbor components.

In computational linguistics, morphology appears in the context of automatic word form recognition. It is based on an on-line lexicon and a morphological parser which (i) relates each word form to its base form (lemmatization) and (ii) characterizes its morpho-syntactic properties (categorization). Automatic word form recognition is presupposed by all other rule-based techniques of automatic language analysis, such as syntactic and semantic parsing.

– *Lexicon:* Listing analyzed words

The words of a language are collected and classified in lexicography and lexicology. Lexicography deals with the principles of coding and structuring lexical entries, and is a practically oriented border area of natural language science. Lexicology investigates semantic relations in the vocabulary of a language and is part of traditional philology.

In computational linguistics, electronic lexica combine with morphological parsers in the task of automatic word form recognition. The goal is maximal completeness with fast access and low space requirements. In addition to building new lexica for the purpose of automatic word form recognition, there is great interest in utilizing the knowledge of traditional lexica for automatic language processing ('mining of dictionaries').

– *Syntax:* Science of composing word forms

In communication, the task of syntax is the composition of literal meanings via the composition of word forms (surfaces).[13] One aspect of this is characterizing well-formed compositions in terms of grammatical rules. The other is to provide the basis for a simultaneous semantic interpretation.

In theoretical linguistics, syntactic analysis has concentrated on a description of grammatical well-formedness. The problem with analyzing well-formedness in isolation is that any finite set of sentences may be described by a vast multitude of different grammars. In order to select the one type of description which turns out to be correct in the long run, theoretical linguistics has vainly searched for 'universals' supposed to characterize the 'innate human language faculty.'

A more realistic and effective standard is the requirement that the grammar must be suitable to serve as a component in an artificial cognitive agent communicating in natural language. There, the descriptive and functional adequacy of the grammar may be tested automatically on the full range of natural language data. This presupposes a grammatical algorithm with low mathematical complexity. Further-

[13] Cf. Section 19.1.

more, the algorithm must be input-output equivalent with the mechanism of natural communication.

- *Semantics:* Science of literal meanings

The semantics of natural language may be divided into lexical semantics, describing the literal meaning of words, and compositional semantics, describing the composition of meanings in accordance with the syntax. The task of semantics is a systematic conversion of the syntactically analyzed expression into a semantic representation based on the functor-argument structure underlying the categories of basic and complex expressions.

The beginning of traditional grammar contributed considerably to the theory of semantics, for example Aristotle's distinction between subject and predicate. However, these contributions have been passed on and developed mostly within philosophy of language. Later, the semantics of traditional grammar has not reached beyond the anecdotal.

In theoretical linguistics, semantics has initially been limited to characterizing syntactic ambiguity and paraphrase. Subsequently, logical semantics was applied to natural language: based on a metalanguage, natural language meanings are being defined in terms of truth conditions.

Computational linguistics uses procedural semantics instead of metalanguage-based logical semantics. The semantic primitives of procedural semantics are based on operations of perception and action by the cognitive agent. The semantics is designed to be used by the pragmatics in an explicit modeling of the information transfer between speaker and hearer.

- *Pragmatics:* Science of using language expressions

Pragmatics describes how the grammatically analyzed expressions are used relative to the context of interpretation. Therefore, pragmatics is not part of the grammar proper, but analyzes the *interaction* between the expressions and the context, presupposing the grammatical analysis[14] of the expressions and a suitable description of the context.

In traditional grammar, phenomena of pragmatics have been handled in the separate discipline of rhetoric. This has been an obstacle to integrating the analysis of language structure and language use.

In theoretical linguistics, the distinction between semantics and pragmatics has evolved only haltingly. Because theoretical linguistics has not been based on a functional model of communication pragmatics has served mostly as the proverbial 'wastebasket' (Y. Bar-Hillel 1971, p.405).

In computational linguistics, the need for a systematic theory of pragmatics became most obvious in natural language generation – as in dialogue systems or machine

---

[14] The components phonology, morphology, lexicon, syntax, and semantics are part of the grammar proper because they deal with the structure of word forms, complex expressions, and sentences.

translation, where the system has to decide what to say and how to say it in a rhetorically acceptable way.

That the different approaches of traditional grammar, theoretical linguistics, and computational linguistics use the same set of components to describe the phenomena of natural language – despite their different methods and goals – is due to the fact that the division of phenomena underlying these components is based on different structural aspects, namely *sounds* (phonology), *word forms* (morphology), *sentences* (syntax), literal *meanings* (semantics), and their *use* in communication (pragmatics).

1.3 Methods and applications of computational linguistics

Computational linguistics uses parsers for the automatic analysis of language. The term 'parser' is derived from on the Latin word *pars* meaning part, as in 'part of speech.' Parsing in its most basic form consists in the automatic

1. *decomposition* of a complex sign into its elementary components,
2. *classification* of the components[15] via lexical lookup, and
3. *composition* of the classified components via syntactic rules in order to arrive at an overall grammatical analysis of the complex sign.[16]

Methodologically, the implementation of natural language grammars as parsers is important because it allows to test the descriptive adequacy of formal rule systems automatically and objectively on real data. This new method of verification is as characteristic for computational linguistics as the method of repeatable experiments is for natural science.

Practically, the parsing of natural language may be used in different applications.

1.3.1 PRACTICAL TASKS OF COMPUTATIONAL LINGUISTICS

– Indexing and retrieval in textual databases
Textual databases electronically store texts such as publications of daily news papers, medical journals, and court decisions The user of such a database should be able to find exactly those documents and passages with comfort and speed which are relevant for the specific task in question. The World Wide Web (WWW) may also be viewed as a large, unstructured textual database, which daily demonstrates to a growing number of users the difficulties of successfully finding the information desired. Linguistic methods for optimizing retrieval are described in Section 2.2.

---

[15] By assigning, e.g., the part of speech.
[16] Including assignment of the part of speech.

– Machine translation

Especially in the European Union, currently with eleven different languages, the potential utility of automatic or even semi-automatic translation systems is tremendous. Different approaches to machine translation are described in Sections 2.4 and 2.5.

– Automatic text production

Large companies which continually bring out new products such as engines, video recorders, farming equipment, etc., must constantly modify the associated product descriptions and maintenance manuals. A similar situation holds for lawyers, tax accountants, personnel officers, etc., who must deal with large amounts of correspondence in which most of the letters differ only in a few, well-defined places. Here techniques of automatic text production can help, ranging from simple templates to highly flexible and interactive systems using sophisticated linguistic knowledge.

– Automatic text checking

Applications in this area range from simple spelling checkers (based on word form lists) via word form recognition (based on a morphological parser) to syntax checkers based on syntactic parsers which can find errors in word order, agreement, etc.

– Automatic content analysis

The printed information on this planet is said to double every 10 years. Even in specialized fields such as natural science, law, or economics, the constant stream of relevant new literature is so large that researchers and professionals do not nearly have enough time to read it all. A reliable automatic content analysis in the form of brief summaries would be very useful. Automatic content analysis is also a precondition for concept-based indexing, needed for accurate retrieval from textual databases, as well as for adequate machine translation.

– Automatic tutoring

There are numerous areas of teaching in which much time is spent on drill exercises such as the more or less mechanical practicing of regular and irregular paradigms in foreign languages. These may be done just as well on the computer, providing the students with more fun (if they are presented as a game, for example) and the teacher with additional time for other, more sophisticated activities such as conversation. Furthermore, these systems may produce automatic protocols detailing the most frequent errors and the amount of time needed for various phases of the exercise. This constitutes a valuable heuristics for improving the automatic tutoring system ergonometrically. It has led to a new field of research in which the 'electronic text book' of old is replaced by new teaching programs utilizing the special possibilities of the electronic medium to facilitate learning in ways never explored before.

– Automatic dialog and information systems

These applications range from automatic information services for train schedules via queries and storage in medical databases to automatic tax consulting.

This list is by no means complete, however, because the possible applications of computational linguistics include all areas in which humans communicate with computers and other machines of this level, today or in the future.

In summary, traditional language sciences may contribute substantially to improve automatic language processing in computational applications. Computers, on the other hand, are an essential tool for improving empirical analysis in linguistics – not only in certain details, but as an accurate, complete, and efficiently functioning theory of language which is realized concretely in terms of unrestricted natural man-machine communication.

1.4 Electronic medium in recognition and synthesis

The expressions and texts of natural language may be realized in different media. The nonelectronic media comprise the *sounds* of spoken language, the *letters* of handwritten or printed language, and the *gestures* of sign language.

Spoken and signed language in its original form has only a fleeting existence. Writing, on the other hand, is the traditional method of storing information more permanently, e.g. on stone, clay, wood, parchment, or paper.

A modern form of storing information is the electronic medium. It codes information abstractly in terms of numbers which are represented magnetically. In contrast to the traditional means of storage, the electronic medium has the advantage of greatest flexibility: the data may be copied, edited, sorted, reformatted, and transferred at will.

The electronic medium may represent language in a realization-dependent or a realization-independent form. The realization-dependent form reproduces accidental properties of *tokens* in a certain medium, such as a tape recording of spoken language, a bitmap of written language, or a video recording of signed language.

The realization-independent form represents language as abstract *types*, coded digitally as electronic sign sequences, e.g. in ASCII.[17] Due to their type character,[18] they may be recognized unambiguously by suitable machines, copied without loss of information, and realized as token surfaces in any imaginable variant in any medium.

In communication, there is a constant transfer between realization-dependent and realization-independent representations (cf. Section 23.2). During *recognition*, the cognitive agent must map realization-dependent representations into the realization-independent ones (d⇒i transfer). During *synthesis*, realization-independent representations must be mapped into realization-dependent ones (i⇒d transfer).

Neither of these directions is trivial to model in computational linguistics, but for different reasons. When building a speaking robot, the challenge with an i⇒d transfer into spoken language (speech synthesis) is to make it sound natural relative to a

---

[17] ASCII stands for *American standard code of information interchange*.

[18] The point is that each instance of such a type is coded identically to any other instance of the same type. For further discussion of the type-token distinction see Section 3.3.

free range of utterance situations. The challenge with an d⇒i transfer from written (optical character recognition) or spoken language (speech recognition) is to correctly interpret tokens from a wide range of different realizations.

The most primitive form of d⇒i transfer leaves recognition to humans. It consists in typing spoken or written language into the computer. This method is still widely in use, such as dictation in the office, transcription of tape recordings in psychology, or electronic typesetting of books which previously existed only in traditional print.

Automatic d⇒i transfer from printed language is based on optical character recognition. Part of an OCR system is a scanner which makes an image of the page as a bitmap – like a camera. Then the OCR software analyzes the image line by line, letter by letter. By comparing the bitmap outline of each letter with stored patterns, the writing is recognized and stored in a form as if it was typed in.

The input to an OCR system may vary widely in font type, font size, and the form of layout. Even within a given document there are head lines, footnotes, tables, and the foot lines of pictures to deal with. Modern OCR systems handle such challenges by means of an initial learning phase in which the user corrects misclassifications by telling the program whether a certain constellation happens to be, e.g., ii or n.

In addition, OCR systems use large dictionaries on the basis of which they decide which of several possible analyses constitutes a legitimate word form. In this manner a high recognition rate is achieved, sufficient for practical use. Depending on the type of machine a page may take between 50 seconds and a few minutes.[19]

The speed of today's OCR systems is quite competitive, especially in light of the fact that the machine does not become tired and that the operation of the scanner can be left to unskilled labor. The most important aspect of language transfer in general, however, is the avoidance of errors. In this respect, the human and the mechanical forms of transfer are equal in that both require proof reading.

Automatic d⇒i transfer from *spoken* language turns out to be considerably more difficult than that from written language. Whereas words in print are clearly separated and use uniformly shaped letters, speech recognition must analyze a continuous stream of sound and deal with different dialects, different pitches of voice, as well as background noises.

The possible applications of a good automatic speech recognition are tremendous, however, because there are many users and many circumstances of use for which a computer interaction based on spoken language would be considerably more user friendly than one based on the key board and the screen. Therefore automatic speech recognition is subject of an intensive worldwide research effort. The projects range from a type-writer capable of interpreting dictation over telephone-based automatic information systems (e.g. for train schedules) to *Verbmobil*.[20]

---

[19] The power of scanners and their OCR software has improved considerably since 1980 while prices have fallen. For these reasons the use of scanners in offices has greatly increased.

[20] Cf. W. Wahlster 1993. *Verbmobil* is intended as a portable computer into which the users can speak in German or Japanese to obtain a spoken English translation. Its use presupposes that the German

The quality of automatic speech recognition should be at least equal to that of an average human hearer. This leads to the following desiderata.

1.4.1 DESIDERATA OF AUTOMATIC SPEECH RECOGNITION

– *Speaker independence*
  The system should understand speech of an open range of speakers with varying dialects, pitch, etc. – without the need for an initial learning phase to adapt the system to one particular user.

– *Continuous speech*
  The system should handle continuous speech at different speeds – without the need for unnatural pauses between individual word forms.

– *Domain independence*
  The system should understand spoken language independently of the subject matter – without the need of telling the system in advance which vocabulary is to be expected and which is not.

– *Realistic vocabulary*
  The system should recognize at least as many word forms as an average human.

– *Robustness*
  The system should recover gracefully from interruptions, contractions, and slurring of spoken language, and be able to infer the word forms intended.

Today's continuous speech systems can achieve speaker independence only at the price of domain dependence. The prior restriction to a certain domain – for example, train schedules, or when and where to meet – has the advantage of drastically reducing the number of hypotheses about the word forms underlying a given sound pattern.[21]

The vocabulary of speaker-independent continuous speech recognition systems is still limited to no more than 1 000 word forms. An average speaker, however, uses about 10 000 words – which in English corresponds to about 40 000 word forms. Her or his passive vocabulary is about three to four times as large. Therefore a speech recognition system for English would have to recognize 120 000 word forms in order to be in the same class as an average speaker.[22] Accordingly, Zue, Cole & Ward[23] estimate that "It will be many years before unlimited vocabulary, speaker-independent continuous dictation capability is realized."

---

and Japanese partners have a passive knowledge of English. In this way, the hearer can understand the output of the system, and the speaker can check whether the system has translated as intended (cf. Section 2.5). The system is limited to the domain of scheduling meetings.

[21] Utilizing domain knowledge is always crucial for inferring the most probable word sequence from the acoustic signal in both, human and artificial speech recognition. The point is that the current domain should not be prespecified by design or have to be preselected by the user. Instead, the system should be domain-independent in the sense that is can determine the current domain automatically.

[22] Based on (i) a training phase to adapt to a particular user and (ii) pauses between each word form, the IBM-VoiceType system can recognize up to 22 000 different word forms.

[23] In R. Cole (ed.) 1998, p. 9

Speech recognition will be fully successful only if the technological side is supplied continuously with small bits of highly specific data from large stores of domain and language knowledge. These bits are needed only momentarily and must be provided very fast in order for the system to work in real time.

Therefore, the crucial question for designing truly adequate speech recognition is:

How should the domain and language knowledge best be organized?

The answer is obvious:

The domain and language knowledge should be organized within a functional theory of language which is mathematically and computationally efficient.

The better natural communication is modeled on the computer, the more effectively speech recognition can be supplied with the necessary bits of information. Conversely, the better the functioning of speech recognition, the easier the d⇒i transfer and thus the supply of knowledge needed for understanding during man-machine dialog.

1.5 Second Gutenberg revolution

The first Gutenberg revolution[24] was based on the technological innovation of printing with movable letters. It allowed to reproduce books inexpensively and in great numbers, making a wealth of information available to a broader public for the first time. This led to a rapid expansion of knowledge which in turn was stored and multiplied in the form of books as well.

The first Gutenberg revolution made more and more information freely available. It therefore became increasingly difficult to *find* the relevant information for a given purpose. Today the accumulated wealth of printed information far exceeds the capacity of a human life span even in narrowly defined subdomains of scientific research.

The second Gutenberg revolution is based on the automatic processing of natural language in the electronic medium. Its purpose is to facilitate access to specific pieces of information in such a way that huge amounts of text can be searched quickly and comfortably on the computer, producing accurate and complete results.

The second Gutenberg revolution is supported by the fact that today's publications originate primarily in the electronic medium, making a d⇒i transfer from the conventional print media unnecessary. In publishing, the traditional media of spoken and

---

[24] Named after Johannes Gutenberg, 1400(?)-1468(?), who invented printing with movable letters in Europe. This technique had been discovered before in Korea, probably around 1234 (D-C. Kim 1981, Y-W. Park 1987). The existence of this early Korean print technique was rediscovered in 1899 by Maurice Courant.

One reason why this technique did not prevail in Korea is the great number of Chinese characters, which are not as suitable for printing as the roughly 40 characters of the Latin alphabet (the Korean alphabet *Hangul* was designed much later in 1446). The other is that in Korea the printing was done by hand only whereas Gutenberg – presumably inspired by the wine presses of his time – integrated the printing blocks into a mechanism which facilitated and sped up the printing of pages, and which could be combined with an engine to run automatically.

printed language have by now become secondary media which are instantiated only when needed.

Even texts which have long existed in the traditional print medium are nowadays being transferred into the electronic medium in order to make them susceptible to the methods of electronic processing. Examples are the complete texts of classical Greek and Latin, the complete Shakespeare, and the Encyclopedia Britannica, which are now available on CD-ROM.

Compared to the printed version of a multi-volume edition, the electronic medium has the advantage of compactness, comfort, and speed. The information can usually be stored on a single CD-ROM. Instead of having to heave several volumes from the shelve and leafing through hundreds of pages by hand in order to find a particular passage, the use of the CD-ROM merely requires typing in the key words.

Given a suitable software, also combinations of words can be searched such as all passages in which the words **painter, Venice,** and **16**th **century** occur within a certain stretch of text. These methods of search can be life-saving, e.g. when a textual database is used for diagnosing a rare disease, or for choosing a particular medication.

Another advantage of the electronic medium is the editing, formatting, and copying of text. In the old days, newspaper articles were put together with mechanical typesetting machines. Information coming in from a wire service had to be typeset from the ticker tape letter by letter. To make room for some late breaking piece of news, the printing plates had to be rearranged by hand.

Today the production of newspapers is done primarily on-line in *soft copy*. Contributions by wire services are not delivered on paper, but by telephone whereby a modem converts the signal into the original layout. Form and contents of the on-line newspaper can be freely reformatted, copied, and edited, and any of these versions can be printed as *hard copies* without additional work.

A newspaper article, like any text, is not merely a sequence of words, but has a structure in terms of header, name of author, date, subsections, paragraphs, etc. In the electronic medium, this textual structure is coded abstractly by means of control symbols.

1.5.1 NEWSPAPER TEXT WITH CONTROL SYMBOLS

```
<HTML>
<HEAD>
<TITLE>9/4/95 COVER: Siberia, the Tortured Land</TITLE>
</HEAD>
<BODY>
<!-- #include "header.html" -->
<P>TIME Magazine</P>
<P>September 4, 1995 Volume 146, No. 10</P>
<HR>
Return to <A href="../../../../../time/magazine/domestic/toc/
950904.toc.html">Contents page</A>
<HR>
```

```
<BR>
<!-- end include -->
<H3>COVER STORY</H3>
<H2>THE TORTURED LAND</H2>
<H3>An epic landscape steeped in tragedy, Siberia suffered
grievously under  communism. Now the world's capitalists covet
its vast riches </H3>
<P><EM>BY <A href="../../../../../time/bios/eugenelinden.html">
EUGENE LINDEN</A>/YAKUTSK</EM>
<P>Siberia has come to mean a land of exile, and the place
easily fulfills its reputation as a metaphor for death and
deprivation. Even at the peak of midsummer, a soul-chilling
fog blows in off the Arctic Ocean and across the mossy tundra,
muting the midnight sun above the ghostly remains of a
slave-labor camp. The mist settles like a shroud over broken
grave markers and bits of wooden barracks siding bleached
as gray as the bones of the dead that still protrude through
the earth in places. Throughout Siberia, more than 20 million
perished in Stalin's Gulag.      ...
```

To be positioned in example 1.5.1, the text was copied electronically from an publication of TIME magazine available on the Internet. The example contains control symbols of the form $<...>$ which specify the formatting of the text in print or
on the screen. For example, $<P>$September 4, 1995 Volume 146, No.
10$</P>$ is to be treated in print as a paragraph, and $<H2>$THE TORTURED
LAND$</H2>$ as a header.

At first, different print shops used their own conventions to mark the formatting
instructions, for which reason the control symbols had to be readjusted each time a
text was moved to another typesetting system. To avoid this needless complication,
the International Standards Organization (ISO) developed the SGML standard.

1.5.2 SGML: *standard generalized markup language.*

> A family of ISO standards for labeling electronic versions of text, enabling both sender and
> receiver of the text to identify its structure (e.g. title, author, header, paragraph, etc.)

Dictionary of Computing, S. 416 (ed. Illingworth et al. 1990)

The SGML language,[25] exemplified in 1.5.1 above, has been adopted officially by
the USA, the European Union, and other countries, and has become widely accepted
by the users. Texts which use SGML for their markup have the advantage that their
formatting instructions can be automatically interpreted by other SGML users. An
easier to use subset of SGML is XML, which is oriented towards handling hypertext.[26]

In addition to the standardized coding of textual building blocks such as header,
subtitle, author, date, table of contents, paragraph, etc., there is the question of how
different types of text, such as articles, theater plays, or dictionaries, should best be
constructed from these building blocks. For example, the textual building blocks of a

---

[25] See C.F. Goldfarb 1990, E. van Herwijnen 1990, 1994.
[26] See S. St.Laurent 1998.

theater play, i.e., the acts, the scenes, the dialog parts of different roles, and the stage descriptions, can all be coded in SGML. Yet the general text structure of a play as compared to an newspaper article or an dictionary entry goes beyond the definition of the individual building blocks.

In order to standardize the structure of different types of texts, the International Standards Organization began in 1987 to develop the TEI-Guidelines. TEI stands for *text encoding initiative* and defines a DTD (*document type definition*) for the markup of different types of text in SGML.[27]

SGML and TEI specify the mark up at the most abstract level insofar as they define the text structure and its building blocks in terms of their function (e.g. header), and not in terms of how this function is to be represented in print (e.g. bold face, 12 pt.). For this reason, texts conforming to the SGML and TEI standards may be realized in any print style of choice.

An intermediate level of abstraction is represented by the formatting systems developed as programming languages for type-setting only a few years earlier. Widely used in academic circles are TEX, developed by D. Knuth, and its macro package LATEX. Since they were first introduced in 1984 they are used by scientists for preparing camera ready manuscripts of research papers and books.

At the lowest level of abstraction are menu-based text processing systems on PCs, such as Winword and WordPerfect. They are initially easy to learn, but their control is comparatively limited and for long documents they are not stable. Also, transferring text from one PC text processing system to another is difficult to impossible.

In summary, SGML and TEI focus on defining the abstract structure of the text, TEX and LATEX focus on control of the print, and PC systems focus on the ease and comfort of the user. Thereby the higher level of abstraction, e.g. SGML, can always be mapped into a lower level, e.g. LATEX. The inverse direction, on the other hand, is not generally possible because the lower level control symbols have no unambiguous interpretation in terms of text structure.

SGML/TEI and TEX/LATEX have in common that their control symbols are placed into the text's source code (e.g. 1.5.1) by hand; then they are interpreted by a program producing the corresponding print. PC systems, on the other hand, are based on WYSIWYG (*what you see is what you get*), i.e., the look of the print is manipulated by the user on the screen whereby the software automatically floods the text's source code with cryptic control symbols.

For authors, the production of camera ready manuscripts on the computer has many practical advantages. With this method, called *desktop publishing* (DTP), the author can shape the form of the publication directly and there are no galley-proofs to be corrected. Also, the time between text production and publication may be shortened, and the publication is much less expensive than with conventional typesetting.

---

[27] A description of TEI Lite and bibliographical references may be found in L. Burnard & C. Sperberg-McQueen 1995.

For linguists, on-line texts have the advantage that they can be analyzed electronically. With most current publications originating in the electronic medium anyhow it is only a question of access and copyright to obtain arbitrarily large amounts of on-line text such as newspapers, novels, or scientific publications in various domains.

One linguistic task is to select texts in such a way that their collection forms a representative and balanced sample of a language at a certain time (corpus construction, cf. Chapter 15). Another task is to analyze the texts in terms of their lexical, morphological, syntactic, and semantic properties. In either case, linguists are not interested in a text because of its content or layout, but as a genuine instance of natural language.

There are many possibilities to process an on-line text for linguistic analysis. For example, using some simple commands one may easily remove all control symbols from the text in 1.5.1 and then transform it into an alphabetical list of word forms.

1.5.3 ALPHABETICAL LIST OF WORD FORMS

10	in	STORY
146	in	suffered
1995	in	sun
20	its	than
4	its	that
a	LAND	The
a	land	the
a	landscape	the
a	like	the
a	LINDEN	the
above	Magazine	the
across	markers	the
and	mean	the
and	metaphor	the
and	midnight	the
and	midsummer	the
Arctic	million	through
as	mist	Throughout
as	more	to
barracks	mossy	to
bits	muting	TORTURED
bleached	No	tragedy
blows	Now	tundra
bones	of	vast
broken	of	Volume
camp	of	wooden
capitalists	of	world's
come	of	An
communism	off	as
Contents	page	at
covet	peak	COVER
dead	perished	EUGENE
death	place	fulfills
deprivation	places	ghostly
earth	protrude	in
easily	remains	Ocean

epic	reputation	over
Even	riches	Return
exile	settles	September
fog	shroud	Siberia
for	Siberia	Stalin's
grave	Siberia	THE
gray	siding	TIME
grievously	slave-labor	under
Gulag	soul-chilling	/YAKUTSK
has	steeped	

In this list, word forms are represented as often as they occur in the text, thus providing the basis for word-form statistics. It would be just as easy, however, to create a unique list in which each word form is listed only once, as for lexical work. Another approach to analyzing an on-line text for linguistic purposes is measuring the co-occurrence of word forms next to each other, based on bigrams and trigrams.

These methods all have in common that they are *letter-based*.[28] They operate with the abstract, digitally coded signs in the electronic medium, whereby word forms are no more than sequences of letters between spaces. Compared to nonelectronic methods – such as type-writing, typesetting, card indices, search by leafing and/or reading through documents, or building alphabetical word lists by hand, – the electronic computation on the basis of letters is fast, precise, and comfortable.

At the same time the letter-based method is limited in as much as any grammatical analysis is by definition outside of its domain. Letter-based technology and grammatical analysis may work closely together, however. By combining the already powerful letter-based technology with the concepts and structures of a functional, mathematically efficient, and computationally suitable theory of language, natural language processing may be greatly improved.

Exercises

Section 1.1

1. Describe different variants of man-machine communication.
2. In what sense is the interaction with a contemporary washing machine a special case of man-machine communication, and why is it not essential to computational linguistics?
3. Why is it difficult to get accustomed to programming languages, as in the interaction with a database?
4. Why is a standard computer a universal machine for processing information, yet at the same time of limited cognitive capacity in comparison to a robot?

---

[28] In the widest sense of word, including numbers and other signs.

5. Describe two different branches of artificial intelligence.
6. What is the principled difference between a robot and a VR system?

Section 1.2

1. Why is the development of talking robots of special interest to the theoretic development of computational linguistics?
2. Compare three different approaches to language analysis and describe their different methods, goals, and applications.
3. What are the components of grammar and what are their respective functions?
4. Why do different approaches to language analysis use the same kinds of components, dividing the phenomena in the same way?

Section 1.3

1. Classify computational linguistics as a science. Explain the notions numerical, nonnumerical, cognitive science, and artificial intelligence.
2. Which sciences are integrated into computational linguistics in an interdisciplinary way?
3. What are the methodological consequences of programming in computational linguistics?
4. What are practical applications of computational linguistics?

Section 1.4

1. What are the different media in which the expressions of natural language can be realized?
2. Explain the distinction between realization-dependent and realization-independed storage of information.
3. Does communication require constant d$\Rightarrow$i and i$\Rightarrow$d transfers only in artificial cognitive agents or also in humans?
4. Explain the notion OCR software.
5. What are the desiderata of automatic speech recognition?
6. What will be required for automatic speech recognition to become ultimately successul?

Section 1.5

1. What is the second Gutenberg revolution and how does it differ from the first?
2. Explain the technological advantages of the electronic medium.
3. Explain the term SGML.
4. What is the role of computers in DTP?
5. Explain the notions *hard copy* and *soft copy*.
6. What are the possibilities and the limitations of a purely technology-based natural language processing?

2. Technology and grammar

Having described possibilities of a purely technological approach to natural language processing in the previous chapter, we turn next to its limitations and ways of overcoming them by using linguistic knowledge. Section 2.1 explains the structures underlying the use of textual databases. Section 2.2 shows how linguistic methods can improve the retrieval from textual databases. Section 2.3 shows how different applications require linguistic knowledge to different degrees in order to be practically useful. Section 2.4 explains the notion of language pairs in machine translation and describes the *direct* and the *transfer* approach. A third approach to machine translation, the *interlingua* approach, as well as computer-based systems for aiding translation are described in Section 2.5.

2.1 Indexing and retrieval in textual databases

A textual database is an arbitrary collection of electronically stored texts. In contrast to a classic, record-based database like example 1.1.1, no structural restrictions apply to a textual database. Thus, the individual texts may be arranged, e.g., in the temporal order of their arrival, according to their subject matter, the name of their author(s), their length, or no principle at all.

The search for a certain text or text passage is based on the standard, letter-based *indexing* of the textual database.

2.1.1 STANDARD INDEXING OF A TEXTUAL DATABASE

> The indexing of a textual database is based on a table which specifies for each letter all the positions (addresses) where it occurs in the storage medium of the database.

The electronic index of a textual database functions in many ways like a traditional library catalog of alphabetically ordered filing cards.

Each filing card contains a *key word*, e.g. the name of the author, and the associated *addresses*, e.g. the shelf where the book of the author may be found. While the filing cards are ordered alphabetically according to their respective key words, the choice of the addresses is free. Once a given book has been assigned a certain address and this address has been noted in the catalog, however, it is bound to this address.

In an unordered library without a catalog, the search for a certain book requires looking through the shelves (linear search). In the worst case the book in question happens to be in the last of them. A library catalog speeds up such searching because it replaces a linear search by specifying the exact address(es) of the physical location. Thus, a book may be found using the alphabetic order of the filing cards irrespective of how the actual locations of the books are arranged.

The electronic index of a textual database uses the letters of the alphabet like the key words of a library catalog, specifying for each letter all its positions (addresses) in the storage medium. The occurrences of a certain word form, e.g. sale, is then computed from the intersection of the position sets of s, a, l, and e. The electronic index is built up automatically when the texts are read into the database, whereby the size of the index is roughly the same as that of the textual database itself.

The search for relevant texts or passages in the database is guided by the user on the basis of words (s)he considers characteristic of the subject matter at hand. Consider for example a lawyer interested in legal decisions dealing with the warranty in used car sales. After accessing an electronic database in which all federal court decisions since 1960 are stored, (s)he specifies the words warranty, sale, and used car. After a few seconds the database returns a list of all the texts in which these words occur. When the user clicks on a title in the list the corresponding text appears on the screen.

The user might well find that not all texts in the query result are actually relevant for the purpose at hand. It is much easier, however, to look through the texts of the query result than to look through the entire database.

Also, the database might still contain texts which happen to be relevant to the subject matter, yet are not included in the query result. Such texts, however, would have to deal with the subject matter without mentioning the query words.

The use of an electronic index has the following advantages over a card index.

2.1.2 ADVANTAGES OF AN ELECTRONIC INDEX

– Power of search
  Because the electronic index of a textual database uses the letters of the alphabet as its keys the database may be searched for any sequence of letters, whereas the keys of a conventional catalogue are limited to certain kinds of words, such as the name of the author.
– Flexibility
  – General specification of patterns
    An electronic index allows searching for patterns. For example, the pattern[1] in.*i..tion matches all word forms of which the first two letters are in, the seventh letter from the end is i and the last four letters are tion, as in inhibition and inclination.

---

[1] A widely used notation for specifying patterns of letter sequences are *regular expressions* as implemented in Unix.

– Combination of patterns

The electronic index allows searching for the combination of several word forms whereby a maximal distance for their co-occurrence may be specified.

Though it is theoretically possible to create a conventional card index for the positions of each letter in a library of books, this would not be practical. For this reason, searching with patterns or the combination of keywords and/or patterns is not technically feasible with a conventional card index.

– Automatic creation of the index structure

The electronic index of a textual database is generated automatically during the reading-in of texts into the database. In a conventional card index, on the other hand, each new key word requires making a new card by hand.

– Ease, speed, and reliability

While an electronic search is done automatically in milliseconds, error free, and complete, a conventional search using a card index requires human labor, is susceptible to errors, and may take anywhere from minutes to hours or days. The advantages of electronic search apply to both, the *query* (input of the search words) and the *retrieval* (output of the corresponding texts or passages).

  – Query

  An electronic database is queried by typing the search patterns on the computer, while the use of a card index requires picking out the relevant cards by hand.

  – Retrieval

  In an electronic database, the retrieved texts or passages are displayed on the screen automatically, while use of a conventional card index requires going to the library shelves to get the books.

The quality of a query result is measured in terms of recall and precision.[2]

2.1.3 DEFINITION OF RECALL AND PRECISION

Recall measures the percentage of relevant texts retrieved as compared to the total of relevant texts contained in the database.

For example: a database of several million pieces of text happens to contain 100 texts which are relevant to a given question. If the query returns 75 texts, 50 of which are relevant to the user and 25 are irrelevant, then the recall is $50 : 100 = 50\%$.

Precision measures the percentage of relevant texts contained in the result of a query.

---

[2] See G. Salton 1989, S. 248.

> For example: a query has resulted in 75 texts of which 50 turn out to be relevant to the user. Then the precision is 50 : 75 = 66.6%.

Experience has shown that recall and precision are not independent of each other, but inversely proportional: a highly specific query will result in low recall with high precision, while a loosely formulated query will result in high recall with low precision.

High recall has the advantage of retrieving a large percentage of the relevant texts from the database. Because of the concomitant low precision, however, the user has to work through a huge amount of material most of which turns out to be irrelevant.

High precision, on the other hand, produces a return most of which is relevant for the user. Because of the concomitant low recall, however, the user has to accept the likelihood that a large percentage of relevant texts remains undiscovered.

Measuring recall is difficult in large databases. It presupposes exact knowledge of all the texts or passages which happen to be relevant for any given query. To obtain this knowledge, one would have to search the entire database manually in order to objectively determine the complete set of documents relevant to the user's question and to compare it with the automatic query result.

Measuring precision, on the other hand, is easy because the number of documents returned by the system in response to a query is small compared to the overall database. The user must only look through the documents in the query result in order to find out which of them are relevant.

In a famous and controversial study, Blair & Maron 1985 attempted to measure the average recall of a leading commercial database system called STAIRS.[3] For this purpose they cooperated with a large law firm whose electronic data comprised 40 000 documents, amounting to a total of 350 000 pages. Because of this substantial, but at the same time manageable size of the data it was possible to roughly determine the real number of relevant texts for 51 queries with the assistance of the employees.

Prior to the study, the employees subjectively estimated an electronic recall of 75%. The nonelectronic verification, however, determined an average recall of only 20% – with a standard deviation of 15.9% – and an average precision of 79.0% – with a standard deviation of 22.2%.

2.2 Using grammatical knowledge

The reason for the surprisingly low recall of only 20% on average is that STAIRS uses only technological, i.e. letter-based, methods. Using grammatical knowledge in addition, recall could be improved considerably. Textual phenomena which resist a technological treatment, but are suitable for a linguistic solution are listed below under the heading of the associated grammatical component.

---

[3] STAIRS is an acronym for *Storage and Information Retrieval System*, a software product developed and distributed by IBM.

2.2.1 PHENOMENA REQUIRING LINGUISTIC SOLUTIONS

– *Morphology*

A letter-based search doesn't recognize words. For example, the search for sell will overlook relevant forms like sold.

Remedy would provide a program for word form recognition which automatically assigns to each word form the corresponding base form. By systematically associating each word form with its base form, all variants of a search word in the database can be found. A program of automatic word form recognition would be superior to the customary method of truncation – especially in languages with a morphology richer than that of English.

– *Lexicon*

A letter-based search doesn't take semantic relations between words into account. For example, the search for car would ignore relevant occurrences such as convertible, pickup truck, station wagon, etc.

A lexical structure which automatically specifies for each word the set of equivalent terms (synonyms), of the superclass (hypernyms), and of the set of instantiations (hyponyms) can help to overcome this weakness, especially when the domain is taken into account.[4]

– *Syntax*

A letter-based search doesn't take syntactic structures into account. Thus, the system doesn't distinguish between, e.g., teenagers sold used cars and teenagers were sold used cars.

Remedy would provide a syntactic parser which recognizes different grammatical relations between, e.g., the subject and the object. Such a parser, which presupposes automatic word form recognition, would be superior to the currently used search for words within specified maximal distances.

– *Semantics*

A letter-based search doesn't recognize semantic relations such as negation. For example, the system would not be able to distinguish between selling cars and selling no cars. Also, equivalent descriptions of the same facts, such as A sold x to B and B bought x from A, could not be recognized.

Based on a syntactic parser and a suitable lexicon the semantic interpretation of a textual database could analyze these distinctions and relations, helping to improve recall and precision.

– *Pragmatics*

According to Blair & Maron 1985, a major reason for the poor recall was the frequent use of context dependent formulations such as concerning our last letter,

---

[4] Some database systems already use *thesauri*, though with mixed results. Commercially available lexica are in fact likely to lower precision without improving recall. For example, in *Websters New Collegiate Dictionary*, the word car is related to vehicle, carriage, cart, chariot, railroad car, streetcar, automobile, cage of an elevator, and part of an airship or balloon. With the exception of automobile, all of these would only lower precision without improving recall.

following our recent discussion, as well as nonspecific words such as problem, situation, or occurrence.

The treatment of these frequent phenomena requires a complete theoretic understanding of natural language pragmatics. For example, the system will have to be able to infer that, e.g., seventeen year old bought battered convertible is relevant to the query used car sales to teenagers.

In order to improve recall and precision, linguistic knowledge may be applied in various different places of the database structure. The main alternatives are whether improvements of the search should be based on preprocessing the query, refining the index, and/or postprocessing the result. Further alternatives are an automatic or an interactive refinement of the query and/or the result, as described below.

2.2.2 LINGUISTIC METHODS OF OPTIMIZATION

A. Preprocessing the query

– Automatic query expansion
(i) The search words in the query are automatically 'exploded' into their full inflectional paradigm and the inflectional forms are added to the query.
(ii) Via a thesaurus the search words are related to all synonyms, hypernyms, and hyponyms. These are included in the query – possibly with all their inflectional variants.
(iii) The syntactic structure of the query, e.g. A sold x to B, is transformed automatically into equivalent versions, e.g. B was sold x by A, x was sold to B by A, etc., to be used in the query.

– Interactive query improvement
The automatic expansion of the query may result in an uneconomic widening of the search and considerably lower precision. Therefore, prior to the search, the result of a query expansion is presented to the user to allow elimination of useless aspects of the automatic expansion and allow for an improved formulation of the query.

B. Improving the indexing

– Letter-based indexing
This is the basic technology of search, allowing to retrieve the positions of each letter and each letter sequence in the database.

– Morphologically-based indexing
A morphological analyzer is applied during the reading-in of texts, relating each word form to its base form. This information is coded into an index which for any given word (base form) allows to find all corresponding (inflected) forms.

– Syntactically-based indexing
A syntactic parser is applied during the reading-in of texts, eliminating morphological ambiguities and categorizing phrases. The grammatical information is coded into an index which allows to find all occurrences of a given syntactic construction.

– Concept-based indexing
The texts are analyzed semantically and pragmatically, eliminating syntactic and semantic ambiguities as well as inferring special uses characteristic of the domain. This information is coded into an index which allows to find all occurrences of a given concept.

C. Postquery processing

– The low precision resulting from a nonspecific formulation of the query may be countered by an automatic processing of the data retrieved. Because the raw data retrieved are small as compared to the database as a whole they may be parsed after the query[5] and checked for their content. Then only those texts are given out which are relevant according to this post query analysis.

The ultimate goal of indexing textual databases is a concept-based indexing founded on a complete morphological, syntactic, semantic, and pragmatic analysis of the texts. This type of indexing promises not only fast search with maximal recall and precision, but also an automatic *classification* of texts. Today's manual classification is not only slow and expensive, but also unreliable. Research on manual classification has shown that two professional classifiers agreed on only 50 percent of a given set of texts.

2.3 Smart versus solid solutions

Which of the alternatives mentioned in 2.2.2 is actually chosen in the design of a textual database depends on the amount of data to be handled, the available memory and speed of the hardware, the users' requirements regarding recall, precision, and speed of the search, and the designer's preferences and abilities. At the same time, the alternatives of 2.2.2 are not independent from each other.

For example, if an improvement of recall and precision is to be achieved via an automatic processing of the query, one can use a simple indexing. More specifically, if the processing of the query explodes the search words into their full inflectional paradigm for use in the search, a morphological index of the database would be superfluous. Conversely, if there is a morphological index, there would be no need for exploding the search words.

Similarly, the automatic expansion of queries may be relatively carefree if it is to be scrutinized by the user prior to search. If no interactive fine-tuning of queries is provided, on the other hand, the automatic expansion should be handled restrictively in order to avoid a drastic lowering of precision.

Finally, the indexing of texts can be comparatively simple if the results of each query are automatically analyzed and reduced to the most relevant cases before output to

---

[5] Operations which are performed while the user is interacting with the system are called *on the fly operations*, in contrast to *batch mode operations*, e.g. building up an index, which are run when the system is closed to public use.

the user. Conversely, a very powerful index method, such as concept-based indexing, would produce results with such high precision that there would be no need for an automatic postprocessing of results.

The different degrees of using linguistic theory for handling the retrieval from textual databases illustrate a more general alternative in the design of computational applications, namely the alternative between *smart* versus *solid* solutions. A classic example of a smart solution is the Eliza program (Weizenbaum 1965), which was intended to illustrate that computer programs may appear to have cognitive abilities that they actually do not have (negative example).[6]

However, there are also positive examples of smart solutions, providing the user with a partial, yet highly welcome improvement by avoiding the difficult, costly, or theoretically unsolved aspects of the task at hand – such as the use of *restricted language* in machine translation.[7] Solid solutions, on the other hand, are based on a complete theoretical and practical understanding of the phenomena involved.[8]

Whether a given problem is suitable for a smart or a solid solution depends much on whether the application requires a perfect result or whether a partial answer is sufficient. For example, a user working with a giant textual database will be greatly helped by a recall of 70%, while a machine translation system with 70% accuracy will be of little practical use.

The two problems differ in that a 70% recall in a giant database is much more than a user could ever hope to achieve with human effort alone. Also, the user never knows which texts the system didn't retrieve.

In translation, on the other hand, the deficits of an automatic system with 70% accuracy are painfully obvious to the user. Furthermore there is an alternative available, namely professional human translators. Because of the costly and time consuming human correction required by today's machine translation the user is faced daily with the question of whether or not the machine translation system should be thrown out altogether in order to rely on human work completely.

Another, more practical factor in the choice between a smart and a solid solution in computational linguistics is the off-the-shelf availability of grammatical components for the natural language in question. Such components of grammar, e.g. automatic word form recognition, syntactic parsing, etc., must be developed independently of any specific applications as part of basic research – solely in accordance with the general criteria of (i) their functional role as components in the mechanism of natural communication, (ii) completeness of data coverage, and (iii) efficiency.

Once these modular subsystems have shown their functional adequacy in real time models of natural communication, they can be put to use in practical applications with

---

[6] As mentioned in the Introduction III, Eliza is based on the primitive mechanics of predefined sentence patterns, yet may startle the naive user by giving the appearance of understanding, both on the level of language and of human empathy.

[7] See 2.5.5, 3.

[8] The alternative between smart and solid solutions will be illustrated with statistically-based (Sections 13.4, 13.5) and rule-based (Chapter 14) systems of word form recognition.

no need for modification, using their standard interfaces. The more such modules become available as ready-made, well-documented, portable, off-the-shelf products for different languages, the less costly will the strategy of solid solutions be in applications. The main reason for the long term superiority of solid solutions, however, is quality. This is because a 70% smart solution is typically very difficult or even impossible to improve to 71%.

2.4 Beginnings of machine translation

The choice between smart and solid solutions is illustrated by different approaches to machine translation. Translation in general requires that the meaning of a given text or utterance in a certain natural language be reconstructed in another language.

On the one hand, translation is a special task. Superficially, it may seem related to bilingual communication because bilingualism obviates the need for translation.[9] However, a bilingual speaker never uses more than *one* language at a time and translation is not included in the everyday repertoire of bilingual speakers and hearers.

On the other hand, translation avoids some of the hardest problems of language production, such as the selection of content, the serialization, the lexicalization, etc., because it starts from a coherent source text that is given in advance. Machine translation tries to utilize these special circumstances in order to translate large amounts of nonliterary text automatically, usually into several different languages at once.

The administration of the European Union,[10] for example, must publish every report, protocol, decree, law, etc., in the 11 different languages of the member states (assuming the membership of 1999). For example, a decree formulated in French under a French EU presidency would have to be translated into the following 10 languages.

French → English	French → Spanish
French → German	French → Portugese
French → Italian	French → Greek
French → Dutch	French → Danish
French → Swedish	French → Finnish

Under a Danish EU presidency, on the other hand, a document may first be formulated in Danish. Then it would have to be translated into the remaining EU languages, resulting in another set of language pairs.

The total number of language pairs for a set of different languages is determined by the following formula.

2.4.1 FORMULA TO COMPUTE THE NUMBER OF LANGUAGE PAIRS

$n \cdot (n-1)$, *where n = number of different languages*

---

[9] Presuming a given pair of languages.

[10] Another example is the United Nations, which generate a volume of similar magnitude.

For example, an EU with 11 different languages has to deal with a total of $11 \cdot 10 = 110$ language pairs.

In a language pair, the source language (SL) and the target language (TL) are distinguished. For example, 'French→Danish' and 'Danish→French' constitute different language pairs. The source language poses the task of correct *understanding* the intended meaning, taking into account the domain and the context of utterance, whereas the target language poses the task of *formulating* the meaning rhetorically correct.[11]

The first attempts at machine translation tried to get as far as possible with the new computer technology, avoiding linguistic theory as much as possible. This resulted in the smart solution of 'direct translation,' which was dominant in the 1950's and -60's.

Direct translation systems assign to each word form in the source language a corresponding form of the target language. In this way one hoped to avoid a meaning analysis of the source text, yet to arrive at translations which are syntactically acceptable and express the meaning correctly.

2.4.2 SCHEMA OF DIRECT TRANSLATION

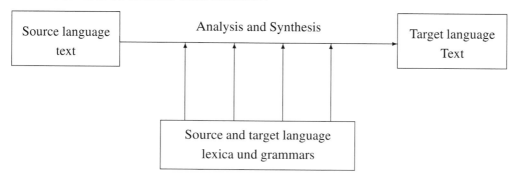

Each language pair requires the programming of its own direct translation system.

Direct translation is based mainly on a differentiated dictionary, distinguishing many special cases for a correct assignment of word forms in the target language. In the source language, grammatical analysis is limited to resolving ambiguities as much as possible, in the target language, to adjusting the word order.

The methodological weakness of direct translation systems is that they do not systematically separate between the source language analysis and the target language synthesis. Consequently one is forced with each new text to add new special cases and exceptions. In this way the little systematic structure present initially is quickly swept away by a tidal wave of exceptions and special cases.

Even though representatives of the direct approach repeatedly asserted in the 1950's that the goal of machine translation, namely

---

[11] For this reason a translation from French into Danish will require a French→Danish dictionary, but hardly a Danish→French dictionary. Because of language specific lexical gaps, idioms, etc., the vocabulary of the two languages is not strictly one to one, for which reason the two dictionaries are not really symmetric.

FULLY AUTOMATIC HIGH QUALITY TRANSLATION (FAHQT)

was just around the corner, their hopes were not fulfilled. Hutchins 1986 provides the following examples to illustrate the striking shortcomings of early translation systems.

2.4.3 EXAMPLES OF AUTOMATIC MIS-TRANSLATIONS

Out of sight, out of mind. ⇒ *Invisible idiot.*
The spirit is willing, but the flesh is weak. ⇒ *The whiskey is alright, but the meat is rotten.*
La Cour de Justice considère la création d'un sixième poste d'avocat général. ⇒ *The Court of Justice is considering the creation of a sixth avocado station.*

The first two examples are apocryphic, described as the result of an automatic translation from English into Russian and back into English. The third example is documented[12] as output of the SYSTRAN system.

To avoid the methodological weaknesses of direct translation, the transfer approach was developed. It is characterized by a modular separation of

– source language analysis and target language synthesis, of
– linguistic data and processing procedures, and of the
– lexica for source language analysis, target language transfer, and target language synthesis.

2.4.4 SCHEMA OF THE TRANSFER APPROACH

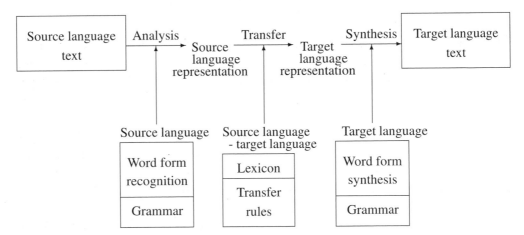

Separating source language analysis and source-target language transfer results in a clearer structure as compared to the direct approach, facilitating debugging and up-scaling of transfer systems. Implementing the different modules independently of each

---

[12] P. Wheeler & V. Lawson 1982.

other and separating the computational algorithm from the language specific data also makes it possible to reuse parts of the software when adding another language pair.

For example, given a transfer system for the language pair A-B, adding the new language pair A-C requires writing new transfer and synthesis modules for language C, but will allow reusing the analysis module of the source language A. Furthermore, if the language specific aspects of the new transfer and synthesis modules are written within a prespecified software framework suitable for different languages, the new language pair A-C should be operational from the beginning.

The three phases of the transfer approach are illustrated below with a word form.

2.4.5 THREE PHRASES OF A WORD FORM TRANSFER *English-German*

1. Source language analysis:

    Unanalyzed surface: `knew`

    Morphological and
    lexical analysis: `(knew (N A V) know)`

    The source language analysis produces the syntactic category (N A V) of the inflectional form (categorization) and the base form **know** (lemmatization).
2. Source-target language transfer:

    Using the base form resulting from the source language analysis, a source-target language dictionary provides the corresponding base forms in the target language.

    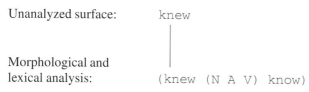

    `know  ⟹   wissen`
    `          kennen`

3. Target language synthesis

    Using the source language category (resulting from analysis) and the target language base forms (resulting from transfer), the desired target language word forms are generated based on target language morphology.

`wußte`	`kannte`
`wußtest`	`kanntest`
`wußten`	`kannten`
`wußtet`	`kanntet`

The transfer of a syntactic structure functions similar to the transfer of word forms. First, the syntactic structure of the source language sentence is analyzed. Second, a corresponding syntactic structure of the target language is determined (transfer). Third, the target language structure is filled with the target language word forms (synthesis) whereby a correct handling of agreement, a domain specific lexical selection, a correct positioning of pronouns, a rhetorically suitable word order, and other issues of this kind must be resolved.

Due to similarities between the direct and the transfer method, they have the following shortcomings in common.

2.4.6 SHORTCOMINGS OF THE DIRECT AND THE TRANSFER APPROACH

– Each language pair requires a special source-target component.
– Analysis and synthesis are limited to single sentences.
– Semantic and pragmatic analysis are avoided, attempting automatic translation without understanding the source language.

Thus, the advantage of the transfer approach over the direct approach is limited to the reusability of certain components, specifically the source language analysis and the target language synthesis for additional language pairs.

2.5 Machine translation today

The importance of language *understanding* for adequate translation is illustrated by the following examples:

2.5.1 SYNTACTIC AMBIGUITY IN THE SOURCE LANGUAGE

1. Julia flew and crashed the air plane.
      Julia (flew and crashed the air plane)
      (Julia flew) and (crashed the air plane)
2. Susan observed the yacht with a telescope.
   Susan observed the man with a beard.
3. The mixture gives off dangerous cyanide and chlorine fumes.
      (dangerous cyanide) and (chlorine fumes)
      dangerous (cyanide and chlorine) fumes

The first example is ambiguous between using the verb fly transitively (someone flies an air plane) or intransitively (something flies). The second example provides a choice between an adnominal and an adverbial interpretation of the prepositional phrase (cf. section 12.5). The third example exhibits a scope ambiguity regarding dangerous. A human translator recognizes these structural ambiguities, determines the intended reading, and recreates the proper meaning in the target language.

  A second type of problem for translation without understanding the source language arises from lexical differences between source and target language:

2.5.2 LEXICAL DIFFERENCES BETWEEN SOURCE AND TARGET

1. The men killed the women. Three days later they were caught.
   The men killed the women. Three days later they were buried.
2. know: wissen savoir
            kennen connaître

3. The watch included two new recruits that night.

When translating example 1 into French, it must be decided whether they should be mapped into ils or elles – an easy task for someone understanding the source language. Example 2 illustrates the phenomenon of a *lexical gap*: whereas French and German distinguish between savoir–wissen and connaître–kennen, English provides only one word, know. Therefore a translation from English into French or German makes it necessary to choose the correct variant in the target language. Example 3 shows a language specific lexical homonymy. For translation, it must decide whether watch should be treated as a variant of clock or of guard in the target language.

A third type of problem arises from syntactic differences between the source and the target language:

2.5.3 SYNTACTIC DIFFERENCES BETWEEN SOURCE AND TARGET

– German:
  Auf dem Hof sahen wir einen kleinen Jungen, der einem Ferkel nachlief.
      Dem Jungen folgte ein großer Hund.

– English:
  In the yard we saw a small boy running after a piglet.
      A large dog followed the boy.
      The boy was followed by a large dog.

German with its free word order can front the dative dem Jungen in the second sentence, providing textual cohesion by continuing with the topic. This cannot be mirrored by a translation into English because of its fixed word order. Instead one can either keep the active verb construction of the source language in the translation, losing the textual cohesion, or one can take the liberty of changing the construction into passive. Rhetorically the second choice would be preferable in this case.

A fourth type of problem is caused by the fact that sequences of words may become more or less stable in a language, depending on the context of use. These fixed sequences range from frequently used 'proverbial' phrases to collocations and idioms.

2.5.4 COLLOCATION AND IDIOM

strong current | high voltage (but: *high current | *strong voltage)
bite the dust | ins Gras beißen (but: *bite the grass | *in den Staub beißen)

For adequate translation, such colloquial and idiomatic relations must be taken into account.

The problems illustrated in 2.5.1–2.5.4 cannot be treated within morphology and syntax alone. Thus, any attempt to avoid a semantic and pragmatic interpretation in machine translation leads quickly to a huge number of special cases. As a consequence, such systems cannot be effectively maintained.

In light of these difficulties, many practically oriented researchers have turned away from the goal of fully automatic high quality translation (FAHQT) to work instead on partial solutions which promise quick help in high volume translation.

2.5.5 PARTIAL SOLUTIONS FOR PRACTICAL MACHINE TRANSLATION

1. *Machine aided translation* (MAT) supports human translators with comfortable tools such as on-line dictionaries, text processing, morphological analysis, etc.
2. *Rough translation* – as provided by an automatic transfer system – arguably reduces the translators' work to correcting the automatic output.
3. *Restricted language* provides a fully automatic translation, but only for texts which fulfill canonical restrictions on lexical items and syntactic structures.

Systems of restricted language constitute a positive example of a smart solution. They utilize the fact that the texts to be translated fast and routinely into numerous different languages, such as maintenance manuals, are typically of a highly schematic nature. By combining aspects of automatic text generation and machine translation, the structural restrictions of the translation texts can be exploited in a twofold manner.

First, an on-line text processing system helps the authors of the original text with highly structured schemata which only need to be filled (text production). Second, the on-line text system accepts only words and syntactic constructions for which correct translations into the various target languages have been carefully prepared and implemented (machine translation).

The use of restricted language may be compared to the use of a car. To take advantage of motorized transportation, one has stay on the road. In this way one may travel much longer distances than one could on foot. However, there are always places a car cannot go. There one can leave the car and continue by walking.

Similarly, due to their automatic input restrictions, systems of restricted language provide reliable machine translation which is sufficiently correct in terms of form and content. If the text to be translated does not conform to the restricted language, however, one may switch off the automatic translation system and look for a human translator.

Besides these smart partial solutions, the solid goal of fully automatic high quality translation (FAHQT) for nonrestricted language has not been abandoned. Today's theoretical research concentrates especially on the interlingua approach, including knowledge-based systems of artificial intelligence. In contrast to the direct and the transfer approach, the interlingua approach does not attempt to avoid semantic and pragmatic interpretation from the outset.

The interlingua approach uses a general, language-independent level called the interlingua. It is desigend to represent contents derived from different source languages in a uniform format. From this representation, the surfaces of different target languages are generated.

2.5.6 SCHEMA OF THE INTERLINGUA APPROACH

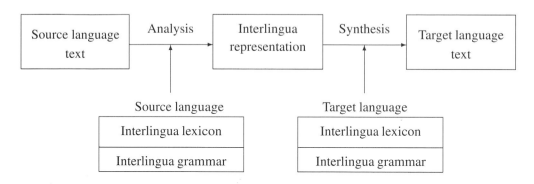

An interlingua system handles translation in two independent steps. The first step translates the source language text into the interlingua representation (analysis). The second step maps the interlingua representation into the targe language (synthesis).

It follows from the basic structure of the interlingua approach that for $n(n-1)$ language pairs only $2n$ interlingual components are needed (namely n analysis and n synthesis modules) – in contrast to the direct and the transfer approach which require $n(n-1)$ components. Thus, as soon as more than three languages ($n > 3$) are involved the interlingua approach has a substantial advantage over the other two.

The crucial question, however, is the exact nature of the interlingua. The following interlinguas have been proposed:

- an artificial logical language,
- a semi-natural language like Esperanto which is man-made, but functions like a natural language,
- a set of semantic primitives common to both, the source and the target language, serving as a kind of universal vocabulary.

Closer inspection shows, however, that these proposals have not yet resulted in theoretically and practically acceptable results. Existing interlingua systems are highly experimental, usually illustrating theoretical principles by translating tiny amounts of data by means of huge systems.

The experimental character of these attempts is not surprising because a general solution to interlingua translation may almost be equated with modeling the mechanism of natural language communication. After all, interlingua translation requires (i) a language-independent representation of cognitive content in the interlingua, (ii) the automatic translation of the natural source language into the language-independent interlingua, and (iii) the automatic generation of the natural target language from the interlingua.

And conversely: as soon as natural communication has been modeled on the computer in a general way, fully automatic high quality translation (FAHQT) is in close reach. At the same time all the other application of computational linguistics mentioned in 1.1.2, such as man-machine communication in natural language, a concept-based indexing of textual databases with a maximal recall and precision, etc., can be provided with solid solutions on the basis of available off the shelve modules.

These practical applications are one reason why the SLIM theory of language aims from the outset at modeling the mechanism of natural language communication in general. Thereby verbal and nonverbal contents are represented alike as concatenated propositions, defined as sets of bidirectional *proplets* in a classic network database. This new format is not only suitable for modeling production (cf. Chapter 23) and interpretation (cf. Chapter 24) in natural man-machine communication, but also as a universal interlingua.

Exercises

Section 2.1

1. Explain the notions recall and precision using the example of database containing 300 texts relevant for a given query, whereby 1000 texts are retrieved, of which 50 turn out to be relevant.
2. What are the weaknesses of a purely technology-based indexing and retrieval in textual databases?
3. Give examples in which truncation leads to the retrieval of irrelevant word forms and relevant word forms are missed.
4. Read Chapter 11, *Language Analysis and Understanding*, in Salton 1989 (pp. 377-424). Give a written summary of 3–5 pages of the linguistic methods for improving information retrieval described there.

Section 2.2

1. Which components of a textual database system are susceptible to linguistically-based optimization?
2. What is the difference between on the fly processing and batch mode processing? Illustrate the difference using the examples query expansion, indexing, and processing of a query result.
3. Why is high quality indexing suited better for fast search than preprocessing the query or postprocessing the retrieved data?

4. What are the costs of high quality indexing as compared to preprocessing the query or postprocessing the retrieved data?

Section 2.3

1. What are the different possibilities of improving retrieval from a textual database, and how are they connected with each other?
2. Describe the different advantages and disadvantages of smart versus solid solutions in applications of computational linguistics.
3. Which kinds of applications are not suitable for smart solutions?
4. Call up the Eliza program in the Emacs editor with 'meta-x doctor' and check it out. Explain why the Eliza program is a smart solution. What is the function of grammatical components in Eliza?

Section 2.4

1. Describe the differences between the direct and the transfer approach to machine translation.
2. What is the ultimate goal of machine translation?
3. In 1995, the EU was expanded from 12 to 15 member states, increasing the number of different languages from 9 to 11. How many additional language pairs resulted from this expansion?
4. Which components of a transfer system can – and which cannot – be reused? Explain your answer with the example of six language pairs for the languages English, French, and German. Enumerate the components necessary in such a system.

Section 2.5

1. Which phenomena of language use make an understanding of the source language necessary for adequate translation?
2. It is sometimes pointed out that English has no word corresponding to German **Schadenfreude**. Does this mean in your opinion that the corresponding concept is alien to speakers of English and cannot be expressed? Provide two further examples of *lexical gaps* relative to language pairs of your choice.
3. Provide two examples of collocations.
4. Where in machine translation could one use off-the-shelve components of grammar?
5. Which linguistic insights could be gained from building such a system of controlled language?
6. Why is machine translation with controlled language an example of a smart solution?
7. What is an interlingua?
8. How many additional components are needed when adding three new languages to an interlingua system?

3. Cognitive foundation of semantics

Modeling the mechanism of natural communication in terms of a general and computationally efficient theory has a threefold motivation in computational linguistics. Theoretically, it requires discovering how natural language actually works – surely an important problem of general interest. Methodologically, it provides a unified, functional viewpoint for developing the components of grammar on the computer and allows objective verification of the theoretical model in terms of its implementation. Practically, it serves as the basis for solid solutions in advanced applications.

The mechanism of natural communication is described in Chapters 3–6 in terms of constructing a robot named CURIOUS. The present chapter lays the ground by describing how CURIOUS perceives and cognitively processes its immediate environment. The result is a preliminary version of the robot functioning without language, but suitable for adding a language component in Chapter 4.

Section 3.1 describes the cognitive abilities of CURIOUS in relation to its task environment. Section 3.2 explains how CURIOUS recognizes simple geometric objects. Section 3.3 defines the notions of the internal *context* and its *concepts*. Section 3.4 describes how the analysis of the task environment results in the automatic derivation of contextual I-propositions. Section 3.5 integrates the components of CURIOUS into a functioning system and defines a program for controlling the buildup and update of the robot's internal representation of its external task environment.

3.1 Prototype of communication

The question of how the natural languages function in communication may seem complicated because there exist so many different ways of using language. Consider the following examples:

3.1.1 VARIANTS OF LANGUAGE COMMUNICATION

- two speakers are located face to face and talk about concrete objects in their immediate environment
- two speakers talk on the telephone about events they experienced together in the past

– a merchant writes to a company to order merchandise in a certain number, size, color, etc., and the company responds by filling the order
– a newspaper informs about a planned extension of public transportation
– a translator reconstructs an English short story in German
– a teacher of physics explains the law of gravitation
– a registrar issues a marriage license
– a judge announces a sentence
– a conductor says: Terminal station, everybody please get off.
– a sign reads: Do not step on the grass!
– a professor of literature interprets an expressionistic poem
– an author writes a science fiction story
– an actor speaks a role

These different variants are not per se an insurmountable obstacle to designing a general model of communication. They only require finding a basic mechanism which works for all of them while able to accommodate their respective differences.

The SLIM theory of language proceeds on the hypothesis that there is a basic prototype which includes all essential aspects of natural communication. This prototype is defined as follows.

3.1.2 PROTOTYPE OF COMMUNICATION

The basic prototype of natural communication is the direct face to face discourse of two partners talking about concrete objects in their immediate environment.

Possible alternatives to 3.1.2 would be approaches which take, e.g., (i) complete texts or (ii) the signs of nature, such as smoke indicating fire, as their basic model.

The prototype hypothesis is proven in two steps. First, a robot is described which allows nonrestricted natural man-machine communication within the basic prototype. Second, it is shown that all the other variants in 3.1.1 are special cases or extensions which can be easily integrated into the cognitive structure of the robot.

Realizing the prototype of communication as a functioning robot requires an exact definition of the following components of basic communication:

3.1.3 THREE COMPONENTS OF THE COMMUNICATION PROTOTYPE

– Specification of the *task environment*[1]
– Structure of the *cognitive agent*
– Specification of the *language*

The task environment of the robot CURIOUS is a large room with a flat floor. Distributed randomly on the floor are objects of the following kind:

---

[1] The notion *task environment* was introduced by A. Newell & H. Simon 1972. The robot-internal representation of the task environment is called the *problem space*.

3.1.4 OBJECTS IN THE WORLD OF CURIOUS

– triangles (scalene, isoceles, etc.)
– quadrangles (square, rectilinear, etc.)
– circles and ellipses

These objects of varying sizes and different colors are elements of the real world.

The robot is called CURIOUS[2] because it is programmed to constantly observe the state of its task environment. The task environment keeps changing in unforeseeable ways because the human 'wardens' remove objects on the floor, add others, or change their position in order to test CURIOUS' attention.

CURIOUS knows about the state of its task environment by exploring it regularly. To avoid disturbing the objects on the floor, CURIOUS is mounted on the ceiling. The floor is divided into even sized fields which CURIOUS can visit from above.

The basic cognition of CURIOUS includes an internal map divided into fields corresponding to those on the floor and a procedure indicating its current external position on the internal map. Furthermore, CURIOUS can specify a certain goal on its internal map and then adjust its external position accordingly.

When CURIOUS finds an object while visiting a certain field, the object is analyzed and the information is stored, for example, Isoceles red triangle in field D2. By systematically collecting data of this kind for all fields, CURIOUS is as well-informed about its task environment as its human wardens.

3.2 From perception to recognition

The first crucial aspect of this setup is that the task environment of CURIOUS is an *open* world: the objects in the task environment are not restricted to a fixed, predefined set, but can be processed by the system even if some disappear and new ones are added in unpredictable ways.

The second crucial aspect is that the task environment is part of the *real* world. Thus, for the proper functioning of CURIOUS a nontrivial form of reference must be implemented, allowing the system to keep track of external objects.

The cognitive functioning of CURIOUS presupposes the real external world as given.[3] Its internal representations do not attempt to model the external world completely, but are limited to properties necessary for the intended interaction with the external world, here the perception and recognition of two-dimensional geometric objects of varying colors.

The performance of the system is evaluated according to the following criteria.

---

[2] CURIOUS is an advanced variant of the *color reader* described in CoL, p. 295 ff.

[3] This is in accordance with the approach of nouvelle AI, which proceeds on the motto *The world is its own best model*. See Section 1.1.

3.2.1 TWO CRITERIA TO EVALUATE CURIOUS

– Measuring the active and reactive behavior (behavior test)
– Measuring the cognitive processing directly (cognition test)

The behavior test is the conventional method of observing actions and reactions of cognitive agents in controlled environments. If only behavior tests are available – as is normally the case with natural cognitive agents – the examination of cognitive functions is limited.[4]

The cognition test consists in evaluating the cognitive performance of a system directly. This kind of test presupposes that the internal states can be accessed and accurately interpreted from the outside.

While we can never be sure whether our human partners see the world as we do and understand us the way we mean it, this can be determined precisely in the case of CURIOUS because its cognition may be accessed directly. Thus, the problem of *solipsism* may be overcome in CURIOUS.

The robot's recognition begins with an unanalyzed internal image of the object in question, e.g. a bitmap representing the outline and the color of a blue square.

3.2.2 INTERNAL BITMAP REPRESENTATION OF EXTERNAL OBJECT

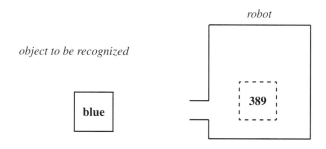

Inside the robot, the blue square is represented as a bitmap outline whereby the color appears as the electromagnetic frequency measured, i.e. 389 nm. Just as an OCR-system (cf. Section 1.4) analyzes bitmap structures to recognize letters, CURIOUS recognizes the form of objects in its task environment by matching their bitmap structures with corresponding patterns.[5]

The recognition of geometric forms may be viewed as a three step process. First, a suitable program approximates the bitmap outline with movable bars resulting in a reconstructed pattern. Second, the reconstructed pattern is logically analyzed in terms

---

[4] Behavior tests with humans may include the use of language by interviewing the subjects about their experience. This however, (i) introduces a subjective element and (ii) is not possible with all types of cognitive agents.

[5] A classic treatment of artificial vision is D. Marr 1982. For a summary see J.R. Anderson 1990[2], p. 36 ff. More recent advances are described in the special issue of *Cognition*, Vol. 67, 1998, edited by M.J. Tarr & H.H.Bülthoff .

of the number of corners, their angles, the length of the edges, etc. Third, the logical analysis is classified in terms of an abstract concept.

This process is illustrated in 3.2.3 with the example of a right-angled triangle.

3.2.3 ANALYSIS OF AN INTERNAL BITMAP REPRESENTATION

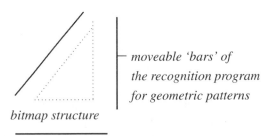

moveable 'bars' of
the recognition program
for geometric patterns

bitmap structure

Moving the edges to match the bitmap outline of a triangle results in a reconstructed pattern. This is logically analyzed as an area enclosed by three lines, two of which form a right angle. Finally, the logical analysis is classified in terms of an abstract concept resulting in the recognition of a right-angled triangle.[6]

The recognition of colors works in the same way. The rough data – corresponding to the bitmap outline – are the electromagnetic frequencies provided by the video camera. The first step of analysis consists in measuring the frequency (reconstructed pattern). The second step consists in associating the value with a color interval (logical analysis). The third step consists in classifying the interval in terms of a color concept resulting in the recognition of the color.

When a system like CURIOUS is alert it is anchored in its task environment by means of its perception and recognition. This means that the relevant aspects of its task environment are represented internally and constantly updated. The current internal representation is called the (nonverbal)[7] *context* of a cognitive agent.

3.2.4 DEFINITION OF THE CONTEXT

The context of a cognitive agent CA at a given point of time t includes

1. the total of all current cognitive parameter values CA_t,

---

[6] For the sake of conceptual simplicity, the reconstructed pattern, the logical analysis, and the classification are described here as separate phases. In practice, these three aspects may be closely interrelated in an incremental procedure. For example, the analysis system may measure an angle as soon as two edges intersect, the counter for corners may be incremented each time a new corner is found, a hypothesis regarding a possible matching concept may be formed early so that the remainder of the logical analysis is used to verify this hypothesis, etc.

[7] In the following, we will avoid the term '(non)verbal' as much as possible because of a possible confusion with the part of speech 'verb.' Instead of 'nonverbal cognition' we will use the term 'context-based cognition.' Instead of 'verbal cognition' we will use the term 'language-based cognition.'

2. the logical analyses of the parameter values and their combinations (reconstructed patterns),
3. the conceptual structures used to classify the reconstructed patterns and
   their combinations.

The cognitive processing of CURIOUS described so far illustrates the difference between perception and recognition. The raw data provided by the video camera are
what CURIOUS perceives. Their classification with respect to geometric shape and
color constitute what CURIOUS recognizes.

3.3 Iconicity of formal concepts

A cognitive agent without language interacts with the world in terms of recognition
and action. Recognition is the process of transporting structures of the external world
into the cognitive agent. Action is the process of transporting structures originating
inside the cognitive agent into the world. Recognition and action are related by means
of inferences on the data structures stored inside the cognitive agent.

Recognition and action alike may be analyzed in terms of three basic notions,
namely (i) concept types, (ii) parameter values, and (iii) concept tokens. The type-
token distinction was introduced by the American philosopher and logician C. S.
PEIRCE (1839–1914).[8]

Within the cognition of CURIOUS, the token of a certain square is represented as
follows.

3.3.1 I-CONCEPT$_{loc}$ OF A square (TOKEN)

$$
\begin{bmatrix}
\text{edge 1: 2cm} \\
\text{angle 1/2: } 90^0 \\
\text{edge 2: 2cm} \\
\text{angle 2/3: } 90^0 \\
\text{edge 3: 2cm} \\
\text{angle 3/4: } 90^0 \\
\text{edge 4: 2cm} \\
\text{angle 4/1: } 90^0
\end{bmatrix}_{loc}
$$

Such a parameter analysis is called an I-concept$_{loc}$ because it is a token *instantiating*
a certain type. The instantiation has edges of length 2cm. The feature *loc* specifies
when and where the token was recognized.

---

[8] An example of a token is the actual occurrence of a sign at a certain time and a certain place, for
example the now following capital letter A. The associated type, on the other hand, is the abstract
structure underlying all actual and possible occurrences of this letter. Realization-dependent differences between corresponding tokens, such as size, font, place of occurrence, etc., are not part of the
associated type.

Different tokens like 3.3.1 which differ only in the length of their edges may be expressed jointly in terms of the following type:

3.3.2 DEFINITION OF THE M-CONCEPT square (TYPE)

$$\begin{bmatrix} \text{edge 1: } \alpha \text{ cm} \\ \text{angle 1/2: } 90^0 \\ \text{edge 2: } \alpha \text{ cm} \\ \text{angle 2/3: } 90^0 \\ \text{edge 3: } \alpha \text{ cm} \\ \text{angle 3/4: } 90^0 \\ \text{edge 4: } \alpha \text{ cm} \\ \text{angle 4/1: } 90^0 \end{bmatrix}$$

In 3.3.2 the length of the edges is represented by the variable α.[9] Such a structure is called an M-concept because it defines a pattern which may be used for *matching*.[10]

The M-concept 3.3.2 is applicable to squares of any size including those which the system will encounter in the future. In the same manner other concepts like *triangle*, *rectangle*, *square*, *pentagon*, *right-angled*, *equilateral*, etc., may be defined explicitly in terms of parameter constellations, variables, and constants.[11]

M-concepts are usually defined for subsets of the available parameters and their possible values. For example, the concept 3.3.2 applies only to a certain constellation of visual parameter values. Other parameters, e.g. color values, are disregarded by the M-concept square. It is possible, however, to define elementary concepts which apply to a multitude of different parameters. For example, *situation* could be defined as the elementary M-concept which matches the totality of current parameter values.

3.3.3 DEFINITION: M-CONCEPT

An M-concept is the structural representation of a characteristic parameter constellation whereby certain parameter values are defined as variables.

In cognitive systems without language, elementary M-concepts may be defined only for those notions for which the corresponding parameters have been *implemented*. For

---

[9] Instances of the same variable in a concept must all take the same value. Strictly speaking, 3.3.2 would thus require an operator – for example a quantifier – binding the variables in its scope. We use sketchy definitions of tokens and types for the sake of simplicity and in order to avoid discussing the different advantages and disadvantages of logical versus procedural semantics (cf. Chapters 19–22).

[10] In context-based cognition, M-concepts are matched with parameter values, resulting in I-concepts$_{loc}$ during recognition and realizing I-concepts$_{loc}$ during action (cf. 3.3.5). In language-based cognition, M-concepts acquire a secondary function as literal meanings attached to word surfaces. During communication, the M-concepts of language are matched with corresponding I-concepts$_{loc}$ of the context (cf. 4.2.3, 23.2.1).

[11] One may also conceive of handling M-concepts connectionistically.

example, the concepts for **warm** and **cold** may be defined in CURIOUS only after (i) the system has been equipped with suitable sensors for temperature and (ii) the resulting measurements have been integrated into the conceptual structure of the system.[12]

The process of matching an M-concept onto relevant parameter values results in an I-concept$_{loc}$.

3.3.4 DEFINITION: I-CONCEPT$_{loc}$

> An I-concept$_{loc}$ results from successfully matching an M-concept onto a corresponding parameter constellation at a certain space-time location.

For any given M-concept there may exist arbitrarily many corresponding I-concepts$_{loc}$. These differ (i) in the instantiations of their variables and (ii) in the specific space-time values *loc* of their occurrence. The space-time values are provided by the agent's internal clock and spatial orientation system (cf. 3.4.2), and are written into the internal database together with the other values of the I-concept$_{loc}$.

The functioning of parameter values, M-concepts, and I-concepts$_{loc}$ in recognition and action, respectively, may be characterized schematically as follows.

3.3.5 CONTEXTUAL RECOGNITION AND ACTION

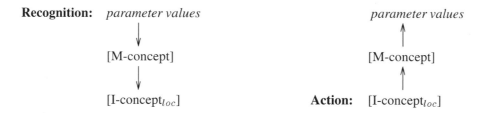

Recognition begins with the incoming parameter values onto which M-concepts are matched. A successful matching results in a corresponding I-concept$_{loc}$. In this way the parameter constellations of the internal context are individuated into I-concepts$_{loc}$. From the total of parameter values at a given moment, certain constellations are picked out and designated as I-concepts$_{loc}$ for, e.g., *triangle*, *red*, *warm*, etc.

Action begins with a certain I-concept$_{loc}$ which is to be realized as a specific outgoing parameter constellation by means of a corresponding M-concept. Consider for example a robot equipped with a gripping device wanting to pick up a glass. For this, a certain I-concept$_{loc}$ is realized as a gripping action (token) with the help of the M-concept (general procedure, type). Thereby distance, size, firmness, etc., of the object are determined via the robot's recognition and integrated into the planned action.

The cognitive formation of tokens (I-concepts$_{loc}$) in recognition and action requires the prior existence of the types (M-concepts). In an artificial system, the M-concepts

---

[12] Elementary M-concepts may be combined into complex M-concepts using procedural variants of logical operators, e.g. & or ¬.

are programmed by its designers. In a natural system, e.g. a human or a higher animal, some M-concepts have become innate in the course of evolution, while others evolve in the interaction with the environment as the result of learning.

In either case the natural evolution of M-concepts can be analyzed as an abstraction over logically similar parameter constellations, whereby the accidental aspects of constellations are represented by variables. Thus the types originate as the formation of classes over sets of similar raw data of perception or unconscious action. Only after the formation of types can individual tokens be instantiated on the basis of these types.

A set of similar parameter values may result in an abstract type for different reasons: frequent occurrence, rare occurrence, simple logical (e.g. geometrical) structure, pleasant or unpleasant nature, etc. For example, geometric shapes (visual parameter values) which are logically similar in that they consist of three straight, intersecting lines may be summarized as an abstract M-concept (type) which speakers of English happen to call **triangle**. Once the M-concept is available it can be applied to newly encountered parameter values resulting in individualized I-concepts$_{loc}$ of triangles. Correspondingly, electromagnetic frequencies which are similar in that they fall into the same narrow range may result in the abstract M-concept (type) of a certain color.[13]

Our analysis of M-concept formation and M-concept use is an *iconic* theory. The notion of iconicity derives from the classical greek word for image (ikon). Intuitively iconicity means the following:

3.3.6 ASPECTS OF ICONICITY

- The parameter values of the internal context are images insofar as they reflect the corresponding structures of the real world.
- The reconstructed patterns (I-concepts$_{loc}$) are images of parameter values because they are logical analyses of parameter values.
- The M-concepts of the internal context are images insofar as they (i) originate as abstractions over similar parameter constellations and (ii) characterize associated classes of reconstructed patterns.

That the recognition of CURIOUS is iconic follows quite simply from the requirement that the system should perceive its task environment *correctly*. For example, if an external triangle is represented internally as the outline of a quadrangle, then the system doesn't work right.

---

[13] Our abstract description of visual recognition is compatible with the neurological view. For example, after describing the neurochemical processing of photons in the eye and the visual cortex, D.E. Rumelhart 1977, p. 59f. writes:

> These features which are abstracted from the visual image are then to be matched against memorial representations of the possible patterns, thus finally eliciting a name to be applied to the stimulus pattern, and making contact with stored information pertaining to the item.

The same holds for action. Assume that CURIOUS is equipped with a gripping device and wants to move an object from field A2 into field A4. If the external action of the gripping device is not iconic with respect to the internal presentation of the movement in question and the object is moved instead into field A5, then the system obviously does not function correctly.

In the literature, the properties and the status of iconicity have long been controversial and are, e.g., at the core of the debate between the naturalists and the conventionalists in ancient Greece (cf. Section 6.4). The often fierce opposition to iconic constructs[14] stems from misunderstanding iconicity in the sense of naive little pictures. That this is not meant at all is illustrated by the examples 3.3.1 (parameter constellation, I-concept$_{loc}$) and 3.3.2 (M-concept), which use an abstract coding. The notion of an icon must be understood in the mathematical sense of a *homomorphism* (cf. Section 21.3), i.e. as a structure-preserving representation.

For example, if the photograph of a familiar face is scanned into the computer and ultimately coded as a sequence of zeros and ones, this coding is iconic in our sense because the original image can be recreated from it on the screen. This point of view was taken by the English philosopher John LOCKE (1632–1704), who said that a person seeing a tree has an image of that tree in his head – which is shown by the fact the person can later make a drawing.

There are basically three arguments against iconicity, which seem to be as popular as they are misguided. The first one is based on the claim that if one were to surgically search the brain of a person who has seen a tree, one would not find such an image.

In reply we point out that this method is much too crude. After all, in the analogous case of a computer it is not certain that a thorough investigation of the hardware would discover an image scanned-in before – even though the computer is a man-made machine and as such completely understood. Furthermore, modern research has shown that the optical cortex does indeed exhibit *'iconic representations'* such as "the line, edge, and angle detectors discovered by Hubel & Wiesel (1962) and the iconic or sensory memories proposed by Sperling (1960) and Neisser (1967)"[15] from which the internal representations of the images are built up.

The second argument was used in the famous controversy among the British empiricists regarding the (non)existence of abstract ideas. Locke took the view that recognition is based on abstract concepts or 'ideas' in the head of the perceiving person. He said that a person recognized, e.g., a triangle on the basis of an internal concept (idea) of a triangle.

---

[14] For example, C.K. Ogden & I.A. Richards 1923 call the use of icons or images in the analysis of meaning 'a potent instinctive belief being given from many sources' (p. 15) which is 'hazardous,' 'mental luxuries,' and 'doubtful' (p. 59). In more recent years, the idea of iconicity has been quietly rehabilitated in the work of W. Chafe 1970, P. Johnson-Laird 1983 (cf. p. 146,7), T. Givón 1985 (cf. p. 189), J. Haiman 1985a,b and others.

[15] S. Palmer 1975.

One generation later this analysis was attacked by George BERKELEY (1685–1753). Berkeley, a bishop by profession,[16] regarded Locke's approach as naive and tried to reduce it to absurdity by asking what *kind* of triangle the concept should be *exactly* : isoceles, scalene, right-angled?

CURIOUS, however, shows Berkeley's argument to be a fallacy. CURIOUS uses an abstract concept of a triangle based on a formal definition involving three straight, intersecting lines forming three angles which together add up to 180 degrees. This concept is realized procedurally as part of a pattern recognition program which may be demonstrated to recognize *all* possible kinds of triangles (cf. 3.2.3).[17]

The third argument is the homunculus-argument, which was used by David HUME (1711–1776) two generations after Locke. The homunculus-argument goes like this: if there are pictures in the mind, then there must be someone to see them. Yet postulating a little man (homunculus) in the head to see the images would not do because the little man would have images in his little head in turn, requiring another homunculus, and so on. Since postulating a homunculus is of no help to understand the interpretation of images, the images themselves are concluded to be superfluous.

CURIOUS, however, shows Hume's argument to be a fallacy as well. CURIOUS uses two kinds of iconic structures: M-concepts and I-concepts$_{loc}$. Neither of them is intended to be seen by any CURIOUS-internal homunculus. Instead the M-concepts are matched onto parameter values whereby their external origin ('Urbild') is classified and instantiated as an I-concept$_{loc}$ (see 3.3.5).

3.4 Contextual I-propositions

In order to perform its task, CURIOUS must (i) move through its environment, (ii) analyze each current field, (iii) represent the objects found there internally, and (iv) integrate the results into a correct cognitive representation of the overall situation. To this purpose, the I-concepts$_{loc}$ derived are combined into elementary *propositions*.

In accordance with the classic view since ARISTOTLE (384–322 B.C.), propositions are simple representations of what is. Propositions are so general and abstract that they have been regarded as both, the states of real or possible worlds and the meanings of language sentences.

Propositions are built from three basic kinds of elements, called *functors*, *arguments*, and *modifiers*. An elementary proposition consists of one functor which com-

---

[16] C.S. Peirce 1871 writes about Berkeley:

> Berkeley's metaphysical theories have at first sight an air of paradox and levity very unbecoming to a bishop. He denies the existence of matter, our ability to see distance, and the possibility of forming the simplest general conception; while he admits the existence of Platonic ideas; and argues the whole with a cleverness which every reader admits, but which few are convinced by.

[17] The program may even be expanded to recognize bitmap outlines with imprecise or uneven contours by specifying different degrees of granularity. Cf. J.L. Austin's 1962 example *France is hexagonal.*

bines with a characteristic number of arguments. Modifiers are optional and may apply to functors as well as to arguments.

In the world, the functors are the (intrapropositional) *relations*, the arguments are the *objects* (in the widest sense), and the modifiers are the *properties*. These basic elements constitute a simple ontology which is intended here for a general representation of cognitive states – in contradistinction to other possible ontologies such as for modeling aspects of the world from the viewpoint of physics (based on atoms, gravity, etc.), biology (based metabolism, reproduction, etc.), or economy (based on markets, inflation, interest rates, etc.).

In the natural languages, the functors are the one-, two-, or three-place *verbs* (cf. 16.2.2), the arguments are the *nouns*, and the modifiers are the *adjectives* and *adverbs* (whereby adjectives modify nouns and adverbs modify verbs). The verbs, nouns, and adjective-adverbials are also called the content words of a language and form the open word classes (cf. 13.1.8).

3.4.1 THE THREE ELEMENTS OF BASIC PROPOSITIONS

	logic	*world*	*language*
1.	functor	relation	verb
2.	argument	object	noun
3.	modifier	property	adjective-adverbial

Elementary propositions can combine with operators such as negation or be concatenated by operators such as conjunction. In the natural languages, operators are represented by the *function words* of the closed word classes (cf. 13.1.8). In the world, operators are realized in part by *extrapropositional relations*.

CURIOUS represents the contextual analysis of its task environment and its planned actions by means of propositions. These are called I-propositions because they are built from I-concepts$_{loc}$ (cf. 3.3.4). The following example illustrates an automatic analysis of a situation in which CURIOUS finds a triangle and a square in field A2.

3.4.2 AN EXAMPLE OF TWO CONTEXTUAL PROPOSITIONS

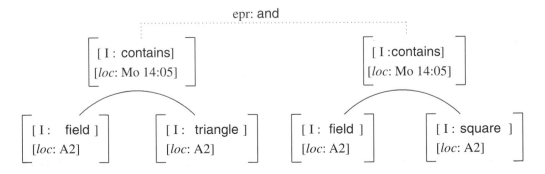

The elementary I-propositions in 3.4.2 may be paraphrased as field contains triangle and field contains square. They are connected by the conjunction and whereby the attribute epr stands for *extrapropositional relation*.[18]

The objects, relations, and properties recognized by CURIOUS are represented as feature structures. The I-concept$_{loc}$ of objects (nouns) specifies in its *loc*-feature the *place* at which the system encountered the object.[19] The I-concept$_{loc}$ of intrapropositional relations (verbs) specifies in its *loc*-feature the *time* at which the system determined the relation between objects.[20] The I-concept$_{loc}$ of properties (modifiers) has a *loc*-feature the value of which agrees with that of the modified.

The coherence of contextual propositions and their connections follows from the coherence of the external world which they reflect. This coherence is maintained by the system-internal algorithm which interprets the input parameters by automatically constructing elementary propositions and combining them into subcontexts.

On the one hand, the I-propositions in the database of CURIOUS relate concretely to spatio-temporal reality via the *loc*-features of their objects and relations. Thereby aspects of the external real world as well as aspects of internal sensations may be represented. This is because the method depicted in 3.3.5 is just as suitable for analyzing internal parameters (such as hunger or the loading state of the on-board battery) as for analyzing external parameters (such as visual perception).[21]

On the other hand, the I-propositions constitute an autonomous system-internal abstract representation which may be processed independently of the spatio-temporal reality. For example, the I-propositions may represent information not only about current, but also about past and possible future states of the task environment, which can be compared with each other, etc. The abstract processing of I-propositions (inferencing, see Section 24.5) may result in actions, however, which again relate meaningfully to the spatio-temporal reality.

The contextual aspects of the external world, e.g. the program of a washing machine, the number of planets in the solar system, the atomic structure of the elements, the colors, etc., are inherently nonlinguistic in nature. To call these structures a 'language' is inappropriate because it would stretch the notion of a language beyond recognition.

The essentially nonlinguistic nature of the external originals holds also for their internal representations in the cognitive agent. Higher nontalking animals like a dog are able to develop types (M-concepts), to derive I-concepts$_{loc}$, to combine them into elementary I-propositions, to concatenate these into subcontexts, and to draw inferences.

---

[18] In Chapters 23 and 24 the intuitive format of 3.4.2, which uses an arc and a dotted line to indicate intra- and extrapropositional relations, respectively, is replaced by indices in the abstract format of a network database.

[19] For example, in the feature [*loc*: A2], the value A2 stands for a certain field in the task environment of CURIOUS.

[20] For example, in the feature [*loc*: Mo 14:05], the value stands for Monday, five minutes after 2 p.m.

[21] In analytic philosophy, internal parameters – such as an individual tooth ache – have been needlessly treated as a major problem because they are regarded as a 'subjective' phenomenon which allegedly must be made objective by means of indirect methods such as the *double aspect* theory. See in this connection the treatment of propositional attitudes in Section 20.3, especially footnote 9.

These cognitive structures and procedures are not explicitly defined as a language, however, but evolve solely as physiologically grown structures.

The contextual structures of a nontalking natural cognitive agent acquire a (description) language aspect only, if and when they are analyzed theoretically. Corresponding artificial systems, on the other hand, usually begin with a language-based definition which is then realized in terms of the hard- and software of the implementation. However, even in artificial systems the language aspect may be completely ignored once it is up and running: on the level of its machine operations the cognitive procedures of a nontalking robot are just as nonlinguistic as those of a corresponding natural agent.

The correlation between the nonlanguage and the language level in the description a nontalking natural cognitive agent and its artificial model may be described schematically as follows.

3.4.3 ARTIFICIAL MODELING OF NATURAL COGNITION

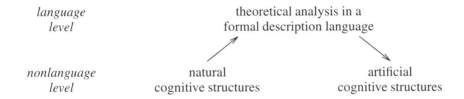

The point is that modeling the context representation of a robot in terms of I-propositions based on feature structures and defining the procedures operating on these context structures in terms of a formal grammar is not in conflict with the essentially nonverbal character of these phenomena. Instead feature structures and grammatical algorithms are general abstract formalisms which may be used as much for the modeling of nonverbal structures as for the description of natural or artificial languages. Furthermore, once the nonverbal structures and the associated inferences have been implemented as electronic procedures they function without any recourse to the language that was used in their construction.[22]

3.5 Recognition and action

The cognitive representation of the task environment in the database of CURIOUS as well as the cognitive derivation of action schemes and their realization are based on five components which interact in well-defined procedures.

---

[22] See autonomy from the metalanguage in Section 19.4.

3.5.1 SCHEMATIC STRUCTURE OF CONTEXT-BASED COGNITION

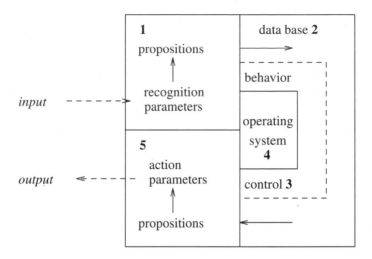

The recognition component 1 analyzes a given field of CURIOUS's task environment by matching M-concepts onto constellations of incoming parameter values. These are instantiated as I-concepts$_{loc}$ (cf. recognition in 3.3.5), automatically combined into concatenated I-propositions (cf. 3.4.2), and stored in the database component 2.

A planned change of the task environment is realized by means of the action component 5. Thereby I-concepts$_{loc}$ are realized with the help of M-concepts as outgoing parameter constellations (cf. action in 3.3.5). The action schemata are based on I-propositions which are generated in the database component 2.

The storing of recognitions and the planning of actions require the behavior control component 3 and the operating system 4. The control program 3 is illustrated below.

3.5.2 EXAMPLE OF A BEHAVIOR CONTROL PROGRAM

1. Primary analysis of the current task environment:
   a) Move into the start field A1.[23]
   b) Analyze the current field:

      i. Approximate bitmap outline with edge program.
      ii. Measure color value inside the bitmap outline.
      iii. Derive I-proposition.

   c) Write I-proposition at index P-0.1 (present) into the database.
   d) If current field is not D4, move into the next field and enter state b. Otherwise go to 2.
2. Secondary analysis of current task environment (inferences):

---

[23] The formulation in 3.5.2 assumes that the task environment is divided into 16 fields, named A1, A2, A3, A4, B1, B2 etc., up to D4.

    a) Count all triangles, rectangles, squares, red triangles, etc., in the primary analysis P-0.1 and write the result at index P-0.2 into the database.

    b) Compare the current secondary analysis P-0.2 with the previous secondary analysis P-1.2 and write the result (e.g. 'number of red triangle increased by 2') at index P-10.3 into the database.

3. Wait for 10 minutes.

4. Return to state 1.

For simplicity, 3.5.2 is formulated as a sketch which must still be realized in a suitable programming language. Furthermore, notions like 'move into the next field,' 'measure the color value,' or 'wait for 10 minutes,' require technical components (such as a system-internal clock) which realize these notions as corresponding operations. Neither would be a problem with existing technology.

The components of context-based cognition interact closely. For example, the behavior control guides the recognition and ensures that the resulting I-propositions are stored by the operating system at the correct index in the database. Then the behavior control determines the name of the next field and executes the command 'move into the next field' with the help of the operating system and the action component.

In the sense of nouvelle AI, CURIOUS exemplifies the notion of an *autonomous agent* because its symbolic processing is closely connected to its recognition and action.

> Without a carefully built physical grounding any symbolic representation will be mismatched to its sensors and actuators. These groundings provide the constraints on symbols necessary for them to be truly useful.'
>
> R.A. Brooks, 1990, S. 6.

In the sense of classic AI, CURIOUS exemplifies also the notion of a *physical symbol system* :

> The total concept [of a physical symbol system] is the join of computability, physical realizability (and by multiple technologies), universality, the symbolic representation of processes (i.e. interpretability), and finally, symbolic structure and designation.
>
> A. Newell & H. Simon 1975, p. 46

The combination of an autonomous agent and a physical symbol system allow CURIOUS to recognize its task environment and to act in it. In addition, CURIOUS can refer to past and possible states of its task environment and compare them with the current state. These inferences operate meaningfully with internal states which may be completely independent of the current state of the concrete outside world.

As a purely contextual cognitive system, the current version of CURIOUS properly belongs into the domain of AI and robotics. Yet its cognitive mechanism – based on parameter values, M-concepts, I-concepts$_{loc}$, I-propositions, the automatic analysis of artificial perception in the form of concatenated I-propositions and the realization of I-propositions in the form of artificial action – is an essential foundation of language-based (verbal) cognition in computational linguistics.

This is firstly because in talking cognitive agents the contextual (nonverbal) cognitive system functions as the *context* of language interpretation and production. Secondly, in the extension to language, the M-concepts serve an additional function as the literal *meanings* of the sign type symbol. Thirdly, the analysis of context-based cognition is needed in order to explain the phylo- and ontogenetic development of natural language from earlier evolutionary and developmental stages without language.

Exercises

Section 3.1

1. Describe how the uses of natural language in 3.1.1 differ.
2. What are the communication components within the prototype hypothesis?
3. Explain the notions task environment and problem space.
4. Compare the description of natural visual pattern recognition (D.E. Rumelhart 1977, J.R. Anderson 1990) with electronic models (e.g. D. Marr 1982). Bring out differences on the level of hardware and common properties on the logical level between the two types of system. Refer in particular to the Section *Template-Matching Models* in J.R. Anderson 1990, p. 58f.

Section 3.2

1. Which criteria can be used to measure the functional adequacy of CURIOUS?
2. What is the problem of solipsism and how can it be avoided?
3. When would CURIOUS say something true (false)?
4. Describe the SHRDLU system of T. Winograd 1972 (cf. H. Dreyfus 1981.)
5. Why is SHRDLU a closed system – and why is CURIOUS an open system?
6. Does SHRDLU distinguish between the task environment and the problem space?
7. What is the definition of the system-internal *context*. How does it relate to the notions task environment and problem space?

Section 3.3

1. Explain the notions type and token, using the letter A.
2. What is the relation between the parameter values, the M-concept and the I-concept$_{loc}$ of a certain square?
3. How does a type originate in time?

4. Why does a token presuppose a type?
5. What is iconicity? What arguments have been made against it?
6. In what sense is the cognitive theory underlying CURIOUS iconic?

Section 3.4

1. What are the three basic types from which elementary I-propositions are built?
2. Which language categories do the three basic semantic types correspond to?
3. How do the I-propositions of the internal context relate to the external reality?
4. Why do the I-propositions of the internal context form an autonomous system?
5. What is the role of language in the modeling of context-based cognition?
6. Why is the use of a grammar for modeling cognitive processes not in conflict with their essentially contextual (nonverbal) nature?

Section 3.5

1. Describe the schematic structure of CURIOUS. Why are its components self-contained modules and how do they interact functionally?
2. Does CURIOUS fulfill the definition of a physical symbol system?
3. What are the operations of CURIOUS and are they decidable? How are these operations physically realized?
4. Explain how the concepts for *left, right, up, down, large, small, fast, slow, hard, soft, warm, cold, sweet, sour*, and *loud* could be added to CURIOUS.
5. How would you implement the concept for *search* in CURIOUS?
6. Would it be possible to add the command *Find a four cornered triangle* to the behavior control 3.5.2? If so, what would happen? What is the logical status of the notion four cornered triangle?

4. Language communication

This chapter describes the functioning of natural language. To this purpose, the robot CURIOUS – introduced in Chapter 3 as a cognitive agent without language – is equipped with additional components needed for natural language communication between the robot and its wardens.

Section 4.1 describes the components of language-based (verbal) cognition as a phylo- and ontogenetic specialization of the corresponding components of context-based (nonverbal) cognition whereby the language-based components are arranged as a second level above the contextual components. Section 4.2 reconstructs language-based reference as an interaction between the language and the context level. Section 4.3 explains the necessary distinction between the literal meaning$_1$ of language signs and the speaker meaning$_2$ of utterances. Section 4.4 describes Frege's principle, and shows why the phenomena of ambiguity and paraphrase are, properly analyzed, no exceptions to it. Section 4.5 presents the principle of surface compositionality as a strict interpretation of Frege's principle and explains its functional and methodological role within the SLIM theory of language.

4.1 Adding language

Communication between man and machine requires a common language. Today it is still the language which is being adapted to the primitive communication capabilities of current machines. This has the disadvantage that the users have to learn the commands of special programming languages.

Computational linguistics aims at constructing machines which can communicate freely in a preexisting natural language (e.g. English). This requires a modeling of cognitive states and procedures.[1]

---

[1] In traditional grammar and theoretical linguistics (cf. 1.2.1), a modeling of cognition has been avoided. The analysis concentrated instead on the structural properties of expressions such as word forms, sentences, or texts. Thereby dictionaries have been compiled listing the words of a language. Generative syntax grammars have been developed which try to formally distinguish between the well-formed and the ill-formed sentences. And the meaning of sentences has been characterized in logical semantics as a relation between expressions and the world – excluding the cognitive structure of the speaker-hearer. When the goal is to model successful language *use*, however, a detailed functional description of the speaker's and the hearer's cognitive processing cannot be avoided.

In particular, a model of the speaker mode must describe the internal procedures which map meanings into signs of natural language. Correspondingly, a model of the hearer mode must describe the internal procedures which map the signs of natural language into the intended meanings.

Language understanding and language production have in common that they may be divided into two subprocesses, namely (i) sign *processing* and (ii) sign *interpretation*. The hearer mode and the speaker mode differ in that they apply these subprocedures in opposite order and directions.

4.1.1 TWO SUBPROCEDURES OF LANGUAGE USE

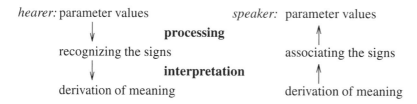

Language understanding begins with sign processing (recognition) as a precondition for sign interpretation. Language production begins with the linguistic interpretation of what is meant as a precondition for sign processing (synthesis).

The processing of a language sign in the speaker-hearer is based on M-forms which characterize the type of the surface, and I-forms$_{loc}$ which represent associated tokens. The linguistic M-forms are counterparts of the contextual M-concepts, while the linguistic I-forms$_{loc}$ are counterparts of the contextual I-concepts$_{loc}$.

In analogy to the analysis of contextual recognition and action in terms of characteristic constellations of parameters, M-concepts, and I-concepts$_{loc}$ (cf. 3.3.5), the linguistic recognition and production (action) of language signs is based on corresponding correlations of parameters, M-forms, and I-forms$_{loc}$.

4.1.2 PROCESSING OF LANGUAGE SIGNS

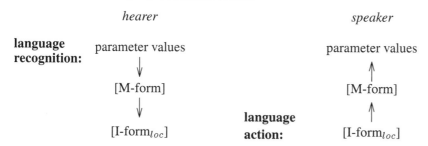

In the hearer mode, language *input* is processed. It consists in incoming parameter values of the acoustic or visual medium which M-forms are applied to. In the case of a successful matching, a corresponding I-form$_{loc}$ is derived and the lexical entry corresponding to the M-form is retrieved.

In the speaker mode, language *output* is processed. It consists in a word form token the surface of which is an I-form$_{loc}$. This surface is realized with the help of the associated M-form as outgoing parameter values in the medium of choice.

Integrating the components of language processing 4.1.2 into the nonverbal version of CURIOUS 3.5.1 results in the following overall structure.

4.1.3 EXPANDED STRUCTURE OF CURIOUS

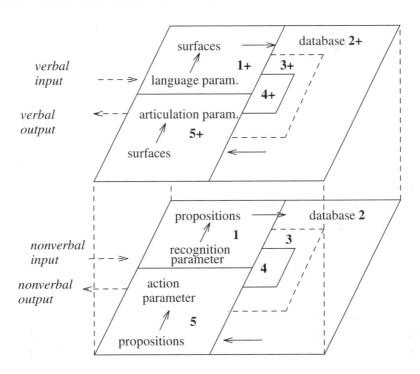

The lower level shows the structure 3.5.1 of the contextual version of CURIOUS, though slanted to allow a three-dimensional overall view. The upper level has the same structure as the lower level and contains the new components of language recognition 1+, language synthesis 5+, as well as counterparts of the components 2, 3, and 4.

Phylo- and ontogenetically the components of the language level may be viewed as *specializations* of the corresponding components of the contextual level. More specifically, component 1+ of word form recognition in 4.1.3 is a special function of the general component 1 for recognizing the task environment: just as CURIOUS recognizes, e.g., a triangle by matching the M-concept onto parameter values, thus forming an I-concept$_{loc}$ (cf. 3.2.3, recognition), the expanded version recognizes a language surface by matching a suitable M-form onto the language-based input parameters resulting in an I-Form$_{loc}$ (cf. 4.1.2, hearer).

Furthermore, component 5+ of language synthesis in 4.1.3 is a special function of the general component 5 for controlling action parameters. Just as CURIOUS can, e.g., change its position from one field to another by realizing the I-concepts$_{loc}$ of a action proposition with the help of the associated M-concepts and its electro motors as output parameter values, it can realize the I-forms of language signs with the help of the associated M-forms and the associated articulation as language-based output parameters (cf. 4.1.2, speaker).

4.2 Modeling reference

The recognition components 1 and 1+ constitute interfaces from external reality to cognition, while the action components 5 and 5+ constitute interfaces from cognition to external reality. Furthermore, the parameters of the contextual recognition and action components 1 and 5 provide the basis for the derivation of M-concepts and I-concepts$_{loc}$ (cf. 3.3.5), while the parameters of the linguistic recognition and action components 1+ and 5+ provide the basis for the derivation of M-forms and I-forms$_{loc}$ (cf. 4.1.2).

At least equally important for contextual and linguistic recognition and action, however, are the contextual and the linguistic database components 2 and 2+. They process the information provided by the recognition components 1 and 1+. Also, from this information they derive action plans which they pass step by step to the action components 5 and 5+ for realization.

The contextual and linguistic databases 2 and 2+ store information in the form of concatenated elementary propositions. The propositions of the linguistic level are composed of M-concepts (types) for which reason they are called M-propositions,[2] while the propositions of the contextual level contain the familiar I-propositions (cf. 3.4.2) consisting of I-concepts$_{loc}$ (tokens).

In the speaker mode, a content is communicated by mapping a subcontext of the database 2 into language. Thereby the I-propositions (tokens) of the subcontext are matched by M-propositions (types) generated in the database 2+. These M-propositions are realized as language surfaces by means of the lexicon and the grammar.

In the hearer mode, a language expression is understood by reconstructing the verbally represented subcontext of the speaker in the hearer's database 2. Thereby the incoming surfaces are assigned literal meanings by the lexicon and combined into M-propositions of the database 2+ by means of the grammar. The speaker's subcontext is reconstructed either by matching these M-propositions onto existing I-propositions in a corresponding subcontext of the database 2 or by storing the M-propositions in the subcontext as new I-propositions.

The matching process underlying language production and language understanding requires that the M-propositions (types) of the language and the I-propositions (tokens) of the context are properly positioned relative to each other. In the most basic

---

[2] In addition to the M-concepts of *symbols* (cf. Section 6.3), M-propositions contain the pointers of *indices* (cf. Section 6.2) as well as *names* (cf. Section 6.4).

way this is enabled in schema 4.1.3 by treating the linguistic database and the contextual database as two parallel layers.

The reconstruction of the communication mechanism in CURIOUS differs from the approach of analytic philosophy. Instead of analyzing communication as an internal cognitive process, analytic philosophy has adopted the viewpoint of a neutral external observer.

4.2.1 AN EXTERNAL VIEW OF REFERENCE

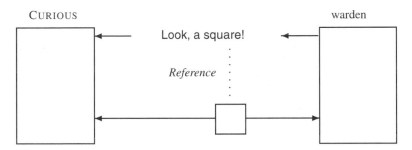

The external observer views the speaker (warden) and the hearer (CURIOUS) recognizing the object of reference in an analogous manner whereby the speaker means by the word **square** the object of reference. This phenomenon of a word (particularly a noun) which *means* an intended object constitutes the basic relation of reference. How does this relation come about?

One thing should be crystal clear: there is nothing in reality corresponding to the dotted line between the word **square** and the referential object in 4.2.1. Any theory trying to *explain* reference functionally by postulating such an external connection between the sign and its referent trivializes reference by relying on a fictional construct.[3]

The SLIM theory of language models reference instead as an internal matching procedure between the literal meanings of language (types) and corresponding contextual referents (I-concepts$_{loc}$, tokens). This procedure is based on the principle of *best match* and is illustrated in 4.2.2 with the sign type symbol, whose literal meaning is defined as an M-concept.

---

[3] Without detracting from their merits in other areas, such a trivial treatment of reference may be found in SHRDLU (T. Winograd 1972) and in Montague grammar (R. Montague 1974). Winograd treats an expression like **blue pyramid** by 'gluing' it once and for all to a suitable object in the toy world of the SHRDLU program. Montague defines the denotation of a predicate like **sleeps** in the metalanguage via the denotation function F for all possible worlds and moments of time. In either case, no distinction is made between the meanings of language expressions and corresponding sets of objects in the world.

   Binding language expressions to their referents in terms of definitions (either in a logical or a programming language) has the short term advantage of (i) avoiding a semantic analysis of language meaning and (ii) treating reference as an external connection – like the dotted line in 4.2.1. The cost for this is high indeed: such systems are in principle limited to being closed systems. See Chapters 19–22, especially Section 20.4, for a detailed discussion.

4.2.2 COGNITIVE 2+1 LEVEL ANALYSIS OF REFERENCE

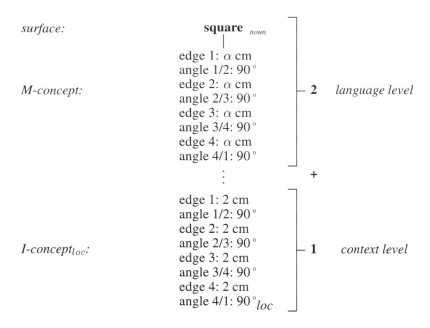

At the language level, the word **square** is lexically analyzed as a fixed constellation of (i) a surface (M-form, here the letter sequence **s**, **q**, **u**, **a**, **r**, **e**), (ii) a category (here the subscript *noun*), and (iii) a literal meaning (here the M-concept 3.3.2). At the contextual level, the referent is defined as an I-concept$_{loc}$.

The relation of reference between the language level and the context level is based on matching the type (M-concept) onto a corresponding token (I-concept$_{loc}$), as indicated by the dotted line. This internal matching of a language meaning (type) and a contextual object (token) possibly never encountered before is the essential cognitive mechanism of reference.

It is based on the three levels of the analyzed surface (syntax), the literal meaning (semantics), and the context. These form a functional [2+1] level structure. The two top levels of syntax and semantics are joined as the '2' in the [2+1] schema because the mechanism of natural language communication requires a *fixed* connection between analyzed surfaces and their literal meanings. In contrast, the internal matching of natural language pragmatics is a *flexible* matching procedure between the level of semantics and the level of context whereby the context is represented by the '1' of the [2+1] schema.[4]

In CURIOUS the fixed connections between the analyzed surfaces and their literal meanings are established by the designer using a programming language. In hu-

---

[4] A closer inspection of the sign types symbol, index, icon, and name in Chapter 6 will show that names constitute a variation to this basic schema because they function in terms of a [1+2] instead of the [2+1] structure (cf. 6.1.4 and 6.4.3).

mans the analogous connections are established by means of conventions which each speaker-hearer has to learn (cf. de Saussure's first law, Section 6.3).

The speaker-hearer may use (tokens of) the same sign to refer to ever new objects in ever varying situations. Assume that two people land on Mars for the first time and both see a rock shaped like a mushroom. If one says Look, a Mars mushroom! the other will understand.

This situation provides no occasion to establish – prior to the utterance – an external relation between the spontaneous ad hoc expression Mars mushroom and the intended referent. Thus, any attempt to explain successful reference in terms of an external relation between the signs and the referential objects, as in 4.2.1, is unrealistic.

Instead, the successful reference of the expression Mars mushroom is based on the analyzed word forms Mars and mushroom, available to speaker and hearer as predefined, internal linguistic entities consisting of a surface, a category, and a minimal literal meaning (analogous to analysis of the word form square in 4.2.2). Furthermore, their context must (i) indicate where they presently are, and (ii) represent the same characteristic rock formation in their respective fields of vision.

4.3 Using literal meaning

Depending on whether or not the relevant context of use is the current task environment, immediate and mediated[5] references are to be distinguished.

4.3.1 IMMEDIATE AND MEDIATED REFERENCE

– *Immediate reference* is the speaker's or the hearer's reference to objects in the current task environment.[6]
– *Mediated reference* is the speaker's or hearer's reference to objects which are not in the current task environment.[7]

Immediate reference and mediated reference have in common that they are based on internal cognitive procedures. They differ only in that in immediate reference the speaker-hearer interacts with the task environment at both, the contextual and the linguistic level. Mediated reference, on the other hand, relates solely to structures of the internal database for which reason the speaker-hearer interacts with the task environment at the linguistic level alone.

---

[5] Translated from German unmittelbar und mittelbar. G. Wahrig 1986 defines unmittelbar as 'ohne örtl. od. zeitl. Zwischenraum' (*without spatial or temporal distance*).

[6] Immediate reference may occur outside the communication prototype 3.1.2, for example, when a hearer finds a note on her desk, saying: Have you found the fresh cookies in the right drawer?

[7] For example, when speaker and hearer talk about the person of J.S. Bach (1685–1750) they refer to a contextual structure for which there is no counterpart in the current real world. Another form of mediated reference is CURIOUS' reference to objects in a state which was current in the past, as in How many red triangles did you find yesterday?

The following schema of immediate reference indicates CURIOUS' linguistic [1] and the contextual [2] interaction with the current task environment.

4.3.2 INTERNAL AND EXTERNAL ASPECTS OF REFERENCE

CURIOUS

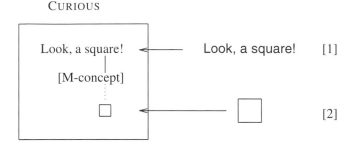

In this analysis, reference is reconstructed as a purely cognitive procedure, without any relation between the external sign [1] and its external referent [2]. Language meanings are treated solely as mental objects in the cognition of the language users.

Such a cognitive analysis of reference has been rejected in analytic philosophy because of concern that it would permit cognitive agents to attach his or her own 'private' meanings to the surfaces of natural language.[8] There is no cause to be worried by the cognitive nature of reference, however. The speakers do not commit the suspected misuse because then their communication would not work.

This can be tested by anyone who decides to use, e.g., the word table for car and vice versa. (S)he should not be surprised if unprepared partners do not understand. Furthermore, if well-meaning partners catch up after a while and join the game, this amounts to a highly local (and usually temporary) change in the convention-based internal connection between certain language surfaces and their meaning concepts.

At any rate, postulating external 'real' relations between language expressions and referents, as in 4.2.1, contributes in no way to making the meanings of language 'objective.' For the functioning of natural language it is sufficient that the surface-meaning connections are established by conventions which are maintained by their constant use within the language community. These conventions must be learned (i.e. internally established) by each and every member of the language community and whoever does not master them has a problem.

In addition to the notion of literal meaning as a property of expression types there is the notion of speaker meaning as a property of utterances, i.e., actions in which tokens of language expressions are being used. These two notions of meaning apply to two different kinds of phenomena.[9] They are equally legitimate and equally necessary to

---

[8] These concerns underlie the laborious arguments guarding against possible accusations of 'psychologism' in the writings of G. Frege, among others.

[9] One might argue that L. Wittgenstein concentrated on the first notion in his early (1921) and on second notion in his late (1953) philosophy. Rather than functionally integrating expression meaning

explain the functioning of natural language. For the sake of clear and brief terminology, the literal meaning of language expressions is called *meaning*$_1$, while the speaker meaning of utterances is called *meaning*$_2$.

The functional connection between meaning$_1$ and meaning$_2$ is described by the first principle of pragmatics, also called PoP-1.[10]

4.3.3 FIRST PRINCIPLE OF PRAGMATICS (POP-1)

The speaker's utterance meaning$_2$ is the use of the sign's literal meaning$_1$ relative to an internal context.

The meaning$_1$ of an expression exists independently of any contextual substructures that might match it. Conversely, the contextual substructures exist independently of any corresponding meaning$_1$ of the language.[11] The meaning$_2$ derived by the hearer is nevertheless called the speaker meaning because the hearer's interpretation is successful only if the speaker's subcontext is reconstructed correctly.[12]

4.4 Frege's principle

A clear distinction between meaning$_1$ and meaning$_2$ allows to strictly maintain Frege's principle in the syntactic and semantic analysis of natural language. This principle is named after GOTTLOB FREGE (1848–1925), mathematician, philosopher, and one of the founders of modern mathematical logic. Though it was not stated explicitly in Frege's writings it may be formulated as follows:

4.4.1 FREGE'S PRINCIPLE

The meaning of a complex expression is a function of the meaning of the parts and their mode of composition.

---

into utterance meaning, as in PoP-1 (4.3.3), Wittgenstein opted to abandon his first approach. See also the discussion of ordinary language philosophy as exemplified by P. Grice in Section 4.5 and the discussion of semantic ontologies in Chapter 20.

[10] A preliminary version of PoP-1 may be found in Hausser 1981, where the distinction between *meaning*$_1$ and *meaning*$_2$ is already used. In CoL, p.271, the first principle is published as one of five principles of pragmatics (see 5.3.4, 5.4.5, 6.1.3, and 6.1.4 below).

[11] For example, the meaning$_1$ of the word square in 4.2.2 is an M-concept which exists independently of any possible referents, either in the internal context or the external task environment. Correspondingly, the square objects in the world and their reflexes in the cognitive agents' context do not depend on the existence of a word with a corresponding meaning$_1$ – as demonstrated by the nonverbal version of CURIOUS in Chapter 3.

This independence of M-concepts and corresponding I-concepts$_{loc}$ holds only for the secondary use of M-concepts as language meanings$_1$ which are lexically bound to the surfaces (M-forms) of symbols, as in 4.2.2. In their primary function as contextual types of certain parameter constellations, on the other hand, M-concepts are the precondition for the derivation of I-concepts$_{loc}$ (cf. 3.3.5).

[12] See the definition of successful communication in 4.5.5 as well as the schemata of language interpretation 5.4.1 and production 5.4.2.

That the meaning of the parts influences the meaning of the whole is demonstrated by the comparison of the syntactically similar sentences a and b:

    a. The dog bites the man
    b. The dog bites the bone

The sentences a and b have different meanings because they differ in one of their parts (i.e. the respective last word).

    That the syntactic composition influences the meaning of the whole is demonstrated by the following sentences a and a', which are composed from exactly the same word forms (parts):

    a. The dog bites the man
    a'. The man bites the dog

The sentences a and a' have different meanings because they differ in their mode of composition (exchange of the nouns **dog** and **man**).

    Frege's principle is intuitively obvious, but it has often been misunderstood. This is caused by a failure to clearly decide whether the notion of 'meaning' in 4.4.1 should be the meaning$_1$ of expressions or the meaning$_2$ of utterances.

    Consider, for example, the meaning$_1$ of **That's beautiful weather!** This expression may be used literally on a sunny summer day or ironically on a dark, wet day in November. In the second case, the meaning$_2$ may be paraphrased as *The weather is disgusting*. Thus, one sign (type) is related in two different utterances to two different contexts of use resulting in two different meanings$_2$.

    If Frege's principle is applied – erroneously – to a meaning$_2$ (e.g. what has been paraphrased as *The weather is disgusting* in the above example), then the meaning of the whole does not follow from the meaning of the parts (i.e. the words **That's beautiful weather**). Examples of this type has been used repeatedly to cast Frege's principle into doubt. In reality, however, these apparent counterexamples are a fallacy based on confusing meaning$_1$ and meaning$_2$.

    If Frege's principle is applied – correctly – to meaning$_1$, then it serves to characterize the functional relation between syntax and semantics: by putting together the word surfaces in the syntax, the associated meanings$_1$ are composed simultaneously on the level of the semantics. Thus, the meaning of complex expressions is derived *compositionally* via the composition of word forms whose surface and meaning$_1$ are in a fixed constellation (as in 4.2.2). A completely different matter is the *use* of a complex meaning$_1$ relative to a context. This aspect lies entirely outside Frege's principle – in the domain of pragmatics, the theory of correlating the meaning$_1$ of language and the context of use.

    Our standard interpretation of Frege's principle implies that two complex expressions with different surfaces must have different meanings$_1$. Correspondingly, two complex expressions with the same surface must have the same meaning$_1$.

4.4.2 STANDARD INTERPRETATION OF FREGE'S PRINCIPLE

$$
\begin{array}{lcccccc}
\text{surface:} & \mathbf{a} & = & \mathbf{a} & \quad & \mathbf{a} & \neq & \mathbf{b} \\
 & \vdots & & \vdots & & \vdots & & \vdots \\
\text{meaning}_1: & \mathbf{A} & = & \mathbf{A} & & \mathbf{A} & \neq & \mathbf{B}
\end{array}
$$

However, there are two apparent exceptions to the standard interpretation, namely syntactic ambiguity and syntactic paraphrase.

A surface is called syntactically ambiguous, if it can be assigned two or more meanings$_1$. For example, good in

They don't know how good meat tastes

can be interpreted as an adverbial modifying **taste** or as an adjective modifying **meat**. These alternative syntactic analyses come out quite clearly in the intonation.

Conversely, different surfaces constitute a set of syntactic paraphrases if their meanings$_1$ turn out to be equivalent. For example, the sentences

The dog bit the man (active)

The man was bitten by the dog (passive)

may be considered to have equivalent meanings$_1$.

As apparent exceptions to the standard interpretation of Frege's principle, syntactic ambiguity and paraphrase may be analyzed superficially as follows.

4.4.3 APPARENT EXCEPTIONS (incorrect analysis)

$$
\begin{array}{lcccccc}
 & \multicolumn{3}{c}{\textbf{ambiguity}} & \multicolumn{3}{c}{\textbf{paraphrase}} \\
\text{surface:} & \mathbf{a} & = & \mathbf{a} & \mathbf{a} & \neq & \mathbf{b} \\
 & \vdots & & \vdots & \vdots & & \vdots \\
\text{meaning}_1: & \mathbf{A} & \neq & \mathbf{A'} & \mathbf{A} & = & \mathbf{B}
\end{array}
$$

In the representation of ambiguity, two different meanings$_1$ A and A' seem to have the same surface, while in the representation of paraphrase two different surfaces seem to have the same meaning$_1$. Both representations seem to disagree with the standard interpretation of Frege's principle.

One important structural aspect which neither 4.4.2 nor 4.4.3 express correctly, however, is that Frege's principle applies to syntactically *analyzed* surfaces and their meaning$_1$. Yet syntactic ambiguity is by its very nature a property of *unanalyzed* surfaces, resulting in the following re-analysis.

4.4.4 SYNTACTIC AMBIGUITY (correct analysis)

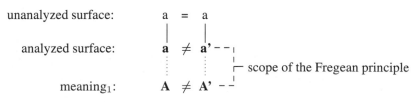

As a property of the unanalyzed surface, syntactic ambiguity is outside of the domain of Frege's principle. Therefore syntactic ambiguity is no exception to its standard interpretation.

Another important aspect which 4.4.3 fails to express is that the meanings$_1$ of different complex surfaces can at best be equivalent, but never identical. In arithmetic, for example, no one in his right mind would express the semantic equivalence of 2+4 and 3+3 in terms of an identical 'underlying form' (e.g. 6). Instead the correct way is to show the equivalence of the respective meaning$_1$ structures.

4.4.5 SYNTACTIC PARAPHRASE

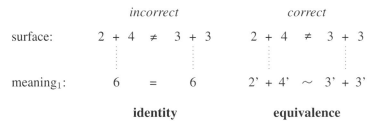

According to the correct analysis, paraphrases are no exception to the standard interpretation of Frege's principle: even though the meanings$_1$ turn out to be equivalent they are not identical.

4.5 Surface compositionality

In its standard interpretation, Frege's principle corresponds to the principle of surface compositionality.[13]

4.5.1 SURFACE COMPOSITIONALITY I (SC-I PRINCIPLE)

> An analysis of natural language is surface compositional if it uses only concrete word forms as the building blocks such that all syntactic and semantic properties of complex expression derive systematically from the syntactic category and the meaning$_1$ of their building blocks.

For linguistic analysis, surface compositionality has the consequence

- methodologically that syntactic analyses are *concrete* because no kind of zero surface or underlying form may be used,
- mathematically that syntactic and semantic analyses may be of *low complexity* (cf. 12.5.7 and 21.5.2, respectively), and
- functionally that the internal matching between meaning$_1$ and context may be extended from single words (cf. 4.2.2) to the systematic syntactic-semantic *combination* of expressions.

The principles of internal matching and surface compositionality – represented by the letters IM and S, respectively, in the SLIM acronym – presuppose each other in the following way. Internal matching is a precondition for a strictly compositional analysis in syntax and semantics because it moves all aspects of meaning$_2$ into the pragmatics where they belong. Conversely, surface compositionality provides the systematically derived meaning$_1$ needed for internal matching with the context.

While functionally integrated into the mechanism of natural language communication, surface compositionality is primarily a methodological standard for ensuring concreteness in the empirical analysis of natural language syntax and semantics. To show the consequences of violating this standard, let us consider two examples, namely generative grammar within nativism and speech act theory within ordinary language philosophy. A third example from yet another tradition is Montague grammar (cf. SCG for a detailed analysis).

Violating surface compositionality: EXAMPLE I

Generative grammar within nativism began with transformational grammar (TG). It dominated theoretical linguistics between 1957 and 1981, and is still influential in current theories[14] – protestations to the contrary notwithstanding. Transformational grammar aims at characterizing the innate linguistic knowledge of the speaker-hearer.

In TG, deep structures defined as context-free constituent structures (cf. Chapter 8) are mapped into surface structures by means of transformations. Since Katz & Postal 1964, deep and surface structure are assumed to be semantically equivalent, i.e. the transformations are 'meaning preserving.'

The correlation between semantically equivalent deep and surface structures in TG is illustrated by the following examples.

4.5.2 EXAMPLES OF 'CLASSICAL' TRANSFORMATIONS

DEEP STRUCTURE: SURFACE STRUCTURE:

Passive:
Peter closed the door   ⇒ The door was closed by Peter

Do-support:
Peter not open the door   ⇒ Peter didn't open the door

Reflexivization
Peter$_i$ shaves Peter$_i$ ⇒ Peter shaves himself

---

[13] The SC-I principle was first described in Hausser 1978. As shown in SCG, it may be interpreted formally as restricting the homomorphism condition. The relation between the SC-I principle and the mathematical notion of a homomorphism is described in Section 21.3, where the informal, intuitive version 4.5.1 is supplemented by the formal variant SC-II (cf. 21.3.5).

[14] A description of the post-transformational systems of GB, LFG, and GPSG may be found in P. Sells 1985. Yet another variant of nativism is HPSG as the continuation of GPSG.

There-insertion
A hedgehog is in the garden   ⇒ There is a hedgehog in the garden

Pronominalization
Peter$_i$ said that Peter$_i$ was tired   ⇒ Peter said that he was tired

Relative clause formation
Peter [Peter was tired]   ⇒ Peter, who was tired

Main clause order in German
Peter die Tür geschlossen hat   ⇒ Peter hat die Tür geschlossen

Object raising[15]
Peter persuaded Jim [Jim sleeps]   ⇒ Peter persuaded Jim to sleep

Subject-raising
Peter promised Jim [Peter sleeps]   ⇒ Peter promised Jim to sleep

The transformational derivation of surface structures from deep structures has no functional role in communication.[16] Instead, transformations are intended to express linguistic generalizations at the level of competence.[17]

Transformations violate surface compositionality because they treat the concrete parts of the surface and their mode of composition as irrelevant for the resulting meaning. The properties distinguishing a surface from its semantically equivalent deep structure are regarded as *syntactic sugar*.

For example, in the passive sentence The door was closed by Peter, the word order (mode of composition) is changed on the way from the deep structure to the surface. Furthermore, the word forms was and by are not treated as meaningful elements in their own right, but smuggled into the surface by transformations (cf. Section 21.3).

From the viewpoint of surface compositionality it is therefore no surprise that the presumed meaning equivalence between active and passive turned out to be spurious. For example, there is general agreement that

Everyone in this room speaks at least two languages

and the corresponding passive sentence

At least two languages are spoken by everyone in this room

differ in meaning. The passive has the dominant reading that there are two languages (e.g. English and French) which everyone speaks. The active only says that each person speaks two languages which may differ from person to person.[18] Because of the

---

[15] A purely lexical treatment of so-called *subject* and *object raising* may be found in SCG, p. 254.

[16] Some followers of Chomsky have recognized this as a weakness. Their attempts to provide transformations in hindsight with a genuine functional role have not been successful, however. N. Chomsky always rejected such attempts as inappropriate for his nativist program (e.g. Chomsky 1965, p.9).

[17] Transformations have later been replaced by similar mechanisms. When asked in 1994 about the frequent, seemingly radical, changes in his theories N. Chomsky pointed out that the 'leading ideas' had never changed. Personal communication by Prof. Dong-Whee Yang, Seoul, Korea 1995. See also Chomsky 1981, p. 3, in the same vein.

lack of meaning equivalence illustrated by these examples, the passive transformation was later abandoned by nativism.

The purpose of transformations was to characterize the innate human language faculty. There is a general law of nature, however, that the form of innate structures follows their function. For example, in an aquatic bird such as a duck the form of the feet is clearly innate. This innate structure can be described in many different ways, but its scientific analysis in biology must explain the relation between the specific form (webbed toes) and the specific function (locomotion in water by paddling).

The same holds for innate cognitive structures. In particular, the human language capacity may be described in many different ways, but its scientific analysis in linguistics must explain the relation between the specific form of natural language surfaces and their specific function in the time-linear coding and decoding of content.

There may be innate structures whose function is not yet discovered, is of seemingly little importance (as the wing pattern of butterflies), or has been lost during evolution (as in the rudimentary pelvis bones of whales). Natural language communication, however, is neither a minor nor a lost function.

Consequently, any structure alleged to be part of the innate human language faculty is implausible, if it cannot be demonstrated to have a functional role in the mechanism of natural communication. We formulate this conclusion as a cognitive variant of *Occam's razor* – a rule of science named after WILLIAM OF OCKHAM (1270–1347).

4.5.3 COGNITIVE VARIANT OF OCCAM'S RAZOR

> Entities or components of grammar should not be postulated as innate if they have no clearly defined function within natural communication.

The cognitive razor applies to transformational grammar as well as all later variants of nativism including LFG[19], GPSG[20], and HPSG.[21] Like transformational grammar, their linguistic generalizations are nonfunctional with respect to communication and inherently in violation of surface compositionality (cf. p. 418f.). It is no accident that their computational complexity turned out to be similar to that of transformational grammar (cf. Section 8.5).

Violating surface compositionality: EXAMPLE II

A second example of a non-surface-compositional approach is the definition of meaning by P. GRICE (1913–1988). It is in the tradition of L.WITTGENSTEIN's (1889–

---

[18] The surface compositional treatment is based on separate derivations of the active and the passive. If they happen to be paraphrases, this may be expressed by establishing semantic equivalence on the level of meanings$_1$. Cf. 4.4.5.

[19] J. Bresnan (ed.) 1982.

[20] G. Gazdar, E. Klein, G. Pullum & I. Sag 1985.

[21] C. Pollard & I. Sag 1987, 1994

1951) ordinary language philosophy and J.L.AUSTIN's (1911–1960) speech act theory, specifically Wittgenstein 1953 and Austin 1962.[22] This philosophical tradition provides an intention-based alternative to the logical definition of meaning in terms of truth conditions (cf. Chapters 19–21).

4.5.4 DEFINITION OF MEANING BY GRICE

> Definiendum: U meant something by uttering x.
> Definiens: For some audience A, U intends his utterance of x to produce in A some effect (response) E, by means of A's recognition of the intention.[23]

The speaker (utterer) U conventionally uses x with a certain intention. Given a token of x, the hearer (audience) A recognizes the intention because A has gotten used to associate x with this particular intention type of U. Knowledge of the type enables the hearer A to recognize the intention of the speaker U because the speaker *habitually intends* for the sentence to have a certain effect.

In this way, the sentence meaning is defined as the type of a convention, while the corresponding utterance meanings are defined as tokens of this type. By reducing sentence meaning to a conventional habit,[24] an explicit representation of literal words meanings and their composition based on syntactic analysis is avoided.

As pointed out repeatedly,[25] however, Grice's construction is not suitable to explain the *evolution* of natural language communication. How are the sentence meanings (types) supposed to originate in the language community so that the corresponding utterance meanings (tokens) can have their intended effect? For the hearer, the type must already exist as a convention in order to be able to recognize a corresponding token. The point is that convention cannot be used to establish a certain type of intention because in order to recognize the first token the type must already be in place.

Another problem for Grice are nonstandard uses of language, such as metaphor or irony. After all, tokens of a purely convention-based sentence meaning can only be recognized by the audience, if they comply with the standard, habitual use.

---

[22] According to S.C. Levinson 1983:227,8,

> there are strong parallels between the later Wittgenstein's emphasis on language usage and language-games and Austin's insistence that "the total speech act in the total speech situation is the *only actual* phenomenon which, in the last resort, we are engaged in elucidating" (1962:147). Nevertheless Austin appears to have been largely unaware of, and probably quite uninfluenced by, Wittgenstein's later work, and we may treat Austin's theory as autonomous.

That Austin, who was 22 years younger than Wittgenstein, was 'largely unaware' of Wittgenstein's writings explains Levinson by their teaching at different universities: Austin was at Oxford while Wittgenstein was at Cambridge.

[23] Cf. P. Grice 1957 and 1968.

[24] This reduction is epitomized by ordinary language philosophy's imprecise and misleading formula

MEANING IS USE.

The SLIM theory of language is also based on use, but on the use of literal expression meanings$_1$ relative to a context, resulting in the speaker's utterance meaning$_2$ (cf. PoP-1, 4.3.3).

[25] See for example J. Searle 1969, p. 44f.

In order to nevertheless handle spontaneous metaphors and other nonstandard uses, Grice proposed another theory. It is based on violating conventions and known as conventional implicature. Conventional implicature formulates common sense principles for making sense of the exception to the rule.[26]

Thus, Grice simultaneously uses two opposite methods for realizing intentions. One is his standard method of *using conventions*, the other is his nonstandard method of *violating conventions*. Neither method is surface compositional because neither is based on the structure of the signs and their literal meaning$_1$.

Conveying intentions by either conforming to habit, or violating it selectively, may succeed for utterer U and audience A if they have been familiar with each other for the better part of a lifetime in highly ritualized situations. For better or worse, however, this is not language communication in the normal sense.

For explaining the interpretation of, e.g., an average news paper editorial with ironic innuendos and metaphoric insinuations, Grice's meaning definition 4.5.4 does not suffice. This is because it does not take into account the differentiated syntactic-semantic structure of natural language signs by means of which an infinite variety of complex meaning structures may be encoded.[27]

Apart from the internal problems of Grice's theory, the notions 'recognition of the intention,' 'producing some effect,' or 'intending for some audience' of definition 4.5.4 are not sufficiently algebraic[28] for a computational model of natural communication. For this, a functional theory of language is required which explains the understanding and purposeful use of natural language in terms of completely explicit, mechanical (i.e. logically-electronic) procedures.

This is a clear and simple goal, but for theories not designed for it from the outset it is practically out of reach. Purely hypothetically it could turn out in hindsight that non-

---

[26] See also J.L. Austin 1962, p. 121 f.

[27] Rather than applying the type-token distinction to conventions, the SLIM theory of language applies it to recognition and action. In contextual recognition, for example, the types arise as classes of similar parameter constellations (cf. 3.3.2). Once a type has evolved in this way it is used to classify corresponding constellations of parameter values, resulting in tokens instantiating the type (cf. 3.3.5).

In the extension to language, the types are used in a secondary function, namely as the meaning$_1$ of symbols. The interpretation of symbols is based on matching these meanings$_1$ (M-concepts) with contextual referent structures (I-concepts$_{loc}$).

The principle of internal matching between types (M-concepts) and tokens (I-concepts$_{loc}$) aims from the outset at handling the spontaneous use of language to express new meanings$_2$ relative to new contexts. Conventions are used only for fixing the relation between the language surfaces and their meaning$_1$ inside the speaker-hearer (in agreement with de Saussure's first law, cf. Section 6.3.)

This straightforward explanation of the primary origin and function of types and tokens on the contextual level and their secondary functioning in the interpretation of language cannot be transferred to Grice's speech act theory. The reason is that the speaker's intentions are not accessible to the hearer's recognition as characteristic parameter constellations – as would be required for the SLIM-theoretic use of the type-token distinction.

[28] Regarding its mathematical properties, a formal definition of Grice's approach has yet to be provided. What such a formalization could look like, however, is indicated by another system of ordinary language philosophy and speech act theory, namely J.R. Searle & D. Vanderweken 1985. Depending on the viewpoint, this formalization is either too imprecise to draw any hard mathematical conclusions at all, or some particular aspect of it may interpreted to demonstrate sky high complexity.

functional theories – such as speech act theory, nativism, behaviorism, structuralism, model theory, etc. – happen to be functionally suitable anyway.

This, however, would constitute a historically unique scenario: nonfunctional theories happen to be of such profound correctness regarding general structure and conceptual detail that they later turn out to be adequate even for the demands of a concretely functioning model – without any need for major corrections. Such stroke of luck has not occurred in linguistics, and is probably without example in the whole history of science.

In conclusion we turn to the notion of *successful communication* within the SLIM theory of language. When talking to another person it is impossible to determine for sure whether the hearer understands the speaker in exactly the way the speaker intended (solipsism, cf. Section 3.2). But the direct access to the internal cognitive processing of a machine designed to communicate in natural language allows an objective reconstruction of this notion.

4.5.5 SUCCESSFUL MAN-MACHINE COMMUNICATION

Let L be an established natural language, SH a human speaker-hearer of L, and CA a cognitive agent (for example, an advanced version of CURIOUS).

– *Successful natural language interpretation*
CA communicates successfully in the hearer mode, if CA understands the L-utterance in the way intended by SH. In technical terms this means that CA correctly recreates the speaker meaning of the L-utterance in its database. The developers of CA can verify the procedure because (i) they themselves can understand the utterance in L and (ii) they can view the interpretation directly in the database of CA.

– *Successful natural language production*
CA communicates successfully in the speaker mode, if CA formulates its intentions in L in a way that SH can understand. This requires technically that CA maps a certain structure in its database into an L-utterance which SH can correctly reconstruct. The developers of CA can verify the procedure because (i) they have direct access to the database structure to be communicated and (ii) they themselves can understand the utterance in L.

The logical structure of the databases in question (cf. 4.1.3) and the procedures of reading-in (hearer-mode) and -out (speaker-mode) are described in more detail in Chapters 22 -24.

Exercises

Section 4.1

1. Which components are required to extend the nonverbal version of CURIOUS presented in Chapter 3 into a robot communicating in natural language?
2. In what sense are the components of language processing in CURIOUS a specialization of its contextual cognitive components?
3. What is the relation between natural language defined as a set of grammatically analyzed expressions, and the cognitive structure of the speaker-hearer using the language? Explain your answer on 3–4 pages.
4. Why does the construction of artificial cognitive agents have a methodological impact on the formation of linguistic theory?

Section 4.2

1. What is the internal aspect of reference?
2. On which basis does the hearer establish reference if the speaker uses an expression not heard before (e.g. Mars mushroom) to refer to an object not seen before?
3. Why is the handling of reference nontrivial in the case of CURIOUS, but trivial in the case of SHRDLU? How does this difference depend on the distinction or nondistinction between task environment and problem space?
4. In what respect can SHRDLU do more than CURIOUS? What would be required to combine the different merits of the two systems?

Section 4.3

1. In what sense is the SLIM theory based on a [2+1] level structure?
2. What is the difference between immediate and mediated reference?
3. Describe the connection between reference and cognitive processing.
4. What is the difference between the speaker-mode and the hearer-mode in natural language communication?
5. Describe the connection between the literal meaning$_1$ of a language expression and the speaker meaning$_2$ of an utterance.
6. Given that the time linearity of natural language signs is represented from left to right, can you motivate why the hearer is placed to the left of the speaker in 4.1.3, 4.3.2, and 5.2.1?

Section 4.4

1. Who was G. Frege and when did he live? Explain the principle that carries his name.
2. Does Frege's principle relate to the speakers' utterance meaning$_2$ or the expressions' literal meaning$_1$?

3. Why are ambiguity and paraphrase apparent exceptions to Frege's principle? Which properties of the analysis allow to eliminate these exceptions? Explain your answer using concrete examples of ambiguity and paraphrase.
4. What is the relation between the principle of surface compositionality and Frege's principle?

Section 4.5

1. Give 9 different examples of transformations and show how they violate surface compositionality.
2. Why can Frege's principle only be applied to the deep structures of transformational grammar, but not to the surface structures?
3. What is a methodological objection to applying Frege's principle to deep structures?
4. Name other areas in which confusing identity and equivalence has led to problems (cf. R. Barcan-Marcus 1960.)
5. Describe the definition of meaning by P Grice and explain why it is not suitable for computational linguistics.
6. Compare the analysis of meaning in the theory of P. Grice and the SLIM theory of language. Explain the different uses of the *type-token* distinction and the different definitions of sentence and utterance meaning in the two theories.
7. What can a concrete implementation of CURIOUS as a talking robot do for an improved understanding of natural language communication?
8. Explain the criteria for successful man-machine communication.
9. Compare the language processing of Eliza, SHRDLU and CURIOUS.

5. Using language signs on suitable contexts

The crucial question for the interpretation of natural language is: how does the relation between the sign and the intended referent come about? The preceding Chapters 3 and 4 investigated this question in the context of designing the talking robot CURIOUS. This design is simplified in that cognition and language are limited to triangles, quadrangles, and circles of various sizes and colors. Nevertheless, CURIOUS models the general functioning of natural language insofar as the system can not only talk about new objects of a known type, but also about situations outside its current task environment, such as past or future situations.

This chapter investigates which principles allow the natural or artificial hearer to *find* the correct subcontext such that the internal matching between the expression used and the subcontext selected results in successful communication as defined in 4.5.4. This leads to a further differentiation of the SLIM theory of language and an extension of CURIOUS to general phenomena of natural language pragmatics.

Section 5.1 compares the structure of CURIOUS with Bühler's organon model and Shannon & Weaver's information theory. Section 5.2 demonstrates with an example of nonliteral use why the precise delimitation of the internal subcontext is crucial for the correct interpretation of natural language signs. Section 5.3 describes how the four parameters defining the origin of signs (STAR-point) allow to infer the subcontexts which are correct for their interpretation. Section 5.4 explains the function of the time-linear order of the signs for production and interpretation as well as de Saussure's second law. Section 5.5 describes the conceptualization underlying language production as an autonomous navigation through the propositions of the internal database.

5.1 Bühler's organon model

A theory of pragmatics analyzes the general principles of purposeful action. It describes how a cognitive agent can achieve certain goals by using certain means in certain situations. Examples of pragmatic problems are the use of a screw driver to fasten a screw, the use of one's legs to go from a to b, the scavenging of the refrigerator in the middle of the night to fix a BLT sandwich and satisfy one's hunger, or the request that someone fix and serve the sandwich.

Depending on whether or not the means employed are signs of language we speak of linguistic and nonlinguistic pragmatics. Just as language recognition and articulation may be analyzed as a phylo- and ontogenetic specialization of contextual (nonverbal) recognition and action, respectively (cf. Section 4.1), linguistic pragmatics may be analyzed as a phylo- and ontogenetic specialization of nonlinguistic pragmatics.

This embedding of linguistic pragmatics into nonlinguistic pragmatics was recognized by PLATO (427(?)–347 BC), who pointed out the *organon* character of language in his dialog Kratylos. In modern times, the tool character (*Werkzeugcharakter*) of language was emphasized by KARL BÜHLER (1879–1963):

> Die Sprache ist dem Werkzeug verwandt; auch sie gehört zu den Geräten des Lebens, ist ein Organon wie das dingliche Gerät, das leibesfremde Zwischending; die Sprache ist wie das Werkzeug ein *geformter Mittler*. Nur sind es nicht die materiellen Dinge, die auf den sprachlichen Mittler reagieren, sondern es sind die lebenden Wesen, mit denen wir verkehren. [Language is akin to the tool: language belongs to the instruments of life, it is an organon like the material instrument, a body-extraneous hybrid; language is – like the tool – a *purposefully designed mediator*. The only difference is that it is not material things which react to the linguistic mediator, but living beings with whom we communicate.]
>
> K. Bühler 1934, p. XXI

Bühler summarized his analysis in terms of the well-known organon model. In addition to the function of language *representation*, the organon model includes the functions of language *expression* and language *appeal*.

5.1.1 BÜHLER'S ORGANON MODEL

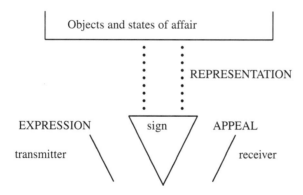

Representation refers to the language-based transfer of information. Expression refers to the way the transmitter produces the sign. Appeal refers to the way the sign affects the receiver beyond the bare content of the sign.

At first glance the relation between the organon model 5.1.1 and the CURIOUS model 4.1.3 is not obvious. This is because the organon model 5.1.1 describes the communication prototype 3.1.2 from the viewpoint of an external observer – like 4.2.1. The CURIOUS model, on the other hand, describes the internal mechanism of natural communication. As a consequence, the organon model is limited to immediate reference, while the CURIOUS model can also handle mediated reference (cf. 4.3.1) to subcontexts of past, future, and other nonactual modalities.

Upon closer investigation, however, the following corelations between the organon model and the CURIOUS model may be established. The function of expression in 5.1.1 is to be located in component 5+ (language synthesis) in 4.1.3. The function of appeal in 5.1.1 is to be located in component 1+ (language recognition) in 4.1.3. The function of representation in 5.1.1 is performed in 4.1.3 by means of lexical, syntactic, and semantic components in the language-based database structure 2+ and interpreted in relation to the contextual database structure 2.

Thus, the organon model and the CURIOUS model are compatible, though the organon model applies only to the communication prototype 3.1.4, and is limited to an external, noncognitive viewpoint. This limitation is reflected in Bühler's terminology: the notions transmitter and receiver are more appropriate to the transmission of signals, as in broadcasting, than to a cognitive modeling of the speaker-hearer.

The mathematical theory of signal transmission was presented in 1949 by C.E. SHANNON and W. WEAVER as *information theory* – 15 years after Bühler's 'Sprachtheorie.' Information theory investigates the conditions under which the transmission of electric and electronic signals is of sufficient quality. Central notions of information theory besides transmitter and receiver are the band width of the channel, the redundancy and relative entropy of the codes, and the noise in the transmission.

The laws of information theory hold also in everyday conversation, but background noises, slurring of speech, hardness of hearing, etc., are not components of the natural communication mechanism. It must be based instead on a model of cognitive agents – which goes far beyond the comparatively primitive structures of a transmitter and a receiver.[1]

5.2 Pragmatics of tools and pragmatics of words

A theory of nonlinguistic pragmatics must describe the structure of the tools, of the objects to be worked on, and the user's strategies of applying a tool to an object in order to realize a certain purpose. Analogously, a theory of linguistic pragmatics requires an explicit definition of meaning$_1$ (tool), of the interpretation context (object to be worked on), and of the strategies for relating a meaning$_1$ and a context such that the intended meaning$_2$ is communicated. Just as a tool is positioned in a specific spot of the object to be worked on and then manipulated in a purposeful way, a suitable meaning$_1$ is positioned relative to a certain subcontext in order to represent it linguistically (speaker mode) or to insert it into the subcontext (hearer mode).

The analogy between an external tool like a screw driver and a cognitive tool like the word **table** shows up in their respective literal and nonliteral uses. While the 'literal

---

[1] Nevertheless there have been repeated attempts to glorify the mechanics of code transmission into an explanation of natural communication. A case in point is U. Eco 1975, whose theory of semiotics within Shannon & Weaver's information theory begins with the example of a buoy which 'tells' the engineer about dangerous elevations of the water table. Other cases are P. Grice's 1957 'bus bell model' and F. Dretsky's 1981 'door bell model.'

use' of a screw driver consists in the fastening and unfastening of screws, there is also an open multitude of 'nonliteral uses' such as punching holes into juice cans, use as a door stop, as a letter weight, etc. Because these nonliteral uses depend on the momentary purpose and the properties of the current context it is impossible to provide a valid enumeration[2] of all the possible uses (i.e. all the different possible instances of $meaning_2$) of a screw driver.

Instead the user infers the most effective application of the tool for each new context and for each new purpose via general principles of pragmatics. These refer to (i) the structure of the tool with its well-known shape and properties of material (constituting the screwdriver's $meaning_1$) and (ii) the current properties of the object to be worked on (constituting the 'context of use').

It is similar with a word like **table**, which may be used not only to refer to proto-typical tables. Assume that the hearer is in a room never seen before in the middle of which there is an orange crate. If the speaker says **Put the coffee on the table!**, the hearer will understand that **table** refers to the orange crate. Given this limited context of use, the minimal $meaning_1$ of the word **table** best fits the structure of the orange crate (*best match*).

5.2.1 NONLITERAL USE OF THE WORD table

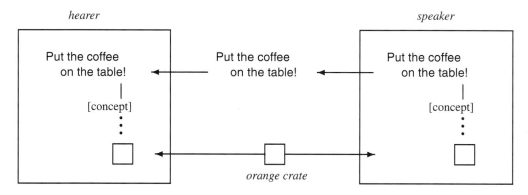

However, if a prototypical table were placed next to the orange crate, the hearer would interpret the sentence differently, putting the coffee not on the orange crate, but on the table. This is not caused by a change in the $meaning_1$ of **table**, but by the fact that the context of use has changed, providing an additional candidate for best match.

The principle of best match can only function properly, if the choice of possible candidates is restricted. Therefore the selection and delimitation of the subcontext is crucial for the successful interpretation of natural language. This leads to a central question of linguistic pragmatics.

> How does the speaker code the selection and delimitation of the used subcon-text into the sign and how can these be correctly inferred by the hearer?

---

[2] This is another problem for the speech act theory of Austin, Grice, and Searle (cf. Section 4.5).

The internal database of a speaker-hearer comprises all the episodical and theoretical facts accumulated in the course of a life time. To correctly find the small, delimited subcontext required for each interpretation of language in this huge database is not trivial. It is particularly challenging in the case of mediated reference (cf. 4.1.3), in which the context of use does not correspond to the current task environment.

5.3 Finding the correct subcontext

As an example of mediated reference consider a post card, i.e., a sign whose the places of origin and of interpretation are far apart in time and space. On a beach in New Zealand on a hot day in February 1999, the 'hearer' Heather is reading a postcard with the statue of liberty on one side and the following text on the other:

5.3.1 POSTCARD EXAMPLE

> New York, December 1, 1998
>
> Dear Heather,
> Your dog is doing fine. The weather is very cold. In the morning he played in the snow. Then he ate a bone. Right now I am sitting in the kitchen. Fido is here, too. The fuzzball hissed at him again. We miss you.
>
> Love,
> Spencer

Which structural properties of the sign enable Heather to select from all her stored knowledge the one subcontext which is correct for the interpretation of this text?

Like all human artifacts the postcard (as a hand-written sign) has a point of origin. In signs this point is defined by the following parameters:

5.3.2 PARAMETERS OF ORIGIN OF SIGNS (STAR-POINT)

1. S = the **S**patial place of origin
2. T = the **T**emporal moment of origin
3. A = the **A**uthor
4. R = the intended **R**ecipient.

The parameters S, T, A, and R have their values defined automatically during production and constitute the STAR-point of a sign.[3] All meaningful utterances have their unique STAR-point which is a necessary property of sign *tokens*.

The word 'point' in STAR-point may be interpreted rather loosely. Even in spoken language the temporal origin is strictly speaking an interval rather than a point. In writing this interval may be considerable length.

---

[3] In some respects the STAR point resembles Bühler's 1934 notion of *origo*. However, the STAR-point is defined as a property of the (token of the) sign whereas the *origo* seems to refer to time-spatial coordinates and gestures of the speaker-hearer during the utterance.

Apart from the parameters of origin of a sign there are the parameters of its interpretation. The latter are called ST-points because they consist of (1) a spatial location S and (2) a moment of time T. Just as the STAR-point is fixed automatically by the origin, an ST-point is defined whenever a sign is interpreted.[4]

While the STAR-point is unique for any given sign (token) and defined once and for all, the number of its possible ST-points is open. For example, a postcard which is lost in the mail may not have any ST-point at all whereas a letter published in the *New York Times* will have many ST-points.

The ST-point is known to each single hearer or reader: it equals his or her current circumstances. Yet the correct interpretation of a sign depends mostly on the knowledge of its STAR-point, which may be difficult to infer.[5] Consider a clay tablet found in an archaeological dig in Mesopotamia. The ST-points of the various interpretation attempts by various scientists make no difference as to what the tablet really means. What is crucial are the place of origin (S), the correct dynasty (T), the writer (A), and the addressee (R) of the clay tablet.[6]

Only face-to-face communication – as in the communication prototype 3.1.2 – provides the hearer with the STAR-point directly. For this reason, the pragmatic interpretation of natural language is especially easy there. Signs not intended for face-to-face communication must ensure their correct contextual positioning by an explicit or implicit specification of their STAR-point.

The role of the STAR-point is described by the second principle of pragmatics.

5.3.3 SECOND PRINCIPLE OF PRAGMATICS (POP-2)

> The STAR-point of the sign determines its primary positioning in the database by specifying the *entry context* of interpretation.

---

[4] Many different constellations between STAR- and ST-point are possible. For example, in someone talking to himself the S- and T-parameters of the STAR- and ST-point are identical. Furthermore, the A-value and the R-value of the STAR-point are equal.

 If two people talk to each other in the situation of the communication prototype, the T-parameter values in the STAR- and ST-point are practically the same, while the S-values are significantly differen. Furthermore, in a dialog the R-value of the STAR-point equals the hearer, while in the case of someone overhearing a conversation, this person will be distinct from the R-value of the STAR-point.

[5] In logical semantics the speaker – and thus the origin of signs – is not formally treated. For this reason, the model-theoretic interpretation can only be defined relative to an (arbitrary) ST-point. This is of no major consequence in the case of eternally true sentences. In the extension of logical semantics to contingent sentences, however, the formal interpretation relative to ST-points is plainly in conflict with the empirical facts of communication.

 Similarly in speech act theory which has attempted to represent the speaker's intention and action in terms of performative clauses like *I request, I declare,* etc. These are treated as part of the type (sentence meaning). Thereby the utterance dependent interpretation of *I*, of the addressee, of the moment of time, and of the place is left untreated. This may seem acceptable for the communication prototype 3.1.2, but for the interpretation of a post card like 5.3.1 a theoretic treatment of the STAR-point as a property of the utterance is unavoidable.

[6] The importance of the STAR-point is also shown by the question of whether the tablet is real or fake. While the glyphs remain unaffected by this question, the different hypotheses regarding authenticity

The need for primary positioning is especially obvious in the case of mediated reference, as in Heather's interpretation of Spencer's postcard.

5.3.4 PRIMARY POSITIONING IN TERMS OF THE STAR-POINT

Heather's cognitive representation:

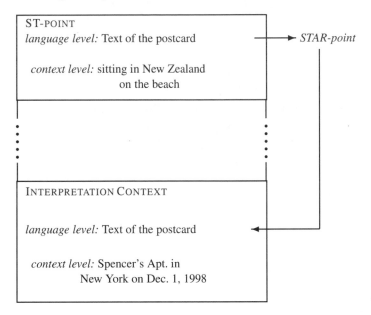

Heather's current situation is stored at the ST-point: sitting on the beach she is looking at the postcard. The signs of the postcard, however, are not matched onto the subcontext of the ST-point, but rather onto a subcontext determined by the STAR-point.

Accordingly, Heather does not object to the statement **the weather is very cold** by pointing to the hot New Zealand summer. Based on the STAR-point she knows that the words of the postcard refer to New York in winter.

Similarly, when the nosy landlady secretly reads Spencer's postcard she is not surprised by **Your dog is doing fine** even though she has no dog. Based on the STAR-point she knows full well that the words of the postcard refer to Heather's dog.

For the hearer, PoP-2 determines the subcontext where the interpretation begins. But it makes no demands on its contents, which depend solely on the person who interprets the sign. A subcontext of interpretation may contain personal memories, factual information, etc., but it may also be completely empty.

In Heather's case, the subcontext *Spencers Apt. in New York on December 1, 1998* is richly filled insofar as Heather is Spencer's friend and knows his apartment from personal experience. In contrast, the corresponding subcontext of Heather's landlady

lead to different interpretations. Another example is an anonymous letter whose threatening quality derives in large part from the fact that it specifies the recipient R without revealing the author A.

is initially a new, empty file. Yet the landlady's interpretation is successful insofar as she reads the postcard text into this file, making sense of it via her acquaintance with Heather (pragmatic anchoring, cf. Section 23.5).

The function of the STAR-point is twofold. On the one hand, it regulates the reference to data structures already present (as in the case of Heather). On the other hand, it is the basis for integrating new information so that it may be retrieved correctly on later occasions (as in the case of the landlady, but also in the case of Heather).

Thus the landlady may smile knowingly when Heather later announces an impending visit from New York. This knowledge is based not only on the text of the postcard, but also in large part on the explicitly specified STAR-point which allowed the landlady to put the content into a correct subcontext.

In addition to its real STAR-point, a sign may also pretend a fictitious one, as in the novel *Felix Krull*, which begins as follows.

> Indem ich die Feder ergreife, um in völliger Muße und Zurückgezogenheit – gesund übrigens, wenn auch müde, sehr müde …
> [While I seize the pen in complete leisure and seclusion – healthy, by the way – though tired, very tired …]

Here the fictitious STAR-point is not even specified explicitly, but filled in inconspicuously, piece by piece, by the author in the course of the text.[7]

Signs which provide their STAR-point explicitly and completely can be correctly interpreted by anyone who speaks the language. This is shown by the possibility of nonintended interpretations, as illustrated by Heather's nosy landlady.

Signs which do not provide their STAR-point, as an undated and unsigned letter, require at least a *hypothesis* of the likely STAR-point for their interpretation. The reader can always provide a tentative STAR-point, but the less is known about the real one, the less (s)he can refer to the data structures coded by the writer.

5.4 Language production and interpretation

The STAR-point of a sign provides the entry context for the speaker's language production and the hearer's interpretation. From there, however, an unlimited variety of other subcontexts may be accessed and traversed. The time-linear coding of the subcontexts into language surfaces during production and their decoding during interpretation is based on the [2+1] level structure (cf. 4.2.2) of the SLIM theory of language.

In language interpretation, the navigation through the subcontexts is controlled by the language signs: the hearer follows the surfaces of the signs, looks up their meanings$_1$ in the lexicon, and matches them with suitable subcontexts.

---

[7] In a novel, the real STAR-point may be of little or no interest to the average reader. Nevertheless it is explicitly specified. The name of the author, Thomas Mann, is written on the cover. The intended recipient is the general readership as may be inferred from the text form 'book.' The time of the first printing and the place of publisher are specified on the back of the title page.

5.4.1 SCHEMA OF LANGUAGE INTERPRETATION (ANALYSIS)

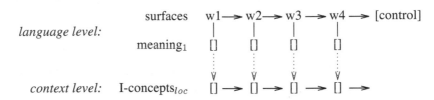

In language production, the navigation control is located in the subcontexts of the speaker. Each contextual concept traversed is matched with the meaning$_1$ of a suitable word form. Utterance of the word form surfaces allows the hearer to reconstruct the speaker's time-linear navigation path.

5.4.2 SCHEMA OF LANGUAGE PRODUCTION (GENERATION)

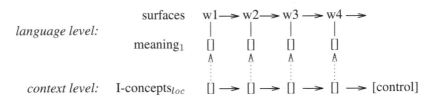

The schemata 5.4.1 and 5.4.2 agree with the view that interpretation ($\downarrow$) and production ($\uparrow$) are inverse vertical procedures. Nevertheless, interpretation and production have their main direction in common, namely a horizontal time-linear structure ($\rightarrow$).

5.4.3 THE TIME-LINEAR STRUCTURE OF NATURAL LANGUAGE SIGNS

The basic structure of natural language signs is their *time-linear order*. This holds for the sentences in a text, the word forms in a sentence, and the allomorphs in a word form.

Time-linear means:

LINEAR LIKE TIME AND IN THE DIRECTION OF TIME.

This formulation may be regarded as a modern version of FERDINAND DE SAUSSURE's (1857–1913) second law.

5.4.4 DE SAUSSURE'S SECOND LAW: *linear character of signs*

SECOND PRINCIPE; CARACTÈRE LINÉAIRE DU SIGNIFIANT.
Le signifiant, étant de nature auditive, se déroule dans le temps seul et a les caractères qu'il emprunte au temps: a) *représente une étendue*, et b) *cette étendue est mesurable dans une seule dimension*: c'est une ligne.
Ce principe est évident, mais il semble qu'on ait toujours négligé de l'énoncer, sans doute parce qu'on l'a trouvé trop simple; cependent il est fondamental et les conséquences en sont

incalculables; son importance est égale à celle de la première loi. Tout le méchanisme de la langue en dépend.

[The designator, being auditory in nature, unfolds solely in time and is characterized by temporal properties: (a) *it occupies an expansion*, and (b) *this expansion is measured in just one dimension*: it is a line.

This principle is obvious, but it seems that stating it explicitly has always been neglected, doubtlessly because it is considered too elementary. It is, however, a fundamental principle and its consequences are incalculable. Its importance equals that of the first law. All the mechanisms of the language depend on it.]

F. de Saussure 1913/1972, p. 103

The failure pointed out so politely by de Saussure in 1913 has continued until today. C- (cf. Chapter 7) and PS-grammars (cf. Chapter 8) are based on two-dimensional trees called *constituent structures* (cf. 8.4.3), treating surface order as the 'problem of linearization.' Only LA-grammar (cf. Chapter 10) uses a time-linear derivation order as the common backbone for production and interpretation of natural language.

The time-linear structure of natural language is so fundamental that a speaker cannot but utter a text sentence by sentence, and a sentence word form by word form. Thereby the time-linear principle suffuses the process of utterance to such a degree that the speaker may decide in the middle of a sentence on how to continue.

Correspondingly, the hearer need not wait until the utterance of a text or sentence has been finished before her or his interpretation can begin. Instead the hearer interprets the beginning of the sentence (or text) without knowing how it will be continued.[8]

In spoken language, the time-linear movement of the navigation must be maintained *continuously* during production and interpretation. If a speaker pauses too long, (s)he ceases to be a speaker. If a hearer stops following the words as they are being spoken and starts to think of something else, (s)he ceases to be a hearer.

The function of time linearity is described by the third principle of pragmatics.

5.4.5 THIRD PRINCIPLE OF PRAGMATICS (PoP-3)

> The matching of word forms with their respective subcontexts is incremental whereby in production the elementary signs follow the time-linear order of the underlying thought path while in interpretation the thought path follows the time-linear order of the incoming elementary signs.

The third principle presupposes the second. The initial step in the interpretation of a natural language sign is to determine the entry context as precisely as possible via the STAR-point (PoP-2). The next step is to match the complex sign word form by word form with a sequence of subcontexts. PoP-3 goes beyond 5.4.3 and de Saussure's second law in that PoP-3 emphasizes the role of the time-linear structure for the incremental *matching* between the signs' meaning$_1$ and the subcontexts within the [2+1] level structure of the SLIM theory of language.

---

[8] Hearers are often able to continue incoming sentences, especially if the speaker is slowly producing well-engrained platitudes describing widespread beliefs. This phenomenon shows again that the principle of possible continuations permeates the interpretation of language as well.

5.5 Thought as the motor of spontaneous production

According to the SLIM theory of language, what is said is a direct reflection of what is thought. Thus, the once famous motto of behaviorism

Thought is nonverbal speech

is turned around into

Speech is verbalized thought.

Behaviorism attempted to reduce thought to speech in order to avoid a definition of thought. The SLIM theory of language, on the other hand, models thought explicitly as an autonomous navigation through the internal database.[9]

The navigation is implemented computationally as a focus point moving through the concatenated propositions. Its momentary direction is controlled by (i) the connections between the propositions traversed, (ii) nonverbal internal influences such as hunger, (iii) nonverbal external influences (sensory input), and (iv) verbal input.

The navigation provides a basic handling of *conceptualization*. This notion is used in conventional systems of language production for choosing the contents to be expressed (what to say), in contradistinction to the notion of *realization* which refers to the way in which a chosen content is represented in language (how to say it).[10]

That the SLIM theory of language treats conceptualization as thought itself, and realization as a simultaneous assignment of language surfaces, is in concord with the fact that spontaneous speech is not based on a conscious choice of content or style. A representative example of spontaneous speech is an eye witness report after a recent shock experience. The horrified description gushes out automatically without any stylistic qualms or hesitation about what to say or how to say it. For better or worse, the same is true for most situations of normal everyday life.

That a person will sometimes speak slowly and deliberately, sometimes fast and emotionally, sometimes formally and politely, and sometimes in dialect or slang, is

---

[9] The model is realized explicitly as the SLIM machine in Chapters 22 – 24. This autonomous cognitive machine has nonverbal and verbal interfaces to the real world. It is able to automatically analyze its environment and to act in it, and to understand and produce language. The cognitive core of a SLIM-machine is its internal database in which concatenated propositions are stored.

The SLIM-theoretic model of thought may be viewed as a formal realization of spreading activation theories in cognitive psychology. Different versions may be found in A.M. Collins & E.F. Loftus 1975, J.R. Anderson & G.H. Bower 1981, J.R. Anderson 1983, D.E. Rumelhart, P. Smolensky, J. McClelland & G.E. Hinton 1986 and others.

[10] Conventional systems of language production are based on the attempt to avoid a general modeling of thought, treating conceptualization instead as a special procedure. It selects and structures the contents to be mapped into natural language. Thereby conceptualization and realization are handled as two separate phases.

The data structure used for language production by conventional systems are sets of constituent structures called a 'tree bank', a set of logical formulas, or similar knowledge representations. These representations have not been designed for a time-linear navigation and are therefore not suitable for it (cf. Section 22.2). Typical problems of conventional production systems are the *extraction*, the *connection*, and the *choice problem* (cf. CoL, p. 111).

a direct reflection of the current navigation through the subcontexts and an indirect reflection of the internal and external factors modifying the state of the database and thus the course of the navigation. Besides current moods and intentions of the cognitive agent, these influences are the relation to the hearer, the presence or absence of a third party, and the events taking place at the time.

The navigation through subcontexts (thought path) is independent of language insofar as it often occurs without a concomitant verbalization.[11] When language is produced, however, it is usually a completely automatic process in which order and choice of the words is a direct reflection of the underlying navigation.

> Speech is irreversible. That is its fatality. What has been said cannot be unsaid, except by adding to it: to correct here is, oddly enough, to continue.
>
> R. Barthes, 1986, p. 76

Even in written language the realization of time-linear navigation shows up at least partially in that the word and sentence order is fixed. For example, Spencer decided to describe events in their temporal order (cf. 5.3.1):

In the morning he played in the snow. Then he ate a bone.

In the medium of writing, the order of the two sentences can be inverted, i.e.,

*Then he ate a bone. In the morning he played in the snow.

but this would destroy the intended interpretation of then.

Alternatively Spencer could have depicted the events as follows:

In the morning Fido ate a bone. Before that he played in the snow..

This anti-temporal sequencing represents an alternative thought path through the contextual database and has the stylistic effect of emphasizing the eating of the bone. Here a later inversion would likewise destroy textual cohesion.

There are also examples in which a later change of sequencing does not destroy cohesion, but 'only' modifies the interpretation:

a. 1. In February, I visited the Equator. 2. There it was very hot. 3. In March, I was in Alaska. 4. There it was very cold.

b. 3. In March, I was in Alaska. 2. There it was very hot. 1. In February, I visited the Equator. 4. There it was very cold.

The four sentences in a and b are the same, but their order is different. The interpretation differs with the orderings because there obviously refers to the subcontext (Alaska/Equator) that was traversed most recently. This holds for both, the reader and the writer. According to our knowledge of the world, example b is highly unlikely, yet a sincere interpretation cannot but navigate a corresponding thought.[12]

---

[11] Whether thought is *influenced* by one's language is another question which has been hotly debated through the centuries. According to the Humboldt-Sapir-Whorf Hypothesis the thought of humans is indeed influenced by their respective languages.

[12] The question of trustworthiness, seriousness, etc. of a sign's author is discussed in CoL, p. 280-1.

In summary, the time-linear structure of natural language underlies written language as much as it underlies spoken language. The only difference is that written signs are more permanent than spoken ones (if we disregard recordings) for which reason

– a writer can correct and/or modify a given text without having to add more sentences,
– a reader can leaf through a text, going forward or backward, without being bound to the linear order of the word forms.

These are additional possibilities, however, which in no way change the essential time-linear structure of written signs.[13]

Exercises

Section 5.1

1. Explain the relation between nonlinguistic and linguistic pragmatics.
2. Who was Karl Bühler, and when did he live?
3. What are the three functions of language specified by Bühler's organon model.
4. What do the CURIOUS model and the organon model have in common and in what way do they differ?
5. Which components of the CURIOUS model 4.1.3 treat *expression* and *appeal*, respectively?
6. Who first presented information theory, what are its tasks, and what are its central notions?
7. Is information theory suitable to explain the mechanism of communication?

Section 5.2

1. How does the view of language as an organon affect the theoretic development of a linguistic pragmatics?
2. Why does PoP-1 presuppose the view of language as a tool?
3. Name and explain two analogies between the use of nonlinguistic and linguistic pragmatics.

---

[13] Even hypertexts, with their various options of continuation, are designed for a time-linear consumption by the reader. There are, however, special types of books, e.g. dictionaries, in which the entries – for the sake of a specialized access – are ordered, e.g., alphabetically (though the definitions of their entries are of a conventional time-linear structure). Characteristically, dictionaries – like telephone book, inventory lists, bank accounts, relational databases, and other non-time-linear language structures – are not regarded as 'normal' texts of natural language.

4. Why is it a central problem of linguistic pragmatics to determine the correct context of use? Is this problem addressed in linguistic mainstream pragmatics (e.g. Levinson 1983)?

Section 5.3

1. Explain the roles of the STAR- and the ST-point in the interpretation of linguistic signs.
2. Why is it necessary for the STAR-point to specify also the intended recipient (example)? Why does this not hold for the ST-point?
3. How is the STAR-point specified in Spencer's postcard?
4. What is the entry context and why is its specification important?
5. Is it possible to interpret a sign relative to an empty subcontext? What would be the purpose?
6. Consider a lawn with the sign **Keep off the grass!** Define the STAR-point of this sign (cf. CoL, p. 280.)
7. The novel *A Clockwork Orange* is written in the first person: **There was me, that is Alex, and my three droogs** ... How does the reader distinguish between the author Antony Burgess and Alex? Explain the notion of an auxiliary STAR-point.

Section 5.4

1. Why is the [2+1] level structure of the SLIM theory of language crucial for the interpretation and production of natural language?
2. What is meant exactly by the notion *time-linear*? How does it differ from the notion *linear time* in complexity theory?
3. Explain the time-linear structure of natural language in terms of the speaker's production and the hearer's interpretation.
4. Who was de Saussure, when did he live, and what is his second law?
5. Why does PoP-3 presuppose PoP-2?
6. Is there a difference between PoP-3 and de Saussure's second law?

Section 5.5

1. What is the difference between spoken and written language regarding production and interpretation? Is written language time-linear as well?
2. What is the difference in the correction of spoken and written language?
3. In the production of language, is there an intermediate state which is not time-linear?
4. What are the three basic problems of production in constituent-structure-based grammars? How are they avoided in the SLIM theory of language?
5. Is the late phylo- and ontogenetic aquisition of reading and writing an argument for or against treating them as central cases of language use?
6. How deliberately are style and topic chosen in spontaneous speech?
7. How does the speaker control what she or he says?

6. Structure and functioning of signs

The semantic and pragmatic functioning of complex expressions is based on individual words. These correspond to different sign types the structure of which is the basis of different reference mechanisms. The classification and analysis of words as sign types belongs traditionally into the philosophical domain of the *theory of signs*.

Section 6.1 shows with an example in telegram style that complex meanings can be transmitted without grammatical composition simply by means of the time-linear order and the $meaning_1$ of the signs, and explains how the iconic, name-based, and indexical reference mechanisms work. Section 6.2 describes the sign type symbol and how it is used to refer by the speaker and the hearer. Section 6.3 describes the sign type index and investigates the phenomenon of repeating reference, i.e., the navigating back to a subcontext recently traversed, especially with third person pronouns. Section 6.4 discusses two exceptions to de Saussure's first law, namely the sign types icon and name. Section 6.5 explains the evolution of icons from pictures and describes the development of modern writing systems as a gradual transition from pictures via visual icons (pictograms) to symbols.

6.1 Reference mechanisms of different sign types

The mechanism of natural language communication as described so far is based on the following structural principles:

1. Handling of language use in terms of an internal matching between $meaning_1$ and a subcontext (PoP-1, Section 4.3).
2. Determining the entrance subcontext by means of the STAR-point of the sign (PoP-2, Section 5.3)
3. Fixing the derivation order in terms of the time-linear sequence of words (PoP-3, Section 5.4).

PoP-1, PoP-2, and PoP-3 are founded so deeply in the basic functioning of natural language communication that their fundamental role shows up in each and every instance of meaningful language production and interpretation. Even though their elementary structure and function are "obvious, ... it seems that stating [them] explicitly has always been neglected, doubtlessly because [they are] considered too elementary" – to borrow and expand de Saussure's formulation quoted in 5.4.4.

Based on PoP-1, PoP-2, and PoP-3 we are now at the point where the individual signs are positioned more or less precisely across the correct internal subcontext. From here our analysis may continue in two different directions. One is to describe the horizontal relation of the individual signs to each other, i.e. the principles of time-linear syntactic-semantic composition. The other is to analyze the vertical relation of individual signs to the subcontext. The correlation of these two structural aspects may be shown schematically as follows.

6.1.1 ALTERNATIVES OF DESCRIPTION

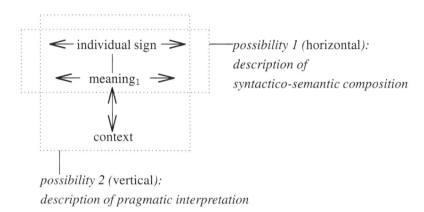

Of these alternatives we choose the second because the vertical pragmatic interpretation is possible even in the case of single signs, as shown by the well-known phenomenon of the one-word sentence. Moreover, the horizontal syntactic-semantic composition is rather complex in the natural languages and requires a separate, detailed analysis. It will be provided in Part III (Morphology and Syntax) and Part IV (Semantics and Pragmatics).

To reduce the – momentarily disregarded – role of syntax in the pragmatic interpretation of complex signs, let us begin with a special type of communication, e.g. the telegram style and similar variants,[1] in which the function of the syntax is minimal:

6.1.2 EXAMPLE WITH MINIMAL SYNTAX

Me up. Weather nice. Go beach. Catch fish. Fish big. Me hungry. Fix breakfast. Eat fish. Feel good. Go sleep.

There is no doubt that this kind of language works. Though it is syntactically simplified, even ungrammatical, it is well-suited to convey nonelementary content. This is based solely on (i) the linear order and (ii) the meaning$_1$ of the individual words. A good theory of language should therefore be able to explain a fully developed syntax as a *refinement* of communication with little or no grammar.

---

[1] A typical example is the simplification in the discourse with foreigners: Me Tarzan, you Jane.

For this, the semantic-pragmatic functioning of individual words must be explained. In referring expressions, analytic philosophy distinguishes the sign types of *symbol*, *index*,[2] and *name*. The SLIM theory of language explains their different reference mechanisms in terms of their characteristic internal structures.

6.1.3 FOURTH PRINCIPLE OF PRAGMATICS (POP-4)

The reference mechanism of the sign type **symbol** is based on a meaning$_1$ which is defined as an M-concept. Symbols refer from their place in a positioned sentence by matching their M-concept with suitable contextual referents (I-concepts$_{loc}$).

The reference mechanism of symbols is called iconic reference. The term iconic is motivated by the matching between M-concepts (types) and corresponding contextual referents (tokens), and should not be misconstrued in the sense of naive little pictures (cf. definition of the M-concept square in 3.3.2 and the example of iconic reference in 4.2.2). Examples of symbols in 6.1.2 are the nouns weather, beech, and fish, the verbs go, catch, and eat, and the adjective-adverbials nice, big, and hungry.

6.1.4 FIFTH PRINCIPLE OF PRAGMATICS (POP-5)

The reference mechanism of the sign type **index** is based on a meaning$_1$ which is defined as a pointer. An index refers by pointing from its place in the positioned sentence into appropriate parameter values.

Indexical reference is illustrated by the adverbs here and now, which point to the spatial and temporal parameter values of an utterance, respectively, and the pronouns I and you, which point to the author and the intended recipient, respectively.
    In contrast to symbols and indices, names have no meaning$_1$.

6.1.5 SIXTH PRINCIPLE OF PRAGMATICS (POP-6)

The reference mechanism of the sign type **name** is based on an act of naming which consists in adding a name-marker to the internal representation of the individual or object in question. Reference with a name consists in matching the name and the corresponding marker.[3]

As an example of name-based reference consider meeting a stranger who introduces his dog as Fido. Afterwards, our internal representation of this particular dog as a

---

[2] This use of the term 'index' as a sign type is distinct from its use in the domain of data bases (cf. 2.1.1). Both uses have in common, however, that they are based on the notion of a pointer.

[3] In analytic philosophy, names have long been a puzzle. Attempts to explain their functioning range from *causal chains* to *rigid designators* (cf. S. Kripke 1972).

contextual referent comprises not only properties like size, color, etc., but also the name-marker **Fido**.[4]

The respective structural basis of iconic, indexical, and name-based reference is illustrated in the following schematic comparison in which the three sign types are used to refer to the same contextual object, i.e. a red triangle.

6.1.6 COMPARING ICONIC, INDEXICAL, AND NAME-BASED REFERENCE

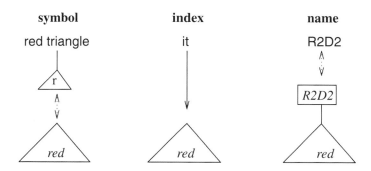

The symbol **red triangle** refers by matching the meaning$_1$ (M-concept) with a suitable object (I-concept$_{loc}$) in a limited range of contextual candidates. The index **it** refers by pointing to the referential object, given that it is compatible with the grammatical number and gender restrictions of the pronoun. The name **R2D2** refers by matching the name surface with an identical marker which was attached to the contextual referent in an act of naming.

Names fit into the basic [2+1] level structure of the SLIM theory of language. However, while symbols and indices establish a fixed connection between the surface and the meaning$_1$, names establish a fixed relation between the name-marker and the contextual referent. Accordingly, in names the matching is not between the meaning$_1$ and the context, but rather between the surface and the marker (cf. Section 6.5).

All three mechanisms of reference must be analyzed as internal, cognitive procedures. This is because it would be ontologically unjustifiable to locate the fixed connections between surface and meaning$_1$ (symbols and indices) and between marker and contextual referent (names) in the external reality.[5]

For explaining the phylo- and ontogenetic development of natural language it is of interest that the basic mechanisms of iconic, indexical, and name-based reference constitute the foundation of nonverbal and preverbal communication as well. Thereby

1. nonverbal iconic reference consists in spontaneously imitating the referent by means of gestures or sounds,
2. nonverbal indexical reference consists in pointing at the referent, and

---

[4] Besides explicit acts of naming, names may also be inferred implicitly. For example, when we observe an unknown person being called by a certain name.

[5] See Section 4.2 as well as the discussion of four possible semantic ontologies in Section 20.4.

3. nonverbal name-based reference consists in pointing at the referent while simultaneously pronouncing a name.

While largely limited to the communication prototype 3.1.2, these nonverbal mechanisms of reference may be quite effective. By avoiding the use of conventionally established surfaces, nonverbal reference allows spontaneous communication in situations in which no common language is available.

The distinction of the *sign types*, i.e. symbol, index, and name, is orthogonal to the distinction between the main *parts of speech*, i.e. noun, verb, and adjective-adverbial, as well as to the corresponding distinction between the basic *elements of propositions*, i.e. argument, functor, and modifier (cf. 3.4.1).

6.1.7 SEVENTH PRINCIPLE OF PRAGMATICS (POP-7)

The sign type *symbol* occurs as noun, verb, and adjective-adverbial. The sign type *index* occurs as noun and adjective-adverbial. The sign type *name* occurs only as noun.

The orthogonal correlation between sign types and parts of speech described in PoP-7[6] may be represented graphically as follows.

6.1.8 RELATION BETWEEN SIGN TYPES AND PARTS OF SPEECH

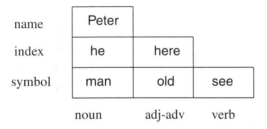

The sign type which is the most general with respect to the parts of speech is the symbol, while the name is the most restricted. Conversely, the part of speech (and, correspondingly, the propositional element) which is the most general with respect to sign types is the noun (object), while the verb (relation) is the most restricted.

6.2 Internal structure of symbols and indices

As signs, the symbols and indices of natural language have an internal structure which is composed of the following components.

---

[6] While CoL has presented preliminary versions of PoP-1 to PoP-5 as the first to fifth principle of pragmatics, respectively, PoP-6 and PoP-7 are defined here for the first time.

6.2.1 INTERNAL COMPONENTS OF THE SIGN TYPES SYMBOL AND INDEX

- The *surface* is constrained by the laws of acoustic articulation (in the original medium of spoken language).
- The *category* reflects the combinatorial properties of the part of speech and the inflectional class to which the sign belongs.
- The *meaning*$_1$ reflects the conceptual structures of the internal context and/or contains characteristic pointers to certain contextual parameters.
- The *glue* connecting surface, category, and meaning$_1$ consists in conventions which must be learned by each member of the language community.

The convention-based nature of the glue, particularly between surface and meaning$_1$, is the subject of de Saussure's first law.

6.2.2 *De Saussure*'S FIRST LAW

PREMIER PRINCIPE; L'ARBITRAIRE DU SIGNE.
Le lien unissant signifiè au signifié est arbitraire, ou encore, puisque nous entendons par signe le total résultant de l'association d'un signifiant à un signifié, nous pouvons dire plus simplement: *le signe linguistique est arbitraire.*
[THE FIRST LAW: ARBITRARINESS OF SIGNS
 The link connecting the designator and the designated is arbitrary; and since we are treating a sign as the combination which results from connecting the designator with the designated, we can express this more simply as: *the linguistic sign is arbitrary.*]

F. de Saussure 1913/1972, p. 100

The arbitrariness of signs may be seen in the comparison of different languages. For example, the symbols **square** of English, **carré** of French, and **Quadrat** of German have different surfaces, but the same meaning$_1$ (i.e. the M-concept defined in 3.3.2). Similarly, the indices **now** of English, **maintenant** of French, and **jetzt** of German have different surfaces, but use the same indexical pointer as their meaning$_1$.

The arbitrariness of signs may also be seen within a single language. For example, the surface (i.e. letter or phoneme sequence) of the symbol **square** contains nothing that might be construed as indicating the meaning$_1$ of an equal-sided quadrangle with 90 degree angles. Similarly, nothing in the surface of indexical **now** might be construed as indicating the meaning$_1$ of a temporal pointer. Thus, the surface-meaning$_1$ relation in the sign types symbol and index is what de Saussure's calls *unmotivated*.

In natural communication, symbols can have the following functions.

6.2.3 POSSIBLE FUNCTIONS OF THE SIGN TYPE SYMBOL

1. *Initial* reference to objects which have not been mentioned so far (cf. 4.3.2).
2. *Repeating* reference to referents which have already been introduced linguistically (cf. 6.3.3 and 6.3.7).
3. *Metaphorical* reference to partially compatible referents (cf. 5.2.1), both in initial and repeating reference.

These different functions are based on

- the minimal meaning$_1$ structure of symbols (M-concepts), and
- the limited selection of compatible referents available in the subcontext.

Consider, for example, the word **table**. The multitude of different possible referents, e.g. dining room tables, kitchen tables, garden tables, writing tables, picnic tables, operating tables, drawing tables, etc., of various brands, sizes, colors, locations, etc., is not part of the meaning$_1$ of this word. The referents arise only at the level of the internal context whereby, e.g., different kinds of tables are distributed over the subcontexts of the cognitive agent in accordance with previous and current experience.

Yet a minimal meaning$_1$ structure of the word **table** is sufficient to refer properly, provided that the subcontext is sufficiently limited over the STAR-point and the time-linear order of the sign. If there are no standard referents, a word like **table** can even refer to an orange crate (cf. 5.2.1). Furthermore, the speaker may differentiate the meaning$_1$ structure of the referring expression by integrating additional symbolic content into the sign, e.g., **the table, the garden table, the green garden table, the small green garden table, the round small green garden table, the round small green garden table near the pool,** etc.

Reference with a symbol will not succeed, if the activated subcontext contains several similar candidates, e.g. several prototypical tables, and the speaker fails to provide additional symbolic characterization or a pointing gesture to single out the one intended. This problem would not only arise for an artificial cognitive agent like CURIOUS, but for a human hearer as well. Cases of uncertain reference are common in daily life. When they occur it is normal for the hearer to request clarification.

In the hearer mode, the M-concept of a symbol may not only be matched onto an existing I-concept$_{loc}$ in a subcontext, but also introduce a new one. For example, when a buyer goes through a house saying, e.g., **living room** in room$_1$, **dining room** in room$_2$, etc., the associated properties are inserted into the respective representations of the hearer-internal context structure. This process differs from naming in that not just markers, but concepts are added to the internal representations of the rooms.

6.2.4 CHARACTERIZING OBJECTS SYMBOLICALLY

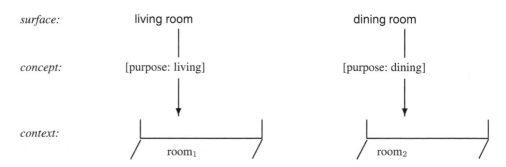

Externally there is nothing changed in room$_1$, but speaker and hearer can leave the house, the town, or the country, and talk without any problem about the living room in the context of their new house.[7]

6.3 Indices for repeating reference

Words exemplifying the sign type index are here, now, I, and you. Their meaning$_1$ consists in pointers at the parameters S, T, A, and R. Thus, these four indices are well-suited to position a sign in accordance with its STAR-point.

In the speaker mode, the characteristic pointers of index words may be easily implemented. Within a robot like CURIOUS, for example, here (S) points to the center of the on-board orientation system, now (T) to the current value of the on-board clock, I (A) to CURIOUS as the center of the internal spatio-temporal orientation system, and you (R) at the current partner in discourse.

The pointer of an indexical word may combine with grammatical and symbolic components of meaning which help to support and refine the process of reference. How much symbolic support is required by an index for successful reference depends on how many potential candidates of reference are provided by the currently activated subcontext.

Index words which contain no additional grammatical or symbolic meaning components are called *pure indices*. If the speaker wishes additional precision or emphasis when using a pure index like here, now, and you,[8] this may be expressed compositionally as in up here, right now, you there, etc.

Examples of a non-pure indices, on the other hand, are the first person pronouns I, me, we, and us in English. They incorporate the symbolic-grammatical distinction between singular (I) and plural (we), and the grammatical distinction between nominative (I, we) and oblique (me, us) case. Thus, even though the pronouns of first person in English all share the same pointer to the parameter A, they differ in their symbolic-grammatical meaning components and are therefore not pure indices.

The pointing area of pronouns of third person, finally, is outside the STAR-point parameters and comprises all objects and persons that have been activated so far and are neither the speaker nor the hearer. Depending on the subcontext, this pointing area may be very large for which reason gender-number distinctions in third person pronouns like he, she, it, they, him, her, them are helpful, but may still not suffice to achieve correct reference to the intended person or object.

Especially in mediated reference, the introduction of third person individuals and objects is therefore done mostly by means of iconic or name-based reference. This has

---

[7] Should the buyer and his/her companion decide later to make room$_2$ into the living room instead, then this requires an explicit verbal agreement to ensure the functioning of communication.

[8] In German, the second person pronoun is not a pure index. For a detailed discussion of the personal pronouns of English in connection with their role as nominal fillers see Section 17.2. The corresponding analysis for German is in Section 18.2.

freed the index of third person pronouns at least partially to serve in another important function, namely repeating reference.

The special task of repeating reference arises in longer sentences and texts when a certain referent has already been introduced and needs to be referred to again. In the SLIM theory of language, repeating reference is analyzed as the return to a contextual object that has been recently traversed.

From the viewpoint of a time-linear coding and decoding of meaning it would be cumbersome, if reference to an object already introduced would always require the use of a complex noun phrase or a name. Instead third person pronouns are ideally suited for a brief, precise, and versatile handling of repeating reference because of their grammatical differentiation and the general nature of their pointing area.

The following example illustrates a case of repeating reference in which a complex symbolic noun phrase is followed by a third person pronoun. As indicated by the arrows, reference of the noun phrase is taken up again by the pronoun.

6.3.1 INDEXICALLY REPEATING REFERENCE

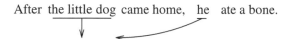

To connect this and the following examples pragmatically beyond their role as linguistic examples in this chapter, let us assume that the little dog is called Fido and that Fido lives in Peter's house together with a big dog named Zach.

Reference to Fido in 6.3.1 is established initially by means of the symbolic meaning$_1$ of the complex noun phrase the little dog. Use of the pronoun he allows repeating reference to Fido with minimal effort. In traditional grammar, the pronominal reference illustrated in 6.3.1 is called *anaphoric*. The 'full' noun phrase preceding the pronoun is called the *antecedent* and is coreferent with the pronoun. In traditional analysis, anaphoric interpretations are limited to individual sentences.

Besides the anaphoric (i.e. reference repeating) interpretation, the pronoun he may also be interpreted noncoreferentially.

6.3.2 INDEXICAL REFERENCE WITHOUT COREFERENCE

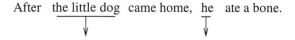

The surface is the same as in 6.3.1. Yet the complex noun phrase and the pronoun refer to different individuals, e.g. Fido and Zach.

According to a surface compositional analysis, the difference between 6.3.1 and 6.3.2 is neither syntactic nor semantic. Instead, it resides in the pragmatics, i.e. the relation between the meaning$_1$ (here specifically the pointer of he) and the context.

Which interpretation is intended by the speaker and which is chosen by the hearer depends solely on the content of the respective subcontexts.

Under certain circumstances it may be desirable to establish initial reference with a third person pronoun and then return to that referent using a (definite) complex noun phrase or a name.[9]

6.3.3 SYMBOLICALLY REPEATING REFERENCE I

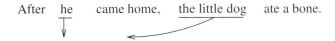

In traditional grammar the pronominal reference illustrated in 6.3.3 is called *cataphoric*. The 'full' noun phrase following the pronoun is called the *postcedent* and is coreferent with the pronoun. Like anaphoric interpretations, cataphoric interpretations are restricted to individual sentences.

Besides the cataphoric (i.e. reference repeating) interpretation, the noun phrase **the little dog** may also be interpreted noncoreferentially.

6.3.4 SYMBOLIC REFERENCE WITHOUT COREFERENCE

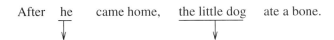

The surface is the same as in 6.3.3, but pronoun and noun phrase refer to different individuals. Without special contextual priming, the interpretation 6.3.4 is in fact much more probable than 6.3.3.

A coreferential interpretation presupposes that the two nominals involved are grammatically and conceptually compatible. For example, if **he** were replaced by **they** in 6.3.1 or 6.3.3, an interpretation of repeating reference to Fido would not be possible because of disagreeing number. Instead an indexical interpretation of the pronoun – as in 6.3.2 or 6.3.4 – would be the only possibility.

A pronoun-based return to the referent of a noun phrase or a noun-phrase-based return to the referent of a pronoun may also be restricted by the sentence structure, as shown by the following examples. They express the intended coreference in terms of equal subscripts (e.g. **Peter**$_k$, **he**$_k$) – a notational convention borrowed from theoretical linguistics.

---

[9] This type of repeating reference is relatively rare. In spoken language, it must be supported by a characteristic intonation. Stylistically it may create a certain expectation in the hearer who is uncertain of the referent of the pronoun and must wait for the following noun phrase to support or specify the interpretation.

6.3.5 SENTENCE STRUCTURE BLOCKING REPEATING REFERENCE

1. Anaphorical positioning:

   After Peter$_i$ came home he$_i$ took a bath.

   Peter$_i$ took a bath after he$_i$ came home.

   %! Near Peter$_i$ he$_i$ sees a snake.[10]

2. Cataphorical positioning:

   After he$_i$ came home Peter$_i$ took a bath.

   %! He$_i$ took a bath after Peter$_i$ came home.[11]

   Near him$_i$ Peter$_i$ saw a snake.

The examples marked with '%!' are grammatically well-formed, but the co-indexed nominals cannot be interpreted as coreferential – despite their closeness of position and despite grammatical and conceptual compatibility.

In communication, repeating reference is needed also across sentence boundaries.

6.3.6 CROSS-SENTENTIAL COREFERENCE

Peter$_k$ wanted to drive into the country. He$_k$ waited for Fido$_i$. When the little dog$_i$ came home he$_k$ was glad.

This example shows that of several coreferential candidates it is not always the one closest to the pronoun or the one in the same sentence which is the most plausible. For example, the third sentence in 6.3.6 taken alone would suggest coreference between the little dog and he. The content of the wider linguistic context, however, makes Peter a more probable antecedent of the second he than the the little dog even though the latter is positioned closer to the pronoun.

Repeating reference with a complex noun phrase is also possible, if the initial reference is based on a proper name[12]:

6.3.7 SYMBOLICALLY REPEATING REFERENCE II

After   Fido   came home,   the little dog   ate a bone.

Like 6.3.3 and 6.3.4, the surface of this example permits both, a coreferential and a non-coreferential interpretation.

Examples like this lead to the generalization that all three sign types may be used for initial reference.

---

[10] G. Lakoff 1968.

[11] R. Langacker 1969.

[12] Instead of a proper name also a noun phrase may be used for the initial reference:

My neighbor went on vacation, but the idiot forgot his credit cards.

In such examples, R. Jackendoff 1972 called the second noun phrase a *pronominal epithet* – which indicates the mistaken view that the noun phrase the idiot was 'pronominalized' via coreference with the antecedent my gardener.

6.3.8 INITIAL REFERENCE ESTABLISHED BY SYMBOL, INDEX, AND NAME

```
                the little dog
     After     he                 came home, he slept.
                Fido
```

Furthermore, all three sign types may be used for repeating reference.

6.3.9 REPEATING REFERENCE USING SYMBOL, INDEX, AND NAME

```
                        the little dog
     After he came home,     he              slept.
                        Fido
```

Besides coreferential interpretations, all of the instances in 6.3.8 and 6.3.9 allow non-coreferential ones as well. Which interpretation is intended by the speaker and chosen by the hearer is a question of pragmatics.

Therefore the noun phrase **the little dog** in 6.3.9 maintains its symbolic character, the pronoun **he** maintains its indexical character, and **Fido** maintains its character as a name – irrespective of whether these signs are used for repeating reference or not. To postulate different morphosyntactic analyses for a sign depending on whether it is used for repeating reference or not would violate surface compositionality.

Unfortunately, nativism has a long tradition[13] of assigning special morphosyntactic analyses to the ana- and cataphoric interpretation of third person pronouns, based on stipulated coreference within the sentence boundary. This is not justified because the personal pronouns form a uniform class not only morphologically and syntactically, but also semantically: they are all of the sign type index. This indexical character is as inherent in pronouns of third person as it is in pronouns of first or second person, regardless of whether such pronouns are used for repeating reference or not.

6.4 Exceptional properties of icon and name

There are two important and interesting exceptions to de Saussure's first law, namely the sign types icon and name. Typical icons are signs with pictorial representations. For example, in the Dutch city of Amsterdam one will find the following street sign.

6.4.1 EXAMPLE OF AN ICON (STREET SIGN)

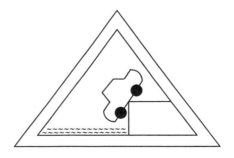

Positioned at parking places between the street and the canal (Gracht) this sign says something like "Attention: danger of driving car into water!"

Foreigners who speek no Dutch and see this sign for the first time can infer the intended meaning₂ directly from the surface of the sign and its positioning – via pragmatic inferences which are obviously not language dependent. Thus the crucial property of icons is that they can be understood spontaneously without the prior learning of a specific language. For this reason, icons are widely used in multilingual applications such as international airports, programming surfaces, manuals, etc.

Icons occur in all media of language such as vision, sound, and gesture. Sound-based icons in natural language are called onomatopoetica. An example in English is the word cuckoo: the sound form of the surface is in a meanigful (motivated, iconic) relation to the sound of the bird in question.

That the meaning₁ of icons may be understood independently of any particular language comes with a price, however. Depending on the medium, there is only a limited number of concepts suitable[14] for iconic representation. For this reason, new icons must be designed and positioned very carefully to ensure proper understanding.

De Saussure viewed onomatopoetica mostly as an exception to his first law.[15] He pointed out, first, that the number of onomatopoetic words is much smaller than 'generally assumed.' Second, they are only partially motivated because they differ from language to language (e.g., bubble bubble in English, glou-glou in French, and gluck-gluck in German). Third, the onomatopoetic word forms are subject to morphological processes for which reason they may become *demotivated* in time.

Regarding the synchronic functioning of a fully developed natural language one may agree with de Saussure that onomatopoetic words are indeed of secondary importance. The main interest of icons as a basic type of sign, however, arises with the question of how natural language developed during evolution. In this context, de Saussure's position must be seen before the background of the age-old controversy between the *naturalists* and the *conventionalists*:

> [The naturalists] maintained that all words were indeed 'naturally' appropriate to the things they signified. Although this might not always be evident to the layman, they would say, it could be demonstrated by the philosopher able to discern the 'reality' that lay behind the appearance of things. Thus was born the practice of conscious and deliberate etymology. The term itself (being formed from the Greek stem *etymo-* signifying 'true' or 'real') betrays its philosophical origin. To lay bare the origin of a word and thereby its 'true' meaning was to reveal one of the truths of 'nature'.

> J. Lyons 1968, p. 4 f.

---

[13] Beginning with R.B. Lees 1960 transformational analysis of reflexive pronouns. A more recent example may be found in C. Pollard & I. Sag 1994.

[14] For example, the sound form cuckoo is considerably more distinctive and thus more suitable for use as an icon than a visual icon for this bird, recognition of which would require above average ornithological knowledge. Conversely, it would be very hard to find an acoustic counterpart to the well-known nonsmoking sign consisting of the schematic representation of a burning cigarette in a circle crossed by a diagonal bar – whereby the latter components are symbolic in character.

[15] Cf. F. de Saussure 1967, p. 81, 82.

The prototypical sign type of naturalists is the icon, that of conventionalists is the symbol.

The naturalists recognized the spontaneous understanding of icons as an important functional aspect of sign-based communication. This insight, however, is not applicable to signs in which the relation between surface and M-concept is based on convention, as in symbols, or in which there is no M-concept at all, as in indices and names.

The conventionalists recognized that most signs have surfaces which – in the terminology of de Saussure – are unmotivated. This insight, however, is no reason to down-play iconic reference as a phenomenon.

According to the SLIM theory of language, the sign types icon and symbol function in terms of the same *iconic* reference mechanism.

6.4.2 COMPARING THE STRUCTURE OF SYMBOL AND ICON

Symbol and icon share the same two-level structure with a fixed connection between the surface and the M-concept. The difference between the two sign types is in their respective *surfaces* and the *kind* of connection between the two levels.

In symbols, the surface is unmotivated and the connection between the levels of the surface and the M-concept is based on the glue of convention. In icons, the surface is motivated and the connection between the two levels is based ideally on their equality.

This analysis explains why icons, but not symbols, can be understood spontaneously, independently of any specific language. Icons use iconic reference *directly* because their meaning$_1$ is apparent in the surface – for which reason icons can be understood without having to learn a convention. Symbols, on the other hand, use iconic reference *indirectly* because the connection between the M-concept and the surface is convention-based – for which reason symbols cannot be understood without the prior learning of the convention.

By relating symbols and icons in a coherent theoretical manner within the [2+1] level structure of the SLIM theory of language, the controversy between the conventionalists and the naturalists is amiably resolved. Furthermore, it is explained why symbols can be used to refer spontaneously to objects never encountered before (as in the Mars mushroom example in Section 4.2): it is because symbols and icons function on the basis of the same iconic mechanism of reference.

Gradual changes in the surface (e.g. simplification) or the M-concept (e.g. semantic reinterpretation) turn icons quasi automatically into symbols. These processes of *de-motivation* destroy the spontaneous understandability of the original icons, but have the advantage of achieving a new level of generality. Once the concept structure has become independent of the surface it is not constrained by the surface medium any more. The expressive power of symbols is much greater than that of icons because symbols may build their meaning$_1$ by combining elementary as well as complex M-concepts from different media and may include logical concepts like negation.

Another exception to de Saussure's first law besides icons are the proper names. Because names have no meaning$_1$[16] (cf. PoP-6 in 6.1.5) their initial choice is essentially free.[17] Once a name has been given, however, it becomes a fixed property of its referent. The effective use of a name requires that it is known to the speaker and the hearer

A name like R2D2 is certainly 'arbitrary' in de Saussure's sense. Names are an exception to his first law, however, because they do not establish a convention-based relation between a surface and a meaning. Rather, in names the matching frontier is directly underneath the surface, in contradistinction to all the other sign types.

6.4.3 COMPARING THE STRUCTURE OF ICON AND NAME

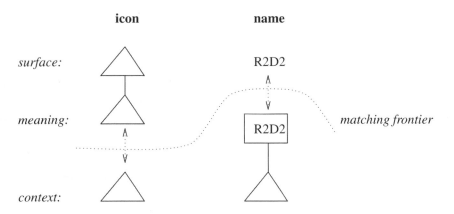

Icons are special signs in that their *meaning doubles as the surface* (↑). The theoretical distinction between the surface and the meaning$_1$ in icons allows their gradual demotivation and the transition to symbols.

Names are special signs in that their *surface doubles as a marker* (↓) on the level of meaning$_1$. The theoretical distinction between the surface and the marker in names allows for reference to different persons having the same name, depending on the context of use.[18]

---

[16] It is possible and common to use symbols as names, e.g. Miller, Walker, Smith, Goldsmith, or Saffire. In such cases, the symbolic meaning may be the basis of word plays – for which reason the choice of a name is an important and sensitive issue –, but it has no function in name-based reference.

[17] Disregarding social traditions like inherited family names. Sign-theoretically the free choice of names is contrasted here with the absence of such a choice in the case of the other signs. See in this connection Section 4.3 as well as L. Wittgenstein 1953 on *private language*.

[18] The crucial problem in the classic AI systems' handling of natural language is the 'gluing' of surfaces to machine internal referents, based on a [−sense, +constructive] ontology. See Section 20.4.

6.5 Pictures, pictograms, and letters

In conclusion let us consider the origin of icons from pictures and the gradual transition from pictures via visual icons (pictograms) to the ideographic,[19] syllabic,[20] and alphabetic writing systems. Pictures are cognitively basic in that they arise at the earliest beginning of mankind and in early childhood.

Though pictures are no signs, they have the sign-theoretically interesting property that they can refer. For example, Heather spontaneously recognizes the Statue of Liberty on Spencer's postcard (cf. 5.3.1). The picture refers to the statue because the viewer establishes a relation to the statue via the picture. This relation is based on the similarity between the picture (here a photograph) and the object shown. If the picture showed something else, e.g. the Eiffel Tower, reference would be to a different object.

Taken by itself, a picture is purely representational. If its positioning is taken into account, however, a picture may be used with a certain communicative intention. For example, the photograph of an ashtray filled with disgusting cigarette stubs placed prominently on the wall of a waiting room in a hospital may be viewed as a mere picture. When the viewer begins to consider why this particular photograph was placed in this particular place, however, the picture turns into an icon. Thereby linguistic and nonlinguistic pragmatics (cf. Section 5.1) are not yet separated.

The iconic use of a picture does not indicate explicitly whether a communicative purpose is intended or not. Moreover, a successful iconic interpretation depends not only on the disposition, but often also on the knowledge of the viewer. For example, paintings from the European middle ages are usually loaded with iconic messages the understanding of which requires a considerable degree of specialized knowledge.[21]

Besides the iconic use of a genuine picture there are icons as a sign type. An icon results from a picture by simplifying it into a basic schema whereby the loss of detail (i) generalizes the content and (ii) explicitly marks communicative purpose. The schematic nature of an icon in conjunction with its relevant positioning invites the viewer to infer the author's intended message.

Using a picture as an icon is like using a rock as a hammer. Modifying the rock to make it better suited for a specialized task is like schematizing a picture into an icon. Just as handle and hitting side of a modified stone (e.g. a paleolithic wedge) are

---

[19] E.g. Chinese.

[20] E.g. Japanese *Kana*.

[21] For example, when an *Annunciation* by the Flemish painter Rogier van der Weyden (1399(?)–1464) shows a glass of water through which there falls a beam of light, this may seem accidental to the uninitiated. A more literate viewer will interpret it as an allegory of the Holy Ghost, however, because it is known that clerical scripture viewed the immaculate conception as similar to light passing through a glass without causing material change.

    This example shows once again that the positioning of signs in general, and icons in particular, is of greatest importance for the interpretation. Positioning the glass with light passing through in a painting of the Annunciation starts thought processes in the educated viewer which go far beyond the purely representational aspect of this part of the picture.

different sides of the same object, the surface (handle) and the meaning of an icon are different sides of the same concept.

In other words, a paleolithic wedge and a random rock differ in that the wedge – but not the rock – provides for a clear distinction between the handling side and the working side. The same is true for an icon in comparison to a picture: only the icon – but not the picture – provides for a distinction between the surface and the meaning level.

This distinction is the basis for an independent development of surfaces and concepts in signs, allowing a process of demotivation – and thus generalization – which considerably improves expressive power.

6.5.1 Transition from picture to icon to symbol

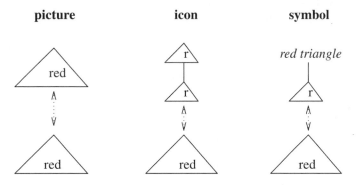

According to this analysis, the iconic reference mechanism originated with pictures. This process of generalization via icons to symbols occurs in all media of language.

In the *acoustic* medium, the development begins with, e.g., imitating the call of bird. The process continues with a simplified icon, e.g. cuckoo, and may end with a demotivated symbol.

In the *gestic* medium, the development begins with imitating a certain activity such as harvesting or threshing grain. Such imitations are often schematized and ritualized into traditional dances. In sign languages the process of demotivating gestures results in true symbols.[22]

In the *visual* medium, the development begins with the picture of, e.g., a bull and continues with a simplified icon. At this point two different kinds of writing systems may evolve. Ideographic writing systems use the icon directly as a pictogram. For example, the meaning 'six bulls' may be expressed in such a system by drawing, painting, or chiseling a sequence of six bull icons.

Letter- or syllable-based writing systems turn the pictogram into a letter by substituting the original meaning with the characteristic sound or syllable of the word in that language. For example, the letter 'A' resembles a bovine head, especially when

---

[22] The notion of a 'symbolic gesture' in the SLIM theory of signs is related to, but pragmatically different from the common use of this term.

turned around to the position it had in its earlier history.[23] The icon evolved into the letter alpha by putting the sound of the corresponding Hebrew word 'lf (= alf) for 'bull' or 'cattle'[24] into the place of the icon's meaning. This rotation principle may be illustrated schematically as follows.

6.5.2 ROTATION PRINCIPLE UNDERLYING TRANSITION FROM ICON TO LETTER

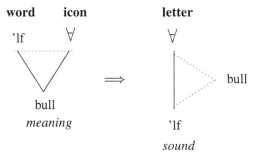

Initially, the word and the corresponding icon stand side by side, referring to the same object. The main bonds are between the word and the referent and between the icon and the referent, producing an equivalence relation between word and icon as a by-product. In the development of the letter, the equivalence relation evolves into the main bond and is established between the visual surface of the icon and the sound surface of the word. The relations to the referent, which motivated the correlation between word and icon in the first place, become obsolete and are forgotten.

The evolution of letters or ideographs as symbols for sounds[25] permits representing natural language in writing. The elementary symbols of the writing system, e.g. the letters, are combined into letter sequences which denote word surfaces. The surfaces in turn denote the meanings$_1$. Aristotle put it this way:

> Spoken words are the signs of mental experience and written words are the signs of spoken words.
>
> Aristotle, DE INTERPRETATIONE, 1

For example, the sequence of letter symbols s, q, u, a, r, and e denotes the surface of the symbol **square**. This surface, which may also be represented acoustically, denotes the concept defined in 3.3.2. The gradual demotivation in the evolution of language symbols – regarding spoken word surfaces as well as letters or signs – is a process of the motivational chain becoming longer and longer until the original motivation is forgotten, irrelevant, or both.

---

[23] The later rotation of the letter was motivated by typographical considerations such as looks (letting the A stand squarely on its 'feet') and maximizing the distinction to other letters.

[24] Cf. T. Givón 1985, pp. 193–5.

[25] Highly developed ideographic writing systems such as Egyptian hieroglyphs or Chinese also make heavy use of the sound form corresponding to the icon denoted in order to overcome the inherent limitations of iconic representation.

Exercises

Section 6.1

1. Explain the functioning of iconic, indexical, and name-based reference.
2. Which indexical words refer to the four parameters of the STAR-point? What is their syntactic category? How pure are they?
3. Explain the use of here in the postcard example 5.3.1. Is it an example of immediate or a mediated reference? Do you require different answers for the speaker and the hearer?
4. How would you implement the meaning$_1$ of yesterday, tomorrow, in three minutes, to my right, to your right, above me, and behind you in CURIOUS?
5. Explain the indexical, symbolic, and grammatical components in the word forms me, my, mine, our, us, ours.
6. Why are Pop-1 – PoP-3 presupposed by PoP-4 – PoP-6?
7. Compare PoP-1 – PoP-7 regarding their empirical content and their functionality.
8. Summarize the 'Universal Grammar' of Chomsky 1981.
9. Make a list of the concrete linguistic universals proposed in Chomsky 1957, 1965, and 1981, comparing their empirical content and their functionality..
10. Make a list of the concrete linguistic universals proposed in GPSG (Gazdar, Klein, Pullum,& Sag 1985), HPSG (Pollard & Sag 1987, 1994), and LFG (Bresnan (ed.) 1982), comparing their empirical content and their functionality.
11. Summarize the linguistic universals that have established within nativism. How do they characterize the innate human language faculty?
12. Why does a coherent treatment of verbal and nonverbal reference in a uniform theoretical framework require a cognitive approach describing the internal states of the agents involved? Give detailed reasons with examples of symbols, indices, and names.
13. Explain nonverbal iconic, indexical, and name-based reference.
14. Describe the language of the bees and explain why it can handle *displaced reference* on the basis of non-verbal indexical and iconic reference (cf. CoL, p. 281f.).

Section 6.2

1. Which four components make up the internal structure of symbols and indices?
2. Describe de Saussure's first law and illustrate it with examples from different languages. Why does it apply to symbols as well as indices?
3. Name three possible functions of symbols, and explain the difference between the speaker and the hearer mode with examples.
4. Why can the meaning$_1$ of symbols be minimal? Are the different shades of the color red part of the meaning of the word red?
5. What is the difference between the index of a textual database (cf. 2.1.1) and the index of a sign? What is their common conceptual root?

Section 6.3

1. Compare the pointing areas of first, second, and third person.
2. What is repeating reference and how can it be achieved in natural language?
3. Explain the anaphoric and the cataphoric use of a third person pronoun.
4. Which syntactic structures block ana- or cataphoric interpretations?
5. Can a third person pronoun refer without an ante- or postcedent?
6. What did Jackendoff mean with his notion *pronominal epithet*?

Section 6.4

1. What is the characteristic property of an icon, and how can it be practically used?
2. Why are icons an exception to de Saussure's first law?
3. Why are onomatopoetic words a kind of icon?
4. How does de Saussure view the role of onomatopoeia in the natural languages?
5. Explain the viewpoints of the naturalists and the conventionalists.
6. What are the specific strength and weakness of the icon as a sign type?
7. Why does demotivation increase the expressive potential of a sign?
8. Would you classify de Saussure's approach as naturalist or conventionalist?

Section 6.5

1. Compare the structure and functioning of icon and name.
2. Why are names an exception to de Saussure's first law?
3. Why is a name like **Peter** less general from a sign theoretic point of view than a symbol like **table** or an index like **here**? What is the great advantage of name-based reference?
4. How does Peirce define the difference between symbols, indices, and icons? Can this definition be used in computational linguistics? (Cf. CoL, p. 276-80.)
5. Why is a picture not a sign? How can a picture be used as a sign?
6. Explain the gradual development of symbols from pictures.
7. What is the difference between ideographic- and letter-based writing systems?
8. Why are letters symbols? What is their meaning$_1$?
9. What is the difference between letter symbols and word symbols?
10. Explain the rotation principles in the evolution of letters from icons.

Part II

Theory of Grammar

7. Generative grammar

Part I has explained the mechanism of natural communication, based on the [2+1] level structure of the SLIM theory of language and different types of language signs. Part II turns to the combinatorial build up of complex signs within the grammar component of syntax. The methods are those of formal language theory, a wide field reaching far into the foundations of mathematics and logic. The purpose here is to introduce the linguistically relevant concepts and formalisms as simply as possible, explaining their historical origin and motivation as well as their different strengths and weaknesses. Formal proofs will be limited to a minimum.

Section 7.1 explains the formal notion of a language as a subset of the free monoid over a finite lexicon. Section 7.2 describes the mathematical, empirical, and computational reasons for using generative grammars in the linguistic description of natural language. Section 7.3 shows that using generative grammar is methodologically a necessary condition for natural language analysis, but not sufficient to ensure that the grammar formalism of choice will be structurally suitable for an empirically successful analysis in the long run. Section 7.4 introduces the historically first generative grammar for natural language, namely categorial grammar or C-grammar. Section 7.5 presents a formal application of C-grammar to a small 'fragment' of English.

7.1 Language as a subset of the free monoid

Formal language theory works with mathematical methods which treat the empirical contents of grammatical analysis and the functioning of communication as neutrally as possible. This is apparent in its abstract notion of language.

7.1.1 DEFINITION OF LANGUAGE

A language is a set of word sequences.

This set (in the sense of set theory) is unordered. Each element of the set, however, is an ordered sequence of words.

The most basic task in defining a formal language is to characterize the grammatically well-formed sequences. Thereby the words are usually treated as simple surfaces with no category, no different word forms, no special base form, and no meaning$_1$.

For example, the words a and b constitute the word set LX = {a, b}, i.e., the lexicon (or alphabet[1]) of a formal language to be defined below. The elements of LX may be combined into infinitely many different word sequences (provided there is no limit on their length). Such an infinite set of all possible sequences is called the *free monoid* over a finite lexicon.

7.1.2 ILLUSTRATION OF THE FREE MONOIDS OVER LX = {a,b}

ε

a, b

aa, ab, ba, bb

aaa, aab, aba, abb, baa, bab, bba, bbb

aaaa, aaab, aaba, aabb, abaa, abab, abba, abbb ...

...

The free monoid over LX is also called the *Kleene closure* and denoted by LX*. The free monoid without the neutral element ε (also called empty sequence or zero sequence) is called the *positive closure* and denoted by LX⁺.[2]

The concept of a free monoid is applicable to artificial and natural languages alike. The lexica of natural languages are comparatively large: if all the different word forms are taken into account, they may contain several million entries. In principle, however, the lexica of natural languages are like those of artificial languages in that they are finite, whereas the associated free monoids are infinite.

Because a free monoid contains all possible word sequences many of them will not be grammatical relative to a given language. In order to filter out the grammatically correct word sequences, a formal criterion is required to distinguish the grammatically well-formed sequences from the ill-formed ones.

In an artificial language, the formal specification of grammatical well-formedness is not difficult. It is defined by those who invented the language. For example, one has defined the artificial language $a^k b^k$ (with $k \geq 1$) as the set of expressions which consist of an arbitrary number of the word a followed by an equal number of the word b. This language is a proper subset of the set of sequences over {a,b} sketched in 7.1.2.

According to the informal description of $a^k b^k$, the following expressions

a b, a a b b, a a a b b b, a a a a b b b b, etc.,

are well-formed. All the other expressions of the free monoid 7.1.2 which do not comply with the structure described, such as

---

[1] In formal language theory, the lexicon of an artificial language is sometimes called the *alphabet*, a word a *letter*, and a sentence a *word*. From a linguistic point of view this practice is unnecessarily misleading. Therefore a basic expression of an artificial or a natural language is called here uniformly a *word* (even if the word consists of only a single letter, e.g. a) and a complete well-formed expression is called here uniformly a *sentence* (even if it consists only of a sequence of one-letter-words, e.g. aaabbb).

[2] In other words: the free monoid over LX equals LX⁺ ∪ {ε}. See M. Harrison 1978, p. 3.

a, b, b a, b b a a, a b a b, etc.,

are not well-formed expressions of the language $a^k b^k$. Analogously to $a^k b^k$ we may invent other artificial languages such as $a^k b^k c^k$, $a^k b^m c^k d^m$, etc.

Not only the free monoid over a finite lexicon contains infinitely many expressions; also the possible languages – as subsets of the free monoid – usually comprise infinitely[3] many well-formed expressions or sentences. Therefore the criterion for filtering a language from a free monoid must be more than just a list.[4]

What is needed instead is a principled structural description of the well-formed expressions which may be applied to infinitely many (i.e. ever new) expression types. In formal language theory these structural descriptions are specified as *recursive rule systems* which were borrowed from logic and called *generative grammars*.

The following example of a generative grammar for the artificial language $a^k b^k$ uses a formalism known today as PS-grammar.[5]

7.1.3 PS-GRAMMAR FOR $a^k b^k$

S → a S b
S → a b

PS-grammar generation is based on substituting the sign on the left of the rule arrow by the signs on the right.

Consider for example the derivation of the string a a a b b b. First the rule S → a S b is applied, replacing the S on the left of the arrow by the signs on the right. This produces the string

a S b .

Next the first rule is applied again, replacing the S in the expression just derived by a S b. This produces the new string

a a S b b.

Finally, the S in this new string is replaced by a b using the second rule, resulting in

a a a b b b.

In this way the string in question may be derived formally using the rules of the grammar 7.1.3, thus proving that it is a well-formed expression of the language $a^k b^k$.

Even though the grammar 7.1.3 uses the finite lexicon {a, b} and a finite number of rules, it generates infinitely many expressions of the language $a^k b^k$. The formal basis for this is the *recursion* of the first rule: the variable S appears both, on the left and on the right of the rule arrow, thus allowing a reapplication of the rule to its own output.

---

[3] That the subsets of infinite sets may themselves be infinite is illustrated by the even numbers, e.g. 2,4,6 ..., which form an infinite subset of the natural numbers 1,2,3,4,5, ... The latter are formed from the finite lexicon of the digits 1,2,3,4,5,6,7,8,9, and 0 by means of concatenation, e.g. 12 or 21.

[4] This is because an explicit list of the well-formed sentences is finite by nature. Therefore it would be impossible to make a list of, e.g., all the natural numbers. Instead the infinitely many surfaces of possible natural numbers are *generated* from the digits via the structural principle of concatenation.

[5] A detailed introduction to PS-grammar is given in Chapter 8.

The second rule of 7.1.3, on the other hand, is an example of a non-recursive rule in PS-grammar.

The generative methodology for analyzing artificial and natural languages is not bound to a particular formalism of grammar. Rather, there are several different formalism to chose from.[6] Within formal language theory, the following *elementary formalisms* have so far been defined.

7.1.4 ELEMENTARY FORMALISMS OF GENERATIVE GRAMMAR

1. Categorial or C-grammar
2. Phrase-structure or PS-grammar
3. Left-associative or LA-grammar

These elementary formalisms differ in the form of their respective categories and rules (cf. 10.1.3, 10.1.4, and 10.1.5) as well as in their conceptual derivation orders (cf. 10.1.6). The formal basis of an elementary formalism is its *algebraic definition*.

7.1.5 ALGEBRAIC DEFINITION

> The algebraic definition of a generative grammar explicitly enumerates the basic components of the system, defining them and the structural relations between them using only notions of set theory.

An algebraic definition of C-grammar is provided in 7.4.2, of PS-grammar in 8.1.1, and of LA-grammar in 10.2.1, respectively.

An elementary formalism requires furthermore that its most relevant mathematical properties have been determined. These are in particular the hierarchy of language classes generated by subtypes of the formalism and their complexity. Also, the formal relations between different elementary formalisms must be established.

Based on the elementary formalisms a multitude of *derived formalisms* have been defined over the years. The reason for developing derived formalisms is the attempt to overcome inherent weaknesses of the underlying elementary formalism.

Some examples of derived formalisms which have enjoyed wider popularity at one time or another are enumerated in 7.1.6 and 7.1.7.

7.1.6 DERIVED FORMALISMS OF PS-GRAMMAR

> Syntactic Structures, Generative Semantics, Standard Theory (ST), Extended Standard Theory (EST), Revised Extended Standard Theory (REST), Government and Binding (GB), Barriers, Generalized Phrase Structure Grammar (GPSG), Lexical Functional Grammar (LFG), Head-driven Phrase Structure Grammar (HPSG)

---

[6] Compare, for example, the PS- and C-grammar analysis of $a^k b^k$ in 7.1.3 and 7.4.4.

7.1.7 DERIVED FORMALISMS OF C-GRAMMAR

> Montague grammar (MG), Functional Unification Grammar (FUG), Categorial Unification Grammar (CUG), Combinatory Categorial Grammar (CCG), Unification-based Categorial Grammar (UCG)

It was even proposed to combine PS- and C-grammar into a derived formalism, as in Cooper grammar, to compensate their respective weaknesses and benefit from their respective strengths.

The mathematical properties of derived formalisms must be characterized just as explicitly and precisely as those of the underlying elementary formalism. At first, derived formalisms seem to have the advantage that the known properties of their respective elementary formalism would facilitate their analysis. In the long run, however, the mathematical analysis of derived formalisms has turned out to be at least as difficult as that of elementary formalisms. Often taking decades, it has frequently been subject to error.[7]

Besides elementary and derived formalisms there also exist *semi-formal systems* of generative grammar, the description of which has not been underpinned by an algebraic definition. Examples are dependency grammar (Tesnière 1959) and systemic grammar (Halliday 1985). In such systems one tries to get by to a certain degree with *conjectures* based on similarities with properties of known formalisms. For example, there is a structural resemblance between dependency grammar and C-grammar.

For systematic reasons, the analysis of mathematical properties and empirical suitability will be concentrated mainly on the elementary formalisms. The question is which of them is suited best (i) for the linguistic analysis of natural language and (ii) for being implemented as automatic parsers. Also, the formal and conceptual reasons of why some elementary formalisms have been producing ever new derived formalisms will be explained.

7.2 Methodological reasons for generative grammar

In contrast to artificial languages like $a^k b^k$ and $a^k b^k c^k$, which were invented for certain purposes (particularly complexity-theoretic considerations), natural languages are given by their respective language communities. This means that the decision of whether or not a natural language expression is grammatically well-formed depends on the language intuition of the native speakers. For example, the grammatical correctness of 7.2.1 as an expression of English is noncontroversial.

---

[7] For example, N. Chomsky originally thought that the *recoverability condition* of deletions would keep transformational grammar decidable (see Section 8.5), which was refuted in the proof by S. Peters & R. Ritchie 1972. G. Gazdar originally thought that the introduction of metarules in GPSG would not increase the originally context-free complexity of his system, which was refuted in the proof by H. Uszkoreit & S. Peters 1986.

7.2.1 GRAMMATICALLY WELL-FORMED EXPRESSION

the little dogs have slept earlier

The following expression, on the other hand, will be rejected as ungrammatical by any competent speaker of English.

7.2.2 GRAMMATICALLY ILL-FORMED EXPRESSION

* earlier slept have dogs little the [8]

The second example resulted by inverting the order of first.

Based on the distinction between well-formed and ill-formed expressions, generative grammars may be written also for natural languages. Just as the laws of physics allow to precompute the location of a celestial body at a certain time, the rules of a descriptively adequate generative grammar should make it possible to *formally* decide for any arbitrary expression whether or not it is grammatically well-formed.

At first glance this goal of theoretical linguistics may look rather academic. In fact, however, the use of generative grammars is indispensable for modern linguistic methodology.

7.2.3 METHODOLOGICAL CONSEQUENCES OF GENERATIVE GRAMMAR

– *Empirical*: formation of explicit hypotheses
   A generative analysis results in a formal rule system which constitutes an explicit hypothesis about which input expressions are well-formed and which are not. Such an explicit hypothesis provides clarity about where the formal grammar is empirically adequate and where it is not – which is an important precondition for an incremental improvement of the empirical description.
– *Mathematical*: determining the formal properties
   Only strictly formalized descriptions allow an analysis of their mathematical properties[9] such as decidability, complexity, and generative capacity. The mathematical properties of a grammar formalism in turn determine whether it is suitable for empirical description and computational realization.
– *Computational*: declarative specification of the parser
   Only a formal rule system may be used as a declarative specification[10] of the parser, characterizing its necessary properties in contrast to accidental properties stemming

---

[8] In linguistics, examples of ungrammatical structures are marked with an asterisk *, a convention which dates back at least to L. Bloomfield 1933.

[9] The mathematical properties of informal descriptions, on the other hand, cannot be investigated because their structures are not sufficiently clearly specified.

[10] Programs which are not based on a declarative specification may still run. However, as long as it is not clear which of their properties are theoretically necessary and which are an accidental result of the programming environment and the programmer's idiosyncrasies, such programs – called hacks – are of little theoretical interest. From a practical point of view, they are difficult to scale up and hard to debug. The relation between grammar systems and their implementation is further discussed in 15.1.

from the choice of the programming environment, etc. A parser in turn provides the automatic language analysis needed for the verification of the individual grammars written within the generative formalism of choice.

Because a good methodology is a precondition for obtaining solid descriptive results the systematic use of generative grammars in natural language analysis has a positive effect also on the practical applications of linguistic analysis.

There may be several formalisms of generative grammar, however, all of which are equally well-defined, yet differently well-suited for the description of natural language. This is because the sole prerequisite for formalizing a theory of grammar is that its constructs are explicit. Thus, while the use of generative grammar is methodologically a necessary condition for investigating the empirical, mathematical, and computational properties of linguistic analyses, it is not sufficient to ensure that the grammar formalism in question will be structurally suitable in these respects.

In order to be suitable empirically, the formalism should have sufficient expressive power to describe all the structures of the language in question. Mathematically, the formalism should have the lowest possible complexity. Computationally, the derivation order of the algorithm should be compatible with an efficient and transparent programming structure. In addition, the formalism should be functionally well-suited to serve as the syntactic component of an artificial cognitive agent which is able to communicate freely and efficiently (i.e. in real time) in natural language.

If a given theory of grammar may be shown to be suboptimal by means of mathematical analysis, this is not the fault of the generative method. Conversely, if no such conclusions can be drawn because of insufficient formalization, this is not a merit of the theory of grammar. Instead, the formalization is a precondition for determining objectively whether a theory of grammar is adequate or not.

7.3 Adequacy of generative grammars

A generative grammar is called descriptively adequate for a given natural language, if it generates all and only the well-formed expressions of the language. An inadequate grammar is either incorrect, incomplete, or both. A grammar is incorrect if it generates expressions which are not well-formed (overgeneration). A grammar is incomplete if there are well-formed expressions it cannot generate (undergeneration).

Let us assume that there is a formal grammar for a natural language we do not speak, e.g. Quechua,[11] and we would like to check whether a certain expression is well-formed. For this, we assign to the word forms their respective categories using the lexicon of our formal Quechua grammar. Then we attempt to derive the resulting sequence of analyzed word forms using the formal grammar rules. The expression in question is well-formed only if such a derivation exists.

---

[11] Quechua is a language of South-American Indians.

However, in natural language such a formal derivation is a reliable criterion of well-formedness only, if the generative grammar is known to be descriptively adequate. As long as there is any doubt about the grammar in question, expressions generated by it must all be presented to native speakers to decide whether the grammar really generated something well-formed or not (correctness). Also, new Quechua expressions provided by the native speakers must all be checked as to whether or not the grammar can generate them (completeness).

Determining whether a given expression is in fact generated by the formal grammar may take long – in fact, infinitely long – even if the check is done automatically on a supercomputer. This is the case whenever the grammar formalism used is of high mathematical complexity. The mathematical complexity of a generative grammar is measured as the maximal number of rule applications (worst case) relative to the length of the input (cf. Section 8.2).

Most current systems of generative grammar for natural language are of high mathematical complexity while attempting to characterize grammatical well-formedness without a functional theory of communication. Thus, the systems in question analyze natural languages inefficiently while being empirically underspecified.

To avoid these shortcomings, the syntactic analysis of natural language should be

- defined *mathematically* as a formal theory of low complexity,
- designed *functionally* as a component of natural communication, and
- realized *methodologically* as an efficiently implemented computer program in which the properties of formal language theory and of natural language analysis are represented in a modular and transparent manner.

These desiderata must pursued simultaneously. After all, what use is a pleasantly designed syntax, if its complexity turns out to be undecidable or exponential? What is the purpose of a mathematically well-founded and efficient formalism, if it turns out to be structurally incompatible with the mechanism of natural communication? How reliable is a mathematically and functionally well-suited grammar, if it has not been computationally verified on realistic, i.e. very large, amounts of data?

7.4 Formalism of C-grammar

Historically, the first generative grammar formalism is categorial grammar or C-grammar. It was invented by the Polish logicians LEŚNIEWSKI 1929 and AJDUKIEWICZ 1935 in order to avoid the Russell paradox in formal language analysis. C-grammar was first applied to natural language by BAR-HILLEL 1953.[12]

The origin of C-grammar in the context of the Russell paradox resulted from the outset in a logical semantic orientation (cf. Section 19.3). Accordingly, the combinatorics

---

[12] A good intuitive summary may be found in P. Geach 1972. See also J. Lambek 1958 and Y. Bar-Hillel 1964, Chapter 14, p. 185-189.

of C-grammar is based on the functor-argument structure of logic. A functor denotes a function which maps suitable arguments into values. A logical function consists of a *name*, a specification of the *domain* (i.e. the set of arguments), a specification of the *range* (i.e. the set of values), and a specification of the *assignment* which associates each argument with at most one value.

7.4.1 STRUCTURE OF A LOGICAL FUNCTION

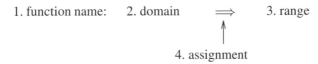

A simple example of a function is the squaring of natural numbers. The name of this function is **square_number**, the domain is the set of natural numbers, the range is likewise the set of natural numbers, and the assignment provides for each possible argument, e.g., 1, 2, 3, 4, 5 ..., exactly one value, i.e., 1, 4, 9, 16, 25 ....[13]

The role of functors and arguments in C-grammar is formally reflected in the definition of its categories and rules. This is shown by the following algebraic definition.[14]

7.4.2 ALGEBRAIC DEFINITION OF C-GRAMMAR

A C-grammar is a quintuple $< W, C, LX, R, CE >$.

1. W is a finite set of word form surfaces.
2. C is a set of categories such that
   a) *basis*
      u and v ϵ C,[15]
   b) *induction*
      if X and Y ϵ C, then also (X/Y) and $(X\backslash Y)$ ϵ C,
   c) *closure*
      Nothing is in C except as specified in (a) and (b).
3. LX is a finite set such that $LX \subset (W \times C)$.
4. R is a set comprising the following two rule schemata:
   $$\alpha_{(Y/X)} \circ \beta_{(Y)} \Rightarrow \alpha\beta_{(X)}$$
   $$\beta_{(Y)} \circ \alpha_{(Y\backslash X)} \Rightarrow \beta\alpha_{(X)}$$
5. CE is a set comprising the categories of *complete expressions*, with $CE \subseteq C$.

---

[13] In contrast, **square_root** is not a function, but called a relation because it may assign more than one value to an argument in the domain. The root of 4, for example, has two values, namely 2 and -2.

[14] A comparable definition of C-grammar may be found in Y. Bar-Hillel 1964, p. 188.

[15] The names and the number of elementary categories (here u and v) are in principle unrestricted. For example, Ajdukiewicz used only one elementary category, Geach and Montague used two, others three.

In accordance with the general notion of an algebraic definition (cf. 7.1.5), the basic components of C-grammar are first enumerated in the quintuple <W, C, LX, R, CE>. Then these components are set-theoretically characterized in clauses 1–5.

More specifically, the set W is defined as a finite, unordered enumeration of the basic surfaces of the language to be described. For example, in the case of the artificial language $a^k b^k$ the set W would contain a and b.

The set C is defined recursively. Because the start elements u and v are in C so are (u/v), (v/u), (u\v), and (v\u) according to the induction clause. This means in turn that also ((u/v)/v), ((u/v)\v), ((u/v)/u), ((u/v)\u), (u/(u/v)), (v/(u/v)), etc., belong to C. C is infinite because new elements can be formed recursively from old ones.

The set LX is a finite set of ordered pairs such that each ordered pair is built from (i) an element of W and (ii) an element of C. Notationally, the second member of the ordered pair, i.e. the category, is written as a subscript to the first as in, e.g., $a_{((u/v)\backslash v)}$. Which surfaces (i.e. elements of W) take which elements of C as their categories is specified in LX by explicitly listing the ordered pairs.

The set R contains two rule schemata. They use the variables α and β to represent the surfaces of the functor and the argument, respectively, and the variables X and Y to represent their category patterns. The first schema combines functor and argument in the order $\alpha\beta$ whereby the Y in the category of the functor α is canceled by the corresponding symbol in the category of argument β. The second schema combines functor and argument in the inverse order, i.e., $\beta\alpha$. This ordering is formally triggered by the backslash '\' (instead of the slash '/') in the category of α. This type C-grammar has been called bidirectional C-grammar by Bar-Hillel because the two rule schemata allow to place a functor either before or behind the argument.

The set CE, finally, describes the categories of those expressions which are considered complete. Depending on the specific C-grammar and the specific language, this set may be finite and specified in terms of an explicit listing, or it may be infinite and characterized by patterns containing variables.

Whether two categorized language expressions may be legally combined depends on whether their categories can be matched onto a suitable rule schema. This process of pattern matching is traditionally handled implicitly in C-grammar because there are only two rule schemata.

7.4.3 IMPLICIT PATTERN MATCHING IN COMBINATIONS OF C-GRAMMAR

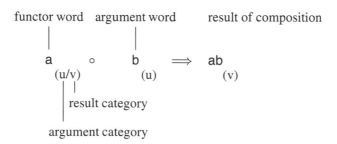

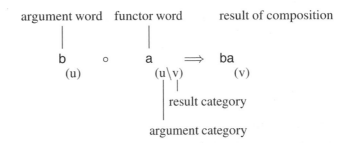

argument word   functor word   result of composition

$$b \quad \circ \quad a \quad \Longrightarrow \quad ba$$
$$(u) \qquad\qquad (u\backslash v) \qquad\qquad (v)$$

result category

argument category

Strictly speaking, however, the rule schemata involved in such combinations should be made visible in terms of explicit pattern matching (cf. Chapters 17 and 18.)

Assuming the rule schemata, the combinatorics is coded into the category structures of C-grammar. Therefore a language is sufficiently specified by defining (i) the lexicon LX and (ii) the set of complete expressions CE. This is illustrated by the following definition of the familiar artificial language $a^k b^k$ (cf. 7.1.3).

7.4.4 C-GRAMMAR FOR $a^k b^k$

$$\text{LX} =_{def} \{a_{(u/v)}, b_{(u)}, a_{(v/(u/v))}\}$$
$$\text{CE} =_{def} \{(v)\}$$

The word a has two lexical definitions with the categories (u/v) and $(v/(u/v))$, respectively, for reasons apparent in the following derivation tree.

7.4.5 EXAMPLE OF $a^k b^k$ DERIVATION, FOR $k = 3$

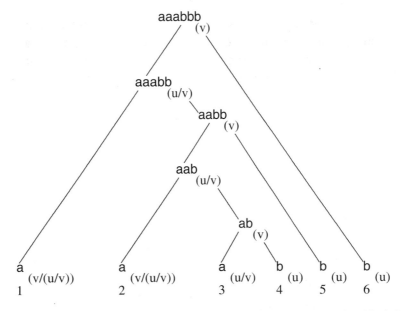

The derivation begins with the combination of words 3 and 4. Only this initial composition uses the categorial reading (u/v) of word a. The result is the expression ab of category (v). This would be a complete expression according to the definition of CE in 7.4.4 provided there were no other words in the input string. Next, word 2 is combined with the result of the last combination, producing the expression **aab** with the category (u/v). This result is in turn combined with word 5, producing the expression **aabb** with the category (v), etc.

In this way complete expressions of category (v) may be derived, each consisting of arbitrarily many words **a** followed by an equal number of words **b**. Thus, the finite definition 7.4.4 generates infinitely many expressions of the artificial language $a^k b^k$.

Structurally, however, C-grammar has the following disadvantages. First, the correct intermediate expressions can only be discovered by trial and error. For example, it is not always obvious where the correct initial composition should take place in the surface of arbitrary inputs. Second, C-grammars require a high degree of lexical ambiguity in order to code alternative word orders into alternative categories.

As a consequence, even experts must often puzzle to find a derivation like 7.4.5 for a given string relative to a given C-grammar (or to show that such a derivation does not exist) – especially in languages not as simple as $a^k b^k$. Correspondingly, an automatic analysis based on a C-grammar is computationally inefficient because large numbers of possible combinations must be tested.

In addition, C-grammar does not permit a time-linear derivation. In 7.4.5, for example, the derivation must begin with word 3 and 4. Due to the characteristic structure of categories and rule schemata in C-grammar it is impossible to design an alternative C-grammar for $a^k b^k$ that can derive arbitrarily long sentences in a time-linear fashion.

7.5 C-grammar for natural language

C-grammar is the prototype of a lexicalist approach: all the combinatorial properties of the language are coded into the categories of its basic expressions. This is as apparent in the C-grammatical definition of the artificial language $a^k b^k$ in 7.4.4 as it is in the following definition of a tiny fragment[16] of English.

7.5.1 C-GRAMMAR FOR A TINY FRAGMENT OF ENGLISH

$\text{LX} =_{def} \{ W_{(e)} \cup W_{(e \backslash t)} \}$, where
$\qquad W_{(e)} = \{ \text{Julia, Peter, Mary, Fritz, Suzy} \dots \}$
$\qquad W_{(e \backslash t)} = \{ \text{sleeps, laughs, sings} \dots \}$
$\text{CE} =_{def} \{ (t) \}$

Compared to 7.4.4 this grammar exhibits two notational modifications.

---

[16] R. Montague used the notion *fragment* to refer to that subset of a natural language which a given formal grammar is designed to handle.

First, the lexical entries are assembled into word classes W_{cat}. This notation, which originated with Montague, makes the writing of lexica simpler because the categories are specified for whole classes rather than for each single word form.

Second, the elementary categories, called u and v in 7.4.2, are renamed as e and t, respectively. According to Montague, e stands for *entity* and t for *truth value*. Thus, the category (e\t) of sleeps, laughs, sings, ... in 7.4.1 is motivated not only syntactically, but also semantically: (e\t) is interpreted as a characteristic function from entities into truth values. The characteristic function of, e.g., sleeps determines the set of sleepers by checking each entity e as to whether the associated truth value t is 1 (true) or 0 (false). Thus, the set denoted by sleeps consists of those entities which the equivalent characteristic function maps into 1.

Because different words like sleep and walk have the same category they have the same domain and the same range. Thus, the difference in their meaning resides solely in their different assignments (cf. 4 in 7.4.1). How to define these different assignments in a non-trivial way (i.e., *not* in terms of explicitly listing ordered pairs in the model definition) is one of the basic problems of model-theoretic semantics.[17]

The semantic interpretation of the C-grammatically analyzed sentence Julia sleeps relative to some model $\mathcal{M}$ is defined as follows. The sentence is true, if the denotation of sleeps maps the denotation of Julia into 1, and false otherwise.

7.5.2 SIMULTANEOUS SYNTACTIC AND SEMANTIC ANALYSIS

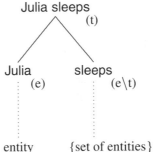

According to the equivalent set-theoretic view, the sentence is true if the denotation of Julia in $\mathcal{M}$ is an element of the set denoted by sleeps in $\mathcal{M}$.

In C-grammar, the categories for natural language expressions are motivated by both, (i) the denotation (semantics) and the (ii) combinatorics (syntax) of an expression. For example, because nouns denote sets they are categorized by Montague[18] as (e/t), i.e., as characteristic functions from entities to truth values, whereby the slant of the slash serves to distinguish nouns syntactically from intransitive verbs.

---

[17] Cf. Section 19.4 and CoL, p. 292-295.

[18] For simplicity and consistency, our notation differs from Montague's in that the distinction between syntactic categories and semantic types is omitted, with arguments positioned before the slash.

Furthermore, because determiners combine with nouns and the resulting noun phrases may be regarded as denoting entities – like proper names – one may categorize the determiner as ((e/t)/e), i.e., as a functor which takes a noun (e/t) to make something like a name (e). Adjectives, on the other hand, take a noun to make a noun, which may be expressed by categorizing adjectives as ((e/t)/(e/t)). From this structure follows the possibility to stack modifiers recursively, e.g. a noun may take an unlimited number of adjectives.

Based on the categorization outlined above, a sentence like **The small, black dogs sleep** may be analyzed as in the following derivation tree.

7.5.3 C-ANALYSIS OF A NATURAL LANGUAGE SENTENCE

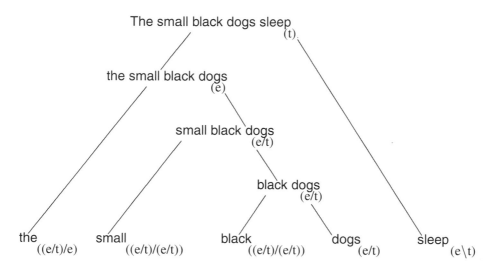

As in the previous examples (cf. 7.4.4 and 7.5.1), the C-grammar analyzing or generating this example consists only of the lexicon LX and the set CE.

7.5.4 C-GRAMMAR FOR EXAMPLE 7.5.3

$$LX =_{def} \{W_{(e)} \cup W_{(e\backslash t)} \cup W_{(e/t)} \cup W_{((e/t)/(e/t))} \cup W_{((e/t)/t)} \}, \text{where}$$
$$W_{(e)} = \{Julia, Peter, Mary, Fritz, Suzy \ldots \}$$
$$W_{(e\backslash t)} = \{sleeps, laughs, sings \ldots \}$$
$$W_{(e/t)} = \{dog, dogs, cat, cats, table, tables \ldots \}$$
$$W_{((e/t)/(e/t))} = \{small, black \ldots \}$$
$$W_{((e/t)/t)} = \{a, the, every \ldots \}$$
$$CE =_{def} \{(t)\}$$

This C-grammar is motivated linguistically by characterizing the semantic functor-argument structure and the syntactic combinatorics at the same time.

In theory, a high degree of motivation and the constraints following from it are generally regarded as desirable in linguistics. However, even though C-grammar is constrained simultaneously by aspects of syntax and semantics, its motivation is only partial. This is because it is not clear how C-grammar and its model-theoretic semantics are supposed to function in natural language communication.[19]

In practice, the structures of C-grammar result in derivations which have the character of problem solving. This disadvantage is aggravated by the fact that C-grammars for larger fragments require extremely high degrees of lexical ambiguity in order to code combinatorial restrictions into the categories.

This is illustrated by the C-grammar 7.5.4, which is not yet adequate for its tiny fragment, but suffers from overgeneration. For example, the combination of $dog_{(e/t)}$ ∘ $Peter_{(e)}$ into the 'sentence' *$dog\ Peter_{(t)}$ is not blocked. Furthermore, there is no proper treatment of the agreement between determiners and nouns, allowing 7.5.3 to generate ungrammatical combinations like *every dogs and *all dog.

In theory, each such difficulty could be handled by a seemingly minor extension of the formalism. In practice, however, the extensions necessary for even moderately sized descriptive work[20] quickly turn the conceptually transparent and mathematically benign elementary formalism of C-grammar into derived systems of high complexity. The need for ever more ad hoc solutions and the concomitant loss of transparency make C-grammar an unlikely candidate for reaching the goal of theoretical linguistics, namely a complete generative description of a natural language.

Exercises

Section 7.1

1. How is the notion of a language defined in formal grammar?
2. Explain the notion of a free monoid as it relates to generative grammar.
3. What is the difference between positive closure and Kleene closure?
4. In what sense can a generative grammar be viewed as a filter?
5. Explain the role of recursion in the derivation of aaaabbbb using definition 7.1.3.
6. What is an algebraic definition and what is its purpose?
7. What is the difference between elementary, derived, and semi-formal formalisms?
8. What is the reason for the development of derived formalisms?

Section 7.2

1. Explain the difference in the well-formedness for artificial and natural languages.
2. Explain characterizing grammatical well-formedness as the descriptive goal of theoretical linguistics.

---

[19] See the discussion of metalanguage-based semantics in Chapters 19 and 20.
[20] For an attempt see SCG.

3. Name three reasons for using generative grammar in modern linguistics.
4. Why is the use of generative grammars a necessary, but not a sufficient condition for a successful language analysis?

Section 7.3

1. Under which circumstances is a generative grammar descriptively adequate?
2. What is meant by the mathematical complexity of a grammar formalism and why is it relevant for practical work?
3. What is the difference between functional and non-functional grammar theories?
4. Which three aspects should be jointly taken into account in the development of a generative grammar and why?

Section 7.4

1. By whom, when, and for which purpose was C-grammar invented?
2. When and by whom was C-grammar first applied to natural language?
3. What is the structure of a logical function?
4. Give an algebraic definition of C-grammar.
5. Explain the interpretation of complex C-grammar categories as functors.
6. Why is the set of categories in C-grammar infinite and the lexicon finite?
7. Name the formal principle allowing the C-grammar 7.4.4 to generate infinitely many expression even though its lexicon and its rule set are finite.
8. Why is the grammar formalism defined in 7.4.4 called *bidirectional* C-grammar?
9. Would it be possible to use C-grammar as the syntactic component of the SLIM theory of language?

Section 7.5

1. Why is C-grammar prototypical of a lexical approach?
2. What is meant by a fragment of a natural language in generative grammar?
3. Explain the relation between a functional interpretation of complex categories in C-grammar and the model-theoretic interpretation of natural language.
4. Explain the recursive structure in the C-grammar 7.5.4.
5. Explain how the semantic interpretation of C-grammar works in principle.
6. Extend the C-grammar 7.5.4 to generate the sentences The man send the girl a letter, The girl received a letter from the man, The girl was sent a letter by the man. Explain the semantic motivation of your categories.
7. Why are there no large scale descriptions of natural language in C-grammar?
8. Why are there no efficient implementations of C-grammar?
9. Why is the absence of efficient implementations a serious methodological problem for C-grammar?
10. Does C-grammar provide a mechanism of natural communication, or would it be suitable as a component of such a mechanism?

8. Language hierarchies and complexity

The second elementary formalism of generative grammar was published in 1936 by the American logician E. Post. Known as *rewrite* or *Post production system*, it originated in the mathematical context of recursion theory and is closely related to automata theory and computational complexity theory.

Rewrite systems were first applied to natural language by N. Chomsky 1957 under the name of *phrase structure grammar*.[1] Based on PS-grammar, Chomsky and others developed a series of derived formalisms initially called transformational grammars.

Section 8.1 provides an algebraic definition of PS-grammar and describes which restrictions on the form of PS-rules result in regular, context-free, context-sensitive, and unrestricted PS-grammars. Section 8.2 explains four basic degrees of complexity and relates them to different types of PS-grammar. Section 8.3 illustrates the notion of generative capacity with applications of PS-grammar to artificial languages. Section 8.4 applies context-free PS-grammar to natural language and defines the linguistic concept of constituent structure. Section 8.5 explains the constituent structure paradox and shows why adding transformations makes the resulting PS-grammar formalism undecidable.

8.1 Formalism of PS-grammar

The algebraic definition of PS-grammar resembles that of C-grammar (cf. 7.4.2) insofar as it enumerates the basic components of the system and characterizes each in terms of set theory.

---

[1] Post's contribution is not mentioned by Chomsky. This is commented by Y. Bar-Hillel 1960 as follows:

> This approach [i.e. rewriting systems] is the standard one for the combinatorial systems conceived much earlier by Post [1936], as a result of his penetrating researches into the structure of formal calculi, though Chomsky seems to have become aware of the proximity of his ideas to those of Post only at a later stage of his work.
>
> Y. Bar-Hillel 1964, p. 103

This is remarkable insofar as Chomsky's thesis advisor Z. Harris and J. Bar-Hillel were in close scientific contact since 1947. Moreover, Bar-Hillel and Chomsky discussed "linguistics, logic, and methodology in endless talks" beginning 1951 (Bar-Hillel 1964, p. 16).

8.1.1 ALGEBRAIC DEFINITION OF PS-GRAMMAR

A PS-grammar is a quadruple $< V, V_T, S, P >$ such that

1. V is a finite set of signs,
2. V_T is a proper subset of V, called *terminal symbols*,
3. S is a sign in V minus V_T, called *start symbol* , and
4. P is a set of rewrite rules of the form $\alpha \rightarrow \beta$, where α is an element of V^+ and β an element of V^*.[2]

The basic components of PS-grammar are the sets V, V_T, and P, plus the start symbol S. The terminal symbols of V_T are the word surfaces of the language. The nonterminal symbols of V minus V_T are called the variables. We will use Greek letters to represent sequences from V^*, upper case Latin letters to represent individual variables, and lower case Latin letters to represent individual terminal symbols.

A PS-grammar generates language expressions by means of rewrite rules whereby the sign sequence on the lefthand side of a rule is replaced by the sign sequence on the right hand side. For example, if $\alpha \rightarrow \beta$ is a rewrite rule in P and γ, δ are sequences in V^*, then

$\gamma\alpha\delta \Rightarrow \gamma\beta\delta$

is a direct substitution of the sequence $\gamma\alpha\delta$ by the sequence $\gamma\beta\delta$. In other words, by applying the rule $\alpha \rightarrow \beta$ to the sequence $\gamma\alpha\delta$ there results the new sequence $\gamma\beta\delta$.

The general format of rewrite rules in PS-grammar suggests systematic restrictions of the following kind (whereby the numbering used here follows tradition):

8.1.2 RESTRICTIONS OF PS-RULE SCHEMATA

0. Unrestricted PS-rules:
   The left hand side and the right hand side of a type 0 rule each consist of arbitrary sequences of terminal and nonterminal symbols.
1. Context-sensitive PS-rules:
   The left hand side and the right hand side of a type 1 rule each consist of arbitrary sequences of terminal and nonterminal symbols whereby the right hand side must be at least as long as the left hand side.
   Example: A B C → A D E C
2. Context-free PS-rules:
   The left hand side of a type 2 rule consists of exactly one variable. The right hand side of the rule consists of a sequence from V^+.
   Examples: A → BC, A → bBCc, etc.[3]

---

[2] V^+ is the positive closure and V^* is the Kleene closure of V (cf. Section 7.2).

[3] Context-free grammar sometimes use so-called 'epsilon rules' of the form A → ε. However, epsilon rules can always be eliminated (cf. J.E. Hopcroft & J.D. Ullman, 1979, p. 90, Theorem 4.3). We specify the right hand side of type 2 rules as a nonempty sequence in order to formally maintain the context-free rules as a special form of the context-sensitive rules.

3. Regular PS-rules:

   The left hand side of a type 3 rule consists of exactly one variable. The right hand side consists of exactly one terminal symbol and at most one variable.[4]

   Examples: A → b, A → bC.

Because the rule types become more and more restrictive – from type 0 to type 3 – the rules of a certain type obey all restrictions of the lower rule types. For example, the regular type 3 rule

   A → bC

complies with the lesser restrictions of the rule types 2, 1, and 0. The context-free type 2 rule

   A → BC

on the other hand, while not complying with the type 3 restriction, complies with the lesser restrictions of the lower rule types 1 and 0. And accordingly for type 1 rules.

The different restrictions on the PS-grammar rule schema described in 8.1.2 result in four different types of PS-grammars. PS-grammars which contain only type 3 rules are called regular, which contain at most type 2 rules are called context-free, which contain at most type 1 rules are called context-sensitive, and which have no restrictions on their rule types are called unrestricted PS-grammars.

These four types of PS-grammar generate in turn four different classes of languages. They are the regular languages generated by the regular PS-grammars, the context-free languages generated by the context-free PS-grammars, the context-sensitive languages generated by the context-sensitive PS-grammars, and the recursively enumerable languages generated by the unrestricted PS-grammars.

The different language classes are properly contained in each other. Thus, the class of regular languages is a proper subset of the class of context-free languages, which in turn is a proper subset of the class of context-sensitive languages, which in turn is a proper subset of the class of recursively enumerable languages.

The differences in the language classes result from differences in the generative capacity of the associated types of PS-grammar. The generative capacity of a grammar type is high, if a corresponding grammar is able not only to recursively generate many formal language structures, but can at the same time *exclude* those which are not part of the language. The generative capacity of a grammar type is low, on the other hand, if the grammar allows only limited control over the structures to be generated.[5]

---

[4] This is the definition of *right linear* PS-grammars. PS-grammars in which the order of the terminal and the nonterminal symbols on the right hand side of the rule is inverted are called *left linear*. Left and right linear grammars are equivalent (cf. Hopcroft & Ullman 1979, p. 219, Theorem 9.2.)

[5] For example, the generative capacity of the PS-grammar 7.1.3 for the artificial language $a^k b^k$ is higher than that of a regular PS-grammar 8.3.2 for the free monoid over $\{a,b\}$ (cf. 7.1.2). The free monoid contains all the expressions of $a^k b^k$, but its regular PS-grammar is unable to *exclude* the expressions which do not belong to the language $a^k b^k$.

8.2 Language classes and computational complexity

Associated with the generative capacity of a grammar type and its language class is the degree of computational complexity, i.e. the amount of computing time and/or memory space needed to analyze expressions of a certain language class. The computational complexity increases systematically with the generative capacity.

In short, different restrictions on a generative rule schema result in

- different *types of grammar* which have
- different *degrees of generative capacity* and generate
- different *language classes* which in turn exhibit
- different *degrees of computational complexity*.

This structural correlation is not limited to PS-grammar, but holds in any well-defined formalism of generative grammar (cf. 8.2.3, 11.5.10).

The complexity of a generative grammar formalism is measured on the basis of an algorithm which implements the grammar formalism as an operational procedure on an abstract automaton (e.g., a Turing machine, a linearly bounded automaton, a push down automaton, a finite state automaton, etc.). The complexity of the algorithm is computed as the number of *primitive operations*[6] required to analyze an arbitrary input expression in the worst possible case (upper bound).[7] Thereby the number of primitive operations is counted in relation to the length of the input.

In elementary formalisms of generative grammar, such as PS-grammar or LA-grammar, complexity is usually determined for well-defined subtypes such as the regular, context-free, context-sensitive, and unrestricted subtypes of PS-grammar, or the C1-, C2-, C3-, B-, and A-LAGs of LA-grammar. Furthermore, the complexity of a language class is equated with that of the associated grammar type. For example, one says *the context-free languages have a complexity of* n^3 because there exists a known algorithm which for any arbitrary context-free PS-grammar can analyze any arbitrary input using at most n^3 primitive operations, whereby n is the length of the input.

The different degrees of computational complexity result in four basic classes.

8.2.1 BASIC DEGREES OF COMPLEXITY

1. *Linear complexity*
   n, $2n$, $3n$, etc.
2. *Polynomial complexity*
   n^2, n^3, n^4, etc.
3. *Exponential complexity*
   2^n, 3^n, 4^n, etc.
4. *Undecidable*
   $n \cdot \infty$

To get an idea of how these complexity degrees affect practical work, consider the Limas corpus[8] of German. The average sentence length in the Limas corpus is 17.54 word forms (including punctuation signs). Thus, a linear $3 \cdot n$ algorithm will require at most 51 operations for the analysis of an average sentence (with $n = 17$), a polynomial n^3 algorithm will require at most $4\,913$ operations, an exponential 3^n algorithm will require at most $127\,362\,132$ operations, and an undecidable algorithm will require at most $17 \cdot \infty \; (= \infty)$ operations.

M.R.Garey & D.S. Johnson 1979 compare the time needed for solving a problem of n^3 (polynomial) and 2^n (exponential) complexity relative to different problem sizes as follows.

8.2.2 TIMING OF POLYNOMIAL VS. EXPONENTIAL ALGORITHMS

time complexity	problem size n		
	10	50	100
n^3	.001 seconds	.125 seconds	1.0 seconds
2^n	.001 seconds	35.7 years	10^{15} centuries

In this example Garey & Johnson use *adding the next word* as the primitive operation of their algorithm.

Of a total of $71\,148$ sentences in the Limas corpus there are exactly 50 which consist of 100 word forms or more, whereby the longest sentence in the whole corpus consists of 165 words.[9] Thus, if we apply the measurements of Garey & Johnson to the automatic analysis of the Limas corpus, an exponential 2^n grammar algorithm, though decidable, could take longer than $1\,000\,000\,000\,000\,000$ centuries in the worst

---

[6] J. Earley 1970 characterizes a primitive operation as "in some sense the most complex operation performed by the algorithm whose complexity is independent of the size of the grammar and the input string." The exact nature of the primitive operation varies from one grammar formalism to the next.

For example, Earley chose the operation of *adding a state to a state set* as the primitive operation of his famous algorithm for context-free grammars (cf. Section 9.3). In LA-grammar, on the other hand, the subclass of C-LAGs uses a *rule application* as its primitive operation (cf. Section 11.4).

[7] We are referring here to time complexity.

[8] As explained in Section 15.3, the Limas corpus was built in analogy to the Brown and the LOB corpus, and contains 500 texts of 2000 running word forms each. The texts were selected at random from roughly the same 15 genres as those of the Brown and LOB corpus in order to come as close as possible to the desideratum of a *balanced* corpus which is *representative* for the whole German language of the year 1973.

[9] These data were provided by Markus Schulze at CLUE (Computational Linguistics at Friedrich Alexander University Erlangen Nürnberg).

case. This amount of time is considerably longer than the existence of the universe and could not be reduced to practical levels by faster machines.

In the case of PS-grammar, the correlation between rule restrictions, grammar types, language classes, and complexity is as follows.

8.2.3 PS-GRAMMAR HIERARCHY OF FORMAL LANGUAGES

rule restrictions	types of PS-grammar	language classes	degree of complexity
type 3	regular PSG	regular languages	linear
type 2	context-free PSG	context-free languages	polynominal
type 1	context-sensitive PSG	context-sensitive lang.	exponential
type 0	unrestricted PSG	rec. enum. languages	undecidable

As an alternative to the PS-grammar hierarchy, also called Chomsky hierarchy, see the LA-grammar hierarchy of formal language classes defined in Chapter 11 (especially 11.5.10 and 11.5.11). The alternative hierarchies of PS- and LA-grammar are compared in Chapter 12.

8.3 Generative capacity and formal language classes

From a linguistic point of view, the question is whether or not there is a type of PS-grammar which generates exactly those structures which are characteristic of natural language. Let us therefore take a closer look at the structures generated by different types of PS-grammar.

The PS-grammar type with the most restricted rules, the lowest generative capacity, and the lowest computational complexity is that of regular PS-grammars.[10] The generative capacity of regular grammar permits the recursive repetition of single words, but without any recursive correspondences.

For example, expressions of the regular language ab^k consist of one a, followed by one, two, or more b. A (right linear) PS-grammar for ab^k is defined as follows.

8.3.1 REGULAR PS-GRAMMAR FOR ab^k (K $\geq$ 1)

$$V =_{def} \{S, B, a, b\}$$
$$V_T =_{def} \{a, b\}$$
$$P =_{def} \{S \rightarrow a\,B,$$
$$B \rightarrow b\,B,$$
$$B \rightarrow b\,\}$$

Another example of a regular language is the free monoid over {a, b} minus the zero-element, which is generated by the following PS-grammar.

8.3.2 REGULAR PS-GRAMMAR FOR {a, b}$^+$

$V =_{def} \{S, a, b\}$
$V_T =_{def} \{a, b\}$
$P =_{def} \{S \rightarrow a\,S,$
$\qquad S \rightarrow b\,S,$
$\qquad S \rightarrow a,$
$\qquad S \rightarrow b\}$

That a regular PS-grammar cannot generate systematic correspondences of arbitrary number is illustrated by the contrast between the already familiar context-free language $a^k b^k$ and the regular language $a^m b^k$.

8.3.3 REGULAR PS-GRAMMAR FOR $a^m b^k$ (K,M $\geq$ 1)

$V =_{def} \{S, S_1, S_2, a, b\}$
$V_T =_{def} \{a, b\}$
$P =_{def} \{S \rightarrow a\,S_1,$
$\qquad S_1 \rightarrow a\,S_1,$
$\qquad S_1 \rightarrow b\,S_2,$
$\qquad S_2 \rightarrow b\}$

The language $a^m b^k$ is regular because the number of a and the number of b in $a^m b^k$ is open – as indicated by the use of two different superscripts ^{m} and ^{k}. The language $a^k b^k$, on the other hand, exceeds the generative capacity of a regular PS-grammar because it requires a correspondence between the number of the a and the number of the b – as indicated by them having the same superscript ^{k}.

The characteristic limitation in the generative capacity of regular PS-grammars follows in an intuitively obvious way from the restrictions on the associated rule type: Because the right hand side of a type 3 rule consists of one terminal and at most one variable it is impossible to recursively generate even pairwise correspondences – as would be required by $a^k b^k$.

The formal proof of the lower generative capacity of regular PS-grammars as compared to the context-free PS-grammars is not trivial, however. It is based on the *pumping lemma* for regular languages,[11] which formally shows that there are languages that may not be generated by regular PS-grammars.

---

[10] The class of regular languages is not part of the hierarchy of LA-grammar, though it may be reconstructed there (CoL, Theorem 3, p. 138). Instead the LA-grammar hierarchy provides the alternative linear class of C1-languages. As shown in Sections 11.5 ff., the class of C1-languages contains all regular languages, all deterministic context-free languages which are recognized by an epsilon-free DPDA, as well as many context-sensitive languages.

[11] Cf. Hopcroft & Ullman 1979, p. 55 ff.

A pumping lemma for a certain language class shows which structures it can have. This is done by explicitly listing the basic structural patterns of the language class such that the infinite set of additional expressions can be pumped, i.e. they can be shown to consist only of repetitions of the basic structural patterns.

The next grammar type in the PS-grammar hierarchy is that of context-free PS-grammars. Examples of context-free languages are $a^k b^k$ (for which a PS-grammar is defined in 7.1.3 and a C-grammar in 7.4.3) and $a^k b^{3k}$ defined below.

8.3.4 CONTEXT-FREE PS-GRAMMAR FOR $a^k b^{3k}$

$$V =_{def} \{S, a, b\}$$
$$V_T =_{def} \{a, b\}$$
$$P =_{def} \{ S \to a S b b b,$$
$$S \to a b b b\}$$

This type of grammar is called context-free because the left hand side of a type 2 rule consists by definition of a single variable (cf. 8.1.2) – without a surrounding 'context' of other signs.[12]

The context-free rule format (see for example 7.1.3 and 8.3.4) results in another restriction on generative capacity: context-free grammars may recursively generate correspondences, but only those of the type *inverse pair* like a b c ... c b a.[13]

This inverse pair structure characteristic of context-free languages shows up clearly in the derivation of context-free expressions. Consider for example the context-free language WW^R whereby W represents an arbitrary sequence of words, e.g. abcd, and W^R stands for the inverse sequence,[14] e.g. dcba.

8.3.5 CONTEXT-FREE PS-GRAMMAR FOR WW^R

$$V =_{def} \{S, a, b, c, d\}, V_T =_{def} \{a, b, c, d\}, P =_{def} \{ S \to a S a,$$
$$S \to b S b,$$
$$S \to c S c,$$
$$S \to d S d,$$
$$S \to a a,$$
$$S \to b b,$$
$$S \to c c,$$
$$S \to d d\}$$

The increased generative capacity of the class of context-free languages as compared to the class of regular languages is associated with an increase in computational

---

[12] This notion of 'context' is peculiar to the terminology of PS-grammar and has nothing to do with the speaker-hearer-internal *context of use* (cf. Chapters 3–6).

[13] Each context-free language is homomorphic with the intersection of a regular set and a semi-Dyck set (Chomsky-Schützenberger Theorem). See M. Harrison 1978, p. 317ff.

[14] The superscript ^{R} in WW^R stands mnemonically for *reverse*.

complexity. While the class of regular languages parses in linear time, the class of context-free languages parses in polynomial time (cf. 8.2.3).

The generative capacity of context-free PS-grammars is still rather limited.[15] As a classic example of a language exceeding the generative capacity of context-free PS-grammar consider $a^k b^k c^k$. Expressions of this language consist of three equally long sequences of a, b, and c, for example,

a b c, a a b b c c, a a a b b b c c c, etc.

The language $a^k b^k c^k$ cannot be generated by a context-free PS-grammar because it requires a correspondence between three different parts – which exceeds the *pairwise* reverse structure of the context-free languages such as the familiar $a^k b^k$ and WW^R.

Another language exceeding the generative capacity of context-free PS-grammars is WW, where W is an arbitrary sequence of words. While the context-free language WW^R defined in 8.3.5 consists of expressions like

aa

abba

abccba

abcddcba

· · ·

which have a pairwise *reverse* structure, the context-sensitive language WW consists of expressions like

aa

abab

abcabc

abcdabcd

· · ·

which do not have a reverse structure. Thus, despite the close resemblance between WW^R and WW, it is simply impossible to write a PS-grammar like 8.3.5 for WW.

While $a^k b^k c^k$ and WW show in an intuitively obvious way that there are languages which cannot be generated by context-free PS-grammar, the formal proof is by no means trivial. As in the case of the regular languages, it is based on a pumping lemma, this time for the context-free languages.[16]

The next larger language class in the PS-grammar hierarchy are the context-sensitive languages, which are generated by PS-grammars using type 1 rules.

---

[15] The class of context-free languages is not part of the hierarchy of LA-grammar, though it may be reconstructed there (CoL, Theorem 4, p. 138). See also Section 11.2, footnote 12. Instead the LA-grammar hierarchy provides the alternative polynomial class of C2-languages. As shown in Section 12.4, the class of C2-languages contains most, though not all, context-free languages, as well as many context-sensitive languages.

[16] See Hopcroft and Ullman, p. 125 ff.

> Almost any language one can think of is context-sensitive; the only known proofs that certain languages are not CSL's are ultimately based on diagonalization.
>
> J.E. Hopcroft and J.D. Ullman 1979, p. 224

Because it is impossible to exhaustively list the basic patterns of practically all structures 'one can think of' there is no pumping lemma for the class of context-sensitive languages.

The structure of a type 1 or context-sensitive rule is specified in 8.3.6:

8.3.6 STANDARD SCHEMA OF CONTEXT-SENSITIVE RULES

$$\alpha_1 A \alpha_2 \rightarrow \alpha_1 \beta \alpha_2, \text{ whereby } \beta \text{ is not the empty sequence.}$$

In PS-grammar, the term *context-sensitive* is interpreted in contrast to the term *context-free*. While a context-free type 2 rule allows nothing but a single variable on the left hand side, a context-sensitive type 1 rule may surround the variable with various terminal symbols. As illustrated in 8.3.6, a type 1 rule is context-*sensitive* because the variable A may be rewritten as β only in the specific 'context' $\alpha_1\_\alpha_2$.

The possibility of specifying a particular environment (context) for the variable on the left hand side of a type 1 rule greatly increases the control and thus the generative power of the context-sensitive PS-grammars. This is illustrated in the following PS-grammar for $a^k b^k c^k$:

8.3.7 PS-GRAMMAR FOR CONTEXT-SENSITIVE $a^k b^k c^k$

$V =_{def} \{S, B, C, D_1, D_2, a, b, c\}$
$V_T =_{def} \{a, b, c\}$
$P =_{def} \{$

$S \rightarrow a\,S\,B\,C,$	*rule 1*
$S \rightarrow a\,b\,C,$	*rule 2*
$C\,B \rightarrow D_1\,B,$	*rule 3a*
$D_1\,B \rightarrow D_1\,D_2,$	*rule 3b*
$D_1\,D_2 \rightarrow B\,D_2,$	*rule 3c*
$B\,D_2 \rightarrow B\,C,$	*rule 3d*
$b\,B \rightarrow b\,b,$	*rule 4*
$b\,C \rightarrow b\,c,$	*rule 5*
$c\,C \rightarrow c\,c\}$	*rule 6*

The rules 3a–3d jointly have the same effect as the (monotonic)
    rule 3 $C\,B \rightarrow B\,C.$
The PS-grammar 8.3.7 uses the rules 3a–3d because the equivalent rule 3 does not comply with the simplifying assumption that only *one* variable on the left hand side of a context-sensitive rule may be replaced.

The crucial function of the context for controlling the three correspondences in the expressions of $a^k b^k c^k$ is shown in the following derivation of a a a b b b c c c. For simplicity, the rules 3a–3d of 8.3.7 are combined into the equivalent rule 3.

8.3.8 DERIVATION OF a a a b b b c c c

	intermediate chains	rules
1.	S	
2.	a S B C	(1)
3.	a a S B C B C	(1)
4.	a a a b C B C B C	(2)
5.	a a a b B C C B C	(3)
6.	a a a b B C B C C	(3)
7.	a a a b B B C C C	(3)
8.	a a a b b B C C C	(4)
9.	a a a b b b C C C	(4)
10.	a a a b b b c C C	(5)
11.	a a a b b b c c C	(6)
12.	a a a b b b c c c	(6)

The high generative capacity of the context-sensitive PS-grammars is based on the possibility of *changing the order* of sequences already derived. The reordering of sequences takes place in the transition from the intermediate sequence 4 to 7.

The possibility of context-sensitively changing the order of sequences provides for a degree of control which is much higher than in context-free PS-grammars. The cost, however, is a high degree of computational complexity. In order to correctly reconstruct automatically which sequence of context-sensitive rule applications resulted in a given expression, a potentially *exponential* number of reordering possibilities must be checked.

This kind of search space is so large that there exists no practical parsing algorithm for the class of context-sensitive PS-grammars. In other words, the class of context-sensitive languages is computationally intractable.

The context-sensitive languages are a proper subset of the recursive languages.[17] The class of recursive languages is not reflected in the PS-grammar hierarchy. This is because the PS-rule schema provides no suitable restriction (cf. 8.1.2) such that the associated PS-grammar class would generate exactly the recursive languages.[18]

A language is recursive if and only if it is decidable, i.e., if there exists an algorithm which can determine in finitely many steps for arbitrary input whether or not the input

---

[17] Hopcroft & Ullman 1979, p. 228, Theorem 9.8.

[18] In the hierarchy of LA-grammar, the class of recursive languages is formally defined as the class of A-languages, generated by unrestricted LA-grammars (cf. 11.2.2). Furthermore, the class of context-sensitive languages is formally defined as the class of B-languages, generated by *bounded* LA-grammars.

belongs to the language. An example of a recursive language which is not context-sensitive is the Ackermann function.[19]

The largest language class in the PS-grammar hierarchy are the recursively enumerable languages, which are generated by unrestricted or type 0 PS-grammars. In unrestricted PS-grammars, the right hand side of a rule may be shorter than the left hand side. This characteristic property of type 0 rules provides for the possibility of *deleting* parts of sequences already generated.

For this reason, the class of recursively enumerable languages is undecidable. The decision of whether or not an expression of a recursively enumerable language is well-formed may thus not just take very long, but forever.[20]

8.4 PS-Grammar for natural language

A simple application of PS-grammar to natural language is illustrated in the following definition. In order to facilitate comparison, it generates the same sentence as the C-grammar 7.5.5.

8.4.1 PS-GRAMMAR FOR EXAMPLE 7.5.4

$$V =_{def} \{S, NP, VP, V, N, DET, ADJ, black, dogs, little, sleep, the\}$$
$$V_T =_{def} \{black, dogs, little, sleep, the\}$$
$$P =_{def} \{ \; S \rightarrow NP \; VP,$$
$$VP \rightarrow V,$$
$$NP \rightarrow DET \; N,$$
$$N \rightarrow ADJ \; N,$$
$$N \rightarrow dogs,$$
$$ADJ \rightarrow little,$$
$$ADJ \rightarrow black,$$
$$DET \rightarrow the,$$
$$V \rightarrow sleep\}$$

The form of this PS-grammar is context-free: it is not yet context-sensitive because the left-hand side of the rules consist of only one variable, and is not regular any more because the right-hand side of some rules contains more than one variable.

Like the C-grammar derivation 7.5.4, the PS-grammar derivation based on definition 8.4.1 can be represented as a tree.

---

[19] Hopcroft & Ullman, p. 175, 7.4.

[20] The class of recursively enumerable languages is not part of the LA-grammar hierarchy, though it may be reconstructed (cf. footnote 15 at the end of Section 11.2).

8.4.2 PS-GRAMMAR ANALYSIS OF EXAMPLE 7.5.4

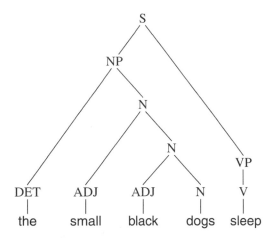

Such trees are called phrase structures in PS-grammar. The category symbols in a phrase structure tree are called nodes. There are two formal relations between nodes: dominance and precedence. For example, the node S dominates the nodes NP and VP in accordance with the rule S → NP VP of the associated grammar 8.4.1. At the same time this rule specifies precedence: the NP node is located in the tree to the left of VP node.

In comparison to C-grammar, which codes the combinatorics of a language into the complex categories of its word forms and uses only two rule schemata for composition, PS-grammar uses only elementary categories the combinatorics of which are expressed in terms of a multitude of rewrite rules. Even the lexicon, which is treated in C-grammar as categorized sets of word forms (cf. 7.5.5), is handled in PS-grammar in terms of rules.

These rules are called terminal rules because they have a terminal symbol (word) on their right-hand side. The remaining rules, called nonterminal rules, generate the PS-grammar sentence frames into which the terminal rules insert various words or word forms.

In addition to the formal differences between C- and PS-grammar, their respective analyses of natural language are linguistically motivated by different empirical goals. The goal of C-grammar is to characterize the *functor-argument structure* of natural language, whereas PS-grammar aims to represent the *constituent structure* of natural language.

Constituent structure represents linguistic intuitions about which parts in a sentence belong most closely together semantically. The principle of constituent structure is defined as a formal property of phrase structure trees.

8.4.3 DEFINITION OF CONSTITUENT STRUCTURE

1. Words or constituents which belong together semantically must be dominated directly and exhaustively by a node.
2. The lines of a constituent structure may not cross (*nontangling condition*).

According to this definition, the following analysis of the sentence the man read a book is linguistically correct.

8.4.4 CORRECT CONSTITUENT STRUCTURE ANALYSIS

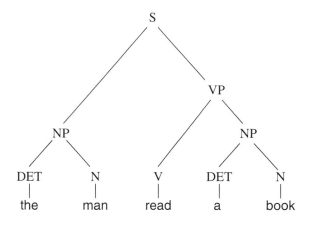

In contrast, the alternative analysis 8.4.5 of the same sentence, while formally possible, violates the principle of constituent structure.

8.4.5 INCORRECT CONSTITUENT STRUCTURE ANALYSIS

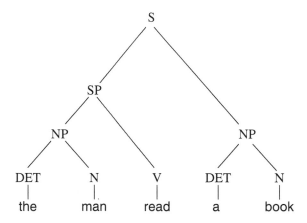

The incorrect analysis violates the constituent structure of the sentence in question because according to the PS-grammarians' intuition read and a book belong semantically together and thus must be dominated directly and exhaustively by a node (as illustrated by the correct analysis 8.4.4).

Historically, the notion of constituent structure is quite recent. It evolved from the *immediate constituent analysis* of the American structuralist L. BLOOMFIELD (1887–1949) and the distribution tests of his student Z. Harris. In Bloomfield's main work *Language* of 1933, immediate constituents do not take center stage, however, and are mentioned on only 4 of 549 pages. They are briefly sketched on pages 161 and 167 using simple sentences and later applied in morphology (op.cit., p. 209/10, 221/2).

> The principle of immediate constituents will lead us, for example, to class a form like gentlemanly not as a compound word, but as a derived secondary word, since the immediate constituents are the bound form -ly and the underlying form gentleman.
>
> L. Bloomfield, *Language*, p. 210

This statement may be translated into the following tree structures:

8.4.6 IMMEDIATE CONSTITUENTS IN PS-GRAMMAR:

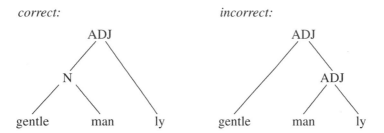

The example gentlemanly is also discussed in Harris 1951 (p. 278–280), where the methodological innovation of distribution tests is introduced.

Distribution tests are realized either as substitution tests or as movement tests. Their goal is to distinguish grammatically correct from grammatically incorrect substitutions or movements.

8.4.7 SUBSTITUTION TEST

correct substitution:

Suzanne has [eaten] an apple

⇓

Suzanne has [cooked] an apple

incorrect substitution:

Suzanne has [eaten] an apple

⇓

* Suzanne has [desk] an apple

The substitution on the left is considered correct because it results in a sentence which is grammatically as well-formed as the original. The substitution on the right is incorrect because it turns a well-formed input sentence into an ungrammatical result.

Analogous considerations hold for movement tests:

8.4.8 MOVEMENT TEST

correct movement:

Suzanne [has] eaten an apple $\implies$ [has] Suzanne eaten an apple (?)

incorrect movement:

Suzanne has eaten [an] apple $\implies$ * [an] Suzanne has eaten apple

For the linguists of American structuralism, the distribution tests were important methodologically in order to objectively support their intuitions about the *correct segmentation* of sentences. The segmentation of sentences and the concomitant hypotheses about more or less closely related subparts were needed in turn to distinguish between linguistically correct and incorrect phrase structures trees.

Such a distinction seemed necessary because for any finite string the number of possible phrase structures is infinite.[21] This constitutes an embarrassment of riches: the possible phrase structures should not all be equally correct linguistically.

The huge variety of possible trees holds primarily for isolated sentences outside a formal grammar. Once the structural principles of a language are known and formulated as a PS-grammar, however, sentences are assigned their phrase structure(s) by the grammar. Compared to the number of possible phrase structure trees for an isolated sentence outside the grammar, the number of trees assigned to it by the grammar for the language in question is usually greatly reduced. Given an unambiguous PS-grammar, for example, there is by definition at most one tree per well-formed sentence.

In the case of context-free artificial languages, the structural principles are sufficiently simple to allow for the definition of adequate formal PS-grammars (e.g., 8.3.1, 8.3.2, 8.3.3, 8.3.4, 8.3.5). When there are several essentially different PS-grammars defined for the same context-free language there is no rational reason to argue about which of them assigns the 'correct' phrase structure trees.

In the case of natural languages, on the other hand, the structural principles are still unknown in PS-grammar. Thus it is an open question which PS-grammar for a single sentence or small set of sentences might turn out to be suited best for the correct extension to cover the whole language. In order to guide the long-term development of PS-grammars for natural language, an empirical criterion is needed.

For this, the intuitive principle of constituent structure was chosen, supported by the associated substitution and movement tests. It has not prevented, however, a constant debate within nativism over which phrase structures for natural language are linguistically correct and why.

---

[21] Even if phrase structures of the form A–B–C...A are excluded, the number of different phrase structure trees still grows exponentially with the length of the input. From a formal point of view such structures are legitimate in context-free phrase structure grammar.

From the viewpoint of formal language theory, the cause of this ultimately fruitless debate is the lack of complete PS-grammars for natural languages – either as elementary or as derived formalisms. That over fifty years of substantially funded research have not resulted in the complete analysis of a single natural language might perhaps be taken as a hint that this approach is not quite optimal.

8.5 Constituent structure paradox

From the viewpoint of the SLIM theory of language, there are several objections to the principle of constituent structure (cf. 8.2.1). First, constituent structure and the distribution tests claimed to support it run counter to the time-linear structure of natural language. Second, the resulting phrase structure trees have no communicative purpose. Third, the principles of constituent structure cannot always be fulfilled.

This is because the conditions 8.2.1 of constituent structure require that the parts which belong together semantically are positioned right next to each other in the natural language surface. Yet there are expressions in natural language – called *discontinuous elements* – in which this is not the case.

For example, there is general agreement that in the sentence

Peter looked the word up.

the discontinuous elements looked and up are more closely related semantically than either the adjacent expressions looked – the word or the word – up.

Discontinuous elements are the structural reason why the principle of constituent structures cannot always be fulfilled in context-free PS-grammars for natural language. This structural problem has been called the *constituent structure paradox*.[22] It is illustrated by the alternative tree structures 8.5.1 and 8.5.2.

8.5.1 VIOLATING THE SECOND CONDITION OF 8.4.3

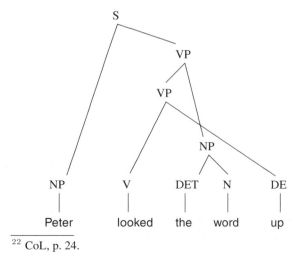

---

[22] CoL, p. 24.

Here the semantically related subexpressions looked and up are dominated directly and exhaustively by a node, thus satisfying the first condition of 8.2.1. The analysis violates the second condition, however, because the lines in the tree cross.

8.5.2 Violating the first condition of 8.4.3

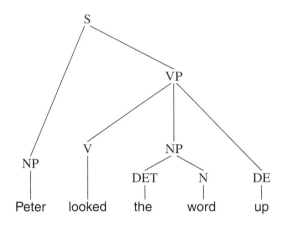

Here the lines do not cross, satisfying the second condition. The analysis violates the first condition, however, because the semantically related expressions looked – up, or rather the nodes V and DE dominating them, are not *exhaustively* dominated by a node. Instead, the node directly dominating V and DE also dominates the NP the word.

From the viewpoint of formal language theory, the constituent structure paradox is caused by the fact that the generative power of context-free PS-grammars is insufficient to handle discontinuous elements in the fashion prescribed by the conditions of 8.2.1. This empirical problem has been known since the early nineteen fifties.[23]

All natural languages have discontinuous elements of one kind or another. In order to nevertheless maintain the principle of constituent structure as much as possible, N. Chomsky 1957 turned Z. Harris' methodologically motivated substitution and movement tests into generative rules which he called transformations and claimed to be innate (Chomsky 1965, p. 47 f.)

A transformation rule takes a phrase structure tree as input and produces a modified phrase structure tree as output. In transformational grammar, many transformations are ordered into a transformational component and applied one after the other to the phrase structure of the input. The input- and output-conditions of a transformation are formally specified as patterns with variables, which are called indexed bracketings.

8.5.3 Example of a formal transformation

$$[[V\ DE]_{V'}\ [DET\ N]_{NP}]_{VP} \Rightarrow [V\ [DET\ N]_{NP}\ DE]_{VP}$$

---

[23] Y. Bar-Hillel writes in 1960 [1964, p. 102] that he abandoned his 1953 work on C-grammar because of analogous difficulties with the discontinuous elements in sentences like He gave it up.

An application of this transformation is illustrated in 8.5.4.

8.5.4 APPLYING THE TRANSFORMATION 8.5.3

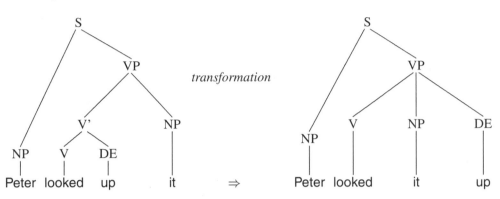

In Standard Theory (ST, Chomsky 1965), the input to the transformational component is generated by a context-free PS-grammar. These 'deep structures' must satisfy the conditions 8.4.3 of constituent structure, but need not correspond to a grammatical sequence (as in the left phrase structure tree of 8.5.4).

If the input pattern (indexed bracketing) of a transformation rule has been matched successfully onto a phrase structure, it is mapped into another phrase structure in accordance with the output pattern of the transformation. After the application of one or more transformations, a surface structure is obtained. The phrase structure of the surface must correspond to a grammatical sequence, but need not fulfill the conditions of constituent structure 8.4.3 (as in the right phrase structure tree of 8.5.4).

The transformation illustrated in 8.5.4 is regarded as 'meaning preserving' – just like the examples in 4.5.2. It is supposed to characterize the innate knowledge of the speaker-hearer without having a communicative function.

From a mathematical viewpoint, a mechanism designed to recursively modify an open set of input structures always results in high degrees of complexity. This has been demonstrated with the PS-grammar 8.3.7 for the simple context-sensitive language $a^k b^k c^k$. Yet while context-sensitive languages are 'only' exponential, transformational grammar is equivalent to a Turing machine, generates the recursively enumerable languages, and is therefore undecidable.

Initially N. Chomsky had hoped to avoid this consequence by imposing a formal restriction called *recoverability of deletions*. According to this condition, a transformation may delete a node only if it can be reconstructed (recovered) via well-defined access to a copy of the node and the subtree dominated by it.

The so-called Bach-Peters-sentences showed, however, that the recoverability of deletions does not always have the desired effect.

8.5.5 EXAMPLE OF A BACH-PETERS-SENTENCE

The man who deserves it will get the prize he wants.

This sentence contains two noun phrases with relative clauses. Each clause contains a pronoun for which the respective other noun phrase serves as the ante- or postcedent (see Section 6.3). Assuming that the pronouns are derived transformationally from full underlying noun phrases which are coreferent with their ante- or postcedent, the deep structure of 8.5.5 will have the following structure.

8.5.6 DEEP STRUCTURE OF A BACH-PETERS-SENTENCE

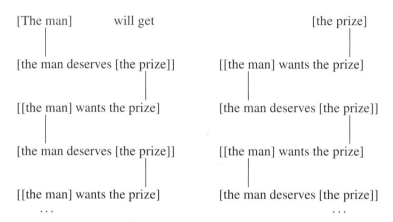

Given the surface of 8.5.5, the transformational algorithm is supposed to produce a deep structure from which this surface may be correctly derived transformationally. Due to the structure of the example, the algorithm will consider bigger and bigger deep structures while observing the recoverability of deletions.

For example, the algorithm will postulate the full noun phrase [the man who deserves it] as deep structure and antecedent of the pronoun he. This deep structure in turn contains the pronoun it, for which the algorithm will postulate the full noun phrase [price which he deserves] as postcedent. This deep structure in turn contains the pronoun he, etc.

This procedure can be continued indefinitely in both relative clauses without ever stopping (halting problem). The formal proof that transformational grammar is undecidable and generates the class of the recursively enumerable languages was established 1969 by Peters & Ritchie[24] and published in 1973.

Later variants of nativism, such as LFG, GPSG, and HPSG, do without transformations, trying to deal with the constituent structure paradox in other ways. In these derived formalisms of PS-grammar, the conditions of constituent structure need only be fulfilled when permitted by the language input. Otherwise (as in cases of discontinuous elements) the intuitions about what is semantically related most closely are

---

[24] In active consultation with N. Chomsky. Personal communication by Bob Ritchie, Stanford 1983.

not expressed in terms of phrase structure trees, but alternatively in terms of feature structures.

In this way, constituent structures have failed to maintain their alleged status as a universal, innate characteristics of human language and as a methodologically meaningful principle. One may therefore well ask why systems like GPSG, LFG, and HPSG continue to hold on to constituent structure. Furthermore, in terms of complexity these later systems are no improvement over transformational grammar: they generate the class of recursively enumerable languages and are undecidable.[25]

Exercises

Section 8.1

1. State an algebraic definition of PS-grammar.
2. What is the difference between terminal and nonterminal symbols in PS-grammar?
3. By whom and when was PS-grammar first invented, under which name, and for which purpose?
4. By whom and when was PS-grammar first used in the description of natural language?
5. Describe the standard restrictions on the rules of PS-grammar.
6. Explain the term generative capacity.

Section 8.2

1. Explain the relation between special types of PS-grammar, formal language classes, and different degrees of complexity.
2. Name the main classes of complexity. Why are they independent of specific formalisms of generative grammar?
3. What is the complexity of language classes in the PS-hierarchy?
4. What is the average sentence length in the Limas corpus?
5. What is the maximal sentence length in the Limas corpus?
6. How much time would an exponential algorithm require in the worst case to parse the Limas corpus?
7. Explain the PS-grammar hierarchy of formal languages.

---

[25] Cf. B. Barton, R.C. Berwick, & E.S. Ristad 1987. If context-free rule loops like $A \to B \to \ldots \to A$ are forbidden, the complexity of LFG is exponential. However, because such loops are formally legal within context-free PS-grammar, this restriction is not really legitimate from the viewpoint of complexity theory. Besides, even an exponential complexity is much too high for computational applications.

8. Which language classes in the PS-grammar hierarchy are of a complexity still practical for computational linguistics?

Section 8.3

1. Define a PS-grammar which generates the free monoid over $\{a, b, c\}$. Classify this language, called $\{a, b, c\}^+$, within the PS-grammar hierarchy. Compare the generative capacity of the grammar for $\{a, b, c\}^+$ with that for $a^k b^k c^k$. Which is higher and why?
2. Where does the term context-free come from in PS-grammar?
3. What kinds of structures can be generated by context-free PS-grammars?
4. Name two artificial languages which are not context-free. Explain why they exceed the generative power of context-free PS-grammars.
5. Define a PS-grammar for $a^k b^{2k}$. Explain why this language fulfills the context-free schema pairwise inverse.
6. Define a PS-grammar for $ca^m dyb^n$. What are examples of well-formed expressions of this artificial language? Explain why it is a regular language.
7. Why would 8.3.5 violate the definition of context-sensitive PS-grammar rules if β was zero?
8. What is a pumping lemma?
9. Why is there no pumping lemma for the context-sensitive languages?
10. Are the recursively enumerable languages recursive?
11. Name a recursive language which is not context-sensitive.

Section 8.4

1. State the definition of constituent structure.
2. Explain the relation between context-free PS-grammars and phrase structure trees.
3. Describe how the notion of constituent structure developed historically.
4. Name two types of distribution tests and explain their role for finding correct constituent structures.
5. Why was it important to American structuralists to segment sentences correctly?

Section 8.5

1. Describe the notion of a discontinuous element in natural language and explain why discontinuous elements cause the constituent structure paradox.
2. How does transformational grammar try to solve the problem caused by discontinuous elements?
3. Compare the goal of transformational grammar and with the goal of computational linguistics.
4. What is the generative capacity of transformational grammar?
5. Explain the structure of a Bach-Peters-sentence in relation to the recoverability of deletions condition. Which mathematical property of transformational grammar was shown with this type of sentence?

9. Basic notions of parsing

This chapter investigates which formal properties make a generative grammar suitable for automatic language analysis and which do not. Thereby context-free PS-grammar and its parsers will be used as the main example.

Section 9.1 describes the basic structure of parsers and explains how the declarative-procedural distinction applies to the relation between generative grammars and parsers. Section 9.2 discusses the relation between context-free PS-grammar and standard C-grammar, and summarizes the relations between the notions of language, generative grammar, subtypes of grammars, subclasses of languages, parsers, and complexity. Section 9.3 explains the principle of type transparency and illustrates with an Earley algorithm analysis of $a^k b^k$ that context-free PS-grammar is not type transparent. Section 9.4 shows that the principle of possible substitutions, on which derivations in PS-grammar are based, does not permit input-output equivalence of PS-grammar with either its parsers or the speaker-hearer. Section 9.5 explains the mathematical, computational, and psychological properties which any empirically adequate generative grammar for natural language must have.

9.1 Declarative and procedural aspects of parsing

Parsers[1] for *artificial* languages are used in computer science for transforming one programming level into another, for example in compilation. Parsers for *natural* languages are used for automatic word form recognition as well as automatic syntactic and semantic analysis. Accordingly, one may distinguish between *morphology parsers*, *syntax parsers*, and *semantic parsers*.

Morphology parsers (cf. Chapter 13–15) take a word form as input and analyze it by (i) segmenting its surface into allomorphs, (ii) characterizing its syntactic combinatorics (categorization), and (iii) deriving the base form (lemmatization). Syntax parsers (cf. Chapter 16–18) take a sentence as input and render as output an analysis of its grammatical structure, e.g. the constituent structure in PS-Grammar or the time-linear derivation in LA-grammar. Semantic parsers (cf. Chapters 22–24) complement the syntactic analyses by deriving associated semantic representations.

---

[1] As explained in Section 1.3, a parser is a computer program which takes language expressions as input and produces some other representation, e.g. a structural analysis, as output.

For modeling the mechanism of natural language communication on the computer, these different types parsers need to be integrated into a functional overall system. Syntax parsers presuppose an automatic word form recognition and thus require a morphology parser of some kind. Semantic parsers presuppose syntactic analysis and thus require a syntax parser.

Common to all kinds of parsers is the distinction between (i) the *structural description* of the expressions to be analyzed and (ii) the *computational algorithm* of the automatic analysis procedure. In modern parsers, these two aspects are separated systematically by treating the structural description by means of a generative grammar which is interpreted and applied by the computational algorithm.

In other words, a modern parser may load one of arbitrary generative grammars G_i, G_j, G_k of a certain formalism (e.g. context-free PS-grammars like 77.1.3, 8.3.1–8.3.5, or C-LAGs like 10.2.2, 10.2.3, 11.5.2, 11.5.3, 11.5.5, 11.5.7, 11.5.8) and analyze a language L_j by interpreting the associated grammar G_j. In the automatic analysis of L_j expressions, a clear distinction is made between the contributions of (i) the grammar G_j and (ii) the parser for the whole class of formal grammars G_i, G_j, G_k.[2]

The separate handling of the grammar and the parsing algorithm corresponds to a distinction which is aspired to in computer science in general, namely the systematic separation of the *declarative specification* and the *procedural implementation* in the solution of a computational problem.

9.1.1 DECLARATIVE & PROCEDURAL ASPECTS IN LINGUISTICS

– The *declarative* aspect of computational language analysis is represented by a generative grammar, written for the specific language to be analyzed within a general, mathematically well-defined formalism.
– The *procedural* aspect of computational language analysis comprises those parts of the computer program which interpret and apply the general formalism in the automatic analysis of language input.

The distinction between the declarative and procedural aspects of a parser is especially clear, if the formal grammar leaves open certain properties which an automatic parser must decide one way or another.

Consider for example the following context-free PS-grammar.

    rule 1: A → B C
    rule 2: B → c d
    rule 3: C → e f

[2] It is not recommended to formulate the rules of the grammar directly in the programming language used. This implicit use of a generative grammar has the disadvantage that the resulting computer program does not show which of its properties are theoretically accidental (reflecting the programming environment or stylistic idiosyncrasies of the programmer) and which are theoretically necessary (reflecting the formal analysis of the language described). Another disadvantage of this approach is that the program works only for a single language rather than a whole subtype of generative grammar and their languages.

The distribution of variables ensures implicitly that in a top-down derivation rule 1 is applied before rule 2 and rule 3. However, the decision of whether rule 2 should be applied before rule 3, or rule 3 before rule 2, or both at once remains open.

For the declarative specification of context-free PS-grammars such a partial rule ordering is sufficient. For a computer program, on the other hand, a complete rule ordering must be decided explicitly – even if it may be irrelevant from a theoretical point of view. Such ordering decisions which go beyond the declarative specification of a formal grammar or which – for reasons of the parsing algorithm – run counter to the conceptual derivation order of the grammar (cf. 9.3.4) are called procedural.

For a given formalism (e.g. context-free PS-grammar) different parsing algorithms may be developed in different programming languages. The result are different procedural realizations which take the same declarative grammars as input. For example, two parsing algorithms (e.g. the Earley algorithm and the CYK algorithm) implemented in two programming languages (e.g. Lisp and C, respectively) will produce results for (i) the same grammar (e.g. 7.1.3 for $a^k b^k$) and (ii) the same language input (e.g. aaabbb) which are identical from a declarative point of view.

For a certain grammar type, however, general parsers are practical only if its complexity is not too high. Therefore, there exist general parsers for the classes of regular and context-free PS-grammars, while no practical parsers can be written for the classes of context-sensitive and a fortiori unrestricted PS-grammars.

9.2 Fitting grammar onto language

Determining where in the PS-hierarchy (cf. 8.2.4) the natural languages might belong is not only of academic interest, but decides whether or not the natural languages may be parsed efficiently within PS-grammar. For reasons of efficiency it would be optimal if the natural languages could be shown to belong into the class of regular languages – because then their PS-grammar analyses could be parsed in linear time.

There are, however, natural language structures in which the surface is obviously context-free rather than regular, e.g., center-embedded relative clauses in German.

9.2.1 CONTEXT-FREE STRUCTURE IN GERMAN

    Der Mann,   schläft.
    (*the man*)   (*sleeps*).
        der die Frau,   liebt,
        (*who the woman*)   (*loves*)
             die das Kind,   sieht,
             (*who the child*)   (*sees*)
                 das die Katze   füttert,
                 (*who the cat*)   (*feeds*)

The structure of this sentence corresponds to the abstract schema 'noun_phrase$_1$ noun_phrase$_2$ ... verb_phrase$_2$ verb_phrase$_1$',

which in turn corresponds to abc ... cba. The structure in 9.2.1 is context-free because there is no grammatical limit on the number of embeddings. Therefore the PS-grammatical analysis of natural language is at least of complexity n^3.

The next question is whether or not natural language can be shown to belong into the PS-grammar class of context-free languages. The answer to this question is less clear than in the case of the regular languages.

N. Chomsky 1957 and 1965 expanded context-free PS-grammar into the derived formalism of transformational grammar by arguing that PS-grammar alone was insufficient to formulate what he considered linguistic generalizations (cf. 4.5.2, 8.5.4). In addition, S. Shieber 1985 presented sentences from Swiss German with the context-sensitive structure of WW (see Section 8.3) which were intended to prove – in analogy to 9.2.1 – that the natural languages are at least context-sensitive.

9.2.2 CONTEXT-SENSITIVE STRUCTURE IN SWISS GERMAN

mer em Hans es huus hälfed aastriiche
we the Hans the house help paint

The formal structure of this example is analyzed as a b a' b' (with a = the Hans, b = the house, a' = help, and b' = paint). This is not context-free because it doesn't have an *inverse* structure – just as in the context-sensitive language WW. If this argument is accepted, the PS-grammatical analysis of natural language is at least context-sensitive. It is thus of exponential complexity and requires billions and billions of centuries for the analysis of longer sentences in the worst case (cf. 8.2.2).

To avoid is implausible conclusion, Harman 1963 and Gazdar 1982 each presented sizable PS-grammar systems intended to show that there are no structures in natural language which could not be handled in a context-free fashion. Harman's proposal came at a time when complexity-theoretic considerations were not widely understood and transformational grammar was pursued with undiminished enthusiasm. Also, Harman did not provide a detailed linguistic motivation for his system.

At the time of Gazdar's proposal, on the other hand, awareness of complexity-theoretic considerations had increased. Also, Gazdar did not employ context-free PS-grammar directly, but used the additional mechanism of *metarules* to define the derived PS-grammar formalism of GPSG (Generalized Phrase Structure Grammar). The purpose of the metarules was to combine huge numbers of context-free PS-rules[3] in order to formulate the linguistic generalizations which he and others considered important in those days.

At the same time, GPSG was hoped to analyze natural languages at a degree of complexity which is computationally practical, i.e., context-free or n^3. However, contrary to Gazdar's original assumption that the use of metarules would not cause an increase

---

[3] "Literally trillions and trillions of rules," S. Shieber, S. Stucky, H. Uszkoreit & J. Robinson 1983.

in complexity, a closer formal investigation proved [4] that GPSG is in fact recursively enumerable and therefore undecidable.

If the natural languages are not context-free – and the majority of theoretical linguists takes this view –, what other formal language class does natural language belong to? By investigating the answer to this question one should remember that the class of context-free languages is the result of using a certain *rule type* (i.e. type 2) of a certain *formalism* (i.e. PS-grammar).

On the one hand, there is no reason why this particular formalism and this particular rule type – resulting in the pairwise inverse structure of context-free languages – should be characteristic for natural language. On the other hand, the context-free languages are the largest class within PS-grammar the mathematical complexity of which is sufficiently low to be practical interest.

The assumption that the natural languages are not context-free implies one of the following two conclusions.

1. PS-grammar is the only elementary formalism of generative grammar, for which reason one must accept that the natural languages are of high complexity and thus computationally intractable.
2. PS-grammar is not the only elementary formalism of generative grammar. Instead, there are other elementary formalisms which define other language hierarchies whose language classes are orthogonal to those of PS-grammar.

In light of the fact that humans process natural language in a highly efficient manner, the first conclusion is implausible. The second conclusion, on the other hand, raises the question of what concrete alternatives there are.

From a historical point of view a natural first step in the search for new formal language classes is to analyze the generative capacity and complexity of C-grammar (Section 7.4). Thereby the easiest strategy is a comparison of the formal properties of C- and PS-grammar.

In such a comparison one of the following three possible relations must hold.

9.2.3 POSSIBLE RELATIONS BETWEEN TWO GRAMMAR FORMALISMS

– *No equivalence*
  Two grammar formalisms are not equivalent, if they generate/recognize different language classes; this means that the two formalisms are of different generative capacity.
– *Weak equivalence*
  Two grammar formalisms are weakly equivalent, if they generate/recognize the same language classes; this means that the two formalisms have the same generative capacity.

---

[4] H. Uszkoreit & S. Peters 1986.

– *Strong equivalence*

Two grammar formalisms are strongly equivalent, if they are (i) weakly equivalent, and moreover (ii) produce the same structural descriptions; this means that the two formalisms are no more than *notational variants*.

For the historical development of modern linguistics it would have been desirable if the elementary formalisms of C- and PS-grammar had turned out to be not equivalent because in this way one would have had a true alternative to the class of context-free languages. In fact, however, it was discovered early on that C-grammar and PS-grammar are *weakly equivalent*.

> The problem arose of determining the exact relationships between these types of [PS-]grammars and the categorial grammars. I surmised in 1958 that the BCGs [Bidirectional Categorial Grammar *á la* 7.4.1] were of approximately the same strength as [context-free phrase structure grammars]. A proof of their equivalence was found in June of 1959 by Gaifman. ... The equivalence of these different types of grammars should not be too surprising. Each of them was meant to be a precise explicatum of the notion *immediate constituent grammars* which has served for many years as the favorite type of American descriptive linguistics as exhibited, for instance, in the well-known books by Harris [1951] and Hockett [1958].

> Y. Bar-Hillel 1960 [1964, p. 103]

That C- and PS-grammar are equivalent only in certain subtypes, namely bidirectional C-grammar and context-free PS-grammar, was eclipsed by the fact that context-free grammars generate the largest language class which is still computationally tractable. That the equivalence between bidirectional C-grammar and context-free PS-grammar is only a weak equivalence did not raise much interest either.

This gave rise to the erroneous belief that PS-grammar and its formal language hierarchy are somehow given by nature and the only fundamental formal system of artificial and natural languages. However, there exists at least one alternative elementary formalism of generative grammar which divides the set of artificial languages into completely different language classes than PS-grammar does.[5]

The general relations between the notions of languages, generative grammars, subtypes of grammars, classes of languages, parsers, and complexity may be summarized as follows.

– *Languages*

Languages exist independently of generative grammars. This is shown not only by the natural languages, but also by artificial languages like $a^k b^m$, $a^k b^k$, $a^k b^k c^k$, $a^k b^k c^k d^k$, $\{a^k b^k c^k\}^*$, WW^R, WW, WWW, etc. Their traditional names characterize the respective languages so well as to allow writing down and recognizing their well-formed expressions.

---

[5] For example, context-free $a^k b^k$ (cf. 7.1.3, 10.2.2) and context-sensitive $a^k b^k c^k$ (cf. 8.3.6, 10.2.3) are classified in LA-grammar as elements of the same linear class of C1-LAGs. Correspondingly, context-free WW^R (cf. 8.3.4, 11.1.5) and context-sensitive WW are classified in LA-grammar as elements of the same polynomial (n^2) class of C2-LAGs.

The definition of an explicit grammar within a given grammar formalism for a natural or artificial language constitutes a second step which usually is not trivial at all. That a given language may be described by different formal grammars of different grammar formalisms is shown by the comparison of the C- and PS-grammar analysis of a^kb^k in 7.4.3 and 7.1.3, respectively.

– *Generative grammars*
  On the one hand, a generative grammar is a general formal framework; on the other, it is a specific rule system defined for describing a specific language within the general framework. For example, PS-grammar as a general formal framework is defined as the quadruple $< V, V_T, S, P >$ with certain additional conditions on its elements. Within this general formalism, specific PS-grammars may be defined for generating specific languages such as a^kb^k.

– *Subtypes of generative grammars*
  Different restrictions on the formal framework of a generative grammar may result in different grammar types. In this way, the subtypes of regular, context-free, context-sensitive, and unrestricted PS-grammars are defined in PS-grammar and the subtypes of C1-, C2-, C3-, B-, and A-LAGs are defined in LA-Grammar. The various restrictions do not exist absolutely, but depend on formal properties of the particular grammar type employed (especially its rule structure.)

– *Language classes*
  The subtypes of a generative grammar may be used to divide the set of possible languages into different language classes. Because the subtypes of a generative grammar depend on the formalism used, the associated language classes do not exist absolutely, but instead reflect the formal properties of the grammar type employed. For example, the pairwise inverse structure characteristic of context-free languages follows directly from the specific restrictions on the rule structure of context-free PS-grammars.
  Nota bene: *languages* exist independently of the formal grammars which may generate them. The *language classes*, on the other hand, do not exist independently, but result from particular restrictions on particular grammar formalisms.

– *Parsers*
  Parsers are programs of automatic language analysis which are defined for whole subtypes of generative grammars (e.g. context-free PS-grammars or the C-LAGs). Thus, the problem with a context-sensitive language like $a^kb^kc^k$ (cf. 8.3.6) is not that one couldn't write an efficient analysis program for it, but rather that no practical parser can be written for context-sensitive PS-grammar in general.

– *Complexity*
  The complexity of a subtype of generative grammar is determined over the number of *primitive operations* needed by an equivalent abstract automaton or parsing program for analyzing expressions in the worst case. The complexity of individual languages is usually determined over the complexity of their respective classes. Because language classes depend on the particular formalism employed, a language

like $a^k b^k c^k$ belongs in PS-grammar into the class of context-sensitive languages, which is of exponential complexity, but in LA-grammar into the class of C1-LAGs, which is of linear complexity.

Besides the complexity of a language in terms of its class, one may also investigate the *inherent* complexity of individual languages. In this case one uses the specific structural properties of the language (independent of any particular grammar formalism) to show how many operations its analysis would require in the worst case on an abstract machine (e.g. a Turing or register machine). For example, languages like 3SAT and Subset Sum (cf. Section 11.4 and 11.5) are inherently complex. Therefore, these languages will be necessarily in a high complexity class (here exponential) in any possible grammar formalism.

The inherent complexity of individual languages is an important tool for determining the minimal complexity of language classes. This form of analysis occurs on a very low level, however, corresponding to machine or assembler code. For this reason, the complexity of artificial and natural languages is usually analyzed at the abstraction level of grammar formalisms, whereby complexity is determined for the grammar type and its language class as a whole.

9.3 Type transparency between grammar and parser

The simplest and most transparent use of a grammar by a parser consists in the parser merely applying the formal grammar mechanism in the analysis of the input. This natural view of the parser as a simple *motor* or *driver* of the grammar was originally intended also in PS-grammar.

> Miller and Chomsky's original (1963) suggestion is really that grammars be realized more or less directly as parsing algorithms. We might take this as a methodological principle. In this case we impose the condition that the logical organization of rules and structures incorporated in the grammar be mirrored rather exactly in the organization of the parsing mechanism. We will call this *type transparency*.
>
> R.C. Berwick & A.S. Weinberg 1984, p. 39.

The following definition specifies the notion of *absolute type transparency* (cf. Berwick & Weinberg 1984, p. 41) in a precise, intuitively obvious, and general way.

9.3.1 DEFINITION OF ABSOLUTE TYPE TRANSPARENCY

– For any given language, parser and generator use the *same* formal grammar,
– whereby the parser/generator applies the rules of the grammar *directly*.
– This means in particular that the parser/generator applies the rules in the *same order* as the grammatical derivation,
– that in each rule application the parser/generator takes the *same input* expressions as the grammar, and

– that in each rule application the parser/generator produces the *same output* expressions as the grammar.

In PS-grammar, it soon turned out that a direct application of the grammar rules by a parser is not possible. The historical reason for this is that Post 1936 developed his production or rewrite system to mathematically characterize the notion of *effective computability* in recursion theory.[6] In this original application, a derivation order based on the substitution of signs by other signs is perfectly natural.

When Chomsky 1957 used the Post production system under the name of PS-grammar for analyzing natural language, he nolens volens inherited its substitution-based derivation order. Because a parser takes terminal strings as input, PS-grammars and their parsers are not *input-output equivalent* – which means that there cannot exist a parser able to apply the rules of PS-grammar directly.

The structural problem of using context-free PS-grammar for parsing is illustrated by the following step by step derivation of a a a b b b based on the PS-grammar 7.1.3 for $a^k b^k$.

9.3.2 *Top-down* DERIVATION OF a a a b b b

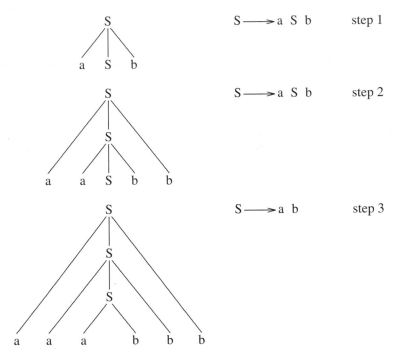

In this substitution-based derivation the variable S on the left-hand side of the rules is repeatedly replaced by the signs on right-hand side of the rules. The derivation begins

---

[6] See for example A. Church 1956, p. 52, footnote 119.

with the variable S and ends when all variables have been replaced by terminal signs. This expansion of variables is called a *top-down* derivation.

A computational procedure cannot apply the derivational mechanism of the PS-grammar directly in the analysis of the input a a a b b b because the PS-grammar inserts terminal symbols always in two different locations into the output string. In contrast to the PS-grammar, the computer program has to deal with the unanalyzed terminal string, and there the structural relation between two locations of arbitrary distance is not at all obvious.

The only possibility to use the grammatical top-down derivation directly for the automatic analysis of input would be a systematic derivation of all outputs (beginning with the shortest), hoping that the string to be analyzed will at some point show up in the set of outputs. In the case of $a^k b^k$ this would succeed fairly easily, but for the whole type of context-free PS-grammar this approach would be no solution. For example, in the case WW^R the number of possible outputs may grow exponentially with their length, for which reason the method of generating all possible strings would quickly become prohibitively inefficient.

Alternatively, one may try to apply the PS-grammar in a *bottom up* derivation.

9.3.3 *Bottom-up* DERIVATION OF a a a b b b

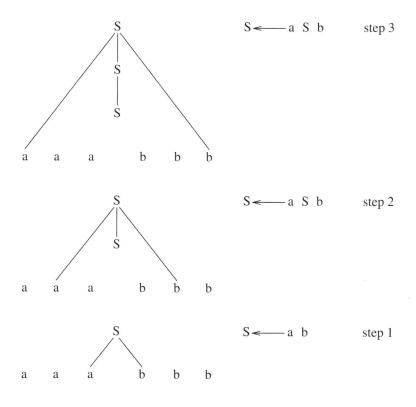

A bottom-up derivation begins with the right hand side of one or more rules and replaces them with the variable of their left-hand side. These variables are matched to the right-hand side of other rules and replaced by their respective left-hand sides. This type of bottom-up derivation may be stopped whenever only one rule is active and its right-hand side is replaced by the start symbol S.

The weakness of a bottom-up derivation in context-free PS-grammar is that it requires the program to find the theoretical center of the pairwise inverse structures. This is a problem because the program has to deal with arbitrary unanalyzed input strings in which the theoretical center is not marked.

In the case of $a^k b^k$, the program could use the border between the a- and b-words to determine the center, but this would be no solution for context-free PS-grammar in general. The language WW^R, for example, contains strings like a a a a a a and a a a a a a a where the center can only be determined via the length. Length, however, is not a general criterion either because there are context-free languages like $a^k b^{3k}$ (cf. 8.3.4) where the center of the derivation is not located at half length of the input.

In order to show how a parser may overcome the difficulties with the logical derivation order of context-free PS-grammar in a generally valid manner, let us consider an analysis within the Earley algorithm, which according to Hopcroft & Ullman 1979, p. 145, "is the most practical, general, context-free recognition and parsing algorithm." The main purpose of the following example 9.3.4 is to illustrate the restructuring of PS-grammar rules necessary for the general parsing of context-free languages.

9.3.4 THE EARLEY ALGORITHM ANALYZING $a^k b^k$

```
.aaabbb

.S
 |            a.aabbb
 |
.ab  -> a.b
.aSb -> a.Sb
        |                aa.abbb
        |
     a.abb   -> aa.bb
     a.aSbb  -> aa.Sbb
        |                   aaa.bbb      aaab.bb
        |
     aa.abbb  -> aaa.bbb  -> aaab.bb-> ...
     aa.aSbbb -> aaa.Sbbb
```

The Earley[7] algorithm uses three operations which are not part of the algebraic definition of PS-grammar (cf. 8.1.1), namely the *predictor-*, *scan-* and *completor*-operation (cf. Earley 1970). With these the input in 9.3.4 is analyzed from left to right, whereby a dot indicates how far the analysis has proceeded.

---

[7] Jay Earley developed his parsing algorithm for context-free PS-grammars in his dissertation at the Computer Science Department of Carnegie Mellon University in Pittsburgh, USA. After this achievement he disappeared from the scientific scene and is now presumed missing.

At the beginning of the analysis, the dot is placed before the the input string

.a a a b b b

which is interpreted as corresponding to the state

.S

As a variable, S cannot have a counterpart in the input string, for which reason the *scan*-operation cannot apply.

However, using the two rules S → a b and S → a S b of the PS-grammar for $a^k b^k$, the *predictor*-operation produces two new states, namely

.a b

.a S b

which are added to the *state set* of the parsing algorithm.

In the next cycle of the algorithm, the dot is moved one position to the right (*completor*-operation), both in the input string, i.e.,

a. a a b b b

and in the states, i.e.,

a. b

a. S b

The *scan*-operation checks whether the sign before the dot in the states corresponds to the sign before the dot in the input string. This is successful in both cases.

In the next cycle, the dot is again moved one position to the right, resulting in the input string

a a. a b b b

and in the states

a b.

a S. b

Because the first state has a terminal symbol preceding the dot, namely b, the *scan*-operation is attempted. This is not successful, however, because the terminal preceding the dot in the input string is an a.

The second state, on the other hand, has an S preceding the dot, to which according to the grammar rules the *predictor*-operation may be applied. This results in the states

a a. b b

a a. S b b,

both of which are shown by another *scan*-operation to fit the input string.

Next the *completor*-operation moves the dot again one position to the right, resulting in the input string

a a a. b b b

and the states

a a b. b

a a S. b b

Applying the *scan*-operation to the first state does not succeed.

However, applying the *predictor*-operation to the second state results in two new states, namely

a a a. b b b
a a a. S b b b

which are both matched successfully onto the input string by another *scan*-operation.

Once more the dot is moved one position to the right, producing the input string

a a a b. b b

and the new states

a a a b. b b
a a a S. b b b

This time the *scan*-operation succeeds on the first state.[8]

Again the dot is moved one position to the right, resulting in the input string

a a a b b. b

and the state

a a a b b. b

Because the state consists of terminal symbols only, only the *scan*-operations can apply. The first of these happens to be successful, resulting in the input string

a a a b b b.

and the state

a a a b b b.

The dot at the end of the input string indicates a successful analysis.

We have seen that the Earley algorithm uses the rules of the grammar, but not directly. Instead, the parsing algorithm disassembles the rules of the grammar successively into their basic elements, whereby the order is determined by the sequence of terminal symbols in the input string – and not by the logical derivation order of the grammatical rule system. Therefore, the relation between the Earley algorithm and the context-free PS-grammars used by it is not type transparent.

For context-free PS-grammar there exists a considerable number of parsers besides the Earley algorithm, such as the CYK algorithm (See Hopcroft & Ullman 1979, p. 139-141) and the chart parser (Kay 1980). On the positive side, these parsers all interpret arbitrary context-free PS-grammars and analyze any of the associated languages. On the negative side, these parsers all lack type transparency.

That parsers like the Earley algorithm, the CYK algorithm, or the chart parser cannot apply the rules of context-free PS-grammar directly, but require huge intermediate structures – called *state sets, tables*, or *charts* – in order to correlate the differing derivation orders of the grammar and the parser has been excused by pointing out that the user, for example the grammar writer, does not notice the procedural routines

---

[8] The second state is also still active at this point because it allows a further *predictor*-operation, resulting in the new states

a a a a. b b b b
a a a a. S b b b b

The subsequent *scan*-operations on the input

a a a b. b b

do not succeed, however.

of the parser. This is only valid, however, if the PS-grammar is already descriptively adequate.

In contrast, if the parser is used in the *development* of descriptively adequate grammars, the lack of type transparency greatly impedes debugging and upscaling of the grammar. For example, if an error occurs because the parser cannot find a legal grammatical structure for well-formed input, then this error must be found in the complex rule system of the PS-grammar.

The number of rules in context-free PS-grammars used for practical applications often exceeds several thousand. Here it would be of great help, if errors could be localized with the help of parser traces. However, because the parser cannot use the rules of the PS-grammar directly the parse trace (protocol of states) is about as unreadable as assembler code (cf. 9.3.4) and therefore of little heuristic value.

One may, of course, write secondary parsers in order to translate the unreadable intermediate structures of the primary parser into a form that can be used for purposes of localizing errors. Compared to a type transparent system like LA-grammar, however, both the construction of the intermediate structures and their reinterpretation for human analysis constitute a essentially superfluous effort the costs of which show up not only in the programming work required, but also in additional computation and increased use of memory.

9.4 Input-output equivalence with the speaker-hearer

The nativist approach has attempted to belittle the structural disadvantages of PS-grammar as a problem of programming, the solution of which was the job of computer science. The real goal was an analysis of the innate language knowledge of the speaker-hearer, and for this the properties in question had no negative effect.

This line of reasoning is not valid, however. Because the form of innate structures follows their function (cf. 4.5.3), a minimal requirement for the description of the innate human language capability is that the grammar formalism used is input-output equivalent with the speaker-hearer.

PS-grammar, however, is just as incompatible with the input-output conditions of the speaker-hearer as it is with those of its parsers, as shown by the following example.

9.4.1 CONTEXT-FREE PS-GRAMMAR FOR A SIMPLE SENTENCE OF ENGLISH

1.	S	→ NP VP	5.	V → read
2.	NP	→ DET N	6.	DET → a
3.	VP	→ V NP	7.	N → book
4.	NP	→ Julia		

In the derivation of the associated analysis tree 9.4.2, rule 1 of 9.4.1 replaces (substitutes) the start symbol S with NP (noun phrase) and VP (verb phrase). Then rule

4 replaces the node NP with the word Julia, and rule 3 replaces the node VP with the nodes V and NP. Then rule 5 replaces the rule V (verb) with the word read, and rule 2 replace the NP with the nodes DET (determiner) and N (noun). Finally, rule 6 replaces the node DET with the word a, and rule 7 replaces the node N with the word book.

9.4.2 PS-GRAMMAR ANALYSIS (*top-down* DERIVATION)

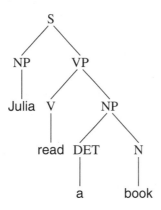

As simple and natural this derivation may seem from a logical point of view, it is nevertheless in clear contradiction to the time-linear structure of natural language. The discrepancy between the input-output conditions of PS-grammars and the use of language by the hearer/reader is illustrated by the following sketch which attempts a time-linear analysis based on the PS-grammar 9.4.1.

9.4.3 ATTEMPT OF A TIME-LINEAR ANALYSIS IN PS-GRAMMAR

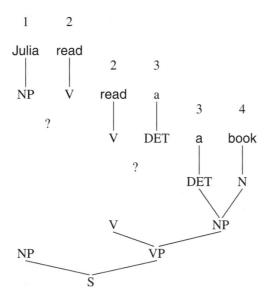

The natural order of reading the sentence Julia read a book is time-linear, i.e. from left to right,[9] one word after the other. After replacing the first two words Julia and read by the nodes NP and V (in accordance with rules 4 and 5 of 9.4.1), the reader's grammar algorithm is looking for a PS-rule to replace the node sequence NP V by a higher node. However, since no such rule is provided by the grammar 9.4.1, the subtree consisting of the first word and its category must be set aside for later use.

Next, the reader's grammar algorithm replaces the third word a by its category DET using rule 6, and attempts to combine the category V of the second word with the category DET of the third word by replacing them by a higher node. Again, no suitable rule is provided by the grammar 9.4.1, and the subtree of the second word must also set aside for later use.

At the end, the reader's grammar algorithm attempts to combine the category of the third word with that of the fourth and final word of the input sentence. Here at last the grammar 9.4.1 provides a suitable rule, namely rule 2, to combine the node sequence DET N into the higher node NP.

Then the reader's grammar algorithm attempts once more to combine the category V of the second word read, which has been set aside for later use, this time with the newly derived NP node. It turns out that this is possible, thanks to rule 3 of the grammar 9.4.1, resulting in the higher node VP. Finally the reader's grammar algorithm attempts once more to combine the category NP of the first word Julia, which has been set aside since the analysis of this four word sentence first began, this time with the newly derived VP node. This time the attempt is successful due to rule 1 and the analysis is completed.

The analysis sketch 9.4.3 shows that a time-linear interpretation is impossible in a constituent-structure-based context-free PS-grammar: the first word of the sentence was the last to be built into the analysis. PS-grammar's lack of type transparency is thus not only an obstacle for the development of systems for the automatic analysis of artificial or natural languages. It is also in conflict with the elementary principle of nature that form follows function – especially innate form.

But what about the standard reply that nativist grammars are *not intended*[10] to describe the operations of the speaker-hearer in communication, that they are rather intended to describe the innate knowledge of the speaker-hearer at a level of abstraction (competence) which is independent of the use of language in concrete communication? This reply cannot convince because in order for a nativist theory to be plausible, its basic structures must be at least *compatible* with the communication mechanism of natural language. And this is impossible with a nativist theory using the formalism of PS-grammar.

---

[9] According to the writing conventions of the Greek-Roman tradition.

[10] N. Chomsky has emphasized tirelessly that it was not the goal of his nativist program to model the communication procedure of the speaker-hearer. See for example Chomsky 1965, p.9.

9.5 Desiderata of grammar for achieving convergence

From a history of science point of view, PS-grammar-based nativism may be evaluated in different ways. By means of continuous revision, nativism has achieved over forty years of holding interest, preserving influence, recruiting new followers, and turning internal strife into a source of restoration rather than decline. On the negative side, it is a textbook example of lacking convergence.[11]

The lack of convergence shows up in at least four ways. First, instead of consolidation there has been a development of ever new derived systems, epitomized by J.D. McCawley's 1982 title *Thirty Million Theories of Grammar*.[12] Second, each time an additional mechanism (e.g. transformations, metarules, etc., regarded as descriptively necessary) was introduced, the mathematical and computational properties were severely degraded as compared to the elementary subformalism of context-free PS-grammar. Third, the description of natural language within PS-grammar has lead continuously to problems of the type descriptive aporia and embarrassment of riches[13] (cf. Section 22.2). Fourth, practical systems of natural language processing pay either only lip service to the theoretical constructs of nativism or ignore them altogether.

There are two main reasons for this lack of convergence. First, nativism is empirically underspecified because it does not include a functional theory of communication. Second, the PS-grammar formalism adopted by nativism is incompatible with the input-output conditions of the speaker-hearer.

Given the longterm consequences of working with a certain formalism,[14] begin-

---

[11] In history of science, a field of research is regarded as developing positively, if its different areas converge, i.e., if improvements in one area lead to improvements in others. Conversely, a field of science is regarded as developing negatively, if improvements in one area lead to a deterioration in other areas.

[12] This number is computed on the basis of the open alternatives within the nativist theory of language. McCawley's calculation amounts to an average of 2 055 different grammar theories a day, or a new theory every 42 seconds for the duration of 40 years (from 1957 – the year in which Chomsky's *Syntactic Structures* appeared – to 1997).

[13] PS-grammar in combination with constituent structure analysis exhibits descriptive aporia in, e.g., declarative main clauses of German such as Peter hat die Tür geschlossen. Because of its discontinuous elements this sentence cannot be provided with a legal constituent structure analysis within context-free PS-grammar (cf. 8.5.1 and 8.5.2).

To resolve this and other problems, transformations were added to context-free PS-grammar (cf. 8.5.3). This resulted in many problems of the type embarrassment of riches. For example, there arose the question of whether the main clauses of German should be derived transformationally from the deep structure of subordinate clauses (e.g. weil Peter die Tür geschlossen hat) or vice versa.

The transformational derivation of main clauses from subordinate clauses was motivated by the compliance of subordinate clauses with constituent structure, while the transformational derivation of subordinate clauses from main clauses was motivated by the feeling that main clauses are more basic (cf. E. Bach 1962 and M. Bierwisch 1963). For a treatment of German main and subordinate clauses without transformations see Chapter 18, especially Section 18.5.

[14] Changing from one formalism to another based on conviction may be costly. It means giving up a carefully maintained status of expert in a certain area, implies revision of sometimes decades of previous work, and may have major social repercussions in one's peer group, both at home and abroad.

However, fields of science are known to sometimes shift in unexpected ways such that established research groups suddenly find their funds, their influence, and their membership drastically reduced.

ning linguists interested (i) in modeling the mechanism of natural communication, (ii) verifying their model computationally, and (iii) utilizing their model in practical applications such as man-machine communication, should select their grammar formalism carefully and consciously. As a contribution to a well-educated choice, the properties of PS-grammar are summarized below.

9.5.1 PROPERTIES OF PS-GRAMMAR

– *Mathematical:*
  Practical parsing algorithms exist only for context-free PS-grammar. It is of a sufficiently low complexity (n^3), but not of sufficient generative capacity for natural language. Extensions of the generative capacity for the purpose of describing natural language turned out to be of such high complexity (undecidable or exponential) that no practical parse algorithm can exist for them.
– *Computational:*
  PS-grammar is not type transparent. This prevents using the automatic traces of parsers for purposes of debugging and upscaling grammars. Furthermore, the indirect relation between the grammar and the parsing algorithm requires the use of costly routines and large intermediate structures.
– *Empirical:*
  The substitution-based derivation order of PS-grammar is incompatible with the time-linear structure of natural language.

Because bidirectional C-Grammar is weakly equivalent to context-free PS-grammar (cf. Section 9.2), C-grammar cannot serve as an alternative. We must therefore look for yet another elementary formalism with the following properties.

9.5.2 DESIDERATA OF A GENERATIVE GRAMMAR FORMALISM

1. The grammar formalism should be mathematically well-defined and thus
2. permit an explicit, declarative description of artificial and natural languages.
3. The formalism should be recursive (and thus decidable) as well as
4. type transparent with respect to its parsers and generators.
5. The formalism should define a hierarchy of different language classes in terms of structurally obvious restrictions on its rule system (analogous – but orthogonal – to the PS-grammar hierarchy),
6. whereby the hierarchy contains a language class of low, preferably linear, complexity the generative capacity of which is sufficient for a complete description of natural language.

When this happens, changing from one formalism to another may become a necessity of opportunistic survival.

In light of these possibilities, the best long term strategy is to evaluate scientific options, such as the choice of grammar formalism, rationally. As in all good science, linguistic research should be conducted with conviction based on the broadest and deepest possible knowledge of the field.

7. The formalism should be input-output equivalent with the speaker-hearer (and thus use a time-linear derivation order).

8. The formalism should be suited equally well for production (in the sense of mapping meanings into surfaces) and interpretation (in the sense of mapping surfaces into meanings).

The following chapters will show that LA-grammar satisfies these desiderata. In contradistinction to the formalisms of C- and PS-grammar, the derivations of which are based on the principle of possible *substitutions*, the derivation of LA-grammar is based on the principle of possible *continuations*.

Exercises

Section 9.1

1. What is the origin of the term parser and what are the functions of a parser?
2. How are morphology, syntax, and semantics parsers related and how do they differ?
3. Describe two different ways of using a generative grammar in a parser and evaluate the alternatives.
4. Explain the notions declarative and procedural. How do they show up in parsers?
5. Is it possible to write different parsers for the same grammatical formalism?

Section 9.2

1. Describe a context-free structure in natural language.
2. Are there context-sensitive structures in natural language?
3. What follows from the assumption that natural language is not context-free? Does the answer depend on which grammar formalism is used? Would it be possible to parse natural language in linear time even if it is context-sensitive?
4. Explain the possible equivalence relations between two formalisms of grammar.
5. How is bidirectional C-grammar related to the types of PS-grammar?
6. Do artificial languages depend on their formal grammars?
7. Do language classes depend on formalisms of grammar?
8. What impact has the complexity of a language class on the possible existence of a practical parser for it?
9. What is the inherent complexity of a language and how is it determined?

Section 9.3

1. Explain the notion of type transparency.
2. For which purpose did Post 1936 develop his production systems?
3. When are a grammar formalism and a parser input-output equivalent?
4. What is the difference between a top-down and a bottom-up derivation in a context-free PS-grammar?
5. Why is it that a context-free PS-grammar is not input-output equivalent with its parsers? Base your explanation on a top-down and a bottom-up derivation.
6. Explain the functioning of the Earley algorithm using an expression of $a^k b^k$. How does the Earley algorithm compensate the substitution-based derivation order of PS-grammar?
7. Explain how the Earley algorithm makes crucial use of the pairwise inverse structure of context-free PS-grammars.
8. Is it possible to parse $a^k b^k c^k$ using the Earley algorithm?
9. Are there type transparent parsers for context-free PS-grammar?
10. Name two practical disadvantages of an automatic language analysis which is not type transparent.

Section 9.4

1. Explain why a nativist theory of language based on PS-grammar is incompatible with the principle *form follows function*, using the notion of input-output equivalence.
2. Demonstrate with an example that the derivation order to PS-grammar is incompatible with the time-linear structure of natural language.
3. Does an additional transformational component (cf. sections 2.4 and 8.5) diminish or increase the incompatibility between the PS-grammar derivation order and the time-linear order of natural language?

Section 9.5

1. Explain the notion of convergence as used in the history of science.
2. How does a lack of convergence show up in the historical development of nativism and what are its reasons?
3. How did McCawley calculate the number in his title *Thirty Million Theories of Grammar*? Does this number indicate a positive development in linguistics?
4. Why can changing to another grammar formalism be costly?
5. Do you see a relation between the problem types 'descriptive aporia' and 'embarrassment of riches,' on the one hand, and the proposal of ever new derived formalisms with high mathematical complexity, on the other?
6. Describe the mathematical, computational, and empirical properties of PS-grammar.
7. Which desiderata must be satisfied by a generative grammar in order to be suitable for a computational analysis of natural language?

10. Left-associative grammar (LAG)

The previous Chapters 7–9 developed the basic notions for analyzing artificial and natural languages within the historical formalisms of C-grammar (1929) and PS-grammar (1935). The following Chapters 10–12 will apply these notions to the third elementary formalism, namely LA-grammar (1985).[1] This comparatively new formalism has not been borrowed from some other field of research to be adapted to natural language analysis, but was developed from the outset as a time-linear, type transparent algorithm that is input-output equivalent with the speaker-hearer.

Section 10.1 explains how the time-linear structure of natural language is modeled by the left-associative derivation order, defines the principle of possible continuations, and shows for C-, PS- and LA-grammar the general connection between rule format and conceptual derivation order. Section 10.2 provides an algebraic definition of LA-grammar. Section 10.3 describes formats for representing time-linear derivations. Section 10.4 illustrates the relation between automatic analysis and automatic generation in LA-grammar using the context-sensitive language $a^k b^k c^k$ and demonstrates the type transparency of LA-grammar. Section 10.5 shows how LA-grammatical analyses of natural language are motivated linguistically.

10.1 Rule types and derivation order

The name LA-grammar is motivated by the left-associative derivation order on which this elementary formalism is based. The notion *left-associative* is known from logic.

> When we combine operators to form expressions, the order in which the operators are to be applied may not be obvious. For example, a + b + c can be interpreted as ((a + b) + c) or as (a + (b + c)). We say that + is *left-associative* if operands are grouped left to right as in ((a + b) + c). We say it is *right-associative* if it groups operands in the opposite direction, as in (a + (b + c)).
>
> A.V. Aho & J.D. Ullman 1977, p. 47

Left- and right-associative bracket structures have the special property that they may be interpreted as regularly increasing.

---

[1] The 'official' publication is Hausser 1992 in the journal *Theoretical Computer Science* (TCS).

10.1.1 INCREMENTAL LEFT- AND RIGHT-ASSOCIATIVE DERIVATION

left-associative:	*right-associative:*

```
      a                                          a
     (a + b)                                   (b + a)
    ((a + b) + c)                           (c + (b + a))
   (((a + b) + c) + d)                   (d + (c + (b + a)))
         · · ·   ⟹                            ⟸   · · ·
```

Of these two regularly increasing structures, the left-associative one corresponds to the traditional direction of Western (Greek-Roman) writing.

When applying the left-associative derivation order to the analysis of language, it is natural to interpret a, b, c... as word forms and + as the concatenation operator. The first word a is a sentence start which is combined with the next word b into the new sentence start (a+b). This result is combined with the next word c into the new sentence start ((a + b) +c), etc. In short, a sentence start is always combined with a next word into a new sentence start until no new next word is available in the input.

The left-associative derivation order is linear in the sense that it regularly adds one word after the next, and it is time-linear because the direction of growth corresponds to the direction of time (cf. 5.4.3). Thus, the left-associative derivation order captures the basic structure of natural language in accordance with de Saussure's second law.

Besides the regular left- and the right-associative bracketing structure there is a multitude of irregular structures, e.g.,

```
 (((a + b) + (c +d)) + e)
 ((a + b) + ((c +d)) + e)
 (a + ((b + c)) + (d + e))
 ((a + (b + c)) + (d + e))
 (((a + b) + c) + (d +e))
         · · ·
```

The number of these irregular bracketings grows exponentially with the length of the string and is infinite, if bracketings like (a), ((a)), (((a))), etc., are permitted.

It is these irregular bracketing structures (and corresponding trees) which C- and PS-grammar generate via the principle of *possible substitutions*. From the large number of possible trees for a terminal chain there follows the central task of linguistic description in C- and PS-grammar, namely to motivate the 'correct' bracketing structure and the 'correct' phrase structure tree as a constituent structure analysis.

LA-grammar, on the other hand, is based on the principle of *possible continuations*, which is formally reflected in the regular left-associative bracketing structure (cf. 10.1.1) and corresponding trees (cf. 10.1.6).

10.1.2 THE PRINCIPLE OF POSSIBLE CONTINUATIONS

Beginning with the first word of the sentence, the grammar describes the possible continuations for each sentence start by specifying the rules which may perform the next grammatical composition (i.e., add the next word).

The time-linear derivation order and the structural characterization of possible continuations is formalized in the specific rule schema of LA-grammar.

10.1.3 SCHEMA OF LEFT-ASSOCIATIVE RULE IN LA-GRAMMAR

$$r_i\text{: cat}_1 \text{ cat}_2 \Rightarrow \text{cat}_3 \text{ rp}_i$$

The rule consists of the name r_i, the category patterns cat_1, cat_2, and cat_3, and the rule package rp_i. The category patterns define a categorial operation which maps a sentence start ss (matched by cat_1) and a next word nw (matched by cat_2) into a new sentence start ss' (characterized by cat_3). The output of a successful application of rule r_i is a *state* defined as an ordered pair (ss' rp_i).

In the next combination, the rules of the rule package rp_i are applied to ss' and a new next word. Because an LA-grammatical analysis starts with the initial words of the input its conceptual derivation order is *bottom-up left-associative* .

To facilitate comparison, the schemata of C- and PS-grammar are restated below.

10.1.4 SCHEMA OF A CANCELING RULE IN C-GRAMMAR

$$\alpha_{(Y|X)} \circ \beta_{(Y)} \Rightarrow \alpha\beta_{(X)}$$

This rule schema combines α and β into $\alpha\beta$ by canceling the Y in the category of α with the corresponding category of β. The result is a tree structure in which $\alpha\beta$ of category (X) dominates α and β. The conceptual derivation order of categorial canceling rules is *bottom-up amalgamating*.

10.1.5 SCHEMA OF A REWRITE RULE IN PS-GRAMMAR

$$A \rightarrow B\,C$$

Replacing the sign A by B and C corresponds to a tree structure in which B and C are immediately dominated by A. The conceptual derivation order is *top-down expanding*.

The characteristic derivation orders following from the different rule schemata have the following schematic representation.

10.1.6 THREE CONCEPTUAL DERIVATION ORDERS

LA-grammar   C-grammar   PS-grammar

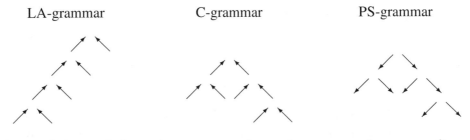

bot.-up left-associative   bottom-up amalgamating   top-down expanding

C- and PS-grammar, being based alike on the principle of possible substitutions, differ only in the direction of their conceptual[2] derivation order. In C-grammar, two categorized expressions are substituted by one categorized expression (bottom-up). In PS-Grammar, one sign is substituted by one, two, or more signs (top-down).

The principle of possible substitutions results in an irregular derivation structure which is reflected in irregular phrase structure trees. From the viewpoint of the SLIM theory of language, a substitution based derivation order is in conflict with the communicative function of language (no input-output equivalence between the grammar and the speaker-hearer) and an obstacle to automatic language analysis (no type transparency between the grammar and the parser).

The principle of possible continuations results in a regular derivation order. The associated tree structure of LA-grammar may seem linguistically uninteresting from the viewpoint of constituent-structure-based C- and PS-grammar. From the viewpoint of the SLIM theory of language, however, it (i) models the fundamental time-linear structure of natural language, (ii) allows input-output equivalence between the grammar and the speaker-hearer, and (iii) results in type transparency between the grammar and the parser.

It follows that the formal notion of constituent structure, defined in terms of phrase structure trees (cf. 8.4.3), has no place in LA-grammar. In the intuitive discussion of natural language examples, however, this notion may be used informally in the more general sense of a *complex grammatical unit*. In this sense, it is an old concept of classical grammar known as a *syntagma*. For example, when we say that in the English declarative sentence

The little dog found a bone.

the verb found is in 'second position,' then we do not mean the position of the second word in the sentence, but rather the position after the first grammatical unit, syntagma, or constituent in a nontechnical sense. The point is that in LA-grammar the positioning of, e.g., the verb in a certain location in the sentence is not based on some substitution or movement, but rather on computing the possible continuations in a strictly time-linear derivation.

10.2 Formalism of LA-grammar

In the following algebraic definition of LA-grammar, we identify positive integers with sets, i.e., $n = \{i \mid 0 \le i < n\}$, for convenience.

---

[2] In C- and PS-grammar, the conceptual derivation order is distinct from the procedural derivation order. For example, the different parsers for context-free PS-grammar use many different derivation orders, some working from left to right,some from right to left, some beginning in the middle (island parsers), some using a left-corner, others a right-corner order, etc. These alternative orders are intended to improve efficiency by utilizing special properties of the language to be analyzed, and belong to the procedural aspect of the respective systems.

10.2.1 ALGEBRAIC DEFINITION OF LA-GRAMMAR

A left-associative grammar (or LA-grammar) is defined as a 7-tuple $<$W, C, LX, CO, RP, ST_S, $ST_F >$, where

1. W is a finite set of *word surfaces*.
2. C is a finite set of *category segments*.
3. LX $\subset$ (W $\times$ C^+) is a finite set comprising the *lexicon*.
4. CO = $(co_0 ... co_{n-1})$ is a finite sequence of total recursive functions from (C^* $\times$ C^+) into C^* $\cup$ $\{\perp\}$,[3] called *categorial operations*.
5. RP = $(rp_0 ... rp_{n-1})$ is an equally long sequence of subsets of n, called *rule packages*.
6. ST_S = $\{(cat_s\ rp_s), ...\}$ is a finite set of *initial states*, whereby each rp_s is a subset of n called start rule package and each $cat_s\ \epsilon\ C^+$.
7. ST_F = $\{(\ cat_f\ rp_f), ...\}$ is a finite set of *final states*, whereby each $cat_f\ \epsilon\ C^*$ and each $rp_f\ \epsilon$ RP.

A concrete LA-grammar is specified by
(i) a lexicon LX (cf. 3),
(ii) a set of initial states ST_S (cf. 6),
(iii) a sequence of rules r_i, each defined as an ordered pair (co_i, rp_i), and
(iv) a set of final states ST_F.

A left-associative rule r_i takes a sentence start *ss* and a next word *nw* as input and tries to apply the categorial operation co_i. If the categories of the input match the patterns of cat_1 and cat_2, the application of rule r_i is successful and an output is produced. The output consists of a pair $(ss'\ rp_i)$, whereby ss' is a resulting sentence start and rp_i is a rule package. If the input does not match the patterns of cat_1 and cat_2, then the application of rule r_i is not successful, and no output is produced.

The rule package rp_i contains all rules which can be applied after rule r_i was successful. A rule package is defined as a set of rule names, whereby the name of a rule is the place number g of its categorial operation co_g in the sequence CO. In practice, the rules are called by more mnemonic names, such as 'rule-g' or 'Fverb+main.'

After a successful rule application, the algorithm fetches a new next word (if present) from the input and applies the rules of the current rule package to the (once new) sentence start and the next word. In this way LA-grammar works from left to right through the input, pursuing alternative continuations in parallel. The derivation stops if there is either no grammatical continuation at a certain point (ungrammatical input, e.g. 10.5.5) or if there is no further next word (complete analysis, e.g. 10.5.3).

---

[3] For theoretical reasons, the categorial operations are defined as total functions. In practice, the categorial operations are defined as easily-recognizable subsets of (C^* $\times$ C^+), where anything outside these subsets is mapped into the arbitrary *"don't care"* value $\{\perp\}$, making the categorial operations total functions.

The general format of LA-grammars is illustrated in 10.2.2 with the context-free language $a^k b^k$, previously described within the frameworks of C-grammar (cf. 7.4.3) and PS-grammar (cf. 7.1.3).

10.2.2 LA-GRAMMAR FOR $a^k b^k$

$LX =_{def} \{[a\ (a)], [b\ (b)]\}$
$ST_S =_{def} \{[(a)\ \{r_1, r_2\}]\}$
$r_1: (X)\quad (a)\ \Rightarrow (aX)\quad \{r_1, r_2\}$
$r_2: (aX)\ (b)\ \Rightarrow (X)\quad \{r_2\}$
$ST_F =_{def} \{[\varepsilon\ rp_2]\}.$

The lexicon LX contains two words, a and b. Each word is an ordered pair, consisting of a surface and a category. The categories, defined as lists of category segments, contain here only a single segment[4] which happens to equal the respective surface.

The initial state ST_S specifies that the first word must be of category (a), i.e., it must be an a. Furthermore, the rules to be applied initially must be r_1 and r_2. Thus, all rules,[5] but not all words, may be used at the beginning of a sentence.

The categorial patterns are based on the sequence variable X, representing zero or more category segments, and the segment constants a and b. Rule r_1 accepts a sentence start of any category (represented by the pattern (X)), and a next word of category (a), i.e. a word a. The result of the categorial operation is expressed by the pattern (a X): an a-segment is added at the beginning of the sentence start category.

Rule r_2 accepts a sentence start the category of which begins with an a-segment (represented by the pattern (aX)) and a next word of category (b), i.e. a word b. The result of the categorial operation is expressed by the pattern (X). It means that an a-segment is subtracted from the beginning of the sentence start category.

The rule package of r_1, called rp_1, contains r_1 and r_2. As long as the next word is an a, r_1 is successful, while r_2 fails because it requires a b as the next word. As soon as the first b is reached in the input, r_2 is successful, while r_1 fails because it requires an a as the next word. The rule package rp_2 contains only one rule, namely r_2. Therefore, once the first b has been added, only b is acceptable as a next word.

An analysis is complete when all the a-segments in the sentence start category have been canceled by b-segments. In other word, the analysis ends after an application of r_2 with an empty sentence start category. This is specified by the final state ST_F of 10.2.2, which requires ε (i.e. the empty sequence) as the final result category.

The following LA-grammar generates the context-sensitive language $a^k b^k c^k$.

10.2.3 LA-GRAMMAR FOR $a^k b^k c^k$

$LX =_{def} \{[a\ (a)], [b\ (b)], [c\ (c)]\}$
$ST_S =_{def} \{[(a)\ \{r_1, r_2\}]\}$
$r_1: (X)\quad (a)\ \Rightarrow (aX)\quad \{r_1, r_2\}$

r$_2$: (aX) (b) ⇒ (Xb) {r$_2$, r$_3$}
r$_3$: (bX) (c) ⇒ (X) {r$_3$}
ST$_F$ =$_{def}$ {[ε rp$_3$]}.

Compared to the corresponding PS-grammar 8.3.7, this LA-grammar is surprisingly simple. Furthermore, the LA-grammars 10.2.2 for context-free $a^k b^k$ and 10.2.3 for context-sensitive $a^k b^k c^k$ resemble each other closely.

In LA-grammar, the relation between the rules and their rule packages defines a finite state transition network (FSN). For example, the FSN underlying the LA-grammar 10.2.3 for $a^k b^k c^k$ has the following form.[6]

10.2.4 THE FINITE STATE BACKBONE OF THE LA-GRAMMAR FOR $a^k b^k c^k$

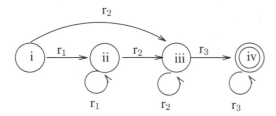

This FSN consists of four states, represented as the circles i – iv. Each state is defined as an ordered pair consisting of a category pattern and a rule package. State i corresponds to the start state in ST$_S$, while the states ii, iii, and iv correspond to the output of rules r$_1$, r$_2$, and r$_3$. State iv has a second circle, indicating that it is a possible final state (cf. definition of ST$_F$ in 10.2.3).

The application of a left-associative rule to an input pair, consisting of a sentence start and a next word, results in a transition. In 10.2.4, the transitions are represented graphically as arrows, annotated with the name of the associated rule. The transitions leading into a state represent the categorial operation of the rule associated with this state. The transitions leading out of a state represent the rule package of the rule common to all the transitions leading into the state.

For example, the transitions leading out of state i are different (namely r$_1$ and r$_2$), corresponding to the rule package of the start state in ST$_S$. The transitions leading into state ii are all the same, representing applications of r$_1$ from different preceding states. Correspondingly, the transitions leading out of state ii are different, corresponding to the rule package of r$_1$. The transitions leading into state iii are all the same, representing applications of r$_2$ from three different preceding states, etc.

An LA-grammar analyzing an input or generating an output navigates through its FSN, from one state to the next, following the transition arrows. Thereby, the generative capacity of LA-grammar resides in the categorial operations of its rules. While

---

[4] In LA-grammars for more demanding languages, the lexical categories consist of several segments.
[5] In more demanding languages, the initial state specifies a proper subset of the grammar rules.
[6] For a more detailed description see CoL, Section 8.2.

the FSN algorithm alone generates only the regular languages, the algorithm of LA-grammar generates exactly the class of recursive languages (cf. Section 11.1).

This enormous generative capacity may be used sparingly, however. For example, the LA-grammars for $a^k b^k$ (cf. 10.2.2) and $a^k b^k c^k$ (cf. 10.2.3) are of the lowest possible complexity: they both parse in linear time.[7]

Thus, there is no border line between the context-free and the context-sensitive languages in LA-grammar. This is a first indication that the formalism of LA-grammar arrives at a hierarchy of language classes different from the formalism of PS-grammar.

10.3 Time-linear analysis

The tree structures of LA-grammar may be displayed equivalently as structured lists.

10.3.1 LA-TREES AS STRUCTURED LISTS

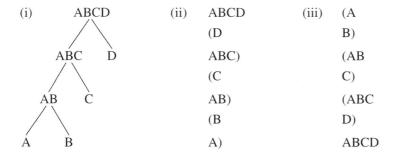

In (i), a left-associative derivation is shown in the form of a tree (cf. 10.1.6). In (ii), the same derivation is shown in the equivalent format of a structured list, whereby sentence start, next word, and resulting sentence start each have their own line. Formats (ii) and (iii) differ only in the direction of a time-linear reading: in (ii) it begins at bottom – in analogy to the tree structure (i) – and in (iii) at the top.

For displaying derivations on the screen, structure (iii) is suited best. This format is shown in 10.3.2, based on the LA-grammar 10.2.2, analyzing the expression aaabbb.

10.3.2 LA-GRAMMAR DERIVATION OF $a^k b^k$ FOR $k = 3$

```
NEWCAT>  a a a b b b

     *START-0
   1
        (A) A
        (A) A
```

---

[7] Because their respective rules have incompatible input conditions (they each take different next words), these two LA-grammars are unambiguous. Unambiguous LA-grammars of this kind are called C1-LAGs and parse in linear time (cf. Chapter 11) .

```
*RULE-1
2
    (A A)  A  A
    (A)  A
*RULE-1
3
    (A A A)  A  A  A
    (B)  B
*RULE-2
4
    (A A)  A  A  A  B
    (B)  B
*RULE-2
5
    (A)  A  A  A  B  B
    (B)  B
*RULE-2
6
    (NIL)  A  A  A  B  B  B
```

This LA-grammar analysis was generated directly as the formatted protocol (trace) of the LA-parser NEWCAT. Because of the absolute type transparency of LA-grammar, the declarative linguistic analysis and the derivation procedure on the computer are merely different realizations of the same left-associative algorithm.

In 10.3.2, the input to be analyzed is typed in after the prompt NEWCAT>. To obtain a better comparison of the input categories, surface and category are printed in inverse order, e.g. (A) A instead of [A (A)].

The first section in 10.3.2 begins with the name of the active rule package *START-0. Then follows the composition number 1, the sentence start (consisting of the first word), the next (i.e. second) word, and finally the name of the rule (RULE-1) which combines the sentence start and the next word. After the next composition number follows the result of the (still) current composition:

```
active rule package:         *START-0
composition number:          1
sentence start:              (A)  A
next word:                   (A)  A
successful rule:             *RULE-1
next composition number:     2
result:                      (A A)  A  A
```

The rule name (cf. 'successful rule') and the resulting sentence start (cf 'result') each have a double function. The rule name simultaneously specifies (i) the rule which is successful in composition n and (ii) the rule package, the rules of which are applied in the next composition $n + 1$. Correspondingly, the result simultaneously represents (i) the output of composition n and (ii) the sentence start of composition $n + 1$.

These double functions in a left-associative derivation are clearly visible in the second composition of 10.3.2:

```
active rule package:         *RULE-1
composition number:          2
```

```
sentence start :              (A A) A A
next word:                    (A) A
successful rule :             *RULE-1
next composition number:      3
result:                       (A A A) A A A
```

The form of the derivation in 10.3.2 is designed to characterize the categorial operations of the rules. Because it only represents successful continuations it constitutes a *depth first* format in LA-grammar. In addition there is also a *breadth first* format, which is used in, e.g., the morphological analysis of word forms (cf. 14.4.2 and 14.4.3). These output formats all have in common that they are protocols of the parser and analyses of the grammar at the same time.

10.4 Absolute type transparency of LA-grammar

The algorithm of LA-grammar is equally suited for parsing and generation. The structural reason for this is the principle of possible continuations. In parsing the next word is provided by the input, while in generation the next word is chosen from the lexicon.

The close relation between parsing and generation is illustrated with the following traces 10.4.1 and 10.4.2. They are based on the same LA-grammar 10.2.3 for the context-sensitive language $a^k b^k c^k$.

An LA-parser loads an arbitrary LA-grammar of the class of C-LAGs. Then arbitrary expressions may be put in to be analyzed. The NEWCAT parser includes an automatic rule counter which at the beginning of each derivation shows the rules attempted in each left-associative composition.

10.4.1 PARSING **aaabbbccc** WITH ACTIVE RULE COUNTER

```
NEWCAT>  a a a b b b c c c
;  1: Applying rules (RULE-1 RULE-2)
;  2: Applying rules (RULE-1 RULE-2)
;  3: Applying rules (RULE-1 RULE-2)
;  4: Applying rules (RULE-2 RULE-3)
;  5: Applying rules (RULE-2 RULE-3)
;  6: Applying rules (RULE-2 RULE-3)
;  7: Applying rules (RULE-3)
;  8: Applying rules (RULE-3)
; Number of rule applications: 14.

    *START-0
    1
        (A) A
        (A) A
    *RULE-1
    2
        (A A) A A
        (A) A
    *RULE-1
    3
```

```
     (A A A) A A A
     (B) B
*RULE-2
4
     (A A B) A A A B
     (B) B
*RULE-2
5
     (A B B) A A A B B
     (B) B
*RULE-2
6
     (B B B) A A A B B B
     (C) C
*RULE-3
7
     (C C) A A A B B B C
     (C) C
*RULE-3
8
     (C) A A A B B B C C
     (C) C
*RULE-3
9
     (NIL) A A A B B B C C C
```

The number of rule applications attempted here is below 2n.

An LA-generator loads likewise an arbitrary LA-grammar of the class of C-LAGs. Then the function 'gram-gen' is called with two arguments: the *recursion factor*[8] of the grammar and a list of words to be used in the generation. In this way, the generation procedure is limited to a certain set of words and to a certain length. The grammatical analysis of generation is also a direct trace, this time of the LA-generator.

10.4.2 GENERATING A REPRESENTATIVE SAMPLE IN $a^k b^k c^k$

```
NEWCAT> (gram-gen 3 '(a b c))

Parses of length 2:
 A B
   2    (B)
 A A
   1    (A A)

Parses of length 3:
 A B C
   2 3    (NIL)
 A A B
   1 2    (A B)
 A A A
   1 1    (A A A)

Parses of length 4:
```

---

[8] CoL, p. 193 ff. In another version, 'gram-gen' is called with the maximal surface length instead of the recursion factor.

```
A A B B
  1 2 2    (B B)
A A A B
  1 1 2    (A A B)
A A A A
  1 1 1    (A A A A)

Parses of length 5:
A A B B C
  1 2 2 3    (B)
A A A B B
  1 1 2 2    (A B B)
A A A A B
  1 1 1 2    (A A A B)

Parses of length 6:
A A B B C C
  1 2 2 3 3    (NIL)
A A A B B B
  1 1 2 2 2    (B B B)
A A A A B B
  1 1 1 2 2    (A A B B)

Parses of length 7:
A A A B B B C
  1 1 2 2 2 3    (B B)
A A A A B B B
  1 1 1 2 2 2    (A B B B)

Parses of length 8:
A A A B B B C C
  1 1 2 2 2 3 3    (C)
A A A A B B B B
  1 1 1 2 2 2 2    (B B B B)

Parses of length 9:
A A A B B B C C C
  1 1 2 2 2 3 3 3    (NIL)
A A A A B B B B C
  1 1 1 2 2 2 2 3    (B B B)

Parses of length 10:
A A A A B B B B C C
  1 1 1 2 2 2 2 3 3    (B B)

Parses of length 11:
A A A A B B B B C C C
  1 1 1 2 2 2 2 3 3 3    (B)

Parses of length 12:
A A A A B B B B C C C C
  1 1 1 2 2 2 2 3 3 3 3    (NIL)
```

This systematic generation begins with well-formed but incomplete expressions of length 2 and represents all well-formed intermediate expressions up to length 12. Complete expressions of the language are recognizable by their result category (NIL).

Each derivation consists of a surface, a sequence of rule names represented by numbers, and a result category. A single derivation is illustrated in 10.4.3.

10.4.3 COMPLETE WELL-FORMED EXPRESSION IN $a^k b^k c^k$

```
A A A B B B C C C
 1 1 2 2 2 3 3 3    (NIL)
```

The surface and the rule name sequence are arranged in such a way that it is apparent which word is added by which rule. The derivation 10.4.3 characterizes a well-formed expression because it corresponds to the final state $(\varepsilon\ rp_3)$, here written as '3 (NIL),' i.e. an element of the set ST_F of the LA-grammar for $a^k b^k c^k$ defined in 10.2.3.

The relation between the LA-grammar defined in 10.2.3, the LA-parser illustrated in 10.4.1, and the LA-generator illustrated in 10.4.2 demonstrates the notion of absolute type transparency (cf. 9.3.1) between a grammar formalism, a parser, and a generator in practice. So far, LA-grammar is the only grammar formalism which achieves absolute type transparency as described by Berwick & Weinberg 1984, p. 41.

10.5 LA-grammar for natural language

Before we turn in the next chapter to the formal properties of LA-grammar (such as language classes, generative capacity, and complexity), let us illustrate the application of LA-grammar to natural language. The purpose is to show the linguistic motivation of LA-grammar analyses in terms of the traditional notions of valency, agreement, and word order on the one hand, and a time-linear derivation on the other.

The formalism of LA-grammar originated in the attempt to implement the C-grammar defined in SCG as a parser (cf. NEWCAT, p. 7). The intuitive relation between C- and LA-grammar is explained in the following comparison of 10.5.1 and 10.5.2.

10.5.1 CONSTITUENT STRUCTURE ANALYSIS IN C-GRAMMAR

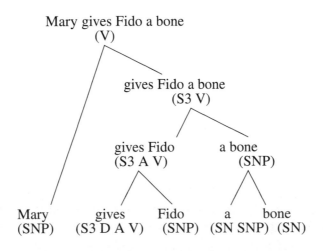

This tree satisfies condition 8.4.3 for constituent structures and could be used in PS- as well as in C-grammar. However, because of its complex categories and the concomitant bias towards a bottom-up derivation it is closer to a C-grammar analysis.

Compared to the algebraic definition of C-grammar 7.4.1, the categories used in 10.5.1 are of an especially simple variety, however. They consist of *lists of category segments* – and not of categorial functor-argument structures based on slashes and multiple parentheses as in traditional C-grammar.

The simplified categories are sufficient to encode the relevant linguistic valency properties. For example, the category (S3 D A V) of gives indicates a verb (V) which takes a nominative of third person singular (S3), a dative (D), and an accusative (A) as arguments. Correspondingly, the category (SN SNP) of the determiner a indicates a functor which takes a singular noun SN to make a singular noun phrase SNP.

The C-grammatical derivation indicated in 10.5.1 may begin by combining gives and Fido, whereby the SNP segment in the category of Fido cancels the D segment in the category of gives, resulting in the intermediate expression gives Fido of category (S3 A V). Next the determiner a and the noun bone must be combined, resulting in the intermediate expression a bone of category (SNP). This cancels the A segment in the category of gives Fido, resulting in the intermediate expression gives Fido a bone of category (S3 V). Finally, the segment SNP in the category of Mary cancels the segment S3 in the category of gives Fido a bone, resulting in a complete sentence.

The lexical categories of the C-grammatical analysis 10.5.1 may be reused in the corresponding left-associative analysis 10.5.2 because their list structure happens to be in concord with the algebraic definition of LA-grammar in 10.2.1.

10.5.2 TIME-LINEAR ANALYSIS IN LA-GRAMMAR

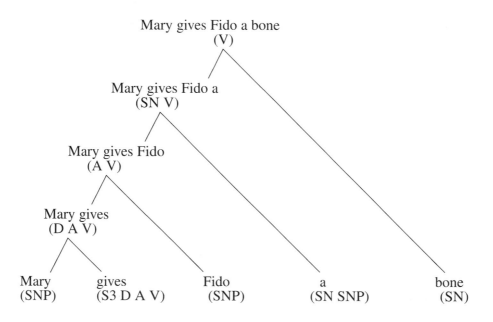

This analysis is based on *LA-E2*, an LA-grammar defined in 17.4.1.

In 10.5.2, the same valency positions are canceled by the same fillers as in 10.5.1. However, the left-associative derivation always combines a sentence start with a next word into a new sentence start, beginning with the first word of the sentence.

The initial combination of Mary and gives is based on the categorial operation

(SNP) (N D A V) $\Rightarrow$ (D A V).

The category segment SNP (for singular noun phrase) of Mary cancels the first segment N (for nominative) of the category of gives. Linguistically speaking, the nominative valency of the verb is filled here by the category of the first word. The agreement between the segments SNP and N is defined in the variable definition of 17.4.1.

The result of this first combination is a sentence start of the category (D A V), indicating an intermediate expression which still needs a D (dative) and an A (accusative) to become a complete sentence. This new sentence start combines with the next word Fido of category (SNP), whereby the category segment SNP cancels the first segment of the sentence start category (D A V).

The result of the second combination is a sentence start of the category (A V), indicating an intermediate expression which still needs an A (accusative) in order to become complete. This new sentence start combines with the next word, i.e. the determiner a of category (SN SNP). The categorial operation of the rule used in this combination has the form

(A V) (SN SNP) $\Rightarrow$ (SN V).

Thus, the result segment SNP of the determiner fills the valency position of the accusative in the category of the sentence start while the argument position SN of the determiner is added to the sentence start category opening a new valency position for a singular noun.

In the next composition of the sentence start Mary gives Fido a and the next word bone, this new valency position is canceled by the category (SN) of the next word. The result is a sentence start of the category (V), for verb. The absence of unfilled valency positions indicates that the expression is now complete. This completeness is only potential, however, because natural language expressions can always be continued – here for example by regularly or because she is so fond of this cute little dog which she picked up in Denver visiting her mother who told her while driving to the cleaners that the Millers had recently moved to Detroit because . . . , etc.

The time-linear analysis 10.5.2 captures the same linguistic intuitions on valency, agreement, and word order as the constituent structure analysis 10.5.1 – which is shown by 10.5.1 and 10.5.2 using the same lexical categories. The difference is that 10.5.1 is based on the principle of possible substitutions while 10.5.2 is based on the principle of possible continuations.

In a constituent structure analysis like 10.5.1, intermediate expressions such as Mary gives or Mary gives Fido a would be considered illegal because they are not supported by the substitution and movement tests on which constituent structure analysis is based. In the left-associative derivation 10.5.2, on the other hand, they are legitimate intermediate expressions because they may be *continued* into a complete well-formed sentence.

Conversely, expressions like **gives Fido** or **gives Fido a bone** will not occur as intermediate expressions in LA-grammar because they cannot be continued into complete well-formed sentences. In C- and PS-grammar, on the other hand, such expressions are used as intermediate expressions because they are supported by the substitution and movement test which motivate C- and PS-grammar intuitively.

The LA-grammar underlying derivation 10.5.2 can be used by a parser directly: the parser reads the sentence in word form by word form, each time applying the rules of the LA-grammar, always combining a sentence start with a next word into a new sentence start. During the parsing procedure the evolving trace of the parse may be displayed directly as the grammatical analysis.

The following automatic analysis illustrates the structured list format used in NEW-CAT and CoL, already familiar from the artificial language analyses in 10.3.2 and 10.4.1 above.

10.5.3 LEFT-ASSOCIATIVE PARSING OF EXAMPLE 10.5.2

```
NEWCAT>  Mary gives Fido a bone \.

    *START
    1
       (SNP) MARY
       (S3 D A V) GIVES
    *NOM+FVERB
    2
       (D A V) MARY GIVES
       (SNP) FIDO
    *FVERB+MAIN
    3
       (A V) MARY GIVES FIDO
       (SN SNP) A
    *FVERB+MAIN
    4
       (SN V) MARY GIVES FIDO A
       (SN) BONE
    *DET+NOUN
    5
       (V) MARY GIVES FIDO A BONE
       (V DECL) .
    *CMPLT
    6
       (DECL) MARY GIVES FIDO A BONE .
```

The final combination adds a full stop which characterizes the expression as a declarative sentence (cf. *LA-E3* defined in 17.5.5).

The parsing analysis 10.5.3 contains not only all the information of the time-linear tree format 10.5.2, but in addition specifies for each left-associative combination the name of the rule involved – which indicates the name of the currently active rule package at the same time (cf. Section 10.3). The structural difference between the linear tree 10.5.2 and the structured list 10.5.3 corresponds to the difference between the equivalent formats (i) and (iii) in 10.3.1.

The next example illustrates the LA-grammar handling of a discontinuous element.

10.5.4 ANALYSIS OF A DISCONTINUOUS ELEMENT

```
NEWCAT>  Fido dug the bone up \.

    *START
    1
       (SNP) FIDO
       (N A UP V) DUG
    *NOM+FVERB
    2
       (A UP V) FIDO DUG
       (SN SNP) THE
    *FVERB+MAIN
    3
       (SN UP V) FIDO DUG THE
       (SN) BONE
    *DET+NOUN
    4
       (UP V) FIDO DUG THE BONE
       (UP) UP
    *FVERB+MAIN
    5
       (V) FIDO DUG THE BONE UP
       (V DECL) .
    *CMPLT
    6
       (DECL) FIDO DUG THE BONE UP .
```

The relation between the discontinuous elements dug and up is represented by the segment UP in the category (N A UP V) of dug. The final position of up is specified over the order of filler positions in the functor category (here N, A, and UP).

The handling of discontinuous elements is a problem only for constituent structure analysis.In contrast, a grammar theory motivated by the basic time-linear structure of natural language and its valency, agreement, and word order properties may treat discontinuous structures in a standard way by coding a certain filler position into the relevant functor category and canceling this position at a later point.

In conclusion consider the handling of an ungrammatical sentence.

10.5.5 LA-ANALYSIS OF UNGRAMMATICAL INPUT

```
NEWCAT>  the young girl give Fido the bone \.

ERROR
Ungrammatical continuation at: "GIVE"
```

```
*START
1
    (SN SNP) THE
    (ADJ) YOUNG
*DET+ADJ
2
    (SN SNP) THE YOUNG
    (SN) GIRL
  *DET+NOUN
3
    (SNP) THE YOUNG GIRL
```

The derivation begins normally, but breaks off after the third word because the grammar does not provide a possible continuation for the current sentence start the young girl and the next word give. The reason is that the category segment SNP does not agree with the nominative segment NOM of give. This is specified in the variable definition of *LA-E2* in 17.4.1.

LA-parsers analyze the grammatical beginning of ungrammatical input as far possible, giving a precise grammatical description of the sentence start. This is of practical use for both, debugging and upscaling of a given LA-grammar. In debugging the break off point and the grammatical structure of the sentence start tell us exactly which rule of the grammar should have fired, what its input categories happen to be, etc. Conversely, if the system accepts input which is ungrammatical (error arising in negative testing), the LA-analysis gives an exact grammatical description of the location where the break off should have been.

In upscaling the new construction may be built step by step from a working sentence start and then led back into the old continuation system. This characteristic property of LA-grammar follows from its time-linear derivation structure and will be used extensively in Part III for developing larger and larger fragments of English (*LA-E1* to *LA-E3* in Chapter 17 and *LA-E3* in Section 23.4) and German (*LA-D1* to *LA-D4* in Chapter 18).

Exercises

Section 10.1

1. What is meant by a 'left-associative grouping of operands?' What other groupings of operands are possible?

2. Which property of natural language is formally modeled by a left-associative derivation order?

3. Compare the principle of possible continuations to the principle of possible substitutions. How do these different principles relate to the intuitive notions of time linearity and constituent structure? Which elementary formalisms are based on which principle?

4. Explain the relation between different rule formats and conceptual derivation orders in C-, PS-, and LA-grammar.

5. What is a syntagma?

Section 10.2

1. Explain the algebraic definition of LA-grammar.

2. Which part of a left-associative rule is used for pattern matching?

3. Which part of a left-associative rule participates in the definition of a finite state transition network?

4. Why do the transitions going into a state all represent the same rule, while the transitions going out of a state all represent different rules?

5. Restate the LA-grammar for $a^k b^k$, explain how it works, and compare it with the corresponding C-grammar 7.4.2 and PS-grammar 7.1.3.

6. Define an LA-grammar for $a^{2k} b^k$.

7. Define an LA-grammar for $a^k b^{2k}$.

8. What are the conditions for a left-associate derivation to terminate?

Section 10.3

1. What is the relation between left-associative tree structures and structured lists? Which format is suited best for computational analysis, and why?

2. Describe the numbered sections in the output of a NEWCAT parse and explain their double function.

3. Why is it remarkable that the LA-grammar for $a^k b^k c^k$ resembles that for $a^k b^k$ so closely? Base your answer on the notions PS-hierarchy, language class, complexity, and parsing algorithm.

4. Define LA-grammars for $\{a^k b^k c^k\}^+$, $a^k b^k c^k d^k$, and $a^k b^k c^k d^k e^k f^k$. What is the PS-grammar language class of these languages? What is their LA-grammar complexity?

Section 10.4

1. What is the relation between an LA-parser, an LA-grammar, and an LA-generator?

2. Why does it make sense to specify the maximal length and the specific words to be used when starting an LA-generator? Which of these restrictions is especially useful for artificial languages, and which for natural languages?

3. Explain how the start of a PS-grammar derivation differs from that in LA-grammar. Does the start symbol of PS-grammar have a counterpart in LA-grammar?

4. Do you see a connection between the formal LA-analysis 10.4.1 and the schema of language understanding in 5.4.1?

5. Do you see a connection between the formal LA-generation 10.4.2 and the schema of language production 5.4.2?

6. Explain the exact properties of type transparency in LA-grammar.

Section 10.5

1. What are the similarities and differences between the C- and the LA-grammar analysis of natural language?

2. How do intermediate expressions differ in a constituent structure analysis and a left-associative analysis of natural language?

3. For which reason is the handling of discontinuous elements a problem for an analysis based on constituent structure, but not for one based on the time-linear structure of natural language?

4. At which point does an LA-parser stop the analysis of an ungrammatical expression?

5. Name three different reasons why an LA-parser may stop the analysis before reaching the end of the input.

11. Hierarchy of LA-grammar

In this chapter, different types of LA-grammar are defined via natural restrictions of its rule system. Then these grammar types are characterized in terms of their generative capacity and computational complexity.

Section 11.1 shows that LA-grammar in the basic, unrestricted form of its algebraic definition generates exactly the recursive languages. Section 11.2 describes possible restrictions on LA-grammar and defines the hierarchy of A-, B-, and C-LAGs. Section 11.3 describes the origin of nonrecursive and recursive ambiguities and their impact on the number of rule applications. Section 11.4 compares the notions of primitive operation used for determining complexity in context-free PS-grammar and in C-LAGs, and explains the relation of the C-LAGs to the automata-theoretic concepts of deterministic and nondeterministic automata. Section 11.5 defines the sub-hierarchy of linear C1-, polynomial C2-, and exponential C3-LAGs within the class of C-LAGs, which differ solely in their degrees of ambiguity.

11.1 Generative capacity of unrestricted LAG

The generative capacity of a grammar formalism follows from its algebraic definition. It specifies the form of the rules which in turn determines which structures can be handled by the formalism and which cannot.

According to the algebraic definition of LA-grammar (cf. 10.2.1), a rule r_i consists of two parts, a categorial operation co_i and the associated rule package rp_i. The categorial operation co_i is defined as a total recursive function.

A function is recursive, if its assignment can be described as a mechanical logical algorithm. A function is total, if each element in the domain is assigned a value in the range. In a partial function, on the other hand, the assignment is undefined for parts of the domain.

Intuitively, the class of total recursive functions represents those structures which may be computed explicitly and completely. Formally, the total recursive functions characterize the class of recursive languages.

A language is recursive if and only if there is an algorithm (i.e. a total recursive function) which can decide for arbitrary input in finitely many steps whether or not the input belongs to the language. The class of recursive languages comprises all languages which are *decidable*. Thus, the recursive languages constitute the largest class the elements of which have a completely specifiable structure.[1]

LA-grammar in its basic, unrestricted form (cf. definition 10.2.1) has the following generative capacity:

11.1.1 GENERATIVE CAPACITY OF UNRESTRICTED LA-GRAMMAR

Unrestricted LA-grammar accepts and generates all and only the recursive languages.

The two parts of this statement are formulated and proven as theorems.[2]

11.1.2 THEOREM 1

Unrestricted LA-grammar accepts and generates *only* the recursive languages.

Proof: Assume an input string of finite length n. Each word in the input string has a finite number of readings (> 0).

Combination step 1: The finite set of start states ST_S and all readings of the first word w_1 result in a finite set of well-formed expressions $WE_1 = \{(ss' \; rp_S) \mid ss' \; \epsilon \; (W^+ \times C^+)\}$.

Combination step n: Combination step k-1, $k > 1$, has produced a finite set of well-formed expressions $WE_k = \{(ss' \; rp_i) \mid i \; \epsilon \; n, \; ss' \; \epsilon \; (W^+ \times C^*)$ and the surface of each ss' has length $k\}$. The next word w_{k+1} has a finite number of readings.

Therefore, the Cartesian product of all elements of WE_k and all readings of the current next word will be a finite set of pairs. Each pair is associated with a rule package containing a finite set of rules. Therefore, combination step k will produce only finitely many new sentence starts. The derivation of this finite set of new sentence starts is decidable because the categorial operations are defined to be total recursive functions.

Q.E.D.

Because all possible left-associative analyses for any finite input can be derived in finitely many steps each of which is decidable there is no halting problem in LA-grammar and associated parsers. Thus LA-grammar satisfies condition 3 of the generative grammar desiderata 9.5.2.

---

[1] In automata theory, the recursive languages are defined as those of which each expression may be recognized by at least one Turing machine in finitely many steps (*'halts on all inputs'*, cf. J.E. Hopcroft & J.D. Ullman 1979, p. 151). The PS-grammar hierarchy does not provide a formal characterization of the recursive languages – in contrast to the regular, context-free, context-sensitive, and recursively enumerable languages, which have both an automata-theoretic and a PS-grammar definition.

[2] CoL, Theorems 1 & 2, p. 134f.

11.1.3 Theorem 2

Unrestricted LA-grammar accepts and generates *all* recursive languages.

Proof:[3] Let L be a recursive language with the alphabet W. Because L is recursive, there is a total recursive function ϱ: $W^* \rightarrow \{0,1\}$, i.e., the characteristic function of L. Let LAG^L be an LA-grammar defined as follows:

The set of word surfaces of LAG^L is W.

The set of category segments C $=_{def}$ W $\cup$ {0,1}.

For arbitrary e, f ϵ W^+, [e (f)] ϵ LX if and only if e = f.

$LX =_{def}$ {[a (a)], [b (b)], [c (c)], [d (d)], ... }

$ST_S =_{def}$ {[(seg_c) {r_1, r_2}]}, where seg_c ϵ {a, b, c, d, ... }

r_1: (X) (seg_c) $\Rightarrow$ (X seg_c)   {r_1, r_2}

r_2: (X) (seg_c) $\Rightarrow$ ϱ (X seg_c)   { }

$ST_F =_{def}$ {[(1) rp_2]}

After any given combination step, the rule package rp_1 offers two choices: application of r_1 to continue reading the input string, or application of r_2 to test whether the input read so far is a well-formed expression of L. In the latter case, the function ϱ is applied to the concatenation of the input categories, which are identical to the input surfaces. If the result of applying r_2 is [(1) rp_2],[4] the input surface is accepted; if it is [(0) rp_2], it is rejected.

Since the categorial operations of LAG^L can be any total recursive function, LAG^L may be based on ϱ, the characteristic function of L. Therefore, LAG^L accepts and generates any recursive language.

Q.E.D.

In LAG^L, the prefinal combination steps serve only to read the surface into the category. At the final combination step, the complete surface is available in the category and is analyzed in one step by a very complex categorial operation defined as the characteristic function ϱ of the language.

Unrestricted LA-grammars are called A-LAGs because they generate *A*ll recursive languages.

11.1.4 Definition of the class of A-LAGs.

The class of A-LAGs consists of unrestricted LA-grammars and generates *all* recursive languages.

In contrast, unrestricted PS-grammars (cf. 8.1.1) generate the set of recursively enumerable languages. Thus, LA-grammar and PS-grammar in the basic unrestricted form of their respective algebraic definitions are not equivalent.

---

[3] This proof was provided by Dana Scott.

[4] I.e., if ϱ maps the category (X seg_c), representing the surface, into the category (1).

11.2 LA-hierarchy of A-, B-, and C-LAGs

The complexity of a grammar formalism – defined as the upper bound on the number of operations required in the analysis of arbitrary input – depends on the following two parameters.

11.2.1 PARAMETERS OF COMPLEXITY

– The *amount* of computation per rule application required in the worst case.
– The *number* of rule applications relative to the length of the input needed in the worst case.

These parameters are independent of each other and apply in principle to any rule-based grammar formalism.

In LA-grammar, the amount parameter depends solely on the categorial operations co_i, while the number parameter is determined completely by the degree of ambiguity. Thus, there are two obvious main approaches to restricting LA-grammars.

11.2.2 MAIN APPROACHES TO RESTRICTING LA-GRAMMAR

R1: Restrictions on the form of categorial operations in order to limit the maximal amount of computation required by arbitrary rule applications.
R2: Restrictions on the degree of ambiguity in order to limit the maximal number of possible rule applications.

Approach *R1* in turn suggests two subtypes of restrictions.

11.2.3 POSSIBLE RESTRICTIONS ON CATEGORIAL OPERATIONS

R1.1: Specifying upper bounds for the *length* of categories;
R1.2: Specifying restrictions on *patterns* used in the definition of categorial operations.

These different ways of restricting categorial operations result in two subclasses of A-LAGs, called B-LAGs and C-LAGs. The B-LAGs are defined in terms of a *linear* **Bound** on the length of categories relative to the length of the input and correspond to the restriction type *R1.1*.

11.2.4 DEFINITION OF THE CLASS OF B-LAGs.

The class of *bounded* LA-grammars, or B-LAGs, consists of grammars where for any complete well-formed expression E the length of intermediate sentence start categories is bounded by $k \cdot n$, where n is the length of E and k is a constant.

The language class generated by the B-LAGs is equal to the class of context-sensitive languages in the PS-grammar hierarchy. The proof[5] is analogous to 11.1.3 and based on corresponding restrictions on linearly bound automata, using the LA-category of the last combination step as a tape. The B-LAGs are a proper subset of the A-LAGs because $CS \subset REC.$[6]

A subclass of the B-LAGs are the C-LAGs, which are defined in terms of restrictions of type *R1.2*. In C-LAGs, the amount of computation required by individual categorial operations co_i is limited by a *Constant*.

If the categorial operations are defined as arbitrary total recursive functions, they do not indicate how much computation they require. However, if the categorial operations are defined in terms of formal patterns, then constant and nonconstant amounts of computation may be distinguished.

In the following rule schemata, the amount of computation required by the categorial operations is constant, independent of length of the input categories.

11.2.5 RULE SCHEMATA WITH CONSTANT CATEGORIAL OPERATIONS

$$r_i: (seg_1...seg_k \; X) \; cat_2 \Rightarrow cat_3 \; rp_i$$
$$r_i: (X \; seg_1...seg_k) \; cat_2 \Rightarrow cat_3 \; rp_i$$
$$r_i: (seg_1...seg_m \; X \; seg_{m+1}...seg_k) \; cat_2 \Rightarrow cat_3 \; rp_i$$

These schemata have in common that the pattern matching of the categorial operation has to check exactly k segments in the *ss* category. These patterns of categorial operations are constant because their patterns check a category always from the outer ends $seg_1 \ldots seg_k$, disregarding an arbitrarily large sequence in the middle of the category.

Categorial operations without this property do not provide a constant upper bound on the amount of computation required. A rule schema with a nonconstant categorial operation is illustrated below.

11.2.6 RULE SCHEMA WITH NONCONSTANT CATEGORIAL OPERATION

$$r_i: (X \; seg_1...seg_k \; Y) \; cat_2 \Rightarrow cat_3 \; rp_i$$

In rules of this form, the pattern matching has to search through an arbitrary number of category segments (represented by X or Y, depending on which side the search begins). Because the length of X and Y may have become arbitrarily long in previous rule applications the amount of computation needed to perform the categorial operation depends on the overall length of the *ss*-category.

LA-grammars the rules of which use only constant categorial operations constitute the class of C-LAGs.

---

[5] CoL, Theorem 5, p. 142.

[6] Hopcroft & Ullman 1979, Theorem 9.8, p. 228. CS stands for the class of context-sensitive languages and REC for the class of recursive languages.

11.2.7 DEFINITION OF THE CLASS OF C-LAGS.

The class of *constant* LA-grammars, or C-LAGs, consists of grammars in which no categorial operation co_i looks at more than k segments in the sentence start categories, for a finite constant k.[7]

The LA-grammar classes considered so far constitute the following hierarchy:

11.2.8 THE HIERARCHY OF A-LAGS, B-LAGS, AND C-LAGS

The class of A-LAGs accepts and generates all recursive languages, the class of B-LAGs accepts and generates all context-sensitive languages, and the class of C-LAGs accepts and generates many context-sensitive, all context-free, and all regular languages.

That all context-free languages are recognized and generated by the C-LAGs is proven[8] on the basis of C-LAGs the categorial operations of which modify only the beginning of the cat_1 pattern and thus correspond to the restrictions on pushdown automata. The class of context-free languages is a *proper* subset of the C-languages because the C-languages contain also context-sensitive languages (e.g. 10.3.3).

The class of regular languages is accepted and generated by C-LAGs in which the length of the categories is restricted by an absolute constant k.[9] The regular, context-free, and context-sensitive languages of the PS-grammar hierarchy have thus been reconstructed in LA-grammar.[10]

---

[7] This finite constant will vary between different grammars.

[8] CoL, Theorem 4, p. 138.

A context-free C-LAG (CF-LAG for short) consists only of rules with the form

r_i: (aX) (b) $\Rightarrow$ (αX) rp_i, with a,b ϵ C and α ϵ C^+

This restriction on the *ss*- and *ss'*-categories corresponds to the working of a PDA which may write not just one symbol but a sequence of symbols into the stack (cf. Hopcroft & Ullman 1979, Chapter 5.2). The following two assertions have to be proven:

1. For each PDA M, a CF-LAG σ may be constructed such that L(M) = L(σ).
This implies CF $\subseteq$ $\mathcal{C}_{cf}$.
2. For each CF-LAG σ, a PDA M may be constructed such that L(σ) = L(M).
This implies $\mathcal{C}_{cf}$ $\subseteq$ CF.

In showing 1 and 2 one has to take into consideration that a CF-LAG uses rules while a PDA uses states, which is not completely the same. Thus it is necessary to provide a constructive procedure to convert states into rules and rules into states – a cumbersome, but not particularly difficult task.

Note with respect to 1 that ε-moves are forbidden in CF-LAGs, but not in PDAs (Hopcroft & Ullman 1979, p. 24). However, there exits for each context-free language a PDA working without ε-moves (Harrison 1978, Theorem 5.5.1) and for an ε-free PDA a CF-LAG may be constructed.

[9] CoL, Theorem 3, p. 138.

[10] Another possibility to modify the generative capacity of LA-grammar consists – at least theoretically – in changing clause 4 of the algebraic definition 10.2.1. For example, if the categorial operations had been defined as arbitrary *partial recursive functions*, then LA-grammar would generate exactly the recursively enumerable languages. This would amount to an increase of generative capacity, making LA-grammar weakly equivalent to PS-grammar. Alternatively, if the categorial operations had been defined as arbitrary *primitive recursive functions*, then it may be shown in analogy to Theorem 2 that

11.3 Ambiguity in LA-grammar

The second main approach to restricting LA-grammar (cf. *R2* in 11.2.2) concerns the number of rule applications. This number is determined by the following factors.

11.3.1 FACTORS DETERMINING THE NUMBER OF RULE APPLICATIONS

The number of rule application in an LA-derivation depends on

1. the length of the input;
2. the number of rules in the rule package to be applied in a certain combination to the analyzed input pair (reading);
3. the number of readings[11] existing at each combination step.

Factor 1 is grammar-independent and used as the length n in formulas characterizing complexity (cf. 8.2.2).

Factor 2 is a grammar-dependent constant. For example, if the largest rule package of an LAG contains five rules, then the maximal number of rule applications in an unambiguous derivation will be at most $5 \cdot n$, where n is the length of the input.

Only factor 3 may push the total number of rule applications beyond a linear increase. In a given left-associative composition, an additional reading comes about when more than one rule in the current rule package is successful on the input. Whether for a given input more than one rule in a rule package may be successful depends on the input conditions of the rules.

Between the input conditions of two rules one of the following relations must obtain:

11.3.2 POSSIBLE RELATIONS BETWEEN INPUT CONDITIONS

1. *Incompatible* input conditions: two rules have incompatible input conditions if there exist no input pairs which are accepted by both rules.
2. *Compatible* input conditions: two rules have compatible input conditions if there exists at least one input pair accepted by both rules and there exists at least one input pair accepted by one rule, but not the other.

---

the resulting kind of LA-grammar generates exactly the primitive recursive languages. This would amount to a decrease in generative capacity as compared to the standard definition.

Using alternative definitions of clause 4 in 10.2.1 is not a good method to obtain different sub-classes of LA-grammar, however. First, alternating the categorial operations between partial recursive, total recursive, and primitive recursive functions is a very crude method. Secondly, the resulting language classes are much too big to be of practical interest: even though the primitive recursive functions are a proper subset of the total recursive functions, the primitive recursive functions properly contain the whole class of context-sensitive languages. Third, the categorial operations have been defined as total recursive functions in 10.2.1 for good reason, ensuring that basic LA-grammar has the maximal generative capacity while still being decidable.

[11] For reasons of simplicity, only syntactic causes of ambiguity are considered here. Lexical ambiguities arising from multiple analyses of words have so far been largely ignored in formal language theory, but are unavoidable in the LA-grammatical analysis of natural language. The possible impact of lexical ambiguity on complexity is discussed in CoL, p. 157 f. and 248 f.

3. *Identical* input conditions: two rules have identical input conditions if it holds for all input pairs that they are either accepted by both rules or rejected by both rules.

Examples of *incompatible* input conditions are (a X) (b) and (c X) (b), as well as (a X) (b) and (a X) (c). If all the rules in a rule package have incompatible input conditions, the use of this rule package cannot be the cause of a syntactic ambiguity.

11.3.3 DEFINITION OF UNAMBIGUOUS LA-GRAMMARS

An LA-grammar is unambiguous if and only if (i) it holds for all rule packages that their rules have *incompatible* input conditions and (ii) there are no lexical[12] ambiguities.

Examples of unambiguous C-LAGs are 10.2.2 for $a^k b^k$ and 10.3.3 for $a^k b^k c^k$.

An example of *compatible* input conditions is (a X) (b) and (X a) (b). Compatible input conditions are the formal precondition for syntactic ambiguity in LA-grammar.

11.3.4 DEFINITION OF SYNTACTICALLY AMBIGUOUS LA-GRAMMARS

An LA-grammar is syntactically ambiguous if and only if (i) it has at least one rule package containing at least two rules with *compatible* input conditions and (ii) there are no lexical ambiguities.

The number of possible readings produced by a syntactically ambiguous LA-grammar depends on its *ambiguity structure*, which may be characterized in terms of the two binary features ±global and ±recursive. We begin with the feature ±global.

In linguistics, a sentence is called syntactically ambiguous if it has more than one reading or structural analysis, as the following example.

11.3.5 +GLOBAL SYNTACTIC AMBIGUITY

Flying air planes can be dangerous.

One reading refers to airborne planes, the other to the activity of piloting. This ambiguity is +global because it is a property of the whole sentence.

Besides +global ambiguity there is also –global (or local) ambiguity in the time-linear analysis of language. For example, reading the sentence 11.3.6 from left to right, there are two readings up to and including the word **barn**.

---

[12] An LA-grammar is lexically ambiguous if its lexicon contains at least two analyzed words with identical surfaces. A nonlinguistic example of a lexical ambiguity is propositional calculus, e. g., (x ∨ y ∨ z) & (...)..., whereby the propositional variables x, y, z, etc., may be analyzed lexically as [x (1)] and [x (0)], [y (1)] and [y (0)], etc. Thereby [x (1)] is taken to represent a true proposition x, and [x (0)] a false one.

While syntactic ambiguities arise in the rule-based derivation of more than one new sentence start, lexical ambiguities are caused by additional readings of the next word. Syntactic and lexical ambiguities can also occur at the same time in an LA-grammar. Furthermore, syntactic ambiguities can be reformulated into lexical ambiguities and vice versa (cf. Hausser 1992, p. 303/4.)

11.3.6 –GLOBAL SYNTACTIC AMBIGUITY

The horse raced by the barn fell.

The first reading interprets raced by the barn as the predicate of the main clause. This initially dominant reading is –global, however, because it is eliminated by the continuation with fell. The second reading interprets raced by the barn as a reduced relative clause and survives as the only reading of the overall sentence. Such examples are called *garden path sentence* because they suggest an initial interpretation which cannot be maintained.

In LA-grammar, the difference between +global and –global ambiguities consists in whether more than one reading survives to the end of the sentence (example 11.3.5) or not (example 11.3.6). Thereby +global and –global readings are treated alike as parallel time-linear derivation branches. This is in contrast to the substitution-based systems of C- and PS-grammar, which recognize only ambiguities which are +global.

The ±global distinction has no impact on complexity in LA-grammar and is made mainly for linguistic reasons. The ±recursive distinction, on the other hand, is crucial for the analysis of complexity because it can be shown that in LA-grammars with nonrecursive ambiguities the maximal number of rule applications per combination step is limited by a grammar-dependent constant (cf. 11.3.7, Theorem 3).

An ambiguity is +recursive, if it originates within a recursive loop of rule applications. In other words, a certain state (cat, rp_i) has several continuations for a given next word, such that one or more of these continuations eventually return into this particular state, thus enabling a repetition of the ambiguity split. Examples of +recursive ambiguities are the C-LAGs for WW^R (cf. 11.5.4) and WW (cf. 11.5.6), which are –global, and for SubsetSum (cf. 11.5.8), which are +global.

An ambiguity is –recursive, if none of the branches produced in the ambiguity split returns to the state which caused the ambiguity. Examples of –recursive ambiguity are the C-LAG for $a^k b^k c^m d^m \cup a^k b^m c^m d^k$ (cf. 11.5.2), which is +global, and the C-LAGs for natural language in Chapter 17 and 18, which exhibit both +global and –global ambiguities.

11.3.7 THEOREM 3

The maximal number of rule applications in LA-grammar with only –recursive ambiguities is

$$(n - (R - 2)) \cdot 2^{(R-2)}$$

for n > (R - 2), where n is the length of the input and R is the number of rules in the grammar.

Proof: Parsing an input of length n requires (n − 1) combination steps. If an LA-grammar has R rules, then one of these rules has to be reapplied after R combination steps at the latest. Furthermore, the maximal number of rule applications in a combination step for a given reading is R.

According to the definition of –recursive ambiguity, rules causing a syntactic ambiguity may not be reapplied in a time-linear derivation path (reading). The first ambiguity-causing rule may produce a maximum of R-1 new branches (assuming its rule package contains all R rules except for itself), the second ambiguity causing rule may produce a maximum of R − 2 new branches, etc. If the different rules of the LA-grammar are defined with their maximally possible rule packages, then after R − 2 combination steps a maximum of $2^{(R-2)}$ readings is reached.

Q.E.D.

Theorem 3 means that in LA-grammars which are not recursively ambiguous the number of rule applications grows only linearly with the length of the input.

11.4 Complexity of grammars and automata

The complexity of a grammar type and its language class is measured in terms of the amount of primitive operations required to process an input in the worst case. Which operation is suitable to serve as the primitive operation and how the number of primitive operations required should be determined, however, is not always obvious.

This holds in particular for grammar formalisms which are not type transparent. Historically, the complexity of PS-grammar types and their language classes was not specified directly in terms of grammar properties, but rather in terms of equivalent abstract automata.[13] These automata each have their own kind of primitive operation.

In order to establish the complexity of the Earley algorithm (cf. 9.3.4), for example, the primitive operation was chosen first and motivated as follows:

> The Griffith and Petrick data is not in terms of actual time, but in terms of "primitive operations." They have expressed their algorithms as sets of nondeterministic rewriting rules for a Turing-machine-like device. Each application of one of these is a primitive operation. We have chosen as our primitive operation the act of adding a state to a state set (or attempting to add one which is already there). We feel that this is comparable to their primitive operation because both are in some sense the most complex operation performed by the algorithm whose complexity is independent of the size of the grammar and the input string.
>
> J. Earley 1970, p. 100

Thus, the complexity statement "context-free PS-grammars parse in n^3" does not apply to context-free PS-grammars directly, but to a certain parsing algorithm which

---

[13] Abstract automata consist of such components as a read/write-head, a write-protected input tape, a certain number of working tapes, the movement of the read/write-head on a certain tape from one cell to another, the reading or deleting of the content in a cell, etc. Classic examples of abstract automata are Turing machines (TM), linearly bounded automata (LBA), pushdown automata (PDA), and finite state automata (FSA). There is a multitude of additional abstract automata, each defined for the purpose of proving various special complexity, equivalence, and computability properties.

takes PS-grammars of this class as input. Accordingly, when Valiant 1975 was able to reduce the complexity of the context-free languages from n^3 to $n^{2.8}$, this was not due to an improvement in PS-grammar, but rather due to Valiant's finding an improved parsing algorithm defined as an abstract automaton.

In contrast, the C-LAGs allow to define the primitive operation (i) directly for the grammar formalism and (ii) transfer it to its parser.

11.4.1 PRIMITIVE OPERATION OF THE C-LAGs

The primitive operation of C-LAGs is a rule application (also counting unsuccessful attempts).

Rule applications are suitable as the primitive operation of C-LAGs because the computation needed by their categorial operations is limited by a constant. That the primitive operation of C-LAGs may simultaneously serve as the primitive operation of its parser is due to the type transparency of LA-grammar.

It follows that the complexity of C-LAGs depends solely on their degree of ambiguity. The linguistic notion of ambiguity is closely related to the automata-theoretic notion of nondeterminism. Before investigating the complexity properties of C-LAGs, let us therefore clarify the distinction between deterministic and nondeterministic automata in its relation to LA-grammar.

An automaton is called deterministic, if each derivation step of its algorithm permits at most one next step. In this sense, all unambiguous LA-grammars (cf. 11.3.3) are deterministic.

An automaton is called nondeterministic, if its algorithm has to chose between several (though finitely many) next steps, or equivalently, has to perform several next steps simultaneously. In this sense, all ambiguous LA-grammars (cf. 11.3.4) are nondeterministic.[14]

A nondeterministic automaton accepts an input if at least one possible path ends in an accepting state. A problem type has the nondeterministic time complexity $\text{NTIME}(f(n))$, if the longest accepted path requires $f(n)$ operations (where n stands for the length of the input). A problem type has the nondeterministic space complexity $\text{NSPACE}(f(n))$, if the longest accepted path requires $f(n)$ memory cells.

NTIME and NSPACE complexity are based on the assumption that the nondeterministic automaton can *guess* the longest accepted path. Thus, NTIME and NSPACE must be seen in light of the fact that there may exist many alternative paths which may have to be computed in order to arrive at a solution, but which do not show up in these complexity measurements.

On a deterministic automaton, on the other hand, a problem type is characterized in terms of the time complexity $\text{DTIME}(f(n))$ and the space complexity $\text{DSPACE}(f(n))$.

---

[14] In PS-grammar, the notions deterministic and nondeterministic have no counterpart.

Because by definition there may exist at most one path on a deterministic automaton DTIME and DSPACE specify the actual amount of time and space units required.

The distinction between the deterministic and the nondeterministic versions of automata has raised the question of whether or not their generative capacity is equivalent. In the case of finite state automata (FSA) it holds for each language L, if L is accepted by a nondeterministc FSA then there exists an equivalent deterministic FSA which also accepts L.[15] A corresponding result holds for Turing machines (TM).[16]

In pushdown automata (PDA), on the other hand, the deterministic and the nondeterministic versions are not equivalent in generative capacity.[17] For example, WW^R is accepted by a nondeterministic PDA, but not by a deterministic one. In the case of linear bounded automata (LBA), finally, it is not known whether the set of languages accepted by deterministic LBA is a *proper* subset of the set of languages accepted by nondeterministic LBA (LBA-problem).

Thus the following relations hold between deterministic and nondeterministic versions of the automata FSA, PDA, LBA, and TM:

DFSA = NFSA
DPSA $\subset$ NDPA
DLBA ? NLBA
DTM = NTM

Automata in which the deterministic and the nondeterministic versions are equivalent raise the additional question as to which degree the space requirement is increased in the transition from the nondeterministic to the deterministic version.

Given the distinction of deterministic (unambiguous) and nondeterministic (ambiguous) LA-grammars, the automata-theoretic notions of complexity, especially NTIME and DTIME, may be applied to LA-grammar. It holds in particular

DTIME complexity of unambiguous C-LAGs = $C \cdot (n - 1)$
NTIME complexity of ambiguous C-LAG   = $C \cdot (n - 1)$,

where C is a constant and n is the length of the input.

This holds because – based on the left-associative derivation structure – each accepted input of length n is analyzed in n–1 combination steps whereby the rule applications of each combination step take the time required for the constant amount of computation C. In other words, unambiguous C-LAGs are of *deterministic linear* time complexity,[18] while ambiguous C-LAGs are of *nondeterministic linear* time complexity.

From a theoretical point of view the applicability of DTIME and NTIME to LA-grammar is of general interest because it facilitates the transfer of results and open

---

[15] M.O. Rabin & D. Scott 1959.

[16] Hopcroft & Ullman 1979, p. 164, Theorem 7.3.

[17] Hopcroft & Ullman 1979, p. 113.

[18] Strictly speaking, unambiguous C-LAGs have an even better complexity than deterministic linear time, namely *real time*.

questions between automata theory and LA-grammar. For example, the question of whether the class of C-LAGs is a *proper* subset of the class of B-LAGs may be shown to be equivalent to the LBA-problem, i.e., the open question of whether or not DLBA is *properly* contained in NLBA (cf. also Section 12.2).

From a practical point of view, however, the automata-theoretic complexity measures have the disadvantage that DTIME applies only to unambiguous LA-grammars and NTIME only to ambiguous ones. Furthermore, the notion of NTIME does not specify the actual amount of computation required to analyze arbitrary new input, but only the amount of computation needed to verify a known result.

In order to employ a realistic method of measuring complexity, which moreover can be applied uniformly to ambiguous and nonambiguous grammars alike, the following complexity analysis of C-LAGs will be specified in terms of the total number of rule applications (relative to the length of the input) required to analyze arbitrary new input in the worst case. This grammatical method of measuring complexity is as simple as it is natural in LA-grammar.

11.5 Subhierarchy of C1-, C2-, and C3-LAGs

Compared to the A- and B-LAGs, the C-LAGs constitute the most restricted class of LAGs, generating the smallest LA-class of languages. Compared to the context-free languages (which are properly contained in the C-languages), however, the class of C-languages is quite large. It is therefore theoretically interesting and practically useful to differentiate the C-LAGs further into subclasses by defining a subhierarchy.

Because the complexity of C-LAGs is measured as the number of rule applications, and because the number of rule applications depends solely on the ambiguity structure, it is natural to define subclasses of C-LAGs in terms of different degrees of ambiguity. These subclasses are called C1-, C2-, and C3-LAGs, whereby increasing degrees of ambiguity are reflected in increasing degrees of complexity.

The subclass with the lowest complexity and the lowest generative capacity are the C1-LAGs. A C-LAG is a C1-LAG if it is not recursively ambiguous. The class of C1-languages parses in linear time and contains all deterministic context-free languages which can be recognized by a DPDA without ε-moves, plus context-free languages with −recursive ambiguities, e.g. $a^k b^k c^m d^m \cup a^k b^m c^m d^k$, as well as many context-sensitive languages, e.g. $a^k b^k c^k$, $a^k b^k c^k d^k e^k$, $\{a^k b^k c^k\}^*$, L_{square}, L^k_{hast}, a^{2^i}, $a^k b^m c^{k \cdot m}$, and $a^{i!}$, whereby the last one is not even an index language.[19]

Examples of unambiguous context-sensitive C1-LAGs are $a^k b^k c^k$ defined in 10.3.3, and the following definition for $a^{2^i} =_{def} \{a^i \mid i \text{ is a positive power of 2}\}$.

---

[19] A C1-LAG for $a^k b^k c^m d^m \cup a^k b^m c^m d^k$ is defined in 11.5.2, for L_{square} and L^k_{hast} in B. Stubert 1993, p. 16 and 12, for $a^k b^k c^k d^k e^k$ in CoL, p. 233, for $a^k b^m c^{k \cdot m}$ in Hausser 1992, p. 296, and for a^{2^i} in 11.5.1. A C1-LAG for $a^{i!}$ is sketched in Hausser 1992, p. 296, footnote 13.

11.5.1 C1-LAG for context-sensitive a^{2^i}

$LX =_{def} \{[a\ (a)]\}$

$ST_S =_{def} \{[(a)\ \{r_1\}]\}$

r_1: (a)   (a) $\Rightarrow$ (aa) $\{r_2\}$

r_2: (aX)   (a) $\Rightarrow$ (Xbb) $\{r_2, r_3\}$

r_3: (bX)   (a) $\Rightarrow$ (Xaa) $\{r_2, r_3\}$

$ST_F =_{def} \{[(aa)\ rp_1],\ [(bXb)\ rp_2],\ [(aXa)\ rp_3]\}.$

This C1-LAG is unambiguous because r_2 and r_3 have incompatible input conditions. A comparison with corresponding PS-grammars for a^{2^i} illustrates the formal and conceptual simplicity of LA-grammar.[20]

A C1-LAG with +global, −recursive ambiguity is 11.5.2 for the context-free language $a^k b^k c^m d^m \cup a^k b^m c^m d^k$. In PS-grammar, this language is called *inherently ambiguous* because there does not exist an unambiguous PS-grammar for it.[21]

11.5.2 C1-LAG for ambiguous $a^k b^k c^m d^m \cup a^k b^m c^m d^k$

$LX =_{def} \{[a\ (a)],\ [b\ (b)],\ [c\ (c)],\ [d\ (d)]\}$

$ST_S =_{def} \{[(a)\ \{r_1, r_2, r_5\}]\}$

r_1: (X)   (a) $\Rightarrow$ (a X) $\{r_1, r_2, r_5\}$

r_2: (a X) (b) $\Rightarrow$ (X) $\{r_2, r_3\}$

r_3: (X)   (c) $\Rightarrow$ (c X) $\{r_3, r_4\}$

r_4: (c X) (d) $\Rightarrow$ (X) $\{r_4\}$

r_5: (X)   (b) $\Rightarrow$ (b X) $\{r_5, r_6\}$

r_6: (b X) (c) $\Rightarrow$ (X) $\{r_6, r_7\}$

r_7: (a X) (d) $\Rightarrow$ (X) $\{r_7\}$

$ST_F =_{def} \{[\varepsilon\ rp_4],\ [\varepsilon\ rp_7]\}$

This C1-LAG contains a syntactic ambiguity in accordance with definition 11.3.4: the rule package rp_1 contains the input compatible rules r_2 and r_5. Nevertheless, the grammar parses in linear time because the ambiguous continuations are not part of a recursion: rp_2 and rp_5 do not contain r_1. In the worst case, e.g. aabbccdd, the grammar generates two analyses based on two parallel time-linear paths which begin after the initial a-sequence.[22]

The type of C-LAG with the second lowest complexity and the second lowest generative capacity are the C2-LAGs. A C-LAG is a C2-LAG if it (i) generates +recursive ambiguities and (ii) its ambiguities are restricted by the single return principle.

---

[20] Hopcroft & Ullman 1979 present the canonical context-sensitive PS-grammar of a^{2^i} on p. 224, and a version as unrestricted PS-grammar on p. 220.

[21] Cf. Hopcroft & Ullman 1979, p. 99–103.

[22] An explicit derivation is given in CoL, p. 154 ff.

11.5.3 THE SINGLE RETURN PRINCIPLE (SRP)

A +recursive ambiguity is single return, if exactly one of the parallel paths
returns into the state resulting in the ambiguity in question.

The class of C2-languages parses in polynomial time and contains certain nondeterministic context-free languages like WW^R and L_{hast}^∞, plus context-sensitive languages
like WW, $W^{k\geq 3}$, $\{WWW\}^*$, and $W_1W_2W_1^RW_2^R$.[23]

An SR-recursive ambiguity is illustrated by the following C2-LAG for WW^R. As
explained in Section 8.3, this language consists of an arbitrarily long sequence W of
arbitrary words, followed by the inverse of this sequence W^R.

11.5.4 C2-LAG FOR CONTEXT-FREE WW^R

$LX =_{def}$ {[a (a)], [b (b)], [c (c)], [d (d)] ... }
$ST_S =_{def}$ {[(seg$_c$) {r$_1$, r$_2$}]}, where seg$_c$ ϵ {a, b, c, d, ... }
r$_1$: (X) (seg$_c$) $\Rightarrow$ (seg$_c$ X) {r$_1$, r$_2$}
r$_2$: (seg$_c$ X) (seg$_c$) $\Rightarrow$ (X) {r$_2$}
$ST_F =_{def}$ {[ε rp$_2$]}

Each time r$_1$ has been applied successfully, r$_1$ and r$_2$ are attempted in the next composition. As soon, however, as r$_2$ was successful, the derivation cannot return to r$_1$
because r$_1$ is not listed in the rule package of r$_2$. Therefore only one branch of the
ambiguity (i.e., the one resulting from repeated applications of r$_1$) can return into the
recursion.

The worst case in parsing WW^R are inputs consisting of an even number of the same
word. This is illustrated in 11.5.5 with the input a a a a a a.

11.5.5 DERIVATION STRUCTURE OF THE WORST CASE IN WW^R

```
rules:              analyses:

2                   a $ a
1 2 2               a a $ a a
1 1 2 2 2           a a a $ a a a
1 1 1 2 2           a a a a $ a a
1 1 1 1 2           a a a a a $ a
1 1 1 1 1           a a a a a a $
```

The unmarked middle of the intermediate strings generated in the course of the derivation is indicated in 11.5.5 by $. Of the six hypotheses established in the course of the
left-associative analysis, the first two are invalidated by the fact that the input string
continues, the third hypothesis correctly corresponds to the input a a a a a a, while

---

[23] A C2-LAG for WW^R is defined in 11.5.4, for L_{hast}^∞ in Stubert 1993, p. 16, for WW in 11.5.6, for
WWW in CoL, p. 215, for $W^{k\geq 3}$ in CoL, p. 216, and for $W_1W_2W_1^RW_2^R$ in 11.5.7.

the remaining three hypotheses are invalidated by the fact that the input does not contain any more words.

A C2-LAG for a context-sensitive language is illustrated by the following definition.

11.5.6 C2-LAG FOR CONTEXT-SENSITIVE WW

$$\text{LX} =_{def} \ \{[\text{a (a)}], [\text{b (b)}], [\text{c (c)}], [\text{d (d)}] \dots \}$$
$$\text{ST}_S =_{def} \ \{[(\text{seg}_c) \ \{r_1, r_2\}]\}, \text{ where seg}_c \ \epsilon \ \{a, b, c, d, \dots \}$$
$$r_1: (\text{X}) \qquad (\text{seg}_c) \ \Rightarrow \ (\text{X seg}_c) \ \{r_1, r_2\}$$
$$r_2: (\text{seg}_c \ \text{X}) \ (\text{seg}_c) \ \Rightarrow \ (\text{X}) \qquad \{r_2\}$$
$$\text{ST}_F =_{def} \ \{[\varepsilon \ rp_2]\}$$

This grammar resembles 11.5.4 except for the result category of r_1: in context-sensitive WW, r_1 defines the category pattern (X seg_c), while in context-free WW^R, r_1 defines the category pattern (seg_c X). The worst case for context-sensitive WW and context-free WW^R is the same, the respective C2-LAGs are both SR-ambiguous, and are both of n^2 complexity. The n^2 increase of readings in the analysis of the worst case of WW^R and WW is clearly visible in 11.5.5.

A C2-LAG of n^3 complexity is the nondeterministic context-sensitive language $W_1W_2W_1^RW_2^R$ (provided by B. Stubert).

11.5.7 C2-LAG FOR CONTEXT-SENSITIVE $W_1W_2W_1^RW_2^R$

$$\text{LX} =_{def} \ \{[\text{a (a)}], [\text{b (b)}]\}$$
$$\text{ST}_S =_{def} \ \{[(\text{seg}_c) \ \{r_{1a}\}], [(\text{seg}_c) \ \{r_{1b}\}]\}, \text{ where seg}_c, \text{seg}_d \ \epsilon \ \{a, b\}$$
$$r_{1a}: (\text{seg}_c) \qquad (\text{seg}_d) \ \Rightarrow \ (\# \ \text{seg}_c \ \text{seg}_d) \quad \{r_2, r_3\}$$
$$r_{1b}: (\text{seg}_c) \qquad (\text{seg}_d) \ \Rightarrow \ (\ \text{seg}_d \ \# \ \text{seg}_c) \ \{r_3, r_4\}$$
$$r_2: (\text{X}) \qquad (\text{seg}_c) \ \Rightarrow \ (\text{X seg}_c) \qquad \{r_2, r_3\}$$
$$r_3: (\text{X}) \qquad (\text{seg}_c) \ \Rightarrow \ (\text{seg}_c \ \text{X}) \qquad \{r_3, r_4\}$$
$$r_4: (\text{X seg}_c) \quad (\text{seg}_c) \ \Rightarrow \ (\text{X}) \qquad\qquad \{r_4, r_5\}$$
$$r_5: (\text{seg}_c \ \text{X} \ \#) \ (\text{seg}_c) \ \Rightarrow \ (\text{X}) \qquad\qquad \{r_6\}$$
$$r_6: (\text{seg}_c \ \text{X}) \quad (\text{seg}_c) \ \Rightarrow \ (\text{X}) \qquad\qquad \{r_6\}$$
$$\text{ST}_F =_{def} \ \{[\varepsilon \ rp_5], [\varepsilon \ rp_6]\}$$

The exact complexity of this language is $\frac{1}{8}n^3 + \frac{1}{4}n^2 + \frac{1}{2}n$. The degree of polynomial complexity depends apparently on the maximal number of SR-recursive ambiguities in a derivation. Thus, a C2-LAG with one SR-recursive ambiguity can be parsed in n^2, a C2-LAG with two SR-recursive ambiguities can be parsed in n^3, etc.

The subclass with the highest complexity and generative capacity are the C3-LAGs. A C-LAG is a C3-LAG if it generates unrestricted +recursive ambiguities. The class of C3-languages parses in exponential time and contains the deterministic context-free language L_{no}, the hardest context-free language HCFL, plus context-sensitive languages like SubsetSum and SAT, which are $\mathcal{NP}$-complete.[24]

A language known to be inherently complex, requiring exponential time irrespective of the algorithm used, is SubsetSum. Its expressions $y\#a_1\#a_2\#a_3\#...\#a_n\#$ are defined such that y, a_1, a_2,..., a_n are all binary strings containing the same number of digits. Furthermore, when viewed as binary numbers presenting the least significant digit first, y is equal to the sum of a subset of the a_i.

11.5.8 C3-LAG FOR SubsetSum.

$$LX =_{def} \{[0\ (0)],\ [1\ (1)],\ [\#\ (\#)]\}$$
$$ST_S =_{def} \{[(seg_c)\ \{r_1, r_2\}]\},\ \text{where } seg_c \in \{0, 1\}$$
$$\qquad seg_c\ \varepsilon\ \{0, 1\}$$

$r_1: (X)$	(seg_c)	$\Rightarrow (seg_c\ X)$	$\{r_1, r_2\}$
$r_2: (X)$	$(\#)$	$\Rightarrow (\#\ X)$	$\{r_3, r_4, r_6, r_7, r_{12}, r_{14}\}$
$r_3: (X\ seg_c)$	(seg_c)	$\Rightarrow (0\ X)$	$\{r_3, r_4, r_6, r_7\}$
$r_4: (X\ \#)$	$(\#)$	$\Rightarrow (\#\ X)$	$\{r_3, r_4, r_6, r_7, r_{12}, r_{14}\}$
$r_5: (X\ seg_c)$	(seg_c)	$\Rightarrow (0\ X)$	$\{r_5, r_6, r_7, r_{11}\}$
$r_6: (X\ 1)$	(0)	$\Rightarrow (1\ X)$	$\{r_5, r_6, r_7, r_{11}\}$
$r_7: (X\ 0)$	(1)	$\Rightarrow (1\ X)$	$\{r_8, r_9, r_{10}\}$
$r_8: (X\ seg_c)$	(seg_c)	$\Rightarrow (1\ X)$	$\{r_8, r_9, r_{10}\}$
$r_9: (X\ 1)$	(0)	$\Rightarrow (0\ X)$	$\{r_5, r_6, r_7, r_{11}\}$
$r_{10}: (X\ 0)$	(1)	$\Rightarrow (0\ X)$	$\{r_8, r_9, r_{10}\}$
$r_{11}: (X\ \#)$	$(\#)$	$\Rightarrow (\#\ X)$	$\{r_3, r_4, r_6, r_7, r_{12}, r_{14}\}$
$r_{12}: (X\ 0)$	(seg_c)	$\Rightarrow (0\ X)$	$\{r_4, r_{12}, r_{14}\}$
$r_{13}: (X\ 0)$	(seg_c)	$\Rightarrow (0\ X)$	$\{r_{11}, r_{13}, r_{14}\}$
$r_{14}: (X\ 1)$	(seg_c)	$\Rightarrow (1\ X)$	$\{r_{11}, r_{13}\ r_{14}\}$

$$ST_F =_{def} \{[(X)\ rp_4]\}$$

This C3-LAG (provided by D. Applegate) copies y into the category. Then the +recursive ambiguity arises by nondeterministically either subtracting or not subtracting each a_i from y. The grammar only enters an accepting state, if the result of the subtraction is zero.

In conclusion let us summarize the different restrictions on LA-grammar, the resulting hierarchy of LA-grammar classes, the associated classes of languages, and their complexity. The LA-hierarchy resembles the PS-hierarchy (cf. 8.1.2) insofar as the subclasses of the LAGs and their languages are defined in terms of increasing restrictions on the rule system.

The language class of the C1-LAGs is a subset of the C2-languages because the –recursive ambiguity structure of the C1-LAGs is more restricted than the SR-

---

[24] A C3-LAG for L_{no} is defined in 12.3.3, for HCFL in B. Stubert 1993, p. 16, for SubsetSum in 11.5.8 and for SAT in Hausser 1992, footnote 19.

recursive ambiguities of the C2-LAGs (siehe 11.5.4). The language class of the C2-LAGs is a subset of the C3-languages because the SR-recursive ambiguity structure of the C2-LAGs is more restricted than the +recursive ambiguities of the C3-LAGs. The language class of the C3-LAGs is a subset of the B-languages because the category patterns of the C-LAGs are more restricted than those of the B-LAGs. The language class of the B-LAGs is a subset of the A-languages because the categories assigned by B-LAGs are restricted in length while those of the A-LAGs are not.

The restrictions on LA-grammar apply directly to the two complexity parameters *amount* and *number* of rule applications (cf. 11.2.1).

11.5.9 Types of restriction in LA-grammar

0. LA-type A: no restriction
1. LA-type B: The length of the categories of intermediate expressions is limited by $k \cdot n$, where k is a constant and n is the length of the input ($R1.1$, amount).
2. LA-type C3: The form of the category patterns results in a constant limit on the operations required by the categorial operations ($R1.2$, amount).
3. LA-type C2: LA-type C3 and the grammar is at most SR-recursively ambiguous ($R2$, number).
4. LA-type C1: LA-type C3 and the grammar is at most –recursively ambiguous ($R2$, number).

The amount and number parameters are controlled via (1) the length of categories assigned to intermediate expressions, (2) the form of the category patterns in the rules, (3) the possible reapplication of rules as defined by the rule packages, and (4) the input compatibility of rules in the same rule package.

The LA-grammar hierarchy is summarized in 11.5.10. Like the PS-grammar hierarchy 8.1.2, it is based on the structural relation between restrictions on the rule system, types of grammar, classes of languages, and degrees of complexity.

11.5.10 LA-grammar hierarchy of formal languages

restrictions	types of LAG	languages	complexity
LA-type C1	C1-LAGs	C1 languages	linear
LA-type C2	C2-LAGs	C2 languages	polynomial
LA-type C3	C3-LAGs	C3 languages	exponential
LA-type B	B-LAGs	B languages	exponential
LA-type A	A-LAGs	A languages	exponential

Of the five different classes of LA-grammar only the class of B-languages occurs also in the PS-grammar hierarchy, namely as the class of context-sensitive languages. The class of A-languages, on the other hand, is properly contained in the class of recursively enumerable languages generated by unrestricted PS-grammars. The classes of C1-, C2- and C3-languages have no counterpart in the PS-grammar hierarchy. However, the PS-grammar class of reqular languages is properly contained in the class of C1-languages, while the class of context-free languages in properly contained in the class of C3-languages.

Furthermore, compared to the restrictions on PS-grammar, which apply only to the form of the rules, the restrictions on LA-grammar apply to the rule system as a whole. This is because in LA-grammar the application of rules is handled explicitly in terms of category patterns and rule packages, whereas in PS-grammar it is handled implicitly in terms of variables contained in the rewrite rules.

Exercises

Section 11.1

1. Explain the notions of *total recursive function* and *partial recursive function.*
2. What is the formal characterization of the class of recursive languages in the hierarchy of PS-grammar?
3. What is the generative capacity of LA-grammar in its basic unrestricted form?
4. Explain the proofs in 11.1.2 and 11.1.3.
5. Describe the difference in generative capacity of unrestricted PS-grammar and unrestricted LA-grammar.

Section 11.2

1. What are possible structural restrictions on LA-grammar?
2. Explain the grammar classes of A-LAGs, B-LAGs and C-LAGs.
3. What is the difference between a constant and a nonconstant categorial operation?
4. Describe the reconstruction of the PS-grammar hierarchy in LA-grammar.
5. Is there a class of LA-grammar generating the recursively enumerable languages?
6. Is there a class of PS-grammar generating the recursive languages?
7. Is there a class of PS-grammar generating the C-languages?

Section 11.3

1. Explain the notions of $\pm$ global ambiguities. What is their relation to the notions of $\pm$ deterministic derivations in automata theory and the notion of ambiguity in PS-grammar.
2. What determines the number of rule applications in an LA-derivation?
3. Describe three different types of input conditions.
4. What is the definition of ambiguous and unambiguous LA-grammars?
5. Explain the notions of $\pm$ recursive ambiguity in LA-grammar.
6. Define the rule packages of a –recursively ambiguous LA-grammar with 7 rules which is maximally ambiguous. How many readings are derived by this grammar after 4, 5, 6, 7, 8, and 9 combination steps, respectively?

Section 11.4

1. Explain the complexity parameters of LA-grammar.
2. How is the elementary operation of C-LAGs defined?
3. Compare Earley's primitive operation for computing the complexity of his algorithm for context-free PS-grammars with that of the C-LAGs.
4. How does Earley compute the amount and number parameters of contextfree PS-grammar?
5. Name four notions of complexity in automata theory and explain them.
6. Explain the relation between the deterministic and the nondeterministic versions of FSA, PDA, LBA, and TM.
7. Explain the application of the notions DTIME and NTIME to the C-LAGs. What is the alternative?

Section 11.5

1. Describe the subhierarchy of the C1-, C2-, and C3-LAGs. What is the connection between ambiguity and complexity?
2. Define a C1-LAG for $a^k b^k c^m d^m \cup a^k b^m c^k d^m$ and explain how this grammar works. Is it ambiguous? What is its complexity?
3. Explain the single return principle.
4. Compare the handling of WW and WW^R in PS-grammar and LA-grammar.
5. Describe the well-formed expressions of SubsetSum and explain why this language is inherently complex.
6. Explain the hierarchy of LA-grammar.
7. Describe the relation between PS-grammar and the nativist theory of language, on the one hand, and between LA-grammar and the SLIM theory, on the other. Would it be possible to use a PS-grammar as the syntactic component of the SLIM theory? Would it be possible to use an LA-grammar as the syntactic component of the nativist theory?

12. LA- and PS-hierarchies in comparison

Having presented two non-equivalent formalisms of generative grammar,[1] each with its own fully developed hierarchy of language and complexity classes, we return in this chapter to the formal properties of natural language. Section 12.1 compares the complexity of the LA- and PS-grammatical language classes. Section 12.2 describes the inclusion relations between language classes in the PS- and the LA-hierarchy, respectively. Section 12.3 defines a context-free language which is a C3-language. Section 12.4 describes the orthogonal relation between the context-free languages and the classes of C-languages. Section 12.5 investigates ambiguity in natural language, and concludes that the natural languages are in the class of C1-LAGs, thus parsing in linear time.

12.1 Language classes of LA- and PS-grammar

A grammar formalism for natural language is like a suit. If it is chosen too small – for example as a regular (type 3) PS-grammar – there are phenomena which cannot be described within its means. If it is chosen too big – for example as a context-sensitive (type 1) PS-grammar –, the characteristic structures of natural language will disappear in a formalism which allows to describe the most complicated artificial structures as easily as the genuinely natural ones.

Furthermore, the 'bigger' a formalism, the more 'expensive' it is in terms of mathematical complexity. Computationally, these costs appear as the time and memory required for the parsing of language expressions.

A formalism may not only simply be too small or too big for the description of natural language, but also too small and too big at the same time – like a pair of trousers which are too short and too wide. For example, the formalism of context-free PS-grammar is too small for the description of natural language, as indicated by the introduction of transformations and similar extensions. At the same time it is too big, as shown by context-free languages like HCFL,[2] the structures of which have no counterpart in natural language.

In principle, there is no theoretical or practical reason why it should not be possible to develop a (sub-)formalism of low generative capacity and complexity which really

---

[1] Namely PS-grammar in Chapters 8 and 9, and LA-grammar in Chapters 10 and 11.
[2] Hardest context-free language, S. Greibach 1973.

fits natural language. It is just that despite great efforts no such subformalism could be found within PS-grammar.

Alternative elementary formalisms are C- and LA-grammar. C-grammar, however, has no fully developed hierarchy with different language classes and degrees of complexity. On the contrary, bidirectional C-grammar is weakly equivalent to context-free PS-grammar (cf. Section 9.2).

Therefore, the only language hierarchy orthogonal to that of PS-grammar is provided by LA-grammar. In 12.1.1 the respective language classes of LA- and PS-grammar are related to the four basic degrees of complexity.

12.1.1 COMPLEXITY DEGREES OF THE LA- AND PS-HIERARCHY

	LA-grammar	PS-grammar
undecidable	—	recursively enumerable languages
exponential	A-languages B-languages C3-languages	context-sensitive languages
polynomial	C2-languages	context-free languages
linear	C1-languages	regular languages

The nonequivalence of the elementary formalisms of LA- and PS-grammar is shown by languages which are in the same class in PS-grammar, but in different classes in LA-grammar, and vice versa. For example, $a^k b^k$ and WW^R are in the same class in PS-grammar (i.e. context-free), but in different classes in LA-grammar: $a^k b^k$ is a C1-LAG parsing in linear time, while WW^R is a C2-LAG parsing in n^2. Conversely, $a^k b^k$ and $a^k b^k c^k$ are in the same class in LA-grammar (i.e. C1-LAGs), but in different classes in PS-grammar: $a^k b^k$ is context-free, while $a^k b^k c^k$ is context-sensitive.

That a language like $a^k b^k c^k$ can be classified into two different language classes (i.e., context-sensitive vs. C1) with different degrees of complexity (i.e., exponential vs. linear) depends on the distinction between the inherent complexity of an individual language and the complexity of its class.

The classification of $a^k b^k c^k$ as context-sensitive in PS-grammar is not because this language is inherently complex, but rather because no lower subclass of PS-grammar happens to fit its structure. It is therefore possible to define an alternative formalism which classifies $a^k b^k c^k$ in a class of linear complexity.

Furthermore, the lower language classes are defined as subsets of the higher language classes. A language like $a^k b^k$, for example, is called context-free in PS-grammar because this is the *smallest* language class containing it. Nominally, however, $a^k b^k$ is also a context-sensitive language because the class of context-free languages is contained in the context-sensitive class. Therefore, a statement like "$a^k b^k c^k$ is a context-sensitive language which in LA-grammar is a C1-LAG parsing in linear time" is no contradiction: because the class of C1-languages is a subset of the class of B-languages, the C1-LAGs are nominally also B-LAGs (and thus context-sensitive).

12.2 Subset relations in the two hierarchies

A language class X is a subset of another language class Y (formally $X \subseteq Y$) if all languages in X are also languages in Y. A language class X is a *proper* subset of another language class Y (formally $X \subset Y$) if X is a subset of Y and in addition there is at least one language in Y which is not an element of X.

The following subset relations hold for the language classes of the PS-hierarchy.

12.2.1 SUBSET RELATIONS IN THE PS-HIERARCHY

regular lang. $\subset$ context-free lang. $\subset$ context-sensitive lang. $\subset$ rec. enum. languages

These subset relations follow from decreasing restrictions on the rewrite rules of PS-grammar (cf. Section 8.2) and are proper subset relations.[3]

The following subset relations hold for the language classes in the LA-hierarchy.

12.2.2 SUBSET RELATIONS IN THE LA-HIERARCHY

C1-languages $\subseteq$ C2-languages $\subseteq$ C3-languages $\subseteq$ B-languages $\subset$ A-languages

These subset relations follow likewise from decreasing restrictions. However, while the B-languages are a proper subset of the A-languages,[4] the proper inclusion of the other classes can only be surmised. In particular, the question of whether the class of C2-languages is a proper subset of the C3-languages corresponds to an unsolved problem of classic automata theory, namely whether $\mathcal{P} \subset \mathcal{NP}$ or $\mathcal{P}=\mathcal{NP}$.

The language class $\mathcal{NP}$ contains all languages which can be recognized in non-deterministic polynomial time, while $\mathcal{P}$ contains all languages which can be recognized in determinist polynomial time.

> The languages recognizable in deterministic polynomial time form a natural and important class, the class $\bigcup_{i \geq 1} \text{DTIME}(n^i)$, which we denote by $\mathcal{P}$. It is an intuitively appealing notion that $\mathcal{P}$ is the class of problems that can be solved efficiently. Although one might quibble that an n^{57} step algorithm is not very efficient, in practice we find that problems in $\mathcal{P}$ usually have low-degree polynomial time solutions.
>
> J.E. Hopcroft & J.D. Ullman 1979, p. 320

---

[3] Hierarchy lemma, Hopcroft & Ullman 1979, p. 228.

[4] See Hopcroft & Ullman 1979, p. 228, Theorem 9.8. The A-LAGs generate the recursive languages while the B-LAGs generate the context-sensitive languages, as shown in 11.1 and 11.2.

A language L is called $\mathcal{NP}$-complete if (i) all languages in $\mathcal{NP}$ can be reduced to L in deterministic polynomial time and (ii) L is in $\mathcal{NP}$. An $\mathcal{NP}$-complete language is designed to represent the worst case of nondeterministic polynomial complexity.

The classic, historically first example of an $\mathcal{NP}$-complete language is SAT, the problem of Boolean *SAT*isfiability.[5] Consider the following Boolean expression.

12.2.3 A WELL-FORMED EXPRESSION IN 3SAT

$$(x \vee \bar{y} \vee \bar{z}) \& (y \vee z \vee u) \& (x \vee z \vee \bar{u}) \& (\bar{x} \vee y \vee u)$$

The sign $\vee$ stands for the logical or (disjunction), the sign & stands for the logical and (conjunction), the letters represent variables for propositions, and the horizontal bar over some of the variables, e.g. $\bar{z}$, stands for negation. 3SAT is a slight simplification of SAT because 3SAT is restricted to conjunctions in which each conjunct consists of a disjunction containing three variables.

The problem of satisfying expressions like 12.2.3 is to find an assignment for the variables which would make the expression true – if such an assignment exists. This problem is inherently complex because the analysis has to keep track of potentially 2^n different assignments. For example, the first variable x may be assigned the values 1 (true) or 0 (false). When the second variable y is encountered, four assignments must be distinguished, namely $(x = 1, y = 1)$, $(x = 1, y = 0)$, $(x = 0, y = 1)$ und $(x = 0, y = 0)$. In other words, each time a new variable is encountered, the number of possible assignments is doubled.

Another example of an $\mathcal{NP}$-complete language is the already familiar Subset Sum. Like 3SAT, Subset Sum is a C3-language, as shown by the definition of the C3-LAG in 11.5.9. Thus, the class of C3-languages contains $\mathcal{NP}$-complete languages. Furthermore, the class of C3-languages is obviously in $\mathcal{NP}$ because C-LAGs verify by definition in nondeterministic linear time – and thus a fortiori in nondeterministic polynomial time.

The C2-languages, on the other hand, are designed to parse in deterministic polynomial time, for which reason they are contained in $\mathcal{P}$. The assumption that the class of C2-languages is not a proper subset of the class of C3-languages would imply

class of C2-languages = class of C3-languages.

This would mean in turn that there exists a C2-LAG for, e.g., Subset Sum. Because all $\mathcal{NP}$-languages can be reduced in deterministic polynomial time to Subset Sum, it would follow that

$\mathcal{P} = \mathcal{NP}$.

The equivalence of $\mathcal{P}$ and $\mathcal{NP}$ is considered improbable, however, which strengthens the guess that the C2-languages are a proper subset of the C3-languages.[6]

---

[5] Cf. Hopcroft & Ullman 1979, p. 324ff.

[6] Assuming that the proper inclusion of C2 $\subset$ C3 could be shown directly (for example by means of a pumping lemma for C2-languages), then this would imply $\mathcal{P} \subset \mathcal{NP}$ only if it can be shown that C2 = $\mathcal{P}$, which is improbable.

Another open question is whether the class of C3-languages is a proper subset of the class of B-languages. Here one may speculate using again the $\mathcal{NP}$-completeness of the C3-LAGs. It is known that the recognition of context-sensitive languages (CS-recognition) is PSPACE-complete (cf. Hopcroft and Ullman 1979, p. 346,7). Because it is improble[7] that a PSPACE-complete problem is in $\mathcal{NP}$ it is also improbable that the class of C-languages is not a proper subset of the set of B-languages.

12.3 Non-equivalence of the LA- and PS-hierarchy

A language which in PS-grammar is context-free, but in LA-grammar is a C3-language, is L_{no} (or *noise*-language) by D. Applegate. L_{no} generates expressions which consist of 0 and 1, and which have the structure $W'\#W^R$. The symbol # separates W' and W^R, W^R is the mirror image of W, and W' differs from W by containing an arbitrary number of additional 0 and 1. These additional words in W' are indistinguishable from those which have a counterpart in W^R and thus function as noise.

A context-free PS-grammar for this language is defined in 12.3.1.

12.3.1 PS-GRAMMAR OF L_{no}

S → 1S1	S → 1S	S → #
S → 0S0	S → 0S	

The rules in the left column generate corresponding words preceding and following #, while the rules in the middle column generate only noise words in W'.

Traditional parsers for context-free languages like the Earley or the CYK algorithm have no difficulty in analyzing L_{no} in n^3. This is because the parser utilizes the basic inverse pair structure of context-free languages. Example 12.3.2 shows a PS-grammar derivation and the corresponding states produced by the Earley algorithm.

12.3.2 PS-GRAMMAR DERIVATION OF 10010#101 IN L_{no}

derivation tree:	generated chains:	states:	
	1S1	1.S1	1S1.
		1.S	
	10S01	0.S0	0S0.
		0.S	
	100S01	0.S0	
		0.S	0S.
	1001S101	1.S1	1S1.
		1.S	
	10010S101	0.S0	
		0.S	0S.
	10010#101	#.	

---

[7] Not only is a PSPACE-complete problem not likely to be in $\mathcal{P}$, it is also not likely to be in $\mathcal{NP}$. Hence the property whose existence is PSPACE-complete probably cannot even be *verified* in polynomial time using a polynomial length 'guess.' M.R. Garey and D.S. Johnson 1979, p. 171.

The Earley algorithm generates only two states for each terminal symbol preceding # in L_{no}, for example '1.S1' and '1.S'. Thus, if # is preceded by k terminal symbols in the input chain, then the algorithm will produce 2k states by the time # is reached.[8]

In contrast, the categorial operations of the C-LAGs reflect the structure of a double ended queue. This structure is well-suited for repetitions of arbitrary number, whereby the repetitions may be modified, e.g. inversed, doubled, halfed, etc.

Parsers in general and C-LAGs in particular are inefficient, however, if the input contains an unknown number of words such that it can only be determined at the end of the analysis whether later words must correspond to earlier words or not. This is the characteristic property of $\mathcal{NP}$-hard languages, i.e., languages which require $\mathcal{N}$ondeterministic $\mathcal{P}$olynomial time for verification and exponential time for analysis.

For LA-grammar, context-free languages like HCFL[9] and L_{no}, on the one hand, and $\mathcal{NP}$-complete context-sensitive languages like 3SAT und Subset Sum, on the other, are structurally similar. These four language have in common that in the first half of the input there may occur arbitrarily many words of which it is not known whether or not they are needed as counterparts in the second half.[10]

The only way a C-LAG can analyze L_{no} is by assigning two interpretations to each word preceding #, one as a 'real' word and one as a 'noise' word. This results in an exponential number of readings for the input chain preceding #, each reading with its own category. For example, if the input is 10010#..., then one reading has the category (10010), representing the hypothesis that all words preceding # are 'real'. Another reading of this input has the category (1001), representing the hypothesis that the last word preceding # is noise, etc.

These hypotheses are generated systematically from left to right.

12.3.3 C3-LAG FOR L_{no}

$LX =_{def} \{[0\ (0)], [1\ (1)], [\#\ (\#)]\}$

$ST_S =_{def} \{[(seg_c)\ \{r_1, r_2, r_3, r_4, r_5\}]\ \}$, where $seg_c, seg_d \in \{0, 1\}$.

$r_1\colon (seg_c)(seg_d) \qquad \Rightarrow \varepsilon \qquad \{r_1, r_2, r_3, r_4, r_5\}$

---

[8] Because L_{no} is a deterministic context-free language, it can be parsed in linear time in PS-grammar. Cf. B. Stubert 1993, p. 71, Lemma 5.1.

[9] That HCFL parses in polynomial time in PS-grammar has several reasons:

1. Context-free PS-grammars use a different method of measuring complexity than C-LAGs. More specifically, the n^3 time complexity for context-free languages in general depends crucially on the use of multi-tape Turing machines. The complexity of C-LAGs, on the other hand, is determined on the basis of the grammars directly.

2. In contrast to abstract automata, no ε-moves are allowed in C-LAGs.

[10] The C-LAG complexity of context-free L_{no} is the same as that of the context-sensitive language L_{no}^3, which generates expressions of the structure W'#W#W'', whereby W' and W'' are noisy versions of W. The LA-grammar of L_{no} is in a higher complexity class than the corresponding PS-grammar because C-LAGs are not designed to utilize the fixed inverse pair structure of the context-free languages and ε-moves are not permitted.

r_2: $(\text{seg}_c)(\text{seg}_d)$ $\quad\quad \Rightarrow (\text{seg}_d)$ $\quad\quad \{r_1, r_2, r_3, r_4, r_5\}$

r_3: $(X)(\text{seg}_c)$ $\quad\quad\quad \Rightarrow (X)$ $\quad\quad\quad \{r_1\ r_2, r_3, r_4, r_5\}$

r_4: $(X)(\text{seg}_c)$ $\quad\quad\quad \Rightarrow (\text{seg}_c\ X)$ $\quad\quad \{r_1\ r_2, r_3, r_4, r_5\}$

r_5: $(X)(\#)$ $\quad\quad\quad\quad \Rightarrow (X)$ $\quad\quad\quad \{r_6\}$

r_6: $(\text{seg}_c\ X)(\text{seg}_c)$ $\quad \Rightarrow (X)$ $\quad\quad\quad \{r_6\}$

$\text{ST}_F =_{def} \{[\varepsilon\ \text{rp}_6]\}$

The +recursive ambiguity of this C3-LAG arises because r_3 and r_4, for example, have (i) compatible (in fact identical) input conditions, (ii) co-occur in rule packages, and (iii) are reapplied in the same analysis path. Rule r_3 ignores the category of the next word, thus treating it as noise. Rule r_4 attaches the category of the next word at the beginning of the new sentence start category, thus ensuring that it will have a counterpart in the second half of the input.

The C3-LAG 12.3.3 for L_{no} is similar to the C3-LAG 11.5.9 for $\mathcal{NP}$-complete **Subset Sum**, where each a_i may be interpreted either as noise or as a 'real' subset. The C3-LAG for **Subset Sum** is context-sensitive, however, because some rules (e.g. r_5) check the beginning and others (e.g. r_4) the end of the sentence start category. The C3-LAG for L_{no} is context-free because its categorial operations all check only the beginnings of sentence start categories.

12.4 Comparing the lower LA- and PS-classes

Context-free PS-grammar has been widely used because it provides the greatest amount of generative capacity within the PS-grammar hierarchy while being computationally tractable. There is general agreement in linguistics, however, that context-free PS-grammar does not properly fit the structures characteristic of natural language. The same holds for computer science, where context-free PS-grammar has likewise turned out to be suboptimal for describing the structures of programming languages.

> It is no secret that context-free grammars are only a first order approximation to the various mechanisms used for specifying the syntax of modern programming languages.[11]

> S. Ginsberg 1980, p.7

Therefore, there has long been a need for an alternative to better describe the syntax of natural and programming languages. Most attempts to arrive at new language classes have consisted in *conservative extensions*, however, which follow context-free PS-grammar too closely. They are based on adding certain mechanisms and result in additional language classes which fit right into the subset relations of the PS-hierarchy. For example, the context-free languages form a proper subset of the tree adjoining languages (TALs),[12] which form a proper subset of the index languages,[13] which in turn form a proper subset of the context-sensitive languages.

---

[11] See also M. Harrison 1978, p. 219ff, in the same vein.

[12] A.K. Joshi et al. 1975.

[13] Hopcroft & Ullman 1979, p. 389 f. A pumping lemma for index languages proved T. Hayashi 1973.

More than five decades of PS-grammar tradition and the long absence of a aubstan-
tial alternative are no valid reasons, however, to regard the PS-grammar hierarchy of
formal languages and its extensions as particularly 'natural.' After all, these language
classes are no more than the result of certain restrictions on a certain formalism.

The context-free languages, for example, are defined in terms of restrictions which
are suggested only by the formalism of rewrite rules (cf. 8.1.2). Similarly, the C1-,
C2-, and C3-languages are defined in terms of restrictions which are suggested only
by the formalim of LA-grammar (cf. 11.5.9). These different kinds of restrictions
result in the language hierarchies of PS- and LA-grammar, respectively, which are
orthogonal to each other.

12.4.1 ORTHOGONAL RELATION BETWEEN C- AND CF-LANGUAGES

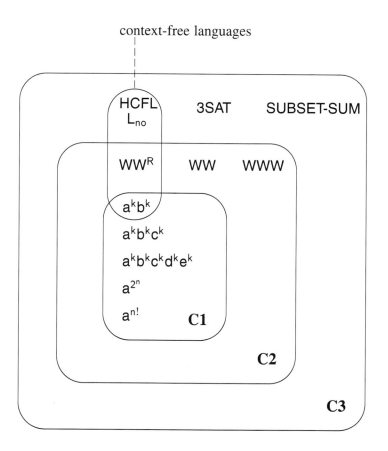

From a pretheoretical point of view one would be inclined to classify the language
$a^k b^k$ with $a^k b^k c^k$, $a^k b^k c^k d^k$, etc., on the one hand, and WW^R with WW on the other.
It is therefore surprising to the untutored that the PS-hierarchy puts $a^k b^k$ and WW^R
into one class (context-free), but $a^k b^k c^k$, $a^k b^k c^k d^k$, etc., with WW into another class
(context-sensitive). The LA-hierarchy is intuitively more natural because there a^k,

$a^k b^k$, $a^k b^k c^k$, $a^k b^k c^k d^k$, etc. are classified together as linear C1-languages, while WW^R and WW, $WW^R W$ and WWW, etc. are classified as C2-languages.

If the distinction between deterministic context-free languages $\mathcal{L}_{dcf}$ and nondeterministic context-free languages $\mathcal{L}_{cf}$ is made within PS-grammar, the orthogonal relation between the PS- and the LA-hierarchy appears even more clearly.

12.4.2 ORTHOGONAL $\mathcal{L}_{dcf}$, $\mathcal{L}_{cf}$, C_1, C_2, AND C_3 CLASSIFICATIONS

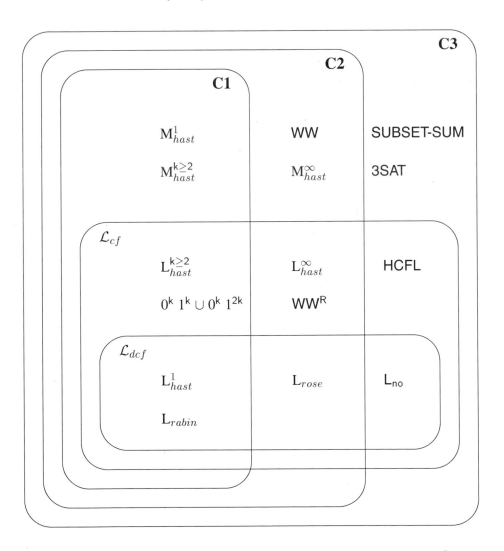

The class of $\mathcal{L}_{dcf}$ cuts across the three subclasses of C in the same way as $\mathcal{L}_{cf}$.

The alternative classifications of LA-grammar provide a new perspective on the theory of formal languages. Furthermore, because the PS-grammar hierarchy may be reconstructed in LA-grammar (cf. Section 11.2), open questions of classic automata theory may be transferred directly to LA-grammar (cf. Section 12.2).

12.5 Linear complexity of natural language

A context-sensitive language which is not a C-language would prove the proper inclusion of the C-languages in the B-languages (cf. 12.2.2). In such a language, the category length would have to grow just within the LBA-definition of context-sensitive languages, but grow faster than the pattern-based categorial operations of the C-LAGs would permit. That this type of language should be characteristic for the structure of natural language is highly improbable.

If the natural languages are contained in the C-LAGs, however, then the following two questions are equivalent:

(i) *How complex are the natural languages?*

(ii) *How ambiguous are the natural languages?*

This is because the C-LAG subclasses differ solely in their degrees of ambiguity.

An utterance is called ambiguous if more than one meaning$_2$ may be derived by the hearer. This may be due to a syntactic, a semantic, or a pragmatic ambiguity. For complexity analysis – being concerned with the combinatorial structure of expressions – only syntactic ambiguity is relevant.

In the SLIM theory of language, a syntactic ambiguity arises if an expresssion is assigned more than one structural analysis. A semantic ambiguity arises if a syntactically unambiguous expression has more than one meaning$_1$. A pragmatic ambiguity is caused by a meaning$_1$ having more than one *use* relative to a given context.

12.5.1 SLIM-THEORETIC ANALYSIS OF AMBIGUITY

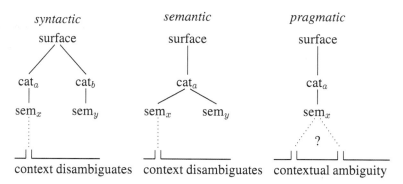

Only syntactic and semantic ambiguities are properties of the type of the expression used, while pragmatic ambiguities are properties of utterances in which tokens of expressions are used.

A syntactic ambiguity is characterized by alternative categories of the surface (here cat$_a$ and cat$_b$), such that each categorial reading has its own meaning$_1$ (here sem$_x$ and sem$_y$). For example, won is syntactically ambiguous between a noun referring to the currency of Korea and the past tense form of the verb to win. If the context does not disambiguate between sem$_x$ and sem$_y$, the syntactic ambiguity will cause the utterance to have more than one meaning$_2$.

A semantic ambiguity is characterized by a surface having only one syntactic analysis, but more than one meaning$_1$. For instance, the surface perch is semantically ambiguous, one literal meaning standing for a kind of fish, the other for a place to roost. Syntactically, however, perch is not ambiguous because both readings are of the category noun.[14] For internal matching, a semantic ambiguity resembles a syntactic one insofar as in either case the expression used has more than one meaning$_1$.

A pragmatic ambiguity consists in alternative uses of one meaning$_1$ relative to a given context. For example, in a context with two equally prototypical tables right next to each other, the utterance of Put the coffee on the table would be pragmatically ambiguous because it is not clear which of the two tables is meant by the speaker (cf. Section 5.2). In contrast to syntactic and semantic ambiguities, a pragmatic ambiguity by its very nature cannot be disambiguated by the context.

A syntactic ambiguity causes an expression to have more than one meaning$_1$. Yet an expression is called semantically ambiguous only if it is syntactically unambiguous (regarding the particular readings in question). Similarly, an utterance is called pragmatically ambiguous only, if the expression used is neither semantically nor syntactically ambiguous. The syntactic or semantic ambiguity of a single word form is also called a lexical ambiguity.

For determining the complexity of natural language expressions, phenomena of pragmatic ambiguity are irrelevant. This is because pragmatic ambiguities do not affect the type of the expression, but arise in the interaction between the semantic interpretation of the expression and the context (cf. Section 4.4).

Phenomena of semantic ambiguity are likewise irrelevant for natural language complexity because semantic ambiguities are by definition associated with syntactically unambiguous surfaces. For example, in light of the fact that the two readings of perch have the same category, it would be superfluous to assign two different syntactic analyses to the sentence The osprey is looking for a perch:

12.5.2 INCORRECT ANALYSIS OF A SEMANTIC AMBIGUITY

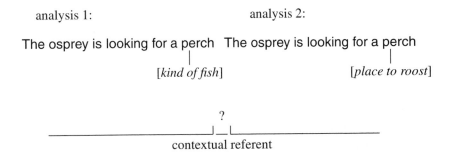

This analysis is misguided because it treats a semantic ambiguity needlessly as a syntactic one.

---

[14] We are ignoring here the denominal verb to perch which stands for sitting on a place to roost.

The linguistically correct analysis treats the surface of the sentence in question as syntactically unambiguous, handling the ambiguity instead semantically by assigning two different meanings$_1$ to **perch**.

12.5.3 CORRECT ANALYSIS OF A SEMANTIC AMBIGUITY

The osprey is looking for a perch

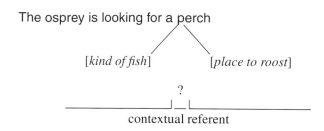

[*kind of fish*]   [*place to roost*]

?

contextual referent

Even though the syntactic derivation is unambiguous, the semantic interpretation provides two different meanings$_1$ which in turn provide for the possibility of two different meanings$_2$ in the interpretation of the sentence relative to a given context. The method used in 12.5.3 is called semantic doubling.[15]

The method of semantic doubling is based on the [2+1] level structure of natural communication. Assigning more than one meaning$_1$ to an analyzed surface realizes the surface compositional insight that it is not always necessary to push semantic distinctions through to the level of syntax.

An ambiguous expression can be analyzed in terms of semantic doubling whenever the distinctions at the semantic level are associated either with *no* (as in 12.5.3) or with a *systematic* distinction (as in 12.5.4 below) at the syntactic level. Thus, syntactic ambiguity may be restricted to cases in which different semantic readings are associated with an unsystematic – and thus unpredictable – syntactic alternative.

As an example of semantic doubling in the case of a systematic syntactic alternative consider prepositional phrases. These generally permit a postnominal and an adverbial interpretation, as shown by the following examples.

12.5.4 MULTIPLE INTERPRETATIONS OF PREPOSITIONAL PHRASES

The man saw the girl with the telescope.
Julia ate the apple on the table behind the tree in the garden.

The first example has two different meaning$_1$ interpretations. On the adverbial reading, the prepositional phrase **with the telescope** modifies the verb **saw**. On the postnominal reading, the prepositional phrase modifies **the girl**.

The second example in 12.5.4 contains three prepositional phrases rather than one, illustrating the theoretical possibility of adding an unlimited number of prepositional phrases. This raises the question of whether or not the number of syntactic readings

---

[15] First proposed in CoL, p. 219-232 and 239-247.

should be doubled each time a new prepositional phrase is added. The following analysis illustrates a (mistaken) syntactic treatment of the alternative interpretations of prepositional phrases.

12.5.5 RECURSIVE PSEUDO-AMBIGUITY

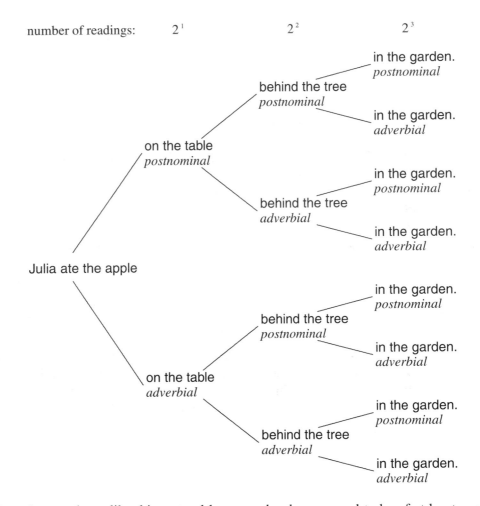

Based on analyses like this, natural language has been argued to be of at least exponential complexity.

For the mechanism of natural communication, however, such a multiplying out of semantic readings in the syntax has no purpose. An analysis like 12.5.5 is possible, but only because any grammatical analysis can be written inefficiently.[16] A good syntactic analysis, however, should aim at finding the absolute minimum of literal meanings sufficient for handling all possible uses.

---

[16] Even for, e.g., unambiguous $a^k b^k$ one may easily write various ambiguous grammars, raising the complexity from linear to polynomial, exponential, or even undecidable.

For communication, an adequate treatment of prepositional phrases requires no more than alternative adverbial and postnominal readings on the *semantic* level. This alternative analysis based on semantic doubling is illustrated in 12.5.6.

12.5.6 CORRECT ANALYSIS WITH *semantic doubling*

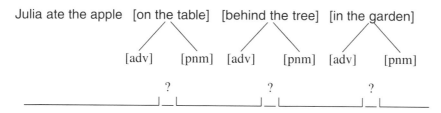

The surface is analyzed as syntactically unambiguous.[17] The semantic interpretation, however, systematically assigns to each prepositional phrase two different literal meanings at the semantic level.

Like 12.5.3, analysis 12.5.6 is based on the [2+1] level structure of natural communication. Analysis 12.5.6 is sufficient for modeling the different meanings$_2$ that may arise in the interpretation relative to different contexts.

The semantic doubling analysis 12.5.6 is more efficient and concrete than the analysis 12.5.5 based on multiplying out the semantic readings in the syntax. While 12.5.5 is of exponential complexity, the alternative analysis 12.5.6 is of the lowest possible complexity, namely linear. This holds for both, the level of syntax and of semantics. The number of alternative meanings$_1$ provided by 12.5.6 for matching with the interpretation context is well below **2n**.

If apparent +recursive syntactic ambiguities of natural language (e.g. 12.5.5) can all be treated as semantic ambiguities (e.g. 12.5.6), then only –recursive ambiguities like 11.3.5 and 11.3.6 remain as candidates for a syntactic treatment. Without +recursive syntactic ambiguities, however, natural languages form a subset of the class of C1-languages.

We formulate this conclusion as the empirical complexity hypothesis for natural language syntax, CoNSyx hypothesis for short.

12.5.7 CoNSyx HYPOTHESIS
(COMPLEXITY OF NATURAL LANGUAGE SYNTAX)

> The natural languages are contained in the class of C1-languages and parse in linear time.

The linear complexity assigned by CoNSyx to natural language is in agreement with the fact that human speakers and hearers usually have no difficulty to produce and

---

[17] Syntactically, the prepositional phrases are categorized as (adv&pnm) using the multicat notation (see Section 15.2).

understand[18] even complicated texts in their native language in real time. The CoNSyx hypothesis will be complemented in Chapter 21 by a corresponding hypothesis for the complexity of semantics, called the CoNSem hypothesis (cf. 21.5.2).[19]

Exercises

Section 12.1

1. Why are the lower classes of a language hierarchy, i.e., those with comparatively low generative capacity, especially interesting for the empirical work in linguistics?
2. Describe the complexity degrees of the subclasses in the LA- and PS-grammar hierarchy.
3. How is the (non)-equivalence of two language classes formally shown?
4. Compare the inherent complexity of $a^k b^k c^k$ and Subset Sum.
5. Which properties of a language determine the language class it belongs to?
6. Explain in what sense the language hierarchies of LA- and PS-grammar are orthogonal.

Section 12.2

1. Compare the inclusion relations in the PS- and the LA-hierarchy.
2. By which method is the proper inclusion of the type 3 language class in the type 2 language class and that of the type 2 language class in the type 1 language class formally proven in PS-grammar?
3. Explain the definition of the languages classes $\mathcal{P}$ and $\mathcal{NP}$.
4. Why is the language 3SAT inherently complex?
5. Which unsolved problems of classic automata theory are related to the open questions of whether $C2 \subset C3$ and whether $C3 \subset B$?

---

[18] At least on the syntactic level.

[19] The structural properties assigned by CoNSyx to natural language syntax may and should be tested empirically in the continuing analysis of various different languages. Possible counterexamples to 12.5.7 would be constructions of natural language with +recursive syntactic ambiguities which do not allow an alternative treatment based on semantic doubling. Given the mechanism of natural communication within the SLIM theory of language, it seems unlikely that such constructions may be found.

Section 12.3

1. In which subclass of the C-languages is L_{no} and why?
2. Write a PS- and an LA-grammar for L_{no}^3, defined as W'#W#W", where W' and W" are noisy versions of W.
3. In which PS-grammar classes are L_{no} and L_{no}^3, respectively?
4. In which LA-grammar classes are L_{no} and L_{no}^3, respectively?
5. Compare the inherent complexity of L_{no} and L_{no}^3 from the viewpoint of the PS-grammar classification of these languages.

Section 12.4

1. What is a conservative extension of context-free PS-grammar? Give two examples and compare their respective classes.
2. Is $a^k b^k c^k d^k e^k f^k$ a TAL? What is the LA-class of this language?
3. What is the structural reason why $a^{i!}$ is not an Index language? Consult Hopcroft & Ullman 1979. What is the LA-class of this language?
4. Describe the relation between the class of context-free languages and the class of C-languages.
5. Why is the hierarchy of LA-Grammar, including the subhierarchy of the C-LAGs, more natural than that of PS-Grammar ?
6. Explain how the orthogonal relation between the LA- and the PS-hierarchy also shows up in the subclasses of deterministic and nondeterministic context-free languages.

Section 12.5

1. Why don't the natural languages fit the class of context-free languages?
2. Why is it likely that the natural languages are a subset of the C-languages?
3. If the natural languages are in the class of C-languages, what is the only possible cause for a higher (i.e. non-linear) complexity degree of natural language?
4. What are the types of ambiguity in natural language?
5. Which types of ambiguity are irrelevant for the complexity of natural language and why?
6. Explain the method of semantic doubling and its consequence on the complexity degree of natural language.
7. Are there +recursive syntactic ambiguities in natural language?
8. Explain why the complexity analysis of natural language depends on the theory of grammar, using examples 9.2.1 and 9.2.2.
9. Explain why the complexity analysis of natural language depends on the theory of language, using the second example of 12.5.4.
10. Explain the CoNSyx hypothesis. How can it be disproved empirically?
11. What is the practical importance of the CoNSyx hypothesis?

Morphology and Syntax

13. Words and morphemes

Part I analyzed natural communication within the SLIM theory of language. Part II presented formal language theory in terms of methodology, mathematical complexity, and computational implementation. With this background in mind, we turn in Part III to the morphological and syntactic analysis of natural language. This chapter begins with the basic notions of morphology.

Section 13.1 describes the principles of combination in morphology, namely inflection, derivation, and composition, as well as the distinction between the open and the closed word classes. Section 13.2 presents a formal definition of the notions morpheme and allomorph. Section 13.3 describes two special cases of allomorphy, suppletion and bound morphemes. Section 13.4 explains the main tasks of automatic word form recognition, namely categorization and lemmatization. Section 13.5 describes the three basic methods of automatic word form recognition, called the word form, the morpheme, and the allomorph method.

13.1 Words and word forms

The words of a natural language are concretely realized as word forms. For example, the English word write is realized as the word forms write, writes, wrote, written, and writing. The grammatical well-formedness of natural language sentences depends on the choice of the word *forms*.

13.1.1 DIFFERENT SYNTACTIC COMPATIBILITIES OF WORD FORMS

> *write
> *writes
> *wrote
> *John has* written *a letter.*
> *writing

In written English, word forms are separated by spaces. For practical purposes, this is sufficient for distinguishing the word forms in a text. Francis & Kučera 1982 define "a graphic word as a string of continuous alphanumeric characters with space on either side; may include hyphens and apostrophes, but no other punctuation marks."

Theoretical linguistics, especially American structuralism, has tried to arrive at a watertight structural definition of the notions word and word form. To establish word forms scientifically, for example in the description of an unknown exotic language, it was proposed to use the method of distribution tests, realized as substitution and movement tests.[1]

In contrast, practical work has used the fact that native speakers have an intuitively clear notion of what the words and word forms of their language are. As observed by E. SAPIR (1884–1939), aborigines who do not read or write can nevertheless dictate in their language word form by word form (E. Sapir 1921, p. 33).

In line with this insight, traditional grammar avoided turning the scientific definition of words and word forms into a major problem and concentrated instead on the classification of what it took as intuitively obvious. The results are used in computational linguistics as the theoretical and empirical basis of automatic word form recognition.

In traditional morphology, the following principles of combination are distinguished.

13.1.2 COMBINATION PRINCIPLES OF MORPHOLOGY

1. *Inflection* is the systematic variation of a word with which it can perform different syntactic and semantic functions, and adapt to different syntactic environments. Examples are learn, learn/s, learn/ed, and learn/ing.
2. *Derivation* is the combination of a word with an affix.[2] Examples are clear/ness, clear/ly, and un/clear.
3. *Composition* is the combination of two or more words into a new word form. Examples are gas/light, hard/wood, over/indulge, and over-the-counter.

These three processes may also occur simultaneously, as in over/indulg/er/s. Furthermore, these processes are *productive* in the sense that a new word like infobahning[3] may be inflected in English as a verb (we infobahn, he infobahn/s, we infobahn/ed, . . .), permit derivations like infobahn/er/s, and may participate in compositions, like pseudo-infobahn/er.

The grammarians of ancient Greece and Rome arranged inflectional word forms into paradigms. We define a word[4] as the set of word forms in its inflectional paradigm.

---

[1] These same tests were also used in the attempt to motivate syntactic constituent structures (cf. 8.4.6, 8.4.7, and 8.4.8).

[2] See the distinction of free and bound morphemes in Section 13.3.

[3] In analogy to autobahning, coined by Americans stationed in Germany after Word War II from 'Autobahn' = highway.

[4] Our terminology is in concord with J. Sinclair 1991:

> Note that a word form is close to, but not identical to, the usual idea of a word. In particular, several different word forms may all be regarded as instances of the same word. So drive, drives, driving, drove, drove, driven, and perhaps driver, drivers, driver's, drivers', drive's, make up ten different word forms, all related to the word drive. It is usual in defining a word form to ignore the distinction between upper and lower case, so SHAPE, Shape, and shape, will all be taken as instances of the same word form.

13.1.3 DEFINITION OF THE NOTION *word*

Word $=_{def}$ {associated analyzed word forms}

According to this definition, a word is an abstract concept which is concretely manifested solely in the associated word forms.[5]

As the name of a word serves its base form. The traditional base form of nouns is the nominative singular, e.g. book, of verbs the infinitive of the present tense, e.g. learn, and of adjective-adverbials the adjective in the positive, e.g. slow.

In LA-grammar, the morphological analysis of word forms is represented as ordered triples consisting of the surface, the syntactic category, and the semantic representation,[6] as in the following example.

13.1.4 EXAMPLE OF AN ANALYZED WORD FORM

[wolves (P-H) wolf]

The surface wolves serves as the key for relating the analyzed word form to corresponding unanalyzed surfaces occurring in texts (cf. 13.4.6). The category (P-H) stands for 'plural non-human' and characterizes the combinatorics of the word form. The semantic representation named wolf applies to the word as a whole (rather than just the word form in question) and serves as the common link between the different forms of the paradigm.[7]

13.1.5 ANALYSIS OF AN INFLECTING WORD

word   *word forms*

wolf $=_{def}$ {[wolf (SN) wolf],
              [wolf's (GN) wolf],
              [wolves (PN) wolf],
              [wolves' (GN) wolf]}

Another term for our notion of a word is *lexeme* (see for example P.H. Matthews 1972, 1974). Terminologically, however, it is simpler to distinguish between *word* and *word forms* than between *lexeme* and *word forms* (or even *lexeme forms*) .

[5] A clear distinction between the notions of *word* and *word form* is not only of theoretical, but also of practical relevance. For example, when a text is said to consist of a '100 000 words' it remains unclear whether the number is intended to refer (i) to the running word forms (tokens), (ii) to the different word forms (i.e., the types, as in a word form lexicon), or (iii) the different words (i.e., the types, as in a base form lexicon). Depending on the interpretation, the number in question may be regarded as small, medium, or large.

[6] LA-grammar uses this format also in the syntactic analysis of natural language.

[7] Semantic properties restricted to specific word forms are also coded in the third position. An example are tense and mood distinctions in German verbs. For example, gibst and gabst have the same combinatorics (category), but differ semantically in their respective tense values.

The different categories in the second position of the analyzed word forms characterize their different combinatorial properties.[8]

Word and word form are distinct also in noninflecting words like and. That the two
may seem to coincide is because there the set of word forms is the unit set.

13.1.6 ANALYSIS OF A NONINFLECTING WORD

word   *word forms*

and $=_{def}$ { [and (cnj) and] }

For reasons of computational efficiency and linguistic concreteness (surface compositionality), the morphological component of the SLIM theory of language takes
great care to assign no more than one category (syntactic reading) per word form
surface whenever possible. This *distinctive* categorization characterizes the combinatorial differences between the concrete surface forms of a word – in contrast to the
traditional *exhaustive* categorization which is based instead on the number of places
in a paradigm schema (see Section 18.2).

The words of a natural language are traditionally divided into the following parts of
speech.

13.1.7 PARTS OF SPEECH

– *verbs*, e.g., walk, read, give, help, teach, . . .

– *nouns*, e.g., book, table, woman, messenger, arena, . . .

– *adjective-adverbials*, e.g., quick, good, low, . . .

– *conjunctions*, e.g., and, or, because, after, . . .

– *prepositions*, e.g., in, on, over, under, before, . . .

– *determiners*, e.g., a, the, every, some, all, any, . . .

– *particles*, e.g., only, already, just. . .

The first three parts of speech are jointly called the *open* classes, whereas the remain-
ders constitute the *closed* classes.

---

[8] In 13.1.5, the category (SN) stands for 'singular noun,' (PN) stands for 'plural noun,' and (GN) stands
for 'genitive noun.' The distinction between non-genitive singulars and plurals is important for the
choice of the determiner, e.g. every vs. all (cf. 17.1.1). Because genitives in English serve only as
prenominal modifiers, e.g. the wolf's hunger, their number distinction need not be coded into the
syntactic category.

13.1.8 OPEN AND CLOSED CLASSES

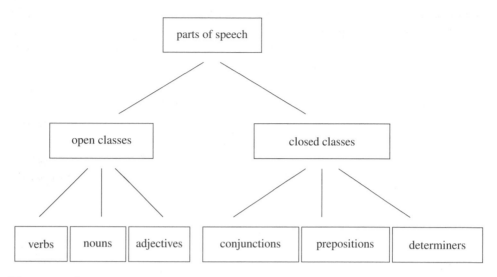

The open classes comprise several 10 000 elements, while the closed classes contain only a few hundred words.

In the open classes, the morphological processes of inflection, derivation, and composition are productive, for which reason it is difficult to specify their elements exactly. Also, the use of words is constantly changing, with new ones entering and obsolete ones leaving the current language. The closed classes, on the other hand, show neither a comparable size and fluctuation, nor are the processes of inflection, derivation, or composition productive.

From the viewpoint of semantic-pragmatic interpretation, the elements of the open classes are also called *content words,* while the elements of the closed classes are also called *function words*. In this distinction, however, the sign type (cf. Chapter 6) must be taken into consideration besides the category. This is because only the *symbols* among the nouns, verbs, and adjective-adverbials are content words in the proper sense. *Indices*, on the other hand, e.g. the personal pronouns he, she, it etc., are considered function words even though they are of the category noun. Indexical adverbs like here or now do not even inflect, forming no comparatives and superlatives. The sign type *name* is also a special case among the nouns.

13.2 Segmentation and concatenation

Because a word may have several different word forms, there arises the question of how many word forms there are for a given set of words. To get a general idea, let us consider 40 000 of the most frequently used elementary base forms of German.[9]

---

[9] The morphology of English happens to be simple compared to, e.g., French, Italian or German. There is little inflection in English. Furthermore, much of composition may be regarded as part of English

Elementary base forms are words which are not derived or composed from other words. Of the 40 000 base forms in question, 23 000 are nouns, 6 000 are verbs, and 11 000 are adjective-adverbials.

German nouns have between 2 to 5 different inflectional surfaces (cf. 14.5.1) – averaging about 4. The regular verbs have about 24 different forms (cf 14.5.3). And the adjective-adverbials have normally 18 different inflectional forms (cf. 14.5.4). These numbers are based on a distinctive categorization as illustrated in 13.1.5.[10]

Using a maximally concrete, surface compositional, distinctive categorization, the lexicon in question would correspond to the following numbers of inflectional forms:

13.2.1 RELATION OF WORDS AND THEIR INFLECTIONAL FORMS

	base forms	inflectional forms
nouns:	23 000	92 000
verbs:	6 000	144 000
adjective-adverbials:	11 000	198 000
	40 000	434 000

According to this estimate, the relation between words and their inflectional forms in German is about 1 to 10 on average.

In addition to the inflectional morphology of German, however, there is also derivational and compositional morphology, allowing the formation of new complex words from old ones. Consider, for example, noun noun composition, such as **Haus/schuh**, **Schuh/haus**, or **Jäger/jäger**, which is of the complexity n^2. This means that from 20 000 nouns 400 000 000 possible compounds of length 2 can be derived (base forms).

Furthermore, noun noun compounds of length 3, such as **Haus/schuh/sohle**, **Sport/schuh/haus**, or **Jäger/jäger/jäger** are of complexity n^3. This means that an additional 8 000 000 000 000 000 (eight thousand trillion) possible words may be formed. Because there is no grammatical limit on the length of noun compounds, the number of possible word forms in German is infinite. These word forms exist potentially because of the inherent productivity of morphology.

---

syntax rather than morphology – in accordance with Francis & Kučera's 1982 definition of a graphic word form cited above. For example, **kitchen table** or **baby wolves** are written as separate words, whereas the corresponding composita in German are written as one word form, e.g. **Küchentisch** and **Babywölfe**. For this reason, some morphological phenomena will be illustrated in this and the following two chapters in languages other than English.

[10] An exhaustive categorization based on traditional paradigm tables would arrive at much higher numbers. For example, the adjective-adverbials of German have 18 inflectional forms per base form according to a distinctive categorization. In contrast, an exhaustive categorization, as presented in the Grammatik-Duden, p. 288, assigns 147 analyzed inflectional forms per base form, whereby the different analyses reflect distinctions of grammatical gender, number, case, definiteness, and comparison, which in most cases are not concretely marked in the surface.

In contradistinction to the possible words forms, the set of actual word forms is finite. There is no limit, however, on the number of 'new' words, i.e. words that have never been used before. These are called neologisms, and coined spontaneously by the language users on the basis of known words and the rules of word formation.

A cursory look through Newsweek or the New Yorker will render English neologisms like the following:

13.2.2 EXAMPLES OF NEOLOGISMS

insurrectionist (inmate)	three-player (set)
copper-jacketed (bullets)	bad-guyness
cyberstalker	trapped-rat (frenzy)
self-tapping (screw)	dismissiveness
migraineur	extraconstitutional (gimmick)

None of these words may be found in a contemporary standard dictionary, yet they have occurred and readers have no problem to understand them.

Because new word forms never observed before are constantly formed morphological analysis should not merely list as many analyzed word forms as possible. Rather, the goal must be a rule-based analysis of potential word forms on demand.

In traditional morphology, word forms are analyzed by disassembling them into their elementary parts. These are called morphemes and defined as the smallest meaningful units of language. In contrast to the number of possible words and word forms, the number of morphemes in a language is finite.

The notion of a morpheme is a linguistic abstraction which is manifested concretely in the form of finitely many allomorphs. The term allomorph is of Greek origin and means "alternative shape." For example, the morpheme wolf is realized as the two allomorphs wolf and wolv.

Just as the elementary parts of the syntax are really the word forms (and not the words), the elementary parts of morphology are really the allomorphs. Accordingly, the definition of morpheme is analogous to that of word.

13.2.3 DEFINITION OF THE NOTION *morpheme*

morpheme $=_{def}$ {associated analyzed allomorphs}

Like the word forms, allomorphs are formally analyzed as ordered triples, consisting of the surface, the category, and the semantic representation. The following examples, based on the English noun wolf, are intended to demonstrate these basic concepts of morphology as simply as possible.[11]

---

[11] Nouns of English ending in -lf, such as calf, shelf, self, etc. form their plural in general as -lves. One might prefer for practical purposes to treat forms like wolves, calves, or shelves as elementary allomorphic forms, rather than combining an allomorphic noun stem ending in -lv with the plural allomorph es. This, however, would prevent us from explaining the interaction of concatenation and allomorphy with an example from English.

13.2.4 FORMAL ANALYSIS OF THE MORPHEME wolf

morpheme	*allomorphs*

wolf $=_{def}$ {[wolf (SN SR) wolf],
                    [wolv (PN SR) wolf]}

The different allomorphs wolf and wolv are shown to belong to the same morpheme by the common semantic representation in the third position. As (the name of) the semantic representation we use the base form of the allomorph, i.e. wolf, which is also used as the name of the associated morpheme.

Some surfaces such as wolf can be analyzed alternatively as an allomorph, a morpheme (name), a word form, or a word (name).

13.2.5 COMPARING MORPHEME AND WORD wolf

morpheme	*allomorphs*		*word*	*word forms*
wolf $=_{def}$	{wolf,		wolf $=_{def}$	{wolf,
	wolv}			wolf/'s,
				wolv/es,
				wolv/es/'}

Other surfaces can be analyzed only as an allomorph, e.g. wolv, or only as a word form, e.g. wolves.

Besides the segmentation into morphemes or allomorphs, a word form surface may also be segmented into the units of its realization medium. Thus, written surfaces may be segmented into letters, and spoken surfaces into syllables or phonemes.

13.2.6 ALTERNATIVE FORMS OF SEGMENTATION

allomorphs:	learn/ing
syllables:	lear/ning
phonemes:	l/e/r/n/i/n/g
letters:	l/e/a/r/n/i/n/g

The syllables lear and ning do not coincide with the allomorphs learn and ing, and similarly in the case of letters and phonemes. While, e.g., syllables are important in automatic speech recognition (cf. Section 1.4), morphological analysis and automatic word form recognition aim at segmenting the surface into morphemes or allomorphs, which are independent[12] of the concrete realization in speaking or writing.

---

[12] In as much as the medium of realization influences the representation of allomorphs (types), there is the distinction between allo*graphs* in written and allo*phones* in spoken language. Allographs are, e.g., happy vs. happi-, allophones the present vs. past tense pronunciation of read.

13.3 Morphemes and allomorphs

The number and variation of allomorphs of a given morpheme determine the degree of regularity of the morpheme and – in the case of a free morpheme – the associated word. An example of a regular word is the verb to **learn**, the morpheme of which is defined as a set containing only one allomorph (compare 13.1.6).

13.3.1 THE REGULAR MORPHEME learn

morpheme   *allomorphs*

learn $=_{def}$ {[learn (N ... V) learn]}

A comparatively irregular word, on the other hand, is the verb to **swim**, the morpheme of which has four allomorphs, namely **swim**, **swimm**,[13] **swam**, and **swum**. The change of the stem vowel may be found also in other verbs, e.g. **sing**, **sang**, **sung**, and is called *ablaut*.

13.3.2 THE IRREGULAR MORPHEME swim

morpheme   *allomorphs*

swim $=_{def}$ {[swim (N ... V1) swim],
                  [swimm (... B) swim],
                  [swam (N ... V2) swim],
                  [swum (N ... V) swim]}

In 13.3.2, the allomorph of the base form is used as the name of the morpheme. Thus, we may say that **swam** is an allomorph of the morpheme **swim**.

Cases in which there is no similarity at all between the allomorphs of a given morpheme are called *suppletion*.

13.3.3 AN EXAMPLE OF SUPPLETION

morpheme   *allomorphs*

good $=_{def}$ {[good (ADV IR) good],
                  [bett (CAD IR) good],
                  [b (SAD IR) good]}

---

[13] This allomorph is used in the progressive **swimm/ing**, avoiding the concatenative insertion of the gemination letter. A psychological argument for handling a particular form non-concatenatively is frequency. Based on speech error data, P.J. Stemberger & B. MacWhinney 1986 provide evidence that the distinction between rote and combinatorial formation is not based only on regularity, but also on frequency, so that even regular word forms can be stored if they are sufficiently frequent.

While the regular comparison in, e.g.,

fast, fast/er, fast/est

uses only one allomorph for the stem, the irregular comparison in, e.g.,

good, bett/er, b/est

uses several.[14] Even in a suppletive form like bett, the associated morpheme is readily available as the third element of the ordered triple analysis.

In structuralism, morphemes of the open and closed classes are called *free* morphemes, in contradistinction to *bound* morphemes. A morpheme is *free* if it can occur as an independent word form, e.g. *book*. Bound morphemes, on the other hand, are affixes such as the prefixes un-, pre-, dis-, etc., and the suffixes -s, -ed, -ing, etc., which can occur only in combination with free morphemes.

The following example is a simplified analysis of the English plural morpheme, which has been claimed to arise in such different forms as book/s, wolv/es, ox/en, and sheep/#.

13.3.4 EXAMPLE OF A BOUND MORPHEME (hypothetical)

morpheme   *allomorphs*

-s $=_{def}$ {[s (PL1) plural],
                 [es (PL2) plural],
                 [en (PL3) plural],
                 [# (PL4) plural]}

In bound morphemes, the choice of the morpheme name, here -s, and the base form of the allomorph, here 'plural,' is quite artificial. Also, postulating the 'zero allomorph' # is in violation of the principle of surface compositionality (see Sections 4.5 and 21.3).

13.4 Categorization and lemmatization

The morphological analysis of an unknown word form consists in principle of the following three steps. First, the unanalyzed surface is disassembled into its basic elements[15] (segmentation). Second, the basic elements are analyzed in terms of their grammatical definitions (lexical look-up). Third, the analyzed basic elements are reassembled on the basis of rules whereby the overall analysis of the word form is derived (concatenation).

Concatenation applies simultaneously to the surface, the category, and the semantic representation, as shown by the following example based on German.

---

[14] For practical purposes, one may analyze good, better, best as basic allomorphs without concatenation.

[15] Depending on the approach, the basic elements of word forms are either the allomorphs or the morphemes.

13.4.1 MORPHOLOGICAL ANALYSIS OF ungelernte

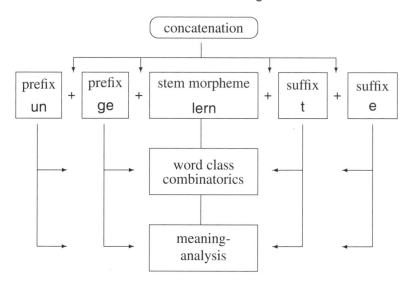

In LA-grammar, the simultaneous concatenation of surface, category, and semantic representation is formally represented by the format of ordered triples.

13.4.2 SCHEMATIC DERIVATION IN LA-GRAMMAR

```
("un" (CAT1) MEAN-a) + ("ge" (CAT2) MEAN-b)
   ("un/ge" (CAT3) MEAN-c) + ("lern" (CAT4) MEAN-d)
      ("un/ge/lern" (CAT5) MEAN-e) + ("t" (CAT6) MEAN-f)
         ("un/ge/lern/t" (CAT7) MEAN-g) + ("e" (CAT8) MEAN-h)
            ("un/ge/lern/t/e" (CAT9) MEAN-i)
```

This schematic[16] analysis goes beyond the structure of 13.4.1 in that it is based on the left-associative derivation order, whereas no derivation order is specified in 13.4.1.

For automatic word form recognition, the following components are required.

13.4.3 COMPONENTS OF WORD FORM RECOGNITION

– *On-line lexicon*

For each element (e.g. morpheme) of the natural language there must be defined a lexical analysis which is stored electronically.

– *Recognition algorithm*

Using the on-line lexicon, each unknown word form (e.g. wolves) must be characterized automatically with respect to categorization and lemmatization:

  – *Categorization*

    consists in specifying the part of speech (e.g. noun) and the morphosyntactic properties of the surface (e.g. plural); needed for syntactic analysis.

---

[16] For simplicity the categories and meanings of the different word form starts and next morphemes are represented as CATn and MEAN-m in 13.4.2.

– *Lemmatization*
   consists in specifying the correct base form (e.g. wolf); provides access to the corresponding lemma in a semantic lexicon.

The formal structure of an on-line lexicon is similar to that of a traditional dictionary. It consists of alphabetically ordered lemmata of the following structure:

13.4.4 BASIC STRUCTURE OF A LEMMA

[surface   (lexical description)]

The lemmata are arranged in the alphabetical order of their surfaces. The surfaces serve as keys which are used for both, the ordering of the lemmata during the building of the lexicon and the finding of a certain lemma once the lexicon has been built. The surface is followed by the associated lexical description.

Because traditional and electronic lexica are based on the same basic structure, traditional dictionaries are well-suited for lemmatization in automatic word form recognition, provided they exist on-line and no copy-rights are violated. For example, in Webster's *New Collegiate Dictionary*, the word wolf has the following lemma:

13.4.5 LEMMA OF A TRADITIONAL DICTIONARY (*excerpt*)

[1]**wolf** \'wu̇lf\ *n. pl* **wolves** \'wu̇lvz\ *often attributed* [ME, fr. OE *wulf*; akin to OHG *wolf*, L *lupus*, Gk *lykos*] **1** *pl also* **wolf a:** any of various large predatory mammals (genus *Canis* and exp. *C. lupus*) that resemble the related dogs, are destructive to game and livestock, and may rarely attack man esp. when in a pack – compare COYOTE, JACKAL **b:** the fur of a wolf . . .

The surface is followed by the lexical description, which specifies the pronunciation, the part of speech (n), the plural form in writing and pronunciation, the etymology, and a number of semantic descriptions and pragmatic uses.

The crucial properties of a lemma like 13.4.5 are the quality of the information contained and the structural consistency of its coding. If these are given on-line, the information can be renamed, restructured, and reformatted automatically[17] without losing any of the original information.

The recognition algorithm in its simplest form consists in matching the surface of the unknown word form with the corresponding key of a lemma in the on-line lexicon, thus providing access to the relevant lexical description.

13.4.6 MATCHING A SURFACE ONTO A KEY

word form surface: wolf

   | *matching*

lemma: [wolf (lexical description)]

---

[17] For this, special programming languages like AWK (A.V. Aho, B.W. Kerningham & P. Weinberger 1988) and PERL (L. Wall & R.L. Schwartz 1990) are available.

There exist several computational methods to match a given surface automatically with the proper lemma in an electronic lexicon.[18]

The simplest is a *linear search*, i.e., going sequentially through the list of lemmata until there is a match between the unknown surface and the key. In small lexica (containing up to 50 lemmata) this method is well-suited. Possible applications are the formal languages of Part II, where each surface must be assigned a category by way of lexical lookup.

The lexica of natural language are considerably larger, however, containing between 20 000 and 1 000 000 entries, depending on their purpose. Even more importantly, most words are related to several word *forms* which must be categorized and lemmatized. Because in the natural languages

- the number of word forms is considerably larger than the number of words, at least in inflectional and agglutinating languages, and
- the lexical lemmata normally define words rather than word forms,

it is best to handle categorization and lemmatization, on the one hand, and access to the lemma, on the other, in two separate steps.

13.4.7 TWO-STEP PROCEDURE OF WORD FORM RECOGNITION

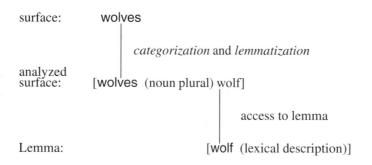

Automatic word form recognition takes place between surface and analyzed surface. It consists of categorization and lemmatization, and is based on a special analysis lexicon. Access to the lemma, containing the semantic representation common to the whole paradigm, takes place in a second step, using a traditional base form lexicon.

13.5 Methods of automatic word form recognition

Possible methods of automatic word form recognition may be distinguished as to whether their analysis lexicon specifies *word forms, morphemes,* or *allomorphs.*[19]

---

[18] See A.V. Aho & J.D. Ullman 1977, p. 336–341.

[19] The fourth basic concept of morphology, the *word*, does not provide for a recognition method because words are not a suitable key for a multitude of word forms.

Each of the three methods exhibits a characteristic correlation between the recognition algorithm and the associated analysis lexicon.

The *word form method*[20] uses an analysis lexicon consisting of word forms.

13.5.1 ANALYZED WORD FORM AS LEXICAL LEMMA

[wolves (part of speech: Subst, num: Pl, case: N,D,A, base form: wolf)]

An analysis lexicon of word forms allows for the simplest recognition algorithm because the surface of the unknown word form, e.g. wolves, is simply matched whole onto the corresponding key in the analysis lexicon.

Of the three steps of morphological analysis, namely (i) segmentation, (ii) lexical lookup, and (iii) concatenation, the first and the third are here identity mappings, for which reason the word form method is a border line case of morphological analysis. Also, categorization and lemmatization are handled here solely by the lexical entry.

The word form method may be useful as a quick and dirty method for toy systems, providing lexical lookup without much programming effort. In the long run this method is costly, however, because of the production,[21] the size,[22] and the basic finiteness of its analysis lexicon.

The last point refers to the fact that the word form method is inherently limited to the entries provided by its analysis lexicon. It therefore provides no possibility to recognize neologisms during run-time – unless all *possible* word forms are provided by the analysis lexicon. This, however, is impossible because of the productivity of natural language morphology.

The *morpheme method*[23] uses the smallest analysis lexicon, consisting of analyzed morphemes.[24] Compared to the word form method, it has the further advantage that neologisms may be analyzed and recognized during run-time using a rule-based segmentation and concatenation of complex word forms into their elements (morphemes). The only requirement is that the elements are lexically known and their mode of composition can be handled correctly by the rules.

The disadvantage of the morpheme method is a maximally complex recognition algorithm. The analysis of an unknown surface during run-time requires the steps of (1) segmentation into allomorphs, (2) reduction of the allomorphs to the corresponding

---

[20] Also known as the *full-form* method based on a *full form lexicon*.

[21] It is possible to derive much of the word form lexicon automatically, using a base form lexicon and rules for inflection as well as – to a more limited degree – for derivation and composition. These rules, however, must be written and implemented for the natural language in question, which is costly. The alternative of producing the whole word form lexicon by hand is even more costly.

[22] The discussion of German noun noun composita in Section 13.2 has shown that the size of a word form lexicon attempting to be complete may easily exceed a trillion word forms, thus causing computational difficulties.

[23] A prototype is the KIMMO-system of *two-level morphology* (K. Koskenniemi 1983).

[24] In light of the morpheme definition 13.2.3, a morpheme lexicon consists strictly speaking of analyzed base form allomorphs.

morphemes, (3) recognition of the morphemes using an analysis lexicon, and (4) the rule-based concatenation of the morphemes to derive the analyzed word form.

In case of the word form wolves, for example, step (1) consists in the segmentation into the allomorphs wolv and es. Step (2) reduces these to the corresponding morpheme surfaces wolf and s, enabling lexical lookup as step (3). In step (4), the resulting analyzed morphemes are concatenated by means of grammatical rules which derive the morphosyntactic properties of the word form as a whole, including categorization and lemmatization.

13.5.2 SCHEMA OF THE MORPHEME METHOD

surface: wolves
                  | | *segmentation*
allomorphs: wolv/es
                  ⇓   ⇓   *reduction*
morphemes: wolf+s   *base form lookup* and *concatenation*

Conceptually, the morpheme method is related to transformational grammar. Allomorphs are not treated as fully analyzed grammatical entities, but exist only as the quasi-adulterated surface reflections of the 'underlying' morphemes which are regarded as the 'real' entities of the theory. Concatenation takes place at the level of morphemes – and not at the level of the concretely given allomorphs. For this reason, the morpheme method violates the principle of surface compositionality (S). Also, because the morpheme method tries to compose the morphemes as much as possible (cf. 8.4.4, 8.4.5) as constituents (cf. 8.4.3, 8.4.6), it violates the principle of time-linear derivation order (L) of the SLIM theory of language.

Mathematically and computationally, the morpheme method is of high complexity ($\mathcal{NP}$ complete)[25] because the system must check the surface for *all possible* phenomena of allomorphy. Faced with English valves, for example, the system would have to consider nonexisting *valf+s as a possible underlying morpheme sequence. Only after all potential allomorph-morpheme reductions have been checked for a given surface can concatenation begin.

The *allomorph method*[26] combines the respective advantages of the word form and the morpheme method by using a simple recognition algorithm with a small analysis lexicon. Based on its rule-based analysis, the allomorph method recognizes neologisms during run-time.

The allomorph method uses two lexica, called the elementary lexicon and the allomorph lexicon, whereby the latter is automatically derived from the former by means of allo-rules before run-time. The elementary lexicon consists in (i) the analyzed elementary base forms[27] of the open word classes, (ii) the analyzed elements of the

---

[25] See Section 12.2. The inherent complexity of the morpheme method is shown in detail by Barton, Berwick, & Ristad 1987, p. 115–186, using the analysis of spies/*spy+s* in the KIMMO system.
[26] The allomorph method was first presented in Hausser 1989b.

closed word classes, and (iii) the allomorphs of the affixes[28] as needed in inflection, derivation, and composition.

During run-time, the allomorphs of the allomorph lexicon are available as precomputed, fully analyzed forms[29] (e.g. 13.2.4, 13.3.1, 13.3.2), providing the basis for a maximally simple segmentation: the unknown surface is matched from left to right with suitable allomorphs – without any reduction to morphemes.

13.5.3 SCHEMA OF THE ALLOMORPH METHOD

surface: wolves
                                  | | *segmentation*
allomorphs: wolv/es   *allomorph lookup* and *concatenation*
                                  ⇑   ⇑   *derivation of allomorphs before run-time*
morphemes & allomorphs: wolf s

Concatenation takes place on the level of analyzed allomorphs by means of combi-rules. This method is in concord with the principles of surface compositionality (S) and time-linear derivation order (L) of the SLIM theory of language.

In conclusion, the methods of automatic word form recognition are compared schematically.

13.5.4 SCHEMATIC COMPARISON OF THE THREE BASIC METHODS

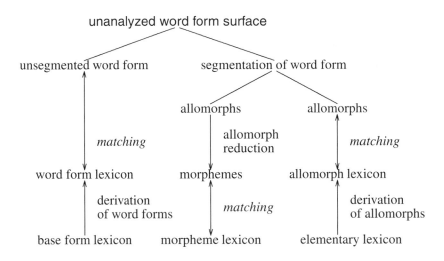

(1) word form method   (2) morpheme method   (3) allomorph method

All three methods are based on matching the input surface with a corresponding key of an analysis lexicon characteristic of the method in question. The first alternative consists in whether the word form surfaces are segmented into their elements during run-time or not. This alternative is relevant not only linguistically, but also of practical consequence for the computational implementation.

If the input surface is not segmented at all, we obtain the word form method in which the input is matched as a whole onto corresponding keys of analyzed word forms. In this case the well-known method of hash tables may be used for lexical lookup, as long as the boundaries of each word form are marked, e.g., by spaces.

If the input surface is segmented, the concrete elements are the allomorphs. To automatically determine the allomorph boundaries – which are not orthographically marked – the method of trie structures is suited best (cf. Section 14.3). The question is whether the allomorphs found in the surface should be reduced to morphemes prior to lexical lookup or whether lexical lookup should be defined for allomorphs.

The first option results in the morpheme method in which both segmentation of the surface into allomorphs and the reduction of the allomorphs into the associated (un-analyzed) morphemes takes place during run-time. The morpheme surfaces obtained in this manner are matched with corresponding keys in an analysis lexicon consist-ing of analyzed morphemes. Then the analyzed morphemes are put together again by grammatical rules to categorize and lemmatize the word form in question.

The second option results in the allomorph method in which the input surface is matched onto fully analyzed allomorphs during run-time. The analyzed allomorphs are generated automatically from an elementary lexicon by allo-rules before run-time. During run-time the analyzed allomorphs need only be concatenated by means of left-associative combi-rules which categorize and lemmatize the input.

Of the three methods, the allomorph method is suited best. It is of low mathematical complexity (linear), describes morphological phenomena of concatenation and allo-morphy in a linguistically transparent, rule-based manner, handles neologisms during run-time, may be applied easily to new languages, is computationally space and time efficient, and can be easily debugged and scaled up. The allomorph method is de-scribed in more detail in the following chapter.

Exercises

Section 13.1

1. Give the inflectional paradigms of man, power, learn, give, fast, and good. Generate new words from them by means of derivation and composition.
2. Call up LA-Morph on your computer and have the above word forms analyzed.
3. Explain the notions word, word form, paradigm, part of speech, and the difference between the open and the closed classes.

---

[27] In the case of irregular paradigms, also the suppletive forms are supplied (cf. 14.1.5).

[28] Thus, no bound morphemes (cf. 13.3.4) are being postulated.

[29] B. MacWhinney 1978 demonstrates the independent status of lexical allomorphs with language acqui-sition data in Hungarian, Finnish, German, English, Latvian, Russian, Spanish, Arabic, and Chinese.

4. Why is it relevant to distinguish between the notions word and word form?
5. What is the role of the closed classes in derivation and composition?
6. Why are only the open classes a demanding task of computational morphology?
7. How do the content and function words relate to the open and the closed classes?

Section 13.2

1. Why is the number of word forms in German potentially infinite?
2. Why is the number of noun noun composita n^2?
3. What do the formal definitions of word and morpheme have in common?
4. What is a neologism?
5. Describe the difference between morphemes and syllables.

Section 13.3

1. What is suppletion?
2. Why is a bound morpheme like -ing neither a member of the open nor the closed classes?
3. What would argue against postulating bound morphemes?

Section 13.4

1. Explain the three steps of a morphological analysis.
2. Why does LA-grammar analyze allomorphs as ordered triples?
3. What are the components of a system of automatic word form recognition.
4. What is a lemma? How do the lemmata of a traditional dictionary differ from those of an on-line lexicon for automatic word form recognition?
5. How does the surface function as a key in automatic word form recognition?
6. Explain the purpose of categorization and lemmatization.
7. What is an analysis lexicon?

Section 13.5

1. Describe three different methods of automatic word form recognition.
2. Why is there no *word method* of automatic word form recognition?
3. Where does the word form method handle categorization and lemmatization?
4. Compare cost and advantage of the word form method.
5. Would you classify the word form method as a smart or as a solid solution?
6. Why is the morpheme method mathematically complex?
7. Why does the morpheme method violate surface compositionality?
8. Why does the morpheme method use surfaces only indirectly as the key?
9. Why is the morpheme method conceptually related to transformational grammar?
10. Why does the allomorph method satisfy the principle of surface compositionality?
11. Why is the run-time behavior of the allomorph method faster than that of the morpheme method?

14. Word form recognition in LA-Morph

The allomorph method has been developed and implemented as a system called LA-Morph. Linguistically, LA-Morph is based on (i) an elementary lexicon, (ii) a set of allo-rules, and (iii) a set of combi-rules. The allo-rules take the elementary lexicon as input and generate from it a corresponding allomorph lexicon before run-time. The combi-rules control the time-linear concatenation of analyzed allomorphs during run-time, resulting in the morphosyntactic analysis of complex word forms.

Section 14.1 explains the allo-rules and defines four degrees of regularity for inflectional paradigms. Section 14.2 gives an overview of allomorphic phenomena in nouns, verbs, and adjective-adverbials of English, and investigates how many allomorphs a morpheme has on average (allomorph quotient of English). Section 14.3 describes the computational method of segmenting word forms into their analyzed allomorphs by means of a trie structure. Section 14.4 explains the combi-rules. Section 14.5 gives an overview of the concatenation patterns of inflectional morphology of English and German.

14.1 Allo-rules

An allo-rule takes a lemma of the elementary lexicon as input and derives from it one, two, or more allomorphs. The input and output is defined in terms of patterns. The basic structure of the allo-rules is as follows:

14.1.1 ABSTRACT FORMAT OF AN ALLO-RULE

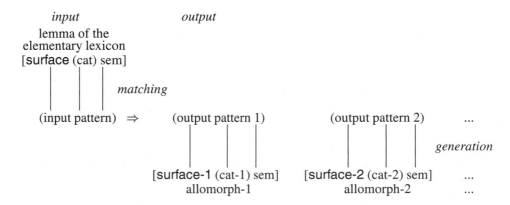

Allo-rules may modify the surface, the category, and the semantic representation of their input.

When the allo-rule component is applied to an elementary lexicon, all its lemmata are run through the allo-rules which are arranged in a fixed order. If a lemma matches the input pattern of an allo-rule, it is accepted and the associated allomorphs are produced. If a lemma is not accepted, it is passed on to the next allo-rule. The last allo-rule is the default rule. Its input pattern accepts any entry and returns it unchanged.

The output of the allo-rules is written sequentially into a file. In this way the elementary lexicon is converted automatically into a lexicon of analyzed allomorphs. Even a lexicon comprising more than 100 000 base forms takes no more than a few seconds to be run through the allo-rule component.

Afterwards the allo-rule component is not applied again until the allo-rules or the elementary lexicon have been modified. The complete description of all morphological regularities and irregularities is a long process, however. Therefore, writing allo-rules for a natural language is a meaningful investment not only for theoretical reasons.

A lemma of an elementary lexicon is illustrated in 14.1.2. The format is that of the LA-Morph version programmed by Hausser & Ellis 1990 (cf. 15.1.3). The list-based categories of this version are close to the LA-grammars for artificial languages.

14.1.2 EXAMPLE OF A BASE FORM LEMMA

```
("derive" (nom a v) derive)
```

The category (nom a v) characterizes the entry as a **v**erb which takes a **nom**inative and an **a**ccusative as arguments.

The allo-rules map the base form lemma into the following analyzed allomorphs:

14.1.3 RESULT OF APPLYING ALLO-RULES TO BASE FORM LEMMA

```
("derive" (sr nom a v) derive)
("deriv" (sr a v) derive)
```

In order to control application of the correct combi-rules, the categories in 14.1.3 have the additional marker 'sr' (semi-regular). The first allomorph is used for the forms derive and derive/s, the second for the forms deriv/ing and deriv/ed.

As an example of a more complicated lexical entry, associated allomorphs, and resulting inflectional forms, consider the analysis of the German verb schlafen (sleep) in 14.1.4, 14.1.5, and 14.1.6, respectively.

14.1.4 BASE FORM ENTRY OF schlafen

```
("schla2fen" (KV VH N GE  {hinueber VS GE } {durch VH A GE }
             {aus VH GE } {ein VS GE }\$ <be VH A GE- >
             <ent VS GE- > <ueber VH A GE- > <ver VH A GE- >)
             schlafen)
```

The first element of this lexical entry is the surface in which the characteristic *ablaut*-variation of schlafen is specified by the marker '2.' The second element is the category[1] (KV ...>) which is rather long and complex because its description includes 4 variants with separable and 4 variants with unseparable prefixes. The third element is the base form *schlafen* without any surface markers.

The allo-rules map the lemma defined in 14.1.4 into the analyzed allomorphs schlaf, schläf, and schlief:

14.1.5 OUTPUT OF ALLO-RULES FOR schlafen

```
("schlaf" (IV V1 VH N GE { hinüber VS GE } { durch VH A GE }
          { aus VH GE } { ein VS GE } $ < be VH A GE- >
          < ent VS GE- > < über VH A GE- > < ver VH A GE- > )
          schlafen)
("schläf" (IV V2 _0 N GE { hinüber VS GE } { durch VH A GE }
          { aus VH GE } { ein VS GE } $ < be VH A GE- >
          < ent VS GE- > < über VH A GE- > < ver VH A GE- > )
          schlafen)
("schlief" (IV V34 _0 N GE { hinüber VS GE } { durch VH A GE }
          { aus VH GE } { ein VS GE } $ < be VH A GE- >
          < ent VS GE- > < über VH A GE- > < ver VH A GE- > )
          schlafen_i)
```

Triggered by the surface marker '2' (cf. 14.1.4), three different allomorphs without surface markers are derived. To ensure application of the correct combi-rules, the first two segments in the category of the base form lemma have been replaced by three new segments in the categories of the respective allomorphs. The remainder of the original category, serving to handle separable and non-separable prefixes, reappears unchanged in the allomorph categories.

---

[1] The category begins with the segment 'KV', specifying the part of speech, verb. The second segment 'VH' indicates that schlafen combines with the auxiliary haben (have) rather than sein (be). The third segment N represents the nominative valency whereby the absence of additional case segments characterizes the verb as intransitive. The fourth segment 'GE' finally indicates that the past participle is formed with ge as in ge/schlaf/en or aus/ge/schlaf/en – in contrast to, e.g., ver/schlaf/en.

Then there follow expressions in curly brackets describing the non-separable prefixes, here hinüberschlafen, durchschlafen, ausschlafen, and einschlafen. These bracketed expressions specify the type of the auxiliary, the type of the past participle, and the valency of each of these variants.

The description of separable prefixes in the base form entry of the stem is necessary for morphological as well as syntactic reasons. The prefix may occur both, attached to the verb as in (i) weil Julia einschlief, and separated from the verb as in (ii) Julia schlief schnell ein. In (ii) the prefix in the category of sleep is used by the syntax to identify ein as part of the verb, which is semantically important. In (i) the prefix in the category is used by the morphology to combine ein and schlief into the word form einschlief.

After the separable prefixes there follow the non-separable prefixes of schlafen marked by angled brackets. Again, the auxiliaries (e.g. hat verschlafen vs. ist entschlafen), the past participle, and the valency structure (e.g. ist entschlafen vs. hat das Frühstück verschlafen) are specified for each prefix variant.

Based on the allomorphs defined in 14.1.4, the combi-rules analyze and generate 9 paradigms with 29 forms each. Thus a single base form entry is mapped into a total of 261 different inflectional forms.

14.1.6 THE WORD FORMS OF schlafen (excerpt)

```
("schlaf/e" (S1 {hinüber}{durch A}{aus}{ein} V) schlafen_p)
("schlaf/e" (S13 {hinüber} {durch A} {aus} {ein} V ) s._k1)
("schlaf/e/n" (P13 {hinüber} {durch A} {aus} {ein} V ) s._pk1)
("schlaf/e/st" (S2 {hinüber} {durch A} {aus} {ein} V ) s._k1)
("schlaf/e/t" (P2 {hinüber} {durch A} {aus} {ein} V ) s._k1)
("schlaf/t" (P2 {hinüber} {durch A} {aus} {ein} V ) s._p)
("schlaf/end" (GER ) schlafen)
("schlaf/end/e" (E ) schlafen)
("schlaf/end/en" (EN ) schlafen)
("schlaf/end/er" (ER ) schlafen)
("schlaf/end/es" (ES ) schlafen)
("schlaf/end/em" (EM ) schlafen)
("schlaf/e/st" (S2 {hinüber} {durch A} {aus} {ein} V ) s._k1)
("schlaf/e/t" (P2 {hinüber} {durch A} {aus} {ein} V ) s._k1)
("schläf/st" (S2 {hinüber} {durch A} {aus} {ein} V ) s._p)
("schläf/t" (S3 {hinüber} {durch A} {aus} {ein} V ) s._p)
("schlief" (S13 {hinüber} {durch A} {aus} {ein} V ) s._i)
("schlief/e" (S13 {hinüber} {durch A} {aus} {ein} V ) s._k2)
("schlief/en" (P13 {hinüber} {durch A} {aus} {ein} V ) s._ik2)
("schlief/est" (S2 {hinüber} {durch A} {aus} {ein} V ) s._ik2)
("schlief/et" (P2 {hinüber} {durch A} {aus} {ein} V ) s._ik2)
("schlief/st" (S2 {hinüber} {durch A} {aus} {ein} V ) s._ik2)
("schlief/t" (P2 {hinüber} {durch A} {aus} {ein} V ) s._i)
("ge/schlaf/en" (H) schlafen)
("ge/schlaf/en/e" (E) schlafen)
("ge/schlaf/en/en" (EN) schlafen)
("ge/schlaf/en/es" (ES) schlafen)
("ge/schlaf/en/er" (ER) schlafen)
("ge/schlaf/en/em" (EM) schlafen)

("aus/schlaf/e" (S1 V) ausschlafen_pk1)
("aus/schlaf/e" (S13 V ) ausschlafen_k1)
("aus/schlaf/en" (P13 A V) ausschlafen_pk1)
   . . .
("aus/schläf/st" (S2 V) ausschlafen_p)
("aus/schläf/t" (S3 V) ausschlafen_p)
   . . .
```

The finite forms of schlafen retain the separable prefixes in their category because they may be needed by the syntax, as in Susanne schlief gestern aus. The non-separable prefixes, on the other hand, are removed from the categories because the associated surfaces, e.g., verschlafe, verschläfst, etc., have no separable variant. For the same reason the separable prefixes are omitted in the categories of non-finite forms.

Depending on whether or not an inflectional paradigm (i) requires exactly one lemma in the elementary lexicon, (ii) requires a special marker in the lemma surface, and (iii) derives more than one allomorph per lemma, four different degrees of regularity may be distinguished.

14.1.7 FOUR DEGREES OF REGULARITY IN LA-MORPH

– *Regular* inflectional paradigm
  The paradigm is represented by one lemma without any special surface markings, from which one allomorph is derived, e.g. learn ⇒ learn, or book ⇒ book.
– *Semi-regular* inflectional paradigm
  The paradigm is represented by one lemma without any special surface markings, from which more than one allomorph is derived, e.g. derive ⇒ derive, deriv, or wolf ⇒ wolf, wolv.
– *Semi-irregular* inflectional paradigm
  The paradigm is represented by one lemma with a special surface marker, from which more than one allomorph is derived, e.g. swlm ⇒ swim, swimm, swam, swum.
– *Irregular* inflectional paradigm
  The paradigm is represented by several lemmata for suppletive allomorphs which pass through the default rule, e.g. go ⇒ go, went ⇒ went, gone ⇒ gone. The allomorphs serve as input to general combi-rules, as in go/ing.

These degrees of regularity may be represented as the following table.

14.1.8 TABULAR PRESENTATION OF THE DEGREES OF REGULARITY

	one lemma per paradigm	lemma without markings	one allomorph per lemma
regular	yes	yes	yes
semi-regular	yes	yes	no
semi-irregular	yes	no	no
irregular	no	no	yes

This structural criterion for the classification of (ir)regularities is of general interest insofar as the characterization of exceptions has always been regarded as a central task of traditional morphology.

14.2 Phenomena of allomorphy

To get a concrete idea of what the allo-rules are needed for, let us consider different instances of allomorphy in English. The following examples of nouns (14.2.1, 14.2.2), verbs (14.2.3, 14.2.4), and adjective-adverbials (14.2.5) are presented in a way which resembles the structure of allo-rules insofar as the (surface of the) input is listed under the key word LEX and the associated output under the key words ALLO1, ALLO2, etc., in a tabular format. They are only a precursor of the actual allo-rules, however, because (i) the structures in question are described by example rather than abstract patterns and (ii) the description is limited to the surfaces.

The *regular* nouns of English occur in four different inflectional forms, namely unmarked singular (book), genitive singular (book/'s), unmarked plural (book/s), and genitive plural (book/s/'). These forms are analyzed and generated concatenatively by combi-rules using the base form and the suffixes 's, s, and ' .

The *semi-regular* nouns of English use different allomorphs for the stem in the singular and the plural form. Like the regular nouns, their different forms are analyzed in terms of concatenation. Thereby markers in the category ensure that each allomorph is combined with the proper suffix, for example wolf/'s, but not *wolv/'s, and wolv/es, but not *wolv/s.

14.2.1 ALLOMORPHS OF SEMI-REGULAR NOUNS

LEX	ALLO1	ALLO2
wolf	wolf	wolv
knife	knife	knive
ability	ability	abiliti
academy	academy	academi
agency	agency	agenci
money	money	moni

The *semi-irregular* nouns of English generate their unmarked plural nonconcatenatively by means of allo-rules alone. The marked forms of singular and plural are handled in a concatenative manner.

14.2.2 ALLOMORPHS OF SEMI-IRREGULAR NOUNS

LEX	ALLO1	ALLO2
analysis	analysis	analyses
larva	larva	larvae
stratum	stratum	strati
matrix	matrix	matrices
thesis	thesis	theses
criterion	criterion	criteria
tempo	tempo	tempi
calculus	calculus	calculi

The *irregular* nouns of English, such as child–children, foot–feet, ox–oxen, and sheep–sheep, are not handled by allo-rules, but rather by two entries in the elementary lexicon, one for the unmarked singular and one for the unmarked plural form. In addition there are pluralitantia like scissors and people which are also handled directly in the elementary lexicon. The marked forms of the irregular nouns such as analysis' are handled by special clauses of the relevant combi-rules.

The *regular* verbs of English occur in four different inflectional forms, namely unmarked present tense and infinitive (learn), marked present tense (learn/s), past tense and past participle (learn/ed), and progressive (learn/ing). In addition there are derivational forms like learn/er as well as their respective inflectional forms such as learn/er/'s, etc. This set of forms is analyzed and generated by combi-rules using the base form and the suffixes s, ed, ing, and er.

The *semi-regular* verbs of English use two different allomorphs. Like the regular verbs, their inflectional forms are analyzed in terms of concatenation. Thereby markers in the category ensure that each allomorph is combined with the proper suffixes, e.g. derive/s, but not *derive/ing, and deriv/ing, but not *deriv/s (cf. 14.1.3).

14.2.3 ALLOMORPHS OF SEMI-REGULAR VERBS

LEX	ALLO1	ALLO2
derive	derive	deriv
dangle	dangle	dangl
undulate	undulate	undulat
accompany	accompany	accompani

The *semi-irregular* verbs of English generate their past tense and past participle nonconcatenatively by means of allo-rules alone. The marked form of the present tense and the progressive are handled in a concatenative manner.

14.2.4 ALLOMORPHS OF SEMI-IRREGULAR VERBS

LEX	ALLO1	ALLO2	ALLO3	ALLO4
swIm	swim	swimm	swam	swum
rUN	run	runn	ran	run
bET	bet	bett	bet	bet

Though some allomorphs of run and bet have the same surface, these forms differ in their respective categories. ALLO2 handles the gemination needed for the progressive form.

The *irregular* verbs of English, such as arise–arose–arisen, break–broke–broken, give–gave–given, go–went–gone, or seek–sought–sought are not handled by allo-rules. Instead the suppletive forms are treated in the elementary lexicon, one lemma for the unmarked present tense, one for the past tense, and one for the past participle.

The *regular* adjective-adverbials of English occur in four different inflectional forms, namely unmarked positive (slow), comparative (slow/er), superlative (slow/est), and marked positive (slow/ly). This set of forms is analyzed and generated by combi-rules using the base form and the suffixes er, est, and ly.

The *semi-regular* adjective-adverbials of English are also analyzed in terms of concatenation whereby markers in the category ensure that each allomorph is combined with the proper suffixes, e.g. abl/er, but not *able/er, or free/ly, but not *fre/ly.

14.2.5 ALLOMORPHS OF SEMI-REGULAR ADJECTIVE-ADVERBIALS

LEX	ALLO1	ALLO2
able	able	abl
happy	happy	happi
free	free	fre
true	true	tru

The *semi-irregular* adjective-adverbials are not instantiated in English. The *irregular* adjective-adverbials of English are exemplified by good–better–best–well.

In light of these examples there arises the question of how many word forms of English are based on allomorphic variants. From the viewpoint of linguistics this question is of interest because the number of allomorphs – compared to the number of base forms – is indicative of the relative degree of (ir)regularity of a natural language in the area of morphology. From the viewpoint of programming it is of interest because the additional number of precomputed allomorphs affects the load on the random access memory (RAM).

If the number of allomorphic variants in natural language turned out to be very large, then the alternative morpheme approach (cf. 13.5.2) would have the advantage of computing allomorphs only on demand – for the surface currently analyzed. On the other hand, if this number turned out to be rather small, then the precomputation of allomorphs would require only a modest amount of additional RAM while resulting in a tremendous simplification and speed up of the run-time analysis.

In short, the choice between the allomorph and the morpheme method depends not only on theoretical and technical considerations, but also on the empirical facts of natural language, namely the number of allomorphs relative to the number of morphemes. This numerical correlation is called the allomorph quotient of a natural language.

14.2.6 DEFINITION OF THE ALLOMORPH QUOTIENT

> The allomorph quotient is the percentage of additional allomorphs relative to the number of base form entries.[2]

Because the allomorph method automatically derives all the possible allomorphs to a given elementary lexicon its application to a certain natural language provides all the data needed for the computation of its allomorph quotient.

Assume we want to apply LA-Morph to a natural language not previously handled, e.g. English. The first step is to look for a traditional lexicon (cf. 13.4.3) that is legally available on-line. Using computational methods, its grammatical information can easily be 'massaged' into a format suitable for the application of allo-rules.

Once the format of the traditional lexicon has been adjusted, and the allo- and combi-rules have been written for the new language, the system may be tested on free

---

[2] That is (the number of allomorphs minus the number of morphemes) divided by (the number of morphemes divided by 100).

text. Thereby many word forms will turn out to have several analyses. For example, if the traditional lexicon used were Webster's New Collegiate Dictionary, then

pseudoclassic
pseudopregnancy
pseudosalt
pseudoscientific

etc.

would have two analyses, in contrast to

pseudogothic
pseudomigrane
pseudoscientist
pseudovegetarian

etc.

which would be unambiguous. This is because the words in the first set happen to be entries in Webster's while those in the second set do not. In order to recognize the highly productive compositions involving the prefix pseudo, the LA-Morph system must provide a general rule-based analysis. As a consequence, word forms like pseudoclassic, which have their own entry in Webster's, are analyzed as ambiguous whereby the second reading stems from the compositional analysis based on the known forms pseudo and classic.

One method to avoid this kind of ambiguity is to remove all non-elementary base forms from the on-line lexicon. This may be done automatically. First all the key word surfaces in the traditional lexicon are parsed by the LA-Morph system. Then all multiple analyses, such as pseudoclassic and pseudo/classic, are collected and their base form entries, here pseudoclassic, are removed.

This approach has the advantage of substantially reducing the size of the elementary lexicon (in German and Italian by about half) while maintaining the original data coverage with respect to categorization and lemmatization. In fact, the data coverage of a LA-Morph system based on an elementary lexicon will be much greater than that of the original lexicon alone because of the general, rule-based handling of derivation and composition.

A second method of eliminating the ambiguities in question leaves the non-elementary base forms in the lexicon and selects the most likely reading after the word form analysis. For example, given the choice between an analysis as an elementary form (e.g. pseudoclassic) and as a complex form (e.g. pseudo/classic), the system would normally take the first. This has the advantage that given the choice between kinship and kin/ship the unlikely compositional interpretation of *water vessel for close relatives* would be filtered out.

The best way is to combine the respective advantages of the two approaches by having two lexica. One is an elementary lexicon which does not contain any non-elementary base forms. It is used for the categorization and lemmatization of word forms.

The other is a base form lexicon of content words. It assigns semantic representations to base forms including composita and derivata established in use. During word form analysis the two lexica are related by matching the result of lemmatization onto a corresponding – if present – key word of the base form lexicon (cf. 13.4.7).

For example, kin/ship resulting from a compositional analysis would be matched onto kinship in the non-elementary base form lexicon, accessing the proper semantic description. In this way, (i) maximal data coverage – including neologisms – is ensured by a rule based analysis, (ii) the possibility of noncompositional meanings is accounted for, and (iii) unnecessary ambiguities are avoided.

LA-Morph systems for which traditional lexica were reduced to elementary lexica in the manner described above have resulted in the following allomorph quotients.

14.2.7 THE ALLOMORPH QUOTIENT OF DIFFERENT LANGUAGES

Italian: 37%
C. Wetzel 1996 reduced a traditional lexicon of 91 800 lemmata into an elementary lexicon of 44 000 lemmata. These base forms are mapped by allo-rules into an analysis lexicon of 60 300 entries. The resulting allomorph quotient of 37% corresponds to an average of 1.37 allomorphs per morpheme.

German: 31%
  O. Lorenz 1996 reduced a traditional lexicon of 100 547 lemmata into an elementary lexicon of 48 422 lemmata. These base forms are mapped by allo-rules into an analysis lexicon of 63 559 entries. The resulting allomorph quotient of 31% corresponds to an average of 1.31 allomorphs per morpheme.

English: 8,97%
Hausser 1989b used an elementary lexicon containing 8 000 of the most frequent base forms from which allo-rules derived an analysis lexicon of 9 500 entries. The resulting allomorph quotient of 19% corresponds to an average of 1.19 allomorphs per morpheme.
J. Leidner 1998 uses an elementary lexicon of 231 000 lemmata which was extracted from several corpora and extended with additional lemmata from machine readable dictionaries. The resulting allomorph quotient of 8.97% corresponds to an average of 1.09 allomorphs per morpheme.

These examples show that the allomorph quotients resulting from the analysis of different languages (and the analysis style of different linguists) are of the same order of magnitude. Furthermore, they are surprisingly low and therefore of no practical consequence for run-time memory.

In addition, it must be taken into account that the analysis lexicon of allomorphs is almost half the size of the original traditional lexicon. With this reduced lexicon the system analyzes not only all key surfaces of the original traditional lexicon, but also all associated inflectional forms, an unlimited number of derivational and compositional forms, as well as their inflectional forms.

Because the phenomenon of allomorphy is limited to the high frequency entries of the lexicon[3] an allomorph quotient will decrease as the size of the lexicon is increased with low frequency items. This is shown by the small and the large system for English cited in 14.2.7. In order to meaningfully compare the allomorph quotient of different languages it should be normalized by basing the calculation on, e.g., the 100 000 most frequent words[4] of the open classes without proper names and acronyms.

14.3 Left-associative segmentation into allomorphs

The construction of an elementary lexicon and the definition of allo-rules is a language specific task which requires the methods and knowledge of traditional linguistics. In order to utilize the analyzed allomorphs computationally, however, there must be a procedure which automatically segments any unknown surface, e.g. wolves, into a fitting sequence of analyzed allomorphs, i.e. wolv/es. This segmentation algorithm is a language independent procedure the implementation of which is a task for computer science.

The automatic segmentation of a given word form surface into its allomorphs should work as follows. If the unknown word form begins with, e.g., W, any allomorphs not beginning with this letter should be disregarded. If the next letter is an O, any allomorphs not beginning with WO should be disregarded, etc.

14.3.1 LEFT-ASSOCIATIVE LETTER BY LETTER MATCHING

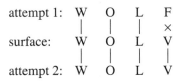

With this method the possible goal WOLF is eliminated as soon as the fourth letter V is encountered. Thus, in the analysis of wolves the allomorph wolf is never even considered. The theoretic relation between wolf and wolv is established solely in terms of the common base form in the third position of their respective analyses (cf. 13.2.4).

If the letter sequence in the unknown surface has been matched successfully by an analyzed allomorph, e.g. wolv, but there are still letters left in the unknown surface,

---

[3] J. Bybee 1985, pp. 208,9.

[4] In order to determine this set, a representative corpus of the language must be used. See Chapter 15.

as in, e.g., **wolves**, the next allomorph is looked for. This method of left-associatively matching allomorphs onto the unknown surface pursues all possible hypotheses in parallel from left to right. It results in a complete segmentation into all possible allomorph sequences.

As an example in which two alternative segmentations are maintained in parallel (ambiguity) consider the German surface **Staubecken**:[5]

14.3.2 ALTERNATIVE SEGMENTATIONS OF A WORD FORM

surface:	Staubecken	Staubecken
segmentation:	Stau/becken	Staub/ecke/n
translation:	*reservoir*	*dust corners*

All that is required for the correct segmentation is to provide the relevant allomorphs **stau**, **staub**, **becken**, **ecke**, and **n**, and to systematically match them from left to right onto the unknown surface. This method works even in writing systems in which only the beginning and end of sentences or even texts are marked, as in the *continua* writing style of classical Latin.

Computationally, the left-associative, parallel, letter-by-letter matching of allomorphs is implemented by means of a trie structure[6] or letter tree. A trie structures functions as an index which allows to find an entry (here an analyzed allomorph) by going through the key word letter by letter from left to right. In LA-Morph the trie structure is built up automatically when the elementary lexicon is run through the allo-rule component.

As an example of a trie structure consider 14.3.3, which shows the storage of the allomorphs **swim**, **swimm**, **swam**, **swamp**, **er**, **ing**, **s**, and **y**.

---

[5] Corresponding hypothetical examples in English are

coverage	grandparent	history	lamp/light	land/s/end
cover/age	grandpa/rent	hi/story	lam/plight	land/send
cove/rage		his/tory		

rampage	rampart	scar/face	sing/able	war/plane
ramp/age	ramp/art	scarf/ace	sin/gable	warp/lane
ram/page	ram/part			

Most of these examples, which were found computationally by C. Wetzel at the CL lab of Friedrich Alexander Universität Erlangen Nünberg (CLUE), are not quite real because noun noun compounds in English are usually formed by separate words, e.g. 'warp lane' or 'ramp art.'

[6] D.E. Knuth 1973, p. 483, attributes the idea of trie structures to R. Briandais 1959. A.V. Aho, J.E. Hopcroft, & J.D. Ullman 1983, p. 163, attribute it to E. Fredkin 1960. The name 'trie' is taken from the middle of the word *retrieval*. According to Aho et al. 1983 "trie was originally intended to be a homonym of 'tree,' but to distinguish these two terms many people prefer to pronounce trie as though it rhymes with 'pie'."

14.3.3 STORING ALLOMORPHS IN A TRIE STRUCTURE

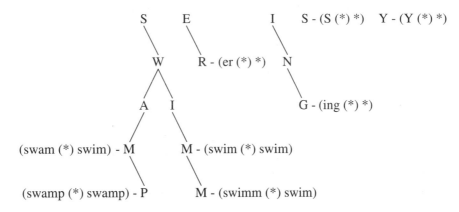

A trie structure consists of several levels, whereby the first level contains all the allomorph-initial letters, the second level contains all the allomorph-second letters, etc. The '/' and ' \ ' connectors between the different levels in 14.3.3 indicate legal continuations from one level to the next. For example, the attempt to find an allomorph analysis for the input SWQ would break off after SW because in 14.3.3 there is no Q underneath the W on the third level.

The lexical analysis of an allomorph is stored at its last letter in the trie structure. Consider for example the search for the analyzed allomorph swim. Because its surface consists of four letters its lexical analysis is stored at the fourth level of the trie structure. The entry is found by going to the letter S on the first level, then to the letter W underneath the S on the second level, etc.

The search is successful if the navigation through the trie structure arrives at a letter with a lexical analysis. When such an analysis is found, three possibilities arise.

– There are no letters left in the surface of the unknown word form, e.g. SWAM. Then the program simply returns the analysis stored at the last letter, here M.
– There are still letters left in the surface of the unknown word form. Then one of the following alternatives applies:
  – The allomorph found so far *is part* of the word form, as swim in SWIMS. Then the program (i) gives the lexical analysis of swim to the combi-rules of the system and (ii) looks for the next allomorph (here s), starting again from the top level of the trie structure.
  – The allomorph found so far *is not part* of the word form, as swam in SWAMPY. In this case the program continues down the trie structure provided there are continuations. In our example, it will find swamp.

Because it becomes apparent only at the very end of a word form which of these two possibilities applies – or whether they apply simultaneously in the case of an ambiguity – they are pursued simultaneously by the program.

The downward search in the trie structure is stopped, if there is an entry and/or there is no continuation to a lower level. The latter can be caused by ungrammatical input (not a possible allomorph of the language) or by an analysis lexicon of allomorphs which is not yet complete.

14.4 Combi-rules

The trie-structure-based segmentation of an unknown surface into analyzed allomorphs in LA-morphology corresponds to lexical lookup in LA-syntax. Similarly, the time-linear composition of allomorphs into word forms in LA-morphology corresponds to syntactic composition of word forms into sentences in LA-syntax.

Because the allomorphs of LA-morphology and the word forms of LA-syntax are similar in structure (ordered triples) their respective time-linear composition is based on the same general rule mechanism of LA-grammar.

14.4.1 COMBI-RULE SCHEMA

$$\text{\textit{input}} \qquad\qquad \text{\textit{output}}$$
$$r_n\text{: (pattern of start) (pattern of next)} \;\Rightarrow\; \text{rp}_n \text{ (pattern of new start)}$$

Each time a rule r_n has mapped an input pair into a new word start, a next allomorph is searched for in the trie structure. The next allomorph and the current word start form a new input pair to which all the rules in the rule package rp_n are applied. This is repeated as long as (i) next allomorphs are being provided by a matching between sections of the word form surface and the continuations in the trie structure (cf. 14.3.3), and (ii) the process of left-associative combinations is not stopped prematurely because no rule in any of the current rule packages is successful.

Combi-rules differ from allo-rules in that they are defined for different domains and different ranges:

> An *allo-rule* takes a lexical entry as input and maps it into one or more allomorphs.
> A *combi-rule* takes a word form start and a next allomorph as input and maps it into a new word form start.

They also differ in that the allo-rules usually modify the surface of their input, whereas the combi-rules combine the surfaces of their input expressions without change. Furthermore, the allo-rules are applied before run-time, whereas the combi-rules apply during run-time.

The combi-rules ensure that

1. the allomorphs found in the surface are not combined into ungrammatical word forms, e.g. *swam+ing or *swimm+s (input condition),

2. the surfaces of grammatical allomorph combinations are properly concatenated, e.g. swim+s ⇒ swims,
3. the categories of the input pair are mapped into the correct result category, e.g. (NOM V) + (SX S3) ⇒ (S3 V),
4. the correct result is formed on the level of semantic interpretation, and
5. after a successful rule application the correct rule package for the next combination is activated.

The structure of morphological analysis in LA-Morph is illustrated in 14.4.2 with the derivation of the word form unduly.

14.4.2 DERIVATION OF unduly IN LA-MORPH

```
1 +u [NIL . NIL]
2  +n [NIL . (un (PX PREF) UN)]
RP:{V-START N-START A-START P-START}; fired: P-START
3    +d [(un (PX PREF) UN) . (d (GG) NIL)]
     +d [NIL . NIL]
4     +u [(un (PX PREF) UN) . (du (SR SN) DUE (SR ADJ-V) DUE)]
RP:{PX+A UN+V}; fired: PX+A
      +u [NIL . NIL]
5 L [(un+du (SR ADJ) DUE) . (l (GG) NIL (ABBR) LITER)]
RP:{A+LY}; fired: none
       +l [(un (PX PREF) UN) . NIL]
       +l [NIL . NIL]
6       +y [(un+du (SR ADJ) DUE) . (ly (SX ADV) LY)]
 RP:{A+LY}; fired: A+LY
("un/du/ly" (ADV) due)
```

This format of a LA-Morph analysis shows the step by step search through the trie structure. At the 2. letter 'n' the allomorph entry un is found, whereby the word form start is NIL. Other allomorphs found are du, l, and ly.

In the analysis we may distinguish between the optional *derivation* and the obligatory *result*. The derivation shows which allomorphs were found, which rules were tried, and which were successful. The derivation is used for debugging and upscaling of the system.

The result, i.e. the content of the last line, is used in applications of LA-Morph, such as syntactic parsing. In such applications the derivation is of little interest and omitted in the output. The result provides the required categorization and lemmatization. For example, in 14.4.2 the word form unduly is categorized as an adverb and lemmatized as due, whereby the negation is coded in the semantic representation stored underneath the lemma.

The combi-rules of LA-Morph ensure that only grammatically correct forms are recognized or generated. As an example consider 14.4.3, in which the legal allomorphs able and ly are attempted to be combined into an illegal word form.

14.4.3 HANDLING OF UNGRAMMATICAL INPUT IN LA-MORPH

```
1 +a [NIL . (a (SQ) A)]
2  +b [NIL . NIL]
3   +l [NIL . (abl (SR ADJ-A) ABLE)]
RP:{V-START N-START A-START P-START}; fired: A-START
4    +e [(abl (SR ADJ) ABLE) . NIL]
     +e [NIL . (able (ADJ) ABLE)]
RP:{V-START N-START A-START P-START}; fired: none
5    +l [(abl (SR ADJ) ABLE) . NIL]
ERROR
Unknown word form: "ablely"
NIL
```

The result characterizes the input *ablely as ungrammatical.

Finally consider the analysis of the simplex **undulate**, which happens to share the first five letters with **unduly** (cf. 14.4.2).

14.4.4 PARSING THE SIMPLEX **undulate**

```
1 +u [NIL . NIL]
2  +n [NIL . (un (PX PREF) UN)]
RP:{V-START N-START A-START P-START}; fired: P-START
3   +d [(un (PX PREF) UN) . (d (GG) NIL)]
    +d [NIL . NIL]
4    +u [(un (PX PREF) UN) . (du (SR SN) DUE (SR ADJ-V) DUE)]
RP:{PX+A UN+V}; fired: PX+A
     +u [NIL . NIL]
5    +l [(un+du (SR ADJ) DUE) . (l (GG) NIL (ABBR) LITER)]
RP:{A+LY}; fired: none
     +l [(un (PX PREF) UN) . NIL]
     +l [NIL . NIL]
6     +a [(un+du (SR ADJ) DUE) . NIL]
      +a [NIL . NIL]
7      +t [(un+du (SR ADJ) DUE) . NIL]
       +t [NIL . (undulat (SR A V) UNDULATE)]
RP:{V-START N-START A-START P-START}; fired: V-START
8       +e [(un+du (SR ADJ) DUE) . (late (ADJ-AV) LATE
                                          (ADV) LATE)]
RP:{A+LY}; fired: none
        +e [(undulat (SR A V) UNDULATE) . NIL]
        +e [NIL . (undulate (SR NOM A V) UNDULATE)]
RP:{V-START N-START A-START P-START}; fired: V-START
("undulate" (NOM A V) UNDULATE)
```

Up to letter 5 this derivation is identical to that of **unduly** in 14.4.2. These hypotheses do not survive, however, and in the end the word form turns out to be a simplex. At letter 8, the allomorph **undulat** (as in **undulat/ing**) is found, but superseded by the longer base form. The fact that the allomorphs **undulat** and **undulate** belong to the same morpheme is expressed solely in terms of their common semantic representation.

14.5 Concatenation patterns

To get a concrete idea of what the combi-rules are needed for, let us consider different instances of inflectional concatenation in English.

14.5.1 CONCATENATION PATTERNS OF ENGLISH NOUNS

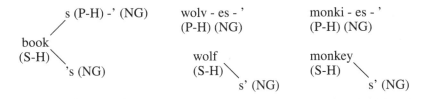

```
          s (P-H) -' (NG)      wolv - es - '          monki - es - '
         /                     (P-H) (NG)             (P-H) (NG)
  book
  (S-H) \                        wolf                   monkey
          's (NG)                (S-H) \                (S-H)    \
                                        s' (NG)                  s' (NG)
```

The category (S-H) stands for a singular nonhuman noun, (NG) for noun genitive, and (P-H) for plural nonhuman noun. The example **book** represents a regular noun. The examples **wolf** and **monkey** represent different types of semi-regular nouns.

14.5.2 CONCATENATION PATTERNS OF ENGLISH VERBS

```
       ing (B *)                 ing (B *)                  ing (B *)
      /                         /                          /
learn - s                 deriv - e - s              apply
(NOM * V) (S3 * V)         (NOM * V) (S3 * V)         (NOM * V)
      \                         \                          \
        ed                        ed                         appli - es (S3 * V)
        (N * V) (HV *)            (N * V) (HV *)                    \
                                                                     ed (N * V) (HV *)
```

The '*' in the categories of 14.5.2 stands for the oblique (i.e. non-nominative) valency positions, which vary in different verbs, e.g. (**sleep** (NOM V) *), (**see** (NOM A V) *), (**give** (NOM D A) *), or (**give** (NOM A TO) *). The V indicates a finite verb form. B and HV represent the non-finite present and past participle, respectively, indicating the combination with a form of **be** or **have**.

The example **learn** in 14.5.2 illustrates the concatenation pattern of a regular verb. The examples **derive** and **apply** represent different instances of semi-regular verbs. In addition to the forms shown above there are derivational forms like **learn/er** and **learn/able**.

14.5.3 CONCATENATION PATTERNS OF ADJECTIVE-ADVERBS

```
          ly (ADV)                       ly (ADV)                        ly (ADV)
         /              able (ADJ)      /            steady (ADJ)       /
quick - er (CAD)        abl - er (CAD)              steadi - er (CAD)
(ADJ)  \                         \                              \
         est (SAD)                 est (SAD)                      est (SAD)
```

The categories (ADJ), (ADV), (CAD), and (SAD) stand for adjective, adverb, comparative adjective, and superlative adjective, respectively. The example quick in 14.5.3 illustrates the concatenation pattern of a regular paradigm, while able and steady represent different instances of semi-regular adjectives.

These examples show that the systematic separation of allomorphy and concatenation results in a highly transparent description of different types of paradigms. The schematic outlines of concatenation patterns like those in 14.5.1, 14.5.2, and 14.5.3, provide the conceptual basis for a systematic development of lexical entries, allorules, and concatenation-rules in new languages. It works equally well for inflectional, derivational, and compositional morphology.

As an example of a more complicated morphology consider the nominal and verbal concatenation patterns of German. They will be referred to in the syntactic analysis of German in Chapter 18,

14.5.4 CONCATENATION PATTERNS OF GERMAN NOUNS

```
                es (-FG)                    es (-FG)                       es (-FG)
               /                           /                              /
schmerz    -e (-FD)          tag                      leib        -e (-FD)
(M-G)                        (M-G)                     (M-G)
               \                           \                              \
                en (P)                      e (MDP-D) -n (PD)              er (P-D) -n (PD)


                s (-FG)                     s (-FG)                        s (-FG)
               /                           /                              /
gipfel                       stachel                   thema
(M-GP-D)\                    (M-G)   \                 (N-G)
                n (PD)                      n (P)
                                                                   themen (P)


                s (-FG)                     s (-FG)
               /                           /
vAter                        auge                      uhu         -s (MGP)
(M-G)                        (N-G)                     (M-G)
               \                           \
                (P) -n (PD)                 n (P)


braten     -s (-FG)          hAnd    -e (P-D) -n (PD)  frau        -en (P)
(M-GP)                       (F)                       (F)


drangsal   -e (P-D) -n (PD)  kenntnis-se (P-D) -n (PD) mUtter - (P-D) -n (PD)

(F)                          (F)                       (F)
```

The purpose of the categories in 14.5.4 is to capture the agreement restrictions of the determiner-noun combination as simply and succinctly as possible. While a traditional paradigmatic analysis always assigns 8 lexical analyses to a German noun (nominative, genitive, dative, and accusative, in the singular and plural) the distinctive categorization of 14.5.4 assigns only one lexical analysis per distinct surface form. Because there are 2 to 5 surface forms per noun, the categorization in 14.5.4 reduces the average number of forms per noun paradigm to less than half.

The category segments are explained and illustrated in the following list.

14.5.5 CATEGORY SEGMENTS OF GERMAN NOUN FORMS

MN	= Masculinum Nominativ	(Bote)
M-G	= Masculinum no Genitiv	(Tag)
-FG	= no Femininum Genitiv	(Tages, Kindes)
-FD	= no Femininum Dativ	(Schmerze, Kinde)
M-NP	= Masculinum no Nominativ or Plural	(Boten)
M-GP	= Masculinum no Genitiv or Plural	(Braten)
MGP	= Masculinum Genitiv or Plural	(Uhus)
M-GP-D	= Masculinum no Genitiv or Plural no Dativ	(Gipfel)
F	= Femininum	(Frau)
N-G	= Neutrum no Genitiv	(Kind)
NG	= Neutrum Genitiv	(Kindes)
ND	= Neutrum Dativ	(Kinde)
N-GP	= Neutrum no Genitiv or Plural	(Leben)
N-GP-D	= Neutrum no Genitiv or Plural no Dativ	(Wasser)
NDP-D	= Neutrum Dativ or Plural no Dativ	(Schafe)
P	= Plural	(Themen)
P-D	= Plural no Dativ	(Leiber)
PD	= Plural Dativ	(Leibern)

The category segments consist of one to four letters whereby the first letter is M, F, -F, N, or P, specifying either gender or plural. The second letter, if present, is a positive or negative specification of case. The third letter, if present, is P, indicating that a singular form is also used for plural. The fourth letter is a case restriction on the plural. The agreement restrictions on determiner-noun combinations in German are described in Section 18.1 and defined in *LA-D2* (18.2.5).

The left-associative concatenation patterns of German verbs is illustrated in 14.5.6 with the semi-irregular forms of **schlafen** (**sleep**). The elementary base forms, analyzed allomorphs, and paradigm forms of this word were presented in 14.1.4, 14.1.5, and 14.1.6, respectively.

14.5.6 A SEMI-IRREGULAR VERB PATTERN OF GERMAN

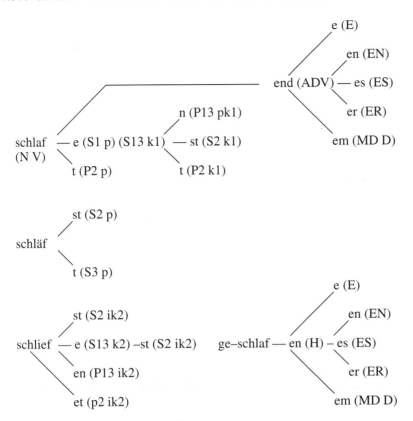

Extensive applications of LA-Morph to different languages with fully developed morphology have shown conclusively that a strictly time-linear[7] analysis works well for concatenation, even in complicated semi-irregular paradigms with prefixes, suffixes, and several allomorphs. Because the time-linear derivation order is also the basis for a complete segmentation of the unknown surface into analyzed allomorphs, segmentation and concatentation work can work incrementally, hand in glove. In addition, a time-linear derivation order corresponds to the empirical situation of word form recognition and synthesis, and is therefore the psychologically most plausible.

---

[7] The popular method of *truncation* removes endings from the right. The goal is to isolate the stem, for which reason this method is also called *stemming*. Because this method does not segment the surface into analyzed allomorphs, but is based on patterns of letters, it works only in morphologically simple languages and is very imprecise even there. Moreover, the right-to-left direction of truncation is in conflict with the basic time-linear structure of natural language production and interpretation.

Exercises

Section 14.1

1. What is the form and function of allo-rules?
2. Why must there be a default rule and where is it ordered among the allo-rules?
3. At which point in the development or application of a LA-Morph system are the allo-rules to be applied?
4. Describe four degrees of regularity in the inflectional paradigms. Which structural properties are the basis for these distinctions?

Section 14.2

1. Describe phenomena of allomorphy in the nouns, verbs, and adjective-adverbials of English.
2. What is the smallest number of allomorphs a morpheme can have?
3. What is the definition of the allomorph quotient?
4. Compare the allomorph quotient of different languages. What does the allomorph quotient say about the word forms of a language?
5. Does the allomorph quotient stay constant when the lexicon is considerably enlarged with low frequency items?
6. Explain the notion of normalizing the allomorph quotients of different natural languages.

Section 14.3

1. How should the segmentation of word forms into allomorphs work conceptually?
2. What is a trie structure and what is its function in automatic word form recognition?
3. What determines the number of levels in a trie structure?
4. Under what conditions is the search for an allomorph in a trie structure discontinued?
5. Under what conditions is the search for an allomorph in a trie structure successful?

Section 14.4

1. What is the relation between the combi-rules of LA-Morph and the rules of LA-grammar in general?
2. Why is neither the word form nor the morpheme method compatible with the algorithm of LA-grammar?
3. What is the function of the combi-rules?
4. What do the allo- and the combi-rules have in common, and how do they differ?
5. Describe the linguistic analysis of a complex word form in LA-Morph.
6. What is shown by the derivation of a LA-Morph analysis? What is this information used for?

7. What is shown by the result of a LA-Morph analysis? What is this information used for?
8. How does LA-Morph handle ungrammatical input?

Section 14.5

1. Describe the concatenation patterns of the inflectional paradigms of English nouns, verbs, and adjective-adverbials.
2. Do left-associative combi-rules show a difference in the combination of a prefix and a stem, on the one hand, and a word start and a suffix, on the other? Illustrate your answer with the example un/du/ly.
3. Which reasons support a left-associative analysis in morphology and which reasons speak against an analysis from right to left, based on truncation and stemming?
4. Is the method of truncation (stemming) a smart or a solid solution?
5. Compare the distinctive categorization of German nouns with a traditional paradigmatic analysis (cf. Section 18.2).

15. Corpus analysis

This chapter describes the application of word form recognition to investigating the frequency distribution of words. Furthermore, the relation between morphological analysis and computational method in automatic word form recognition is used to make some methodological points about grammar components in computational linguistics.

Section 15.1 describes the basic modularity which systems of computational linguistics must satisfy. Section 15.2 presents the method of subtheoretical variants using the example of multicats for lexical ambiguities. Section 15.3 describes the principles of building representative corpora, needed for the testing of automatic word form recognition systems. Section 15.4 explains the frequency distribution of word forms in natural language. Section 15.5 describes the method of statistically-based tagging and evaluates its accuracy in comparison to rule-based systems.

15.1 Implementation and application of grammar systems

The design of a linguistic component such as word form recognition requires the choice of a general *theory of language* in order to ensure a smooth functional interaction with other components of the system, e.g. syntax and semantics. This decision in turn influences another important choice, namely that of a *grammar system*.

A well-defined grammar system consists of two parts.

15.1.1 PARTS OF A GRAMMAR SYSTEM

– Formal algorithm
– Linguistic method

On the one hand, the algorithm and the method must be independent of each other in order to enable the formal definition of the algorithm. This in turn is a precondition for mathematical complexity analysis and a declarative specification of the implementation (cf. Section 9.1).

On the other hand, the algorithm and the method must fit together well. This is because the linguistic method *interprets* the formal algorithm with respect to an area of empirical phenomena, while the formal algorithm provides a procedural *realization* of the linguistic method.

For example, designing a grammar system of automatic word form recognition presented us with the following options:

15.1.2 OPTIONS FOR GRAMMAR SYSTEM OF WORD FORM RECOGNITION

– Formal algorithm:
  C- (Section 7.4), PS- (Section 8.1), or LA-grammar (Section 10.2).
– Linguistic method:
  Word form, morpheme, or allomorph method (cf. Section 13.5).

For empirical, methodological, and mathematical reasons the allomorph method and the algorithm of LA-grammar were chosen, resulting in the grammar system of LA-morphology, implemented as LA-Morph.

The distinction between algorithm and method holds for also in syntactic grammar systems. Transformational grammar uses the algorithm of PS-grammar and the method of constituent structure. Montague grammar uses the algorithm of C-grammar and the method of truth conditions. LA-syntax uses the algorithm of LA-grammar and the method of traditional analysis based on valency, agreement, and word order.

A grammar system should be defined in such a way that it can be *implemented* in different manners and *applied* to different languages. This modularity between grammar system, implementation, and application leads to the following standard.

15.1.3 MINIMAL STANDARD OF WELL-DEFINED GRAMMAR SYSTEMS

A grammar system is well-defined only if it simultaneously allows

1. different *applications* in a given *implementation*, and
2. different *implementations* in a given *application*.

The first point is necessary for empirical reasons. In order to determine how suitable a grammar system is for the analysis of natural language, it must be tested automatically on large amounts of realistic data. This is the simplest, most straightforward, and most effective scientific strategy to show whether or not its *method* brings out a common core of linguistic principles, and whether or not its *algorithm* has sufficient generative capacity while maintaining low mathematical complexity.

The second point is necessary for practical reasons. In various circumstances the computational implementation of a grammar system will have to be replaced by another one – because of new hardware using a different operating system, a changing preference regarding the programming language, changing requirements on the efficiency of the system, changing interface requirements, etc.

Under such circumstances there is no valid theoretical reason why a reimplementation should necessitate changes in its formalism or its applications. On the contrary, different implementations of a grammar system are an excellent way to demonstrate which properties of the grammar system are a *necessary* part of the abstract specification and which result merely from *accidental* properties of the programming.

The methodologically necessary[1] modularity between a grammar system, its implementations, and its applications may be shown schematically as follows.

15.1.4 MODULARITY OF A GRAMMAR SYSTEM

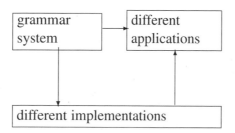

For systems of computational linguistics, this modularity is as important as the functional compatibility[2] of the different components of grammar, and the computational suitability[3] of the algorithm.

Regarding different applications, the grammar system of LA-morphology has so far been used for word form recognition of English, German, Korean, and Italian (cf. 14.2.5), with smaller systems for French, Japanese, and Polish. These experiences have strengthened confidence in the approach of LA-morphology.[4]

Regarding different implementations, LA-morpholgoy has so far been realized as the following computer programs.

15.1.5 DIFFERENT IMPLEMENTATIONS OF LA-MORPHOLOGY

1988 in LISP (Hausser & Todd Kaufmann)
1990 in C (Hausser & Carolyn Ellis)
1992 in C, 'LAMA' (Norbert Bröker)

---

[1] These considerations should be seen in light of the fact that it is tempting – and therefore common – to quickly write a little hack for a given task. A hack is an ad hoc piece of programming code which works, but has no clear algebraic or methodological basis. Hacks are a negative example of a *smart solution*, cf. Section 2.3.

In computational linguistics, as in other fields, little hacks quickly grow into giant hacks. They may do the job in some narrowly defined application, but their value for linguistic or computational theory is limited at best. Furthermore, they are extremely costly in the long run because even their own authors encounter ever increasing problems in debugging and maintenance, not to mention upscaling within a given language or applications to new languages.

[2] To be ensured by the theory of language.

[3] To be ensured by the principle of type transparency, cf. Section 9.3.

[4] Especially valuable for improving the generality of the system was the handling of the Korean writing system, Hangul. The various options for coding Hangul are described in K.-Y. Lee 1994.

1994 in C, 'LAP' (Gerald Schüller)
1995 in C, 'Malaga' (Björn Beutel)

These different implementations are jointly referred to as LA-Morph. Despite differences[5], they share the following structural principles.

- Specification of the allo- (cf. 14.1.1) and the combi-rules (cf. 14.4.1) on the basis of patterns which are matched onto the input.
- Storage of the analyzed allomorphs in a trie structure and their left-associative lookup with parallel pursuit of alternative hypotheses (cf. Section 14.3).
- Modular separation of motor, rule components, and lexicon, permitting a simple exchange of these parts, for example in the application of the system to new domains or languages.
- Use of the same motor and the same algorithm for the combi-rules of the morphological, syntactic, and semantic components during analysis.
- Use of the same rule components for analysis and generation in morphology, syntax, and semantics.

Within the general framework, however, the empirical analysis of large amounts of different types of data has lead to the development of subtheoretical variants.

15.2 Subtheoretical variants

The algebraic definition of LA-grammar in Chapters 10–12 aimed at the simplest, most transparent notation for explaining the complexity-theoretic properties of the algorithm in applications to artificial languages. Applications to natural languages, however, encountered empirical phenomena which suggested modifications of the original notation. These resulted in subtheoretical variants. In contrast to a derived formalism, a subtheoretical variant does not change the component structure and the theoretical status of the original system.

---

[5] The use of LISP in the 1988 version allowed for a quick implementation. It employed the same motor for word form recognition and syntax, and was tested on sizable fragments of English and German morphology and syntax. It had the disadvantage, however, that the rules of respective grammars were defined as LISP functions, for which reason the system lacked an abstract, declarative formulation.

The 1990 C version was the first to provide a declarative specification of the allo- and the combi-rules based on regular expressions (RegExp) in a tabular ASCII format, interpreted automatically and compiled in C.

The 1992 reimplementation aimed at improvements in the pattern matching and the trie structure. It did not get wider use, however, because it did not allow for the necessary interaction between the levels of surface and category, and had omitted an interface to syntax.

The 1994 reimplementation repaired the deficiencies of the 1992 version and was tested on sizable amounts of data in German, French, and Korean. These experiences resulted in many formal and computational innovations.

The 1995 implementation of the Malaga system took a new approach to handling of pattern matching, using attribute-value structures. Malaga provides a uniform framework for simultaneous morphological, syntactic, and semantic parsing and generation.

One phenomenon leading to a modification of the original format were lexical ambiguities in morphology and syntax. Consider, for example, the statistically most frequently used word form of German, the determiner **der**. As indicated in 15.2.1, it has at least three different sets of agreement properties, which may be represented as three different lexical readings.

15.2.1 COMBINATORICS OF THE GERMAN DETERMINER der

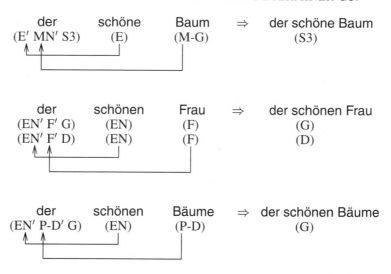

The determiner categories consist of three segments of which the first determines the adjective ending, the second the form of a compatible noun, and the third the result. The result category segment S3 stands for singular 3rd person, G for genitive, and D for dative. The specification of number and person (here S3) is required only in the nominative, which – as subject – agrees in these parameters with the finite verb form (cf. Sections 16.2 and 16.3).

Compared to a traditional paradigm analysis, the categorial analysis in 15.2.1 is designed to avoid unnecessary ambiguities. This is motivated both linguistically (concreteness) and computationally (lower combinatorial complexity). For example, the categories (E) and (EN) of the adjective forms **schöne** and **schönen** do not specify any values for gender, number, case, and definiteness (which would produce 7 and 19 lexical readings, respectively), but only the ending (cf. 14.5.5), the choice of which depends on the determiner rather than the noun, as shown by the following examples.

15.2.2 AGREEMENT OF ADJECTIVE-ENDING WITH DETERMINER

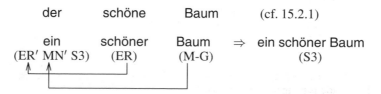

The point is that der schöne Baum and ein schöner Baum have the same case and number (nominative singular) and use the same noun (Baum). Therefore, the grammatical well-formedness of choosing schöner versus schöne depends solely on the choice of an definite (der) versus indefinite (ein) determiner.

The categorization of the noun forms Baum as (M-G), i.e. masculine minus genitive, Bäume as (P-D), i.e. plural minus dative, and Frau as (F), i.e. feminine (cf. 13.1.4), agrees with that of table 14.5.2. This distinctive categorization avoids unnecessary readings and reflects the fact that German nouns either have a grammatical gender, in which case they are singular, or are plural, in which case there is no gender distinction. There is also a possibility of fusing the second and third reading of 15.2.1 by means of the new category segment G&D (for genitive and dative, cf. categorization in 15.2.3), reducing the total number of readings to three.

These reductions are valuable for practical syntax. For example, a traditional, exhaustive categorization of the word forms in der schönen Frauen would require a total of 134 ordered pairs to be checked for their grammatical compatibility, whereas the distinctive categorization requires the checking of only 5 ordered pairs.

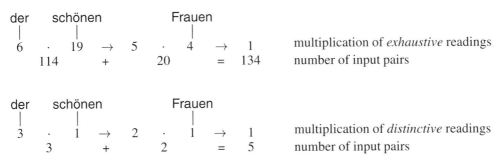

The number behind an arrow represents the number of successful compositions, while that below a product sign represents the number of attempted compositions.

In the end, however, there remains the fact that even in a distinctive categorization the most frequently used word form of German, der, has three different categories.[6] From the viewpoint of computational linguistics, this remarkable fact can be handled either by means of three different lexical entries with equal surfaces, as in 15.2.3, or one lexical entry with one surface and three alternative categories, as in 15.2.4. The latter is the so-called multicat notation first explored in the LAP-system (cf. 15.1.5).

15.2.3 REPRESENTING LEXICAL READINGS VIA DIFFERENT ENTRIES

[der (E′ MN′ S3) DEF-ART]
[der (EN′ F′ G&D) DEF-ART]
[der (EN′ P-D′ G) DEF-ART]

---

[6] See O. Jespersen 1921, p. 341–346.

15.2.4 Representing lexical readings via multicats

[der ((E′ MN′ S3) (EN′ F′ G&D) (EN′ P-D′ G)) DEF-ART]

The multicat solution 15.2.4 has the advantage that only one lexical lookup is necessary. Furthermore, instead of branching immediately into three different parallel paths of syntactic analysis, the alternatives coded in the multicat may be processed in one branch until the result segments come into play. There are many other lexical phenomena, e.g. separable and nonseparable prefixes of German verbs (cf. 14.1.2), for which multicats provide a simplification of lexical description and syntactic analysis.

The use of multicats requires that format and implementation of the combi-rules in morphology and syntax be extended to handle the alternatives coded in the new kind of categories. On the one hand, such an extension of the combi-rules capabilities leads to a version of LA-grammar which differs from the original LAP-system as defined for the algebraic definition. On the other hand, an LA-grammar using multicats can always be reformulated as one in the original format using several lexical readings. Thus, the use of multicats does not change the theoretical status as compared to a corresponding version without multicats.

Another subtheoretical variant of LA-grammar was the definition of categories and rule patterns as attribute-value structures. This system was written by B. Beutel 1995 and is called Malaga. In Malaga the list-based patterns of LAP are replaced by hierarchically structured patterns. The alternative formats are illustrated in 15.2.5 and 15.2.6 for the same schematic LA-rule with the same input and output.

15.2.5 List-based matching (LAP)

$$ss \qquad nw \qquad ss'$$

input-output: (a   b   c   d)   (b)   (a   c   d)

rule pattern: (X   b   Y)   (b) $\Longrightarrow$ (X   Y)

categorial operation

15.2.6 Feature-based matching (Malaga)

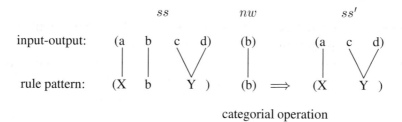

$$ss \qquad\qquad nw \qquad\qquad ss'$$

input-output:
$\begin{bmatrix} mm1 = a \\ mm2 = b \\ mm3 = c \\ mm4 = d \end{bmatrix}$
$\begin{bmatrix} mm5 = b \end{bmatrix}$
$\begin{bmatrix} mm1 = a \\ mm3 = c \\ mm4 = d \end{bmatrix}$

rule pattern:
$\begin{bmatrix} mm2 = b \\ X \end{bmatrix}$
$\begin{bmatrix} mm5 = b \end{bmatrix} \Longrightarrow$
$\begin{bmatrix} X \end{bmatrix}$

categorial operation

The comparison of 15.2.5 and 15.2.6 shows the different formats of list-based and feature-based *categories* (input-output level) as well as the different formats of list-based and feature-based category *patterns* (rule level). The list-based and the feature-based presentations in 15.2.5 and 15.2.6, respectively, are equivalent. In both, the *nw*-segment *b* cancels a corresponding segment in the *ss*.

Malaga[7] uses the subtheoretical variants of multicats and attribute-value structures at the same time. The format of feature structures is particularly useful for the semantic representations of the extended network database described in Chapters 23 and 24. For a theoretical presentation of the lexicon, the morphology, and the syntax of natural language, on the other hand, the original list-based format is more parsimonious, more restricted, and therefore more transparent.

For this reason the list-based notation familiar from the LA-grammars for artificial languages in Part II will also be used in the following analyses of English and German syntax. The explanation of specific syntactic rules will use the graphic correlation between category segments and matching pattern variables illustrated in the list-based notation of 15.2.5.

A derived formalism of LA-grammar was developed by G. Hanrieder 1996. This system, called LAUG (Left-Associative Unification-based Grammar), is motivated by the desire to use unification with typed feature structures. LAUG is implemented in Prolog and designed in the context of an existing application, namely the SUNDIAL (Speech Understanding and DIALogue) project.[8] The specific requirements of the application include (i) the syntactic and semantic parsing of over 12 000 utterances transcribed from more than 1 000 naturally occurring telephone dialogues, (ii) robustness in the parsing of word graph lattices provided by the speech recognition component analyzing the dialogues, and (iii) selection of the 'best path.'

The SUNDIAL application of LAUG provided the first precise comparison with an alternative component based on the formalism of UCG (Unification Categorial Grammar).[9] It turned out that the LAUG implementation was 4 – 7 times faster than the UCG-based alternative. Thus the results of the theoretical complexity analysis comparing LA-grammar to constituent-structure-based alternatives (cf. Part II) were confirmed on the level of a concrete application.

15.3 Building corpora

The empirical testing of a grammar system requires the building of corpora which represent the natural language to be analyzed in terms of a suitable collection of samples. Depending on which aspect of a language is to be documented by a corpus, the

---

[7] The acronym Malaga is self-referential (like GNU), and stands for 'Malaga analyzes left-associative grammars with attributes.'

[8] SUNDIAL was supported by the European Union from 1988–1993, and is currently continued at universities and research institutes in joint projects with industry.

[9] H. Zeevat, E. Klein, & J. Calder 1987.

samples may consist of printed prose, transcribed dialogue, speeches, personal letters, theater plays, etc., either pure or mixed according to some formula. The samples may be collected either diachronically for a longer period of time or strictly synchronically (usually restricted to samples from a certain domain in a given year).

For German, the first comprehensive frequency analysis was presented 1897/8 by Wilhelm Kaeding (cf. H. Meier 1964), intended as a statistical basis for improving stenography. With his small army of 'Zählern' (people who count) Kaeding analyzed texts comprising almost 11 million running word forms (= 20 million syllables) and 250 178 different types.[10] In contrast to the synchronic corpora of today, Kaeding's collection of texts was designed to cover the German language from 1750 to 1890.

The arrival of computers provided a powerful tool for new investigations of this kind. H. Kučera & W.N. Francis 1967 took the lead with the Brown corpus[11] for American English. The Brown corpus comprises 500 texts of 1 014 231 running word forms (tokens) and 50 406 different types.

In 1978 followed the LOB Corpus[12] for British English. Designed as a pendant to the Brown corpus, it consists of 500 texts, about 1 million tokens, and 50 000 types. Both corpora were compiled from texts of the following 15 genres.

15.3.1 TEXT GENRES OF THE BROWN AND THE LOB CORPUS

	Brown	LOB
A Press: reportage	44	44
B Press: editorial	27	27
C Press: reviews	17	17
D Religion	17	17
E Skills, trade, and hobbies	36	38
F Popular lore	48	44
G Belle lettres, biography, essays	75	77
H Miscellaneous (government documents, foundation records, industry reports, college catalogues, industry house organ)	30	38
J Learned and scientific writing	80	80
K General fiction	29	29
L Mystery and detective fiction	24	24
M Science fiction	6	6
N Adventure and western fiction	29	29
P Romance and love story	29	29

---

[10] The size of Kaeding's corpus was thus more than 10 times as large and the number of types twice as large as that of the computer-based 1973 Limas-Korpus (see below).

[11] Named after Brown University in Rhode Island, where Francis was teaching at the time.

[12] The *Lancaster-Oslo/Bergen* corpus was compiled under the direction of Geoffrey N. Leech and Stig Johannson. Cf. K. Hofland and S. Johannson 1982.

R Humour	9	9
Total	500	500

The numbers on the right indicate how many texts were included from the genre in question, indicating slight differences between the Brown and the LOB corpus.

For building the Brown corpus, Kučera & Francis 1967, p. xviii, formulated the following desiderata:

15.3.2 DESIDERATA FOR THE CONSTRUCTION OF CORPORA

1. Definite and specific delimitation of the language texts included, so that scholars using the Corpus may have a precise notion of the composition of the material.
2. Complete synchronicity; texts published in a single calendar year only are included.
3. A predetermined ratio of the various genres represented and a selection of individual samples through a random sampling procedure.
4. Accessibility of the Corpus to automatic retrieval of all information contained in it which can be formally identified.
5. An accurate and complete description of the basic statistical properties of the Corpus and of several subsets of the Corpus with the possibility of expanding such analysis to other sections or properties of the Corpus as may be required.

These desiderata are realized using the mathematical methods of statistics, i.e., the basic equations of stochastics, distributions of independent and dependent frequencies, normalization, computing the margin of error, etc. The goal is to find a theoretical distribution which corresponds to the empirical distribution (invariance of empirical distribution proportions).

The German counterpart to the American Brown Corpus (1967) and the British LOB Corpus (1978) is the Limas Corpus (1973).[13] Like its English pendants, it consists of 500 texts, each containing roughly 2 000 running word forms, amounting to a total of 1 062 624 tokens. Due to the richer morphology of German, the number of types is 110 837, i.e., more than twice than that of the corresponding English corpora.

The selection of genres and amounts selected in 15.3.1 are intended to make the corpora as *representative* and *balanced* as possible.[14] Intuitively these two notions are easy to understand. For example, the editions of a daily newspaper in the course of a year are more representative for a natural language than a collection of phone books or banking statements. And a corpus containing texts from different genres in relative proportions like those indicated in 15.3.1 is balanced better than one which consists of texts from one genre alone.

---

[13] See K. Hess, J. Brustkern & W. Lenders 1983.
[14] Cf. H. Bergenholtz 1989, D. Biber 1994, N. Oostdijk & P. de Haan (eds.) 1994.

It is difficult, however, to *prove* a specific quantification like that of 15.3.1 as 'representative' and 'balanced.' N. Oostdijk 1988, for example, objected to 15.3.1

> that 'genre' is not a well-defined concept. Thus genres that have been distinguished so far have been identified on a purely intuitive basis. No empirical evidence has been provided for any of the genre distinctions that have been made.

A representative and balanced corpus ultimately requires knowledge of which genres are used how often in a given time interval by the language community, in writing and reading as well as speaking and hearing. Because it is next to impossible to realistically determine the correlation of production and reception in written and spoken language in various genres, the building of representative and balanced corpora is more an art than a science. It is based largely on general common sense considerations. Moreover, it depends very much on the purpose for which the corpus is intended.[15]

Today, corpora of 1 million running word forms are considered far too small for valid natural language statistics.[16] A still unsurpassed effort is the British National Corpus (BNC), available on-line since 1995. It has a size of 100 million running word forms. Of these, 89.7 million running word forms and 659 270 types[17] are of written language, while 10.34 million running word forms are of spoken language.

The building and analysis of corpora has been combined with the efforts at standardizing the mark up (SGML, XML, TEI, cf. Section 1.5) of on-line texts into a largely autonomous field. This type of work is located between information science and computational linguistics.

15.4 Distribution of word forms

The value of a corpus does not reside in the content of its texts, but in its quality as a realistic sample of a natural language. The more representative and balanced the

---

[15] The difficulties of constructing balanced and representative corpora are avoided in 'opportunistic corpora,' i.e., text collections which contain whatever is easily available. The idea is that the users themselves construct their own corpora by choosing from the opportunistic corpus specific amounts of specific kinds of texts needed for the purpose at hand. This requires, however, that the texts in the opportunistic corpus are preprocessed into a uniform format and classified with respect to their domain, origin, and various other parameters important to different users.

For example, a user must be able to automatically select a certain year of a newspaper, to exclude certain sections such as private ads, to select specific sections such as the sports page, the editorial, etc. To provide this kind of functionality in a fast growing opportunistic corpus of various different kinds of texts is labor-intensive – and therefore largely left undone.

With most of the corpus work left to the users, the resulting 'private corpora' are usually quite small. Also, if the users are content with whatever the opportunistic corpus makes available, the lexically more interesting domains like medicine, physics, law, etc., will be left unanalyzed.

[16] With today's powerful computer hardware the 'law of big numbers' in stochastics may be accommodated much more easily in corpus building and analysis than in the past.

[17] The following type numbers refer to the surfaces of the word forms in the BNC. The numbers published by the authors of the BNC, on the other hand, refer to tagged word forms (cf. Section 15.5). According to the latter method of counting, the BNC contains 921 073 types.

The strictly surface-based ranking underlying the table 15.4.2 was determined by Marco Zierl at the CL lab of Friedrich Alexander University Erlangen Nürnberg (CLUE).

set of samples, the higher the value of the corpus for, e.g., computing the frequency distribution of words and word forms in the language.

On the most basic level, the statistical analysis is presented as a *frequency* list and an *alphabetic* list of word forms, illustrated in 15.5.1 and 15.5.3, respectively. In the frequency list, the types of the word forms are ordered in accordance with their token frequency. The position of a word form in the frequency list is called its *rank*.

At the beginning of the BNC frequency list, for example, the top entry is the word form the. It occurs 5 776 399 times and comprises 6.4% of the running word forms (tokens). The low end of the list, on the other hand, contains the word forms which occur only once in the corpus. These are called *hapax legomena*.[18] There are 348 145 hapax legomena in the BNC – amounting to 52.8% of its types.

Word forms with the same frequency, such as the hapax legomena, may be collected into *frequency classes* (F-classes). An F-class is defined in terms of two parameters, (i) the characteristic frequency of the F-class types in the corpus and (ii) the number of types with that frequency.

15.4.1 DEFINITION OF THE NOTION FREQUENCY CLASS (F-CLASS)

F-class $=_{def}$ [frequency of types # number of types]

The number of types in an F-class is the interval between the highest and the lowest rank of its elements (cf. 15.4.2).

The notion of an F-class is applicable uniformly to all ranks of the frequency list. The high frequency word forms result in F-classes which contain only a single type, but occur often. In the BNC, for example, the results in the 'single-type' F-class [5 776 399 # 1]. The low frequency words forms, on the other hand, result in F-classes with a large number of types which occur only once. In the BNC, for example, the hapax legomena result in the F-class [1 # 348 145]. The middle of the frequency list results in F-classes containing several or many types with a frequency > 1. For example, the F-class with the frequency 3 happens to be [3 # 39 691] in the BNC.

The number of tokens corresponding to an F-class equals the product of its type frequency and its type number. For example, because there are 39 691 word forms (types) which each occur 3 times in the BNC, the resulting F-class [3 # 39 691] corresponds to a total of 119 073 (= 3 · 39 691) tokens.

There are much fewer F-classes in a corpus than ranks. In the BNC, for example, 655 270 ranks result in 5 301 F-classes. Thus, the number of the F-classes is only 0.8% of the number of ranks. Because of their comparatively small number the F-classes are well suited to bring the type-token correlation into focus.

The frequency distribution in the BNC is illustrated in 15.4.2 for 27 F-classes – 9 at the beginning, 9 the middle, and 9 the end.

---

[18] From Greek, *'said once.'*

15.4.2 TYPE-TOKEN DISTRIBUTION IN THE BNC (*surface-based*)

F-class	start_r	end_r	types	tokens	types-%	tokens-%	
beginning (the first 9 F-classes)							
1 (the)	1	1	1	5776399	0.000152	6.436776	
2 (of)	2	2	1	2789563	0.000152	3.108475	
3 (and)	3	3	1	2421306	0.000152	2.698118	
4 (to)	4	4	1	2332411	0.000152	2.599060	
5 (a)	5	5	1	1957293	0.000152	2.181057	
6 (in)	6	6	1	1746891	0.000152	1.946601	
7 (is)	7	7	1	893368	0.000152	0.995501	
8 (that)	8	8	1	891498	0.000152	0.993417	
9 (was)	9	9	1	839967	0.000152	0.935995	
sums			9	19 648 696	0.001368 %	21.895 %	
middle	(9 samples)						
1000	1017	1017	1	9608	0.000152	0.010706	
2001	2171	2171	1	4560	0.000152	0.005081	tokens
3000	3591	3591	1	2521	0.000152	0.002809	per
3500	4536	4536	1	1857	0.000152	0.002069	type:
4000	5907	5910	4	5228	0.000607	0.005826	1307
4500	8332	8336	5	4005	0.000758	0.004463	801
4750	10842	10858	17	9367	0.002579	0.010438	551
5000	16012	16049	38	11438	0.005764	0.012746	301
5250	44905	45421	517	26367	0.078420	0.029381	51
end	(the last 9 F-classes)						
5292	108154	114730	6577	59193	0.997620	0.065960	9
5293	114731	122699	7969	63752	1.208763	0.071040	8
5294	122700	132672	9973	69811	1.512736	0.077792	7
5295	132673	145223	12551	75306	1.903775	0.083915	6
5296	145224	161924	16701	83505	2.533260	0.093052	5
5297	161925	186302	24378	97512	3.697732	0.108660	4
5298	186303	225993	39691	119073	6.020456	0.132686	3
5299	225994	311124	85131	170262	12.912938	0.189727	2
5300	311125	659269	348145	348145	52.807732	0.387946	1
sums			551 116	1 086 559	83.595012 %	1.210778 %	

For each F-class the associated rank interval is specified (start_r to end_r). The first nine F-classes each contain only one type, for which reason they each cover only 0.000152 % (= 100 : 659 270) of the types, which adds up to 0.001368%. The tokens corresponding to these classes, on the other hand, cover 21.895% of the running word forms of the BNC. The earliest F-class containing more than one type is located between F-classes 3 500 and 4 000. The 9 F-classes at the end of 15.4.2 contain types which occur only 1–9 times in the BNC. Together, they cover only 1.2% of the running word forms, but they comprise 83.6% of the types.

In other words, 16.4% of the types in the BNC suffice to cover 98.8% of the running word forms. Conversely, the remaining 1.2% of the running word forms require 83.6% of the types. This corresponds to the interval between rank 659 270 and 108 155, which represents 551 115 types.

The distribution of type and token frequency characterized in 15.4.2 is found generally in corpora of natural language. It may be shown graphically as follows.

15.4.3 CORRELATION OF TYPE AND TOKEN FREQUENCY

Precentage of tokens

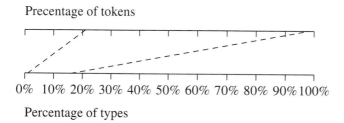

0% 10% 20% 30% 40% 50% 60% 70% 80% 90% 100%

Percentage of types

The fact that 0.001% of the types cover almost 22% of the running word forms, and that 16% of the types cover more than 98% of the running word forms in a corpus, is sometimes misinterpreted as if the small lexica of today's systems of speech recognition (cf. Section 1.4) were quite sufficient for practical purposes. This, however, is a serious mistake because the *semantic significance* increases with decreasing frequency.

For example, the user is helped little by a system of automatic word form recognition which can lexically recognize **the, of, and,** and **to,** but misses on significant BNC hapax legomena like **audiophile, butternut, customhouse,** or **dustheap,** to mention only a few, all listed and analyzed in a traditional lexicon like Webster's *New Collegiate Dictionary.*

In addition, there are many significant words contained in a traditional lexicon, such as **aspheric, bipropellant,** and **dynamotor,** which occur not even once in the BNC, despite its size and its careful design as a representative, balanced corpus. Thus, the word form list of a large corpus may help to update a traditional lexicon, but it should not be expected to be equally or more complete.

The correlation of type and token frequency exemplified in 15.4.3 was described most prominently by GEORGE K. ZIPF (1902–1950) as a general law of nature.[19] According to Zipf 1935, the frequency distribution of types always follows roughly

---

[19] Also known as the *Estoup-Zipf-Mandelbrot Law.* The earliest observation of the phenomenon was made by the Frenchman J.B. Estoup 1916, who – like Kaeding – worked on improving stenography. After doing a statistical analysis of human speech at Bell Labs, E.U. Condon 1928 also noted a constant relation between rank and frequency. B. Mandelbrot 1957, famous for his work on fractals, worked on mathematical refinements of Zipf's Law. Cf. R.G. Piotrovskij et al. 1985

the same pattern, independently of the text size, the text category, or the natural language, and corresponds to the following formula.[20]

15.4.4 ZIPF'S LAW

frequency · rank = constant

This means that the frequency of a given word multiplied by its rank produces a number which is roughly the same as the product of rank and frequency of any other word in the text.[21]

15.4.5 ILLUSTRATION OF ZIPF'S LAW

word form	rank	·	frequency	=	constant
the	1	·	5 776 399	=	5 776 399
and	2	·	2 789 563	=	5 579 126
...					
was	9	·	839 967	=	7 559 703
...					
holder	3 251	·	2 870	=	9 330 370

This example shows that Zipf's law holds only roughly, in the sense of a same order of magnitude. To compensate the curve-like increase and to achieve a better approximation of a constant, logarithmic modifications of the formula have been proposed.[22]

Zipf explained the correlation of frequency and rank he observed in texts as the law of least effort. According to Zipf, it is easier for the speaker-hearer to use a few word forms very frequently than to use the total of all word forms equally often. Zipf observed furthermore that the word forms used most frequently are usually the shortest and that they get longer as their frequency decreases. Zipf also pointed out that the semantic significance of word forms increases with decreasing frequency.

15.5 Statistical tagging

In the beginning of computational corpus analysis, no systems of automatic word form analysis with sufficient data coverage were available. Therefore corpus analysis was initially purely letter-based. The objects investigated were *unanalyzed word forms*, i.e., sequences of letters between spaces in the on-line text.

---

[20] One of the many critiques of the formula on empirical grounds is presented by M. Joos 1936. See also the reply in Zipf 1949.

[21] For simplicity, this example is based on the BNC data. Strictly speaking, however, Zipf's law applies only to individual texts and not to corpora.

[22] The plot of log(frequency) (y axis) versus log(rank) (x axis) approximates a straight line of slope -1.

Most on-line texts contain not only word forms, however, but also a large number of special symbols, such as markups for head lines, footnotes, etc., for example in SGML or some other convention, as well as punctuation signs, quotation marks, hyphens, numbers, abbreviations, etc. In order for the statistical analysis of word forms to be linguistically meaningful, these special symbols must be interpreted and/or eliminated by a preprocessor.

The statistical analysis of properly preprocessed on-line texts results in tables of word forms which specify their frequency relative to the total corpus and usually also relative to its genres. This is illustrated by the frequency list of the Brown-Corpus.

15.5.1 TOP OF BROWN CORPUS FREQUENCY LIST

```
69971-15-500 THE          21341-15-500 IN
36411-15-500 OF           10595-15-500 THAT
28852-15-500 AND          10099-15-485 IS
26149-15-500 TO            9816-15-466 WAS
23237-15-500 A             9543-15-428 HE
```

The entry `9543-15-428 HE`, for example, indicates that the word form `HE` occurs 9 543 times in the Brown corpus, in all 15 genres, and in 428 of the 500 sample texts.

What is missing in 15.5.1, however, is the categorization and lemmatization of the word forms. To fill this gap at least partially, W.N. Francis 1980 developed TAGGIT, a pattern-based system of categorization which required a lot of post-editing. Building from there (cf. I. Marshall 1987, p. 43-5), the CLAWS1-system was developed by R. Garside, G. Leech, & G. Sampson 1987, who tried to induce the categorization from the statistical distribution of word forms in the texts. This statistically-based *tagging* was developed in part for getting better and quicker results in large corpora than with pattern matching.

Statistical tagging is based on categorizing by hand – or half automatically with careful post-editing – a small part of the corpus, called the *core corpus*. The categories used for the classification are called *tags* or *labels*. Their total is called the *tagset*.

The choice of a specific tagset[23] is motivated by the goal to maximize the statistical differentiation of transitions from one word form to the next (bigrams). For this reason the tagging of the BNC core corpus is based on an especially large tagset, called the *enriched (C7) tagset*, which comprises 139 tags (without punctuation marks).

After hand-tagging the core corpus, the probabilities of the transitions from one word form to the next are computed by means of *Hidden Markov Models* (HMMs).[24] Then the probabilities of the hand-tagged core corpus are transferred to the whole corpus using a simplified tagset. In the BNC this *basic (C5) tagset* comprises 61 labels.

---

[23] The consequences of the tagset choice on the results of the corpus analysis are mentioned in S. Greenbaum & N. Yibin 1994, p. 34.

[24] The use of HMMs for the grammatical tagging of corpora is described in, e.g., G. Leech, R. Garside & E. Atwell 1983, I. Marshall 1983, S. DeRose 1988, R. Sharman 1990, P. Brown, V. Della Pietra, et al. 1991. See also K. Church & L.R. Mercer 1993.

15.5.2 Subset of the *basic (C5) tagset*

AJ0 Adjective (general or positive) (e.g. good, old, beautiful)
CRD Cardinal number (e.g. one, 3, fifty-five, 3609)
NN0 Common noun, neutral for number (e.g. aircraft, data, committee)
NN1 Singular common noun (e.g. pencil, goose, time, revelation)
NN2 Plural common noun (e.g. pencils, geese, times, revelations)
NP0 Proper noun (e.g. London, Michael, Mars, IBM)
UNC Unclassified items
VVB The finite base form of lexical verbs (e.g. forget, send, live, return)
VVD The past tense form of lexical verbs (e.g. forgot, sent, lived, returned)
VVG The -ing form of lexical verbs (e.g. forgetting, sending, living, returning)
VVI The infinitive form of lexical verbs (e.g. forget, send, live, return)
VVN The past participle form of lexical verbs (e.g. forgotten, sent, lived, returned)
VVZ The -s form of lexical verbs (e.g. forgets, sends, lives, returns)

Once the whole corpus has been tagged in the manner described, the frequency counts may be based on tagged word forms rather than letter sequences. Thereby different instances of the same surface with different tags are treated as different types. This is shown by the following example, which was selected at random from the tagged BNC list available on-line.[25]

15.5.3 Alphabetical word form list (sample from the BNC)

```
1 activ nn1-np0 1          8 activating aj0-nn1 6
1 activ np0 1              47 activating aj0-vvg 22
2 activa nn1 1             3 activating nn1-vvg 3
3 activa nn1-np0 1         14 activating np0 5
4 activa np0 2             371 activating vvg 49
1 activatd nn1-vvb 1       538 activation nn1 93
21 activate np0 4          3 activation nn1-np0 3
62 activate vvb 42         2 activation-energy aj0 1
219 activate vvi 116       1 activation-inhibition aj0 1
140 activated aj0 48       1 activation-synthesis aj0 1
56 activated aj0-vvd 26    1 activation. nn0 1
52 activated aj0-vvn 34    1 activation/ unc 1
5 activated np0 3          282 activator nn1 30
85 activated vvd 56        6 activator nn1-np0 3
43 activated vvd-vvn 36    1 activator/ unc 1
312 activated vvn 144      1 activator/ unc 1
1 activatedness nn1 1      7 activator/tissue unc 1
88 activates vvz 60        61 activators nn2 18
5 activating aj0 5         1 activators np0 1
```

Each entry in 15.5.3 consists (i) of a number detailing the frequency of the tagged word form in the whole corpus, (ii) the surface of the word form, (iii) the label, and (iv) the number of texts in which the word form was found under the assigned label. The frequency list of the BNC consists of the same entries as those illustrated in 15.5.3, but ordered according to frequency rather than alphabetically.

15.5.3 illustrates the output of the statistical tagger CLAWS4, which was developed for analyzing the BNC and is generally considered one of the best statistical taggers.

[25] Meanwhile, the tagged BNC-lists have been removed from the web.

The error rate[26] of CLAWS4 is quoted by Leech 1995 at 1.7%, which at first glance may seem like a very good result.

This error rate applies to the running word forms, however, and not to the types. If we take into consideration that the last 1.2% of the low frequency tokens requires 83.6% of the types (cf. 15.4.2, 15.4.3), an error rate of 1.7% may also represent a very bad result – namely that about 90% of the types are not analyzed or not analyzed correctly. This conclusion is born out by a closer inspection of the sample 15.5.3.

The first surprise is that of 38 entries, 27 are categorized more than once namely `activ` (2), `activa` (3), `activate` (3), `activated` (7), `activating` (6), `activation` (2), `activator` (2), and `activators` (2). Thereby, `activ` – which is a typing error – is classified alternatively as `nn1-np0` and `np0`, neither of which makes any sense linguistically. The classification of `activate` as `np0` is also mistaken from the viewpoint of traditional English dictionaries. The typing error `activatd` is treated as well-formed and labeled `nn1-vvb`. In `activation.` the punctuation sign is not removed by the preprocessor, yet it is labeled `nn0`. In `activation/` and `activator/` the '/' is not interpreted correctly and they are labeled `unc` (for *unclassified*), whereby the identical entries for `activator/` are counted separately.

In addition to a high error rate, the frequency counts of the BNC analysis are impaired be a weak preprocessor. For example, by treating different numbers as different word forms, as in

```
1 0.544 crd 1
1 0.543 crd 1
1 0.541 crd 1
```

58 477 additional types are produced, which is 6.3% of the labeled BNC types. In addition, there are more than 20 000 types resulting from numbers preceded by measuring units like &pound (11 706) and &dollar (8 723). Furthermore, word-form-initial hyphens and sequences of hyphens are counted as separate word forms, producing another 4 411 additional types, which are moreover labeled incorrectly.

Thus, statistical labeling increases the number of types from 659 270 surfaces to 921 074 labeled BNC forms. A proper preprocessing of numbers and hyphens would reduce the number of surface types by an additional 83 317 items to 575 953. All in all, statistical tagging increases the number of types in the BNC by at least 37.5%.

The BNC tagging analysis is a typical example of the strenghts and weaknesses of a smart solution (cf. Section 2.3). On the positive side there is a comparatively small effort as well as robustness, which suggest the use of statistical taggers at least in the preparatory phase of a building a solid system of automatic word form recognition.

[26] Unfortunately, neither G. Leech 1995 nor L. Burnard 1995 specify what exactly constitutes an error in tagging the BNC. A new project to improve the tagger was started in June 1995, however. It is called *'The British National Corpus Tag Enhancement Project'* and its results were originally scheduled to be made available in September 1996.

On the negative side there are the following weaknesses, which show up even more clearly in languages with a morphology richer than English.

15.5.4 WEAKNESSES OF STATISTICAL TAGGING

1. The categorization is too unreliable to support rule-based syntactic parsing.
2. Word forms can be neither reduced to their base forms (lemmatization) nor segmented into their allomorphs or morphemes.
3. The overall frequency distribution analysis of a corpus is distorted by an artificial inflation of types.

It is in the nature of statistical tagging that the classification of a surface is based solely on statistical circumstances.[27] Thus, if it turns out that a certain form has been classified incorrectly by a statistical tagger, there is no way to correct this particular error. Even if the tagger is successfully improved as a whole, its results can never be more than probabilistically-based conjectures.

The alternative solid solution is a rule- and lexicon-based system of automatic word form recognition. If a word form is not analyzed or not analyzed correctly in such a system, then the cause is either a missing entry in the lexicon or an error in the rules. The origin of the mistake can be identified and corrected, thus solidly improving the recognition rate of the system.

In addition, the rule-based analysis allows to extend frequency analysis from letter based word forms to words and their parts. This is of interest in connection with hapax legomena (i.e. about half of the types). For statistics, the hapax legomena constitute the quasi unanalyzable sediment of a corpus. A rule-based approach, in contrast, cannot only analyze hapax legomena precisely, but reduce their share at the level of elementary base forms by half. In the Limas corpus, for example, word forms like **Abbremsung**, **Babyflaschen**, and **Campingplatz** are hapax legomena, but their parts, i.e., **bremse**, **baby**, **flasche**, **camping**, and **platz**, occur more than once.

Exercises

Section 15.1

1. What is the definition of a grammar system?
2. What is the function of the formal algorithm in a grammar system?

---

[27] It is for this reason that a surface may be classified in several different ways, depending on its various environments in the corpus.

3. Why does a grammar system require more than its formal algorithm?
4. Why must a grammar system be integrated into a theory of language?
5. Explain the methodological reason why a grammar system must have an efficient implementation on the computer.
6. Why is a modular separation of grammar system, implementation, and application necessary? Why do they have to be closely correlated?
7. What differences exist between various implementations of LA-Morph, and what do they have in common?

Section 15.2

1. Explain the linguistic motivation of a distinctive categorization using examples.
2. What are multicats and why do they necessitate an extension of the basic algorithm of LAG?
3. Compare list-based and attribute-based matching in LA-Morph.
4. What motivates the development of subtheoretical variants?
5. Why is the transition to a new subtheoretical variant labor intensive?
6. What is the difference between a subtheoretical variant and a derived formalism?

Section 15.3

1. For which purpose investigated Kaeding the frequency distribution of German?
2. What is a representative, balanced corpus?
3. List Kučera & Francis' desiderata for constructing corpora.
4. Explain the distinction between the types and the tokens of a corpus.

Section 15.4

1. Describe the correlation of type and token frequency in the BNC.
2. What is the percentage of hapax legomena in the BNC?
3. In what sense are high frequency word forms of low significance and low frequency word forms of high significance?
4. What is Zipf's law?

Section 15.5

1. What motivates the choice of a tagset for statistical corpus analysis?
2. Why is it necessary for the statistical analysis of a corpus to tag a core corpus by hand?
3. What is the error rate of the statistical BNC tagger CLAWS4? Does it refer to types or tokens? Is it high or low?
4. Why does statistical tagging substantially increase the number of types in a corpus? Are these additional types real or spurious?
5. Is statistical tagging a smart or a solid solution?
6. What is the role of the preprocessor for the outcome of the statistical analysis of a corpus? Explain your answer using concrete examples.

16. Basic concepts of syntax

In this and the following two chapters, the grammar component of syntax is described. Section 16.1 analyzes the structural border between the grammar components of morphology and syntax. Section 16.2 discusses the role of valency in the syntactic-semantic composition of natural languages. Section 16.3 explains the notions of agreement and the canceling of valency positions by compatible valency fillers. Section 16.4 demonstrates the handling of free word order with an LA-grammar for a small fragment of German. Section 16.5 demonstrates the handling of fixed word order with an LA-grammar for a corresponding fragment of English.

16.1 Delimitation of morphology and syntax

The grammar component of morphology is limited to the analysis of individual word forms, but provides the basis for their syntactic composition in terms of their morphosyntactic categorization. The grammar component of syntax is limited to characterizing the composition of word forms into complex expressions, but depends on their morphological analysis.

Thus, the structural boundary between the morphology and syntax coincides with the boundaries between the word forms in the sentence surface: everything within a word form boundary is in the domain of morphology, while everything dealing with the composition of word forms is in the domain of syntax.

16.1.1 CORRELATION OF LA-MORPHOLOGY AND LA-SYNTAX

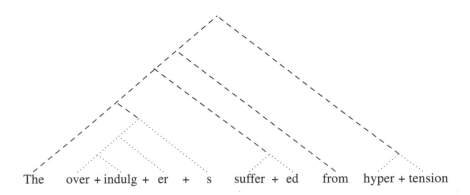

The tree structures of LA-morphology and LA-syntax both satisfy the SLIM-theoretic principles of surface compositionality (S) and time-linear composition (L). However, the time-linear adding of a complex word form like over+indulg+er+s in the syntax presupposes its prior time-linear composition in morphology.

Thus, even though LA-morphology and LA-syntax are based on the same structural principles, their components are separated in a modular fashion[1] because their respective time-linear compositions occur in different phases. This particular correlation of syntactic and morphological composition is in concord with the traditional view according to which complex morphological structures originate historically as "frozen syntactic structures" (H. PAUL 1920[v], Chapter 19).

The separation of LA-morphology and LA-syntax poses no problem even in borderline cases such as idioms. Critical cases can be handled either lexico-morphologically or syntactically because both components show the same incremental composition of surfaces and meanings₁. Which grammar component is appropriate should be determined by the structural properties of the phrase in question.

For example, the phrase over-the-counter should be handled in the lexicon because of its use as a unit and the lack of internal variation, expressed orthographically by the hyphenation. The variant without the hyphens, on the other hand, should be handled in the syntax.

In German, a phrase like gang und gäbe[2] may at first glance seem to belong into syntax because the phrase is written as three separate words. However, because gäbe is like a bound morpheme which may not occur independently, the phrase should be handled in the lexicon. Technically, this may be achieved by an internal representation as gang_und_gäbe, treating the phrase as a single word form for the purposes of syntax.

A syntactic treatment is generally motivated in idioms which (i) retain their compositional meaning as an option, (ii) are subject to normal variations of word order, and (iii) exhibit internal inflectional variation. As an example, consider German seinen Hut nehmen, which literally means *to take one's hat* with the idiomatic meaning of *stepping down from office*.

Because this phrase obeys word order variations, as in nachdem Nixon seinen Hut genommen hatte, and shows internal inflectional variation, as in Nixon nahm seinen Hut, a compositional treatment in the syntax is the most appropriate. The idiomatic use can be handled by marking the paraphrase *stepping down from office* in the semantic representation of nehmen. Whenever this verb is used, it is checked whether the object is Hut. If this is the case, pragmatics is provided with the paraphrase *stepping down from office* as an option.

---

[1] This is in contrast to the nativist variant of 'generative semantics.' It used transformations to derive word form surfaces from syntactic deep structures, e.g. persuade from *cause to come about to intend* (G. Lakoff 1972, p. 600).

[2] This phrase is commented in G. Wahrig's dictionary as "Adj.; nur noch in der Wendung ... *das ist so üblich* [≺ mhd. *gäbe* „annehmbar"; zu geben]".

The nature of the morphology-syntax relation may be illuminated further by looking at different types of natural languages. This topic belongs into the domain of language typology, where synthetic languages with a rich morphology are distinguished from analytic languages with a rudimentary morphology.

According to L. Bloomfield 1933, modern Chinese is an analytic language in which each word (form) is either a morpheme consisting of one syllable, or a compound, or a phrase word. A synthetic language, on the other hand, is Eskimo in which long chains of morphemes are concatenated into a single word form such as [a:wlis-ut-iss?ar-si-niarpu-na] *I am looking for something suitable for a fish-line* (op.cit., p. 207).

For Bloomfield the distinction between synthetic and analytic is relative, however, insofar as one language can be more synthetic than another in one respect, yet more analytic in another. As an alternative schema of classification he cites the isolating, agglutinative, polysynthetic, and inflectional types of morphology:

> Isolating languages were those which, like Chinese, used no bound forms; in agglutinative languages the bound forms were supposed to follow one another, Turkish being the stock example; polysynthetic languages expressed semantically important elements, such as verbal goals, by means of bound forms, as does Eskimo; inflectional languages showed a merging of semantically distinct features either in a single bound form or in closely united bound forms, as when the suffix ō in a Latin form like amō 'I love' expresses the meanings 'speaker as actor,' 'only one actor,' 'action in present time,' 'real (not merely possible or hypothetical) action.' These distinctions are not co-ordinate and the last three classes were never clearly defined.
>
> L. Bloomfield 1933, p. 208

Despite these difficulties in terminology and classification, there remains the fact that some natural languages compose meaning$_1$ mainly in the syntax (e.g. Chinese) and others mainly in morphology (e.g. Eskimo). This alternative exists also within a given natural language. For example, in English the complex concept denoted by the word form **overindulgers** may roughly be expressed analytically as **people who eat and drink too much.**

That meaning$_1$ may be composed in morphology as well as in syntax is no reason against their modular separation into two different components. A reason for their separation, on the other hand, is that they are being based on different principles of combination. Those of morphology are inflection, derivation, and composition (cf. 13.1.2), whereas those of syntax are stated below.

16.1.2 COMBINATION PRINCIPLES OF SYNTAX

1. *Valency* (cf. Section 16.2)
2. *Agreement* (cf. Section 16.3)
3. *Word order* (cf. Section 16.4 for German and 16.5 for English)

In addition to these traditional combination principles there is the higher principle of the left-associative (time-linear) derivation order, which underlies morphological and syntactic composition alike – both in language production (speaker mode) and language interpretation (hearer mode).

16.2 Valency

The notions valency carrier and valency filler go back to the French linguist L. TES-
NIÈRE 1959, who borrowed them from chemistry. The valency positions of a carrier
must be filled, or canceled, by compatible fillers in order for an expression to be syntactically and semantically complete.

Prototypical valency carriers in the natural languages are the verbs. The verb give,
for example, is a three-place carrier because each act of giving necessarily requires (i)
a giver, (ii) a receiver, and (iii) something that is given. This is the valency structure
of the verb give.

The valency structure of a carrier is a basic, inherent, lexical property.[3] For example,
one may say Suzy has given already in the context of a charity. Despite the missing
direct and indirect object, it follows semantically that there is *something* that Suzy
gave and *someone* who received whatever Suzy gave.

From the viewpoint of elementary propositions (cf. 3.4.2), valency carriers realize
functors and valency fillers realize arguments in language. In addition there is the
third basic type of (optional) modifiers. The systematic composition of valency carriers, valency fillers, and modifiers results in the functor-argument structure of natural
language. It serves simultaneously as the basis of (i) syntactic combinatorics and (ii)
semantic interpretation.

In LA-grammar, the structure of valency carriers is coded by means of composite
syntactic categories, defined as lists of category segments. For example, the English
verb form ate is analyzed as follows.

[ate (N$'$ A$'$ V) eat]

The last segment (here V) is interpreted as the *result segment*, while the segments
marked by $'$ (here N$'$ and A$'$) are interpreted as *valency positions*. The category (N$'$ A$'$
V) represents a V-expression (i.e., results in a finite verbal expression) which requires
a nominative N and an accusative A as valency fillers in order to be complete.

LA-grammar and C-grammar have in common that they describe the functor-
argument structure of language in terms of valency positions which are canceled by
appropriate fillers, whereby canceled positions disappear from the carrier category.[4]
Compared to two rules of C-grammar, however, the rule structure of LA-grammar is

---

[3] In addition to the basic valency structure of a word, natural languages allow special uses with secondary valency structures. For example, the English verb to sleep is normally intransitive, but in the
sentence This yacht sleeps 12 people it is used transitively. Thereby, the change to the secondary
valency structure is accompanied by a profound change in the meaning of the word, which may be
paraphrased as provide bunks for.

There is also the inverse process of reducing the number of basic valency positions, called *Detran-
sitivierung* in German. For example, give in German has a secondary use as a one-place – or arguably
two-place – valency carrier, as in Heute gibt es Spaghetti (= today it gives spaghetti). Again, the
secondary valency structure is accompanied by profound changes in the original word meaning.

These phenomena of *transitivization* and *detransitivization* do not call a systematic analysis of
valency structure into doubt. On the contrary, they throw additional light onto the fundamental importance of the valency structure in the analysis of natural language.

considerably more flexible and differentiated (see for example 16.4.8). This is why the categories of LA-grammar may have a simpler form than those of C-grammar.

The different categorial and derivational structures in C- and LA-grammar are illustrated in 16.2.1 with simplified[5] analyses (see also Section 10.5).

16.2.1 CARRIERS, FILLERS, AND MODIFIERS IN CG AND LAG

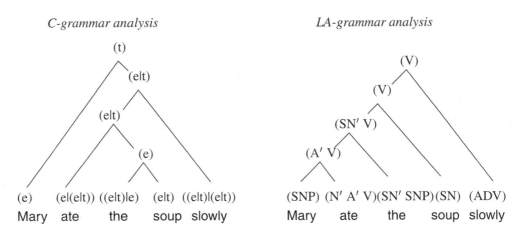

In C-grammar, the basic handling of the functor-argument structure is formally elegant. However, even the smallest fragments of C-grammar for natural language are prohibitively complicated (cf. Section 7.5) because the syntactic combinatorics are handled directly in terms of the semantic functor-argument structure. Moreover, the interpretation of the categorial functor-argument structure in C-grammar is incompatible with a time-linear derivation order.

In LA-grammar, the main task of the syntax is the time-linear building up and down of valency positions in order to specify possible continuations at each point in the derivation. The hierarchical functor-argument structures are derived indirectly by means of separate semantic clauses which interpret the syntactic rules (cf. 21.4.3 as well as Chapters 23 and 24).

The LA-grammar-specific coding of valency positions and result segments in various different valency carriers of German is illustrated by the following examples.

16.2.2 EXAMPLES OF DIFFERENT VALENCY CARRIERS IN GERMAN

– the one-place verb form **schläfst** (*sleep*):

---

[4] In contrast, PS-grammar provides no formal means to account for the functor-argument structure of language at the level of syntax. This is reflected in the formal nature of the respective categories, which are atomic in PS-grammar, but of a composite structure in C- and LA-grammar.

[5] The C-analysis is simplified in that it does not represent alternative orders of functor and argument by means of different slashes. Also, ungrammatical combinations are not excluded by the categorization. The LA-analysis is simplified in that no rules and agreement conditions are specified.

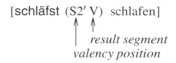

The valency position S2′ indicates that this verb form requires a nominative of 2nd person singular. The result segment V shows that after filling the valency position there results a complete verbal expression (sentence).

– the two-place verb form **liest** (*read*):

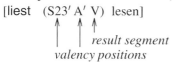

The valency position S23′ indicates that this verb form requires a nominative of 2nd or 3rd person singular. The additional valency position A′ indicates that this form requires an additional accusative in order to result in a complete sentence. Specification of the case ensures that ungrammatical oblique fillers, such as a dative or a genitive, are excluded. For example, **Der Mann liest einen Roman** is grammatical while *Der Mann liest einem Roman is not.

– the two-place verb form **hilft** (*help*):

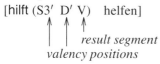

The oblique valency position D′ indicates that this form requires a dative (in addition to the nominative). Non-dative valency fillers, as in *Der Mann half die Frau, are not grammatical.

– The three-place verb form **gebt** (*give*):

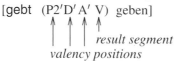

The valency position P2′ indicates that this form requires a nominative of 2nd person plural. The oblique valency positions D′ and A′ indicate the need for additional dative and accusative valency fillers.

– The three-place verb form **lehrte** (*taught*):

The valency position S13′ indicates that this verb form requires a nominative of 1st or 3rd person singular. The two oblique valency positions A′ indicate the need for two additional accusative valency fillers. Because of this valency structure **Der Vater lehrte mich das Schwimmen** is grammatical while *Der Vater lehrte mir das Schwimmen** is not.

– The one-place preposition nach (*after*):

[nach (D′ ADP) nach]

           ↑   ↑

         | *result segment*

      valency position

The preposition *nach* requires a dative noun phrase as argument, as indicated by the valency position D′. The result segment ADP (for ad-phrase) expresses that filling the valency position results in an expression which functions as an adverbial or postnominal modifier (cf. 12.5.4–12.5.6).

As shown by the last example, there exist valency carriers besides the verbs in the natural languages, for example prepositions and determiners (cf. 15.2.1).

In LA-grammars for natural languages, the syntactic categories have at least one category segment – in contrast to LA-grammars for artificial languages, which also use empty lists as categories, especially in final states (e.g. 10.2.2, 10.3.3, 11.5.2, 11.5.4, 11.5.7). Syntactic categories consisting of one segment can serve only as valency fillers or as modifiers, e.g.

[Bücher (P-D) buch]	(*books*)
[ihm (D) er]	(*him*)
[gestern (ADV) gestern]	(*yesterday*)

Valency carriers may also function as valency fillers using their result segment, e.g V, as the filler segment. In this case, the segments representing valency positions are attached at the beginning of the category resulting from the composition.

16.3 Agreement

The second combination principle of natural language syntax besides valency is agreement. Agreement violations are among the most obvious and most serious grammatical mistakes one can make in natural language, even though the intended meaning of the utterance is usually not really jeopardized.

16.3.1 AGREEMENT VIOLATION IN ENGLISH

    *Every girls need a mother.

Agreement interacts with valency in that a potential valency filler can only cancel a structurally compatible valency position.

As a simple example of the functional interaction between a valency carrier and compatible fillers consider the following analysis of the sentence he gives her this, which shows the successive filling (and canceling) of valency positions in a left-associative derivation.

16.3.2 A SIMPLE ANALYSIS IN LA-SYNTAX

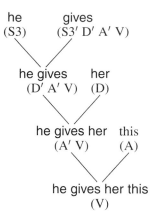

The left-associative bottom-up derivation is represented in this and the following examples as a tree structure growing downward from the terminal string (input).[6] The step by step canceling of valency positions is specified in the category of the sentence start, which shows at each step how many valency positions remain to be filled.

A maximally simple LA-grammar for the derivation 16.3.2 is *LA-plaster*.[7] On the positive side, it is written just like the LA-grammars for artificial languages in Chapters 10–12. On the negative side, it is consciously primitive insofar as no variables are used in the rule patterns.

16.3.3 AN LA-GRAMMAR FOR 6.3.2 (*LA-plaster*)

$LX =_{def}$ { [he (S3) *], [her (D) *], [this (A) *], [gives (S3' D' A' V) *]}

$ST_S =_{def}$ { [(S3) {MAIN+FV}] }
MAIN+FV: (S3) (S3' D' A' V) $\Rightarrow$ (D' A' V) {FV+MAIN1}
FV+MAIN1:   (D' A' V) (D) $\Rightarrow$ (A' V)   {FV+MAIN2}
FV+MAIN2:   (A' V) (A) $\Rightarrow$ (V)   { }
$ST_F =_{def}$ { [(V) rp$_{FV+MAIN2}$] }

For simplicity, the lexicon LX is defined as a word form lexicon. In this way the grammar is independent from a component of automatic word form recognition, but limited to the extremely small vocabulary of LX which comprises only the word forms occurring in the example 16.3.2.[8]

---

[6] This format corresponds to the derivation structure (iii) in 10.3.1 as well as the automatic parsing analyses 10.3.2, 10.4.1, 10.5.3, 10.5.4, and 10.5.5. The equivalent format of LA-trees growing upward from the terminal string, illustrated in 10.5.2 and 16.2.1, on the other hand, has served only for the comparison with corresponding constituent structures, and is of no further use.

[7] In German, this grammar is called LA-Beton. The English translation of Beton as *concrete* has a misleadingly positive ring, hence the name *LA-Plaster*.

[8] The third position of the lemmata is indicated by a '*' because the base forms are not referred to by the rules.

According to the start state ST_S, the derivation must begin with a word of category (S3) and the rule MAIN+FV (main constituent plus finite verb). The input condition of this rule requires that the sentence start must be of category (S3), which matches he, and that the next word must be of category (S3′ D′ A′ V), which matches gives. The categorial operation of MAIN+FV cancels the S3′ position in the valency carrier. The result category (D′ A′ V) characterizes a sentence start which needs a dative and an accusative in order to complete the semantic relation expressed by the verb. The rule package of MAIN+FV activates the rule FV+MAIN1.

According to the input condition of FV+MAIN1, the sentence start must be of category (D′ A′ V), which matches he gives, and the next word must be of category (D), which matches her. The categorial operation of FV+MAIN1 cancels the D′ position in the sentence start. The result category (A′ V) characterizes the new sentence start he gives her as requiring an accusative in order to complete the semantic relation expressed by the verb. The rule package of FV+MAIN1 activates the rule FV+MAIN2

According to the input condition of FV+MAIN2, the sentence start must be of category (A′ V), which matches he gives her, and the next word must be of category (A), which matches this. The categorial operation of FV+MAIN2 cancels the A′ position in the sentence start. The result category (V) characterizes the new sentence start he gives her this as an expression which does not have any open valency positions. FV+MAIN2 activates the empty rule package

The final state ST_F characterizes the output of FV+MAIN2 as a complete sentence of the 'language' described by *LA-Plaster*. At this point the derivation in 16.3.3 is completed successfully.

In an extended grammar, the sentence start he gives her this could be continued further, for example by a concluding full stop, an adverb like now, a subordinate clause like because she asked for it, or a conjunction like and she smiled happily. Thus, whether a given sentence start is a complete expression or not requires not only that its analysis constitutes a final state, but also that the surface has been analyzed completely.

Even though *LA-Plaster* lacks the flexibility and generality LA-grammars for natural languages normally strive for, it is fully functional as a parser, strictly time-linear, and descriptively adequate insofar as it accepts 16.3.2. Furthermore, because the input and output patterns of the rules refer to categories, the grammar would analyze many additional sentences, if the lexicon LX were extended with additional word forms of the categories (S3), (D), (A), and (S3′ D′ A′ V), respectively.

In all three combination steps of 16.3.2 there is agreement between the filler segment and the canceled valency position. In the first combination, S3′ is canceled by S3, in the second, D′ by D, and in the third, A′ by A. Thus, agreement is formally handled here as the *identity* of the filler segment and the associated valency position. Violations of identity-based agreement show up clearly in the input categories.

16.3.4 EXAMPLE OF AN ERROR IN IDENTITY-BASED AGREEMENT

$$\begin{array}{ccccc} \text{I} & + & \text{gives} & \Rightarrow & \text{Error: ungrammatical continuation} \\ \text{(S1)} & & \text{(S3}'\text{ D}'\text{ A}'\text{ V)} & & \end{array}$$

That the first person pronoun I does not agree with **gives** in 16.3.4 is obvious on the level of the respective categories because the filler segment S1 has no corresponding valency segment in the carrier category (S3$'$ D$'$ A$'$ V).

16.4 Free word order in German (*LA-D1*)

The third combination principle of natural language syntax besides valency and agreement is word order. Just as natural languages may be distinguished typologically with respect to their relative degree of morphological marking (specifying case, number, genus, tense, verbal mood, etc.), they may also be distinguished typologically with respect to their relative degree of word order freedom.

Investigating the typology of different languages, J. GREENBERG 1963 observed that grammatical functions may be coded either in their morphology or in their word order. Furthermore, languages with a rich morphology have a free word order, while languages with a simple morphology have a fixed word order.

This is because the richer the morphology, the fewer grammatical functions are left for the word order to express – allowing it to be free.[9] Conversely, the simpler the morphology, the more grammatical functions must be expressed by means of the word order – thus limiting its freedom.

A language with a relatively free word order is German. The position of the verb is fixed, but the order of fillers and modifiers is free. This is illustrated by a declarative main clause with a three-place verb: the finite verb is fixed in second position, while the three fillers allow 6 variations of word order.

16.4.1 WORD ORDER VARIATIONS IN A DECLARATIVE MAIN CLAUSE

Der Mann gab der Frau den Strauß.
(the man gave the woman the bouquet.)
Der Mann gab den Strauß der Frau.
(the man gave the bouquet the woman.)
Der Frau gab der Mann den Strauß.
(the woman gave the man the bouquet.)
Der Frau gab den Strauß der Mann.
(the woman gave the bouquet the man.)
Den Strauß gab der Mann der Frau.
(the bouquet gave the man the woman.)

---

[9] This freedom is utilized for other aspects of communication, such as textual cohesion.

Den Strauß gab der Frau der Mann.
(the bouquet gave the woman the man.)

These all translate into English as the man gave the woman the bouquet. Which variant is chosen by the speaker depends on the purpose and the circumstances of the utterance context.[10]

In contrast to the grammatically correct examples in 16.4.1, the following example 16.4.2 is ungrammatical because it violates the verb second rule of declarative main clauses in German.

16.4.2 WORD ORDER VIOLATION IN GERMAN

*Der Mann der Frau gab einen Strauß.
(the man the woman gave the bouquet.)

In LA-grammar, the free word order of German is formally captured by allowing the canceling of arbitrary valency positions in the carrier category, and not just the currently first segment.

16.4.3 FREE CANCELING OF VALENCY POSITIONS IN A CARRIER OF GERMAN

$$
\begin{array}{llll}
\text{Der Mann} & + & \text{gab} & \Rightarrow & \text{Der Mann gab} \\
\text{(S3)} & & \text{(S3' D' A' V)} & & \text{(D' A' V)}
\end{array}
$$

$$
\begin{array}{llll}
\text{Der Frau} & + & \text{gab} & \Rightarrow & \text{Der Frau gab} \\
\text{(D)} & & \text{(S3' D' A' V)} & & \text{(S3' A' V)}
\end{array}
$$

$$
\begin{array}{llll}
\text{Den Strauß} & + & \text{gab} & \Rightarrow & \text{Den Strauß gab} \\
\text{(A)} & & \text{(S3' D' A' V)} & & \text{(S3' D' V)}
\end{array}
$$

The arrow indicates the canceling of a valency position (marked by ′) by a suitable (here identical) valency filler. Which of the available valency positions is to be canceled is decided here by the case-marked filler. In all three examples, the resulting sentence start can be continued into a complete well-formed sentence, e.g. Den Strauß gab der Mann der Frau.

How can *LA-Plaster* (cf. 16.3.3) be generalized in such a way that the free word order characteristic of German declarative main clauses is handled correctly? The obvious first step is to fuse FV+MAIN1 and FV+MAIN2 of *LA-Plaster* into one rule.

---

[10] Given the isolation of these variants in the context of a linguistic example, some may seem less natural than others. The reason for this, however, is not grammar, but that for some it is more difficult to imagine utterance situations suitable for their topic-comment structures.

16.4.4 GERMAN LA-GRAMMAR WITH PARTIAL FREE WORD ORDER

LX $=_{def}$ { [er (S3) *], [ihr (D) *], [das (A) *], [gab (S3′ D′ A′ V) *]}
Variable definition: $np\ \varepsilon$ {D, A}, with np' correspondingly D′ or A′
                       x, y = .?.?.?.? (i.e. an arbitrary sequence up to length 4)

ST$_S$ $=_{def}$ { [(S3) {MAIN+FV}] }
MAIN+FV: (S3) (S3′ D′ A′ V)$\Rightarrow$ (D A V) {FV+MAIN}
FV+MAIN: (x np' y V) (np) $\Rightarrow$ (x y V) {FV+MAIN}
ST$_F$ $=_{def}$ { [(V) rp$_{FV+MAIN}$] }

This extension of *LA-Plaster* translates the English word form lexicon LX of 16.3.3
into German. Furthermore, 16.4.4 uses the variables x, y, and np to fuse the rules
FV+MAIN1 and FV+ MAIN2 of 16.3.3. The variables x and y are *unspecified*, standing for a sequence of zero, one, two, three, or four arbitrary category segments, while
np is a *specified* variable, the range of which is explicitly limited to the category segments D and A.

   The variables in a rule are assigned values by matching the rule's input patterns
onto the input expressions. Thereby the value for a given variable must be the same
throughout the rule, i.e., the variable is *bound*. Consider the following example:

16.4.5 FV+MAIN MATCHING A NEXT WORD ACCUSATIVE

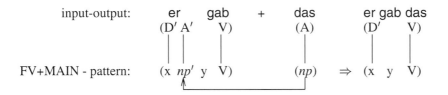

On the level of the input, the sentence start has category (D′ A′ V) and the next word
has the category (A). Therefore, the next word variable np at the level of the rule
is bound to the value A. Accordingly, everything preceding the A′ in the sentence
start category at the input level is matched by (and assigned to) the variable x – here
the sequence <D′> – while everything between the A′ and the V is matched by (and
assigned to) the variable y – here the empty sequence < >. These bound variables,
together with the constant V, are used for building the output category – here (D′ V).
The categorial operation cancels the segment A′ in the sentence start category because
the variable np' is not specified in the output category pattern.

   The new rule FV+MAIN of 16.4.4 generates and analyzes not only the variant er
gab das (ihr), but also er gab ihr (das) (i.e. the one sentence of *LA-Plaster*, though
this time in German). The reason is that the input pattern of FV+MAIN accepts an
accusative (cf. 16.4.5) as well as a dative (cf. 16.4.6) after the verb.

16.4.6 FV+MAIN MATCHING A NEXT WORD DATIVE

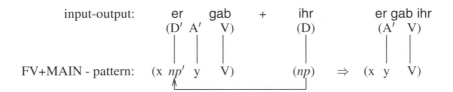

The variable *np* is bound to the category segment D for the duration of this rule application because the next word category is (D). Everything before the D′ in the sentence start is represented by x, while everything between the D′ and the V is represented by y. Thus, x is bound to the empty sequence < >, while y bound to the sequence <A′>. With these bound variables the output of the rule application is specified.

Another important aspect of FV+MAIN in 16.4.4 – as compared to 16.3.3 – is the fact that FV+MAIN calls up itself via its rule package. This provides for a possible recursion which can proceed as long as the categorial states of the input permit. For example, the output of 16.4.6 fits the input pattern of FV+MAIN one more time:

16.4.7 REAPPLICATION OF FV+MAIN

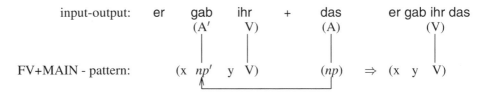

After this second application of FV+MAIN, the category is not suitable for further applications. Thus the content of the current rule package provides options, but whether these options are realized or not is controlled by the content of the current categories.

It is only a short step from the tentative LA-grammar 16.4.4 to an LA-grammar which handles the syntactic structures in question simply and adequately.[11]

16.4.8 GERMAN LA-GRAMMAR WITH FREE WORD ORDER (*LA-D1*)

LX $=_{def}$ { [er (S3) *], [ihr (D) *], [das (A) *], [gab (S3′ D′ A′ V) *]}

Variable definition: *np* ε {S3, D, A}, with *np*′ correspondingly S3′, D′ or A′

                   x, y = .?.?.?.? (i.e. an arbitrary sequence up to length 4)

---

[11] *LA-D1* and the following LA-grammars for natural languages represent the categories and rule patterns in the form of lists, similar to the algebraic definition 10.2.1. Current applications of LA-grammar use the Malaga-System (cf. 15.2.6), which represents categories and rule patterns as feature-value structures. For development and application, Malaga is as powerful as it is comfortable. For a principled theoretical description, however, the list-based notation is more parsimonious and transparent.

$\mathrm{ST}_S =_{def} \{ [(np) \{MAIN+FV\}] \}$
MAIN+FV: $(np) \ (x \ np' \ y \ V) \ \Rightarrow \ (x \ y \ V) \ \{FV+MAIN\}$
FV+MAIN: $(x \ np' \ y \ V) \ (np) \ \Rightarrow \ (x \ y \ V) \ \{FV+MAIN\}$
$\mathrm{ST}_F =_{def} \{ [(V) \ \mathrm{rp}_{FV+MAIN}] \}$

Compared to the tentative LA-grammar 16.4.4, the range of the variable *np* is extended to S3. While the rule FV+MAIN is the same as in 16.4.4, the rule MAIN+FV has a new formulation: its input patterns now resemble that of FV+MAIN except that the patterns for the sentence start and the next word are in opposite order.

The required verb second position is ensured in *LA-D1* by the start state which activates only one initial rule, MAIN+FV, and by the fact that MAIN+FV is not called by any other rule. The otherwise free word order of *LA-D1* is based on the variables in the input patterns of MAIN+FV and FV+MAIN. Using the word form lexicon specified in 16.4.8, the following sentences can be analyzed/generated:

16.4.9 WORD ORDER VARIANTS OF *LA-D1*

er gab ihr das	das gab er ihr	ihr gab er das
er gab das ihr	das gab ihr er	ihr gab das er

LA-D1 works also with an extended lexicon, containing for example the following additional lemmata:

[ich (S1) *], [du (S2) *], [wir (P1) *], [schlafe (S1′ V) *], [schläfst (S2′ V) *], [schläft (S3′ V) *], [schlafen (P1′ V) *], [lese (S1′ A′ V) *], [liest (S2′ A′ V) *], [las (S3′ A′ V) *], [helfe (S1′ D′ V) *], [hilfst (S2′ D′ V) *], [half (S3′ D′ V) *], [lehre (S1′ A′ A′ V) *], [lehrst (S2′ A′ A′ V) *], [lehrt (S3′ A′ A′ V) *], [gebe (S1′ D′ A′ V) *], [gibst (S2′ D′ A′ V) *].

For practical work, this ad hoc way of extending the lexicon is cumbersome, suggesting the use of a general component for automatic word form recognition instead (cf. Chapters 13–15). For the theoretical explanation of certain syntactic structures, however, the definition of small word form lexica has the advantage that only the entries actually used in the examples need to be specified, and that their categories may be simplified[12] to the properties needed for the syntactic structures at hand.

After the indicated extension of the word form lexicon and extending the range of *np* to include the segments S1 and S2, *LA-D1* will also accept, e.g., ich **schlafe**:

16.4.10 IDENTITY-BASED SUBJECT-VERB AGREEMENT IN GERMAN

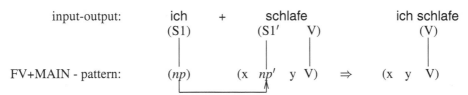

---

[12] Thus, multicats (cf. 15.2.4) and feature structures (cf. 15.2.6) – which are not really necessary for the syntactic properties to be explained here – can be omitted.

This example shows that the identity-based handling of agreement in *LA-D1* works correctly for nominatives in different persons and numbers, and verbs with different valency structures. Ungrammatical input like *ich schläfst will not be accepted:

16.4.11 AGREEMENT VIOLATION IN GERMAN

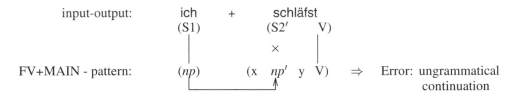

The handling of free word order in German is illustrated by the following derivation of a declarative main clause with a topicalized dative (see 16.3.2 for comparison).

16.4.12 DERIVATION IN *LA-D1* (identity-based agreement)

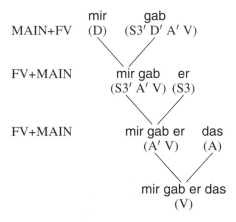

LA-D1 handles agreement between the nominative and the finite verb in all possible person-number combinations. Only valency fillers of the correct case are accepted. Sentences with superfluous valency fillers, e.g. Susanne schläft das Buch, are rejected. If the rule name FV+MAIN is added to the (rule package of the) start state, *LA-D1* is extended to yes/no-interrogatives, e.g. gab er ihr diesen. The tiny two-rule grammar of *LA-D1* thus handles 18 different word order patterns of German.[13]

If the lexicon is extended to lexical ambiguities which lead to syntactic ambiguities – as in sie mag sie, then these are presented correctly as multiple analyses by the parser. Because LA-grammar computes possible continuations *LA-D1* can be used equally well for analysis and generation, both in the formal sense of Section 10.4 and the linguistic sense of language production described in Section 5.4. Because *LA-D1* has no recursive ambiguities it is a C1-LAG and parses in linear time.

---

[13] To a one-place verb there correspond 2 basic sentence patterns, one declarative and one interrogative, to a two-place verb altogether 4, and to a three-place verb altogether 12.

LA-D1 exemplifies a syntactic *grammar system* (cf. Section 15.1) called LA-syntax. As an algorithm, LA-syntax uses C-LAGs. As a linguistic method, LA-syntax uses traditional syntactic analysis in terms of valency, agreement, and word order – with the additional assumption of a strictly time-linear derivation order.

16.5 Fixed word order in English (*LA-E1*)

As in all natural languages, the verbs of English have an inherent valency structure. For example, by assigning the verb form **gave** the category (N′ D′ A′ V) we express that this carrier requires a nominative, a dative, and an accusative to complete the semantic relation expressed. Compared to German, however, English has a fixed word order and a simple morphology.

LA-grammar describes the fixed word order properties of English by canceling the valency positions in the order in which they are listed in the category.[14]

16.5.1 FIXED CANCELING OF VALENCY POSITIONS IN A CARRIER OF ENGLISH

$$
\begin{array}{lcccl}
\text{Peter} & + & \text{gave} & \Rightarrow & \text{Peter gave} \\
\text{(SNP)} & & \text{(N' D' A' V)} & & \text{(D' A' V)}
\end{array}
$$

$$
\begin{array}{lcccl}
\text{Peter gave} & + & \text{Mary} & \Rightarrow & \text{Peter gave Mary} \\
\text{(D' A' V)} & & \text{(SNP)} & & \text{(A' V)}
\end{array}
$$

$$
\begin{array}{lcccl}
\text{Peter gave Mary} & + & \text{books} & \Rightarrow & \text{Peter gave Mary books} \\
\text{(A' V)} & & \text{(PN)} & & \text{(V)}
\end{array}
$$

In contrast to the German example 16.4.3, only the currently first position in the carrier category may be canceled. This simple principle results in a fixed word order. It is formalized by the rules of *LA-E1*, an English counterpart to *LA-D1*.

16.5.2 ENGLISH LA-GRAMMAR WITH FIXED WORD ORDER (*LA-E1*)

LX $=_{def}$ { [Peter (SNP) *], [Mary (SNP) *], [books (PN) *],
          [gave (N′ D′ A′ V) *]}

Variable definition: $np\ \varepsilon$ {SNP, PN}, $np'\ \varepsilon$ {N′, D′, A′},

x = .?.?.?.? (i.e. an arbitrary sequence up to length 4)

ST$_S$ $=_{def}$ { [(x) {NOM+FV}] }

---

[14] Accordingly, the variant of **gave** used in, e.g., John gave the book to Mary is based on the alternative lexical category (N′ A′ TO′ V).

NOM+FV: $(np)\ (np'\ x\ V) \Rightarrow$ (y V) {FV+MAIN}
FV+MAIN: $(np'\ x\ V)\ (np) \Rightarrow$ (y V) {FV+MAIN}
$ST_F =_{def}$ { [(V) rp$_{FV+MAIN}$] }

The valency fillers are represented by the variable *np* and the valency positions by the variable *np'*. According to the variable definition of 16.5.2, *np* is defined to range over the category segments SNP or PN, while *np'* defined to range over N', D', or A'.

In *LA-E1*, the categorial pattern of the valency carrier is (np' y V). The valency position to be canceled is decided by the carrier: it is the currently first position.[15] This does normally not result in agreement conflicts because English noun phrases – except for a few pronouns – are not morphologically marked for case. For example, in **Peter gave Mary Fido** or **The man gave the woman a dog** there is no possibility of ambiguity or a case-based agreement violation.

The usual absence of case markings in English does not imply lexical ambiguity, which would be highly inefficient. According to the principle of surface compositionality, it implies instead that most English noun phrases have no case, as expressed by the filler segments SNP (singular noun phrase) and PNP (plural noun phrase). In as much as case is needed for semantic purposes, it is *assigned* to the English fillers by the respective valency positions they cancel.

The *definition-based* handling of agreement[16] and the canceling of valency positions in the order prescribed by the carrier category is illustrated by the following derivation.

16.5.3 DERIVATION IN *LA-E1* (definition-based agreement)

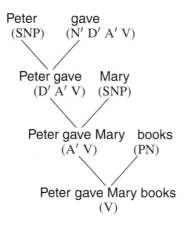

---

[15] In *LA-D1*, the categorial pattern of the valency carrier is (x *np'* y V). The valency position to be canceled is decided by the fillers: their case marking, e.g. **der Mann, dem Mann, den Mann**, determine more or less unambiguously which of the available valency positions in the carrier they agree with – and which they may thus cancel.

[16] For a complete description of the agreement restrictions of English see the variable definition of *LA-E2* in 17.4.1.

This derivation is the English counterpart to the German derivations 16.3.2 and 16.4.9 with their *identity-based* agreement and free word order.

LA-D1 and *LA-E1* demonstrate that the elementary formalism of C-LAGs handles different instances of agreement as well as variations and restrictions of word order on a high level of generality, in linear time, and without a counterintuitive proliferation of rules, lexical readings, or additional components like transformations.[17] The efficiency, flexibility, and parsimony of LA-grammar is due to its novel technique of matching variable-based rule patterns onto categorially analyzed input expressions using a strictly time-linear order.

Exercises

Section 16.1

1. Explain the boundary between the components of morphology and syntax. Does surface compositionality play a role in their delimitation?
2. What do morphology and syntax have in common? How are they related in the long term history of a language?
3. Why are idioms a border line phenomenon between morphology and syntax? Why don't they pose a problem for linguistic analysis? Explain different examples.
4. Describe some characteristics of synthetic and analytic languages. Which other types of language are mentioned by Bloomfield, and how are they defined?
5. Compare the combination principles of morphology with those of syntax.

Section 16.2

1. Explain the valency structure of the verb give.
2. What are secondary valency structures and how can they be handled?
3. What is meant by *transitivization* and *detransitivization*?
4. Do secondary valency structures call the concept of valency into question?
5. Are there other valency carriers besides verbs in natural language?

---

[17] The elementary formalisms of C- and PS-grammar require additional lexical readings or rule sequences for each additional constellation of agreement and each additional variant of word order. To improve the handling of word order, additional mechanisms like transformations or metarules were built into PS-grammar. To improve the handling of agreement, the mechanism of unification was added to PS- and C-grammar. These extensions have in common in that they increase mathematical complexity, but not the understanding of natural language communication.

6. What is the semantic relation/difference between valency carriers and modifiers? How are they handled in categorial grammar, and why is this analysis insightful? Why is it not transferred explicitly into LA-syntax?

7. Explain how syntactic well-formedness is treated as a functional effect rather than a primary object of description within the SLIM theory of language.

8. Explain the interpretation of the category (a$'$ b$'$ c$'$ d) in terms of valency positions and result segment.

9. Are the syntactic principles of valency and agreement contained in the formal algorithm of C-LAGs, or do they constitute something additional?

Section 16.3

1. What is agreement, and how can it be violated?

2. How do valency and agreement interact?

3. What is identity-based agreement?

4. Write a derivational structure like 16.3.2 for the sentence **he sees her**. Write a variant of *LA-Plaster* (16.3.3) for it.

5. What is good about *LA-Plaster*, and in which respect is it deficient?

Section 16.4

1. What relation between word order variation and morphology has been observed?

2. What is the word order of declarative main clauses in German?

3. Explain the use of variables in the rule patterns of LA-grammar.

4. What is necessary for a recursive rule application, and how is it stopped?

5. Explain why **er ihr gab das** is not accepted by *LA-D1*. Go through the derivation and show where and why it fails.

6. Why does adding the rule name FV+MAIN to the rule package of the start state result in an extension of *LA-D1* to German yes/no-interrogatives?

7. The sentence start **Der Mann gab** of category (D$'$ A$'$ V) may be continued with a dative or an accusative. What decides which case is chosen? Consider the difference between speaker and hearer mode in your answer.

8. Which algorithm and which method defines the grammar system LA-Syntax?

Section 16.5

1. How is the difference between the free word order of German and the fixed word order of English reflected in the rule patterns of *LA-D1* and *LA-E1*, respectively?

2. Why would it be inefficient to use identity-based agreement in English, and what is the alternative?

3. Why are *LA-D1* and *LA-E1* surface compositional, time-linear, and type transparent?

4. What is the complexity of *LA-D1* and *LA-E1*, respectively?

5. Explain the inherent difficulties of basic C- and PS-grammar with the handling of (i) word order variation and (ii) agreement.

6. Write the C-grammar *C-D1* and the PS-grammar *PS-D1* which should each handle the same 18 sentence patterns and associated agreement phenomena as *LA-D1*. Test the three grammars as parsers and make sure by means of a suitable list of test examples that they neither over- nor undergenerate. Explain whether the three grammars are strongly or weakly equivalent. Compare your experiences with the three grammar formalisms regarding transparency, elegance, and efficiency.

7. Explain the extensions of basic C- and PS-grammar commonly used for handling word order variation and agreement. What are the mathematical and computational properties of these derived formalisms?

17. LA-syntax for English

This chapter extends the simple LA-syntax for English called *LA-E1* to the time-linear derivation of complex valency fillers, complex verb forms, and yes/no-interrogatives. The emphasis is on explaining the descriptive power of categorial operations in combination with the finite state backbone defined by rule packages.

Section 17.1 describes the left-associative derivation of complex noun phrases in pre- and postverbal position, showing that the categorial patterns of the rules involved are the same in either position. Section 17.2 develops a system for distinguishing nominal fillers to handle agreement with the nominative and the oblique valency positions in the verb. Section 17.3 shows that the system works also for the nominative positions in the auxiliary to be. Section 17.4 formalizes the intuitive analysis as *LA-E2*, an extension of *LA-E1*, and describes the finite state backbone of the system. Section 17.5 explains why upscaling the LA-syntax for a natural language is simple, and demonstrates the point by expanding *LA-E2* to yes/no-interrogatives in *LA-E3*.

17.1 Complex fillers in pre- and postverbal position

For the sake of simplicity, the LA-syntax[1] for English has so far used only elementary valency fillers like Peter, Mary, and books. The now following extension to complex nominal fillers like every child or all children focusses on the time-linear derivation of complex valency fillers positioned before and after their valency carrier.

The formal treatment of complex noun phrases in English requires a description of their *internal* and *external agreement*. Internal agreement consists in restrictions which apply to the grammatical composition of complex noun phrases from smaller parts. External agreement consists in restrictions which apply to the combination of (elementary or complex) nominal fillers and verbal valency carriers.[2]

---

[1] See the definition of *LA-E1* defined in 16.5.2, which illustrates the handling of word order in basic declarative main clauses of English.

[2] One could argue that another possible aspect of internal agreement in noun phrases is the choice between the relative pronouns who vs. which. This choice, however, is not really an instance of grammatical agreement in English, but rather motivated semantically. It expresses whether the referent is considered human or not, just as the choice of the pronouns he/she vs. it. For example, in some contexts a dog will be referred to as it or which. In other contexts, however, the dog may be referred to as he, she, or who. Because the choice of relative pronouns depends on the speaker's viewpoint the human/nonhuman distinction is not coded into the syntactic category of nouns. For an alternative

Internal determiner-noun agreement in English is restricted in terms of *number*. For example, **every child** is grammatical, while ***every children** is not. Furthermore, the noun phrases resulting from a determiner-noun combination must be marked for number because they have an external agreement restriction with the finite verb. For example, **every child sleeps** is grammatical, while ***all children sleeps** is not.

Using an identity-based handling of agreement (cf. Section 16.3), these restrictions may be represented by the following categories of determiners and nouns:

17.1.1 DETERMINER AND NOUN CATEGORIES OF ENGLISH

categories	*surfaces*	*examples of lemmata*
singular and plural determiners:		
(SN′ SNP)	a, an, every, the	[a (SN′ SNP) *]
(PN′ PNP)	all, several, the	[all (PN′ PNP) *]
singular and plural nouns:		
(SN)	man, woman, book, car	[woman (SN) *]
(PN)	men, women, books, cars	[men (PN) *]

According to this categorization, the derivation of a noun phrase is based on canceling the argument position, i.e. SN′ or PN′, in the determiner with an identical noun segment, resulting in a singular (SNP) or plural (PNP) noun phrase.[3]

If a complex noun phrase like **the girl** precedes the verb, the time-linear derivation requires that the noun phrase be first put together and then combined with the verb. This is illustrated in the following sample derivation.

17.1.2 COMPLEX NOUN PHRASE BEFORE THE VALENCY CARRIER

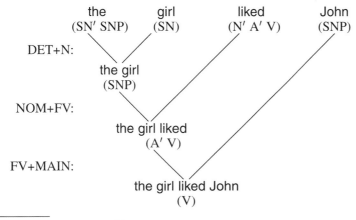

treatment see CoL, p. 366.

The rules DET+N, NOM+FV, and FV+MAIN used in 17.1.2 will now be explained by showing the matching between the input categories and the rule patterns. These rules, together with DET+ADJ and AUX+NFV to be discussed below, constitute the grammar *LA-E2* formally defined in 17.4.1.

The preverbal application of DET+N illustrated in 17.1.2 is based on the following categorial operation:

17.1.3 PREVERBAL APPLICATION OF DET+N

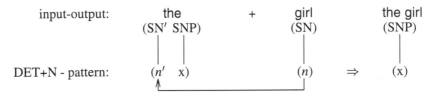

Next NOM+FV applies, which is based on the following categorial operation:

17.1.4 APPLICATION OF NOM+FV TO COMPLEX NOMINATIVE NP

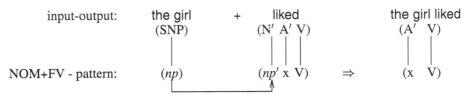

The derived noun phrase **the girl** of category (SNP) cancels the nominative position in the valency carrier, resulting in the sentence start **the girl liked** of category (A′ V).

Finally, the rule FV+MAIN fills the accusative position with the name **John**, resulting in a sentence start of category (V) – with no open valency positions left.

17.1.5 FV+MAIN ADDING ELEMENTARY OBJECT NP

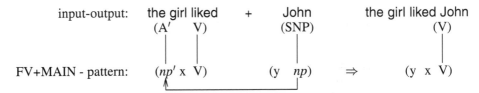

In contrast, if a complex noun phrase follows the verb, the parts of the noun phrase must be added step by step. This is required by the time-linear derivation order. As an example, consider the complex noun phrase **a boy** in the following derivation.

---

[3] For the sake of simplicity, the determiner **the** is treated in 17.1.1 as two lexical entries, one for singular and one for plural. The same would hold for, e.g., **no, my, your, his, her, our, their**. Alternatively one might try an alternative analysis which assigns only one category accepting both singular and plural noun arguments. This, however, would cause syntactic complications because the number of the result is determined by the *filler*. Given the analogous situation with German **der** (cf. Section 15.2), we rather eliminate the potentially adverse effects of lexical ambiguity in high frequency items by using multicats (cf. 15.2.4) in the actual implementation.

17.1.6 COMPLEX NOUN PHRASE AFTER VALENCY CARRIER

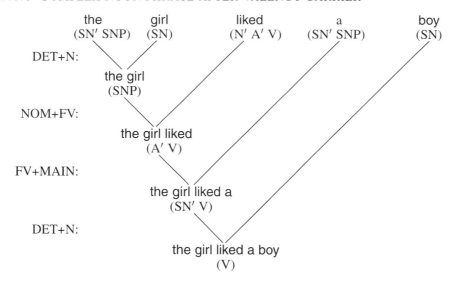

The sentence start **the girl liked** of category (A′ V) and the determiner **a** serve as input to FV+MAIN. The combination **the girl liked + a** results in a legitimate intermediate expression because the resulting sentence start may be continued into a complete well-formed sentence.

17.1.7 FV+MAIN ADDING BEGINNING OF COMPLEX OBJECT NP

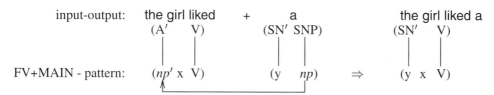

The rule pattern of FV+MAIN is the same as in 17.1.5. Here, it results in the sentence start **the girl liked a** of category (SN′ V). The nominal segment SN′ at the beginning of the result category specifies the number of the following noun. It satisfies the input condition of the rule DET+N to be applied next.

17.1.8 POSTVERBAL APPLICATION OF DET+N

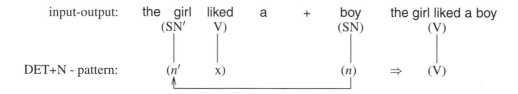

The rule pattern of DET+N is here the same as in 17.1.3.

A simple extension of derived noun phrases is the addition of one or more adjectives between the determiner and the noun. In English, adjectives are morphologically unmarked and there is no agreement between either the determiner and the adjective(s) or between the adjective(s) and the noun. Unlike C-grammar (cf. 16.2.1), the semantic function of adjectives as modifiers of the noun is not reflected in the category.

The LA-rule for adding adjectives is called DET+ADJ. It requires that the sentence start category begins with a noun segment SN$'$ or PN$'$ (like DET+N) and that the next word, i.e. the adjective, has the category (ADJ).

17.1.9 DET+ADJ RECURSIVELY ADDING ADJECTIVES

input-output: the   +   beautiful   the beautiful + young   ...
               (SN$'$ SNP)   (ADJ)   (SN$'$ SNP)   (ADJ)   (SN$'$ SNP)

DET+ADJ - pattern:$(n'$ x)   (ADJ) $\Rightarrow$ $(n'$ x)   (ADJ) $\Rightarrow$ $(n'$ x)

DET+ADJ produces an output category which is the same as the sentence start category – here (SN$'$ SNP). Thus the rule may apply to its own output with no categorially induced upper limit (in contrast to 16.4.5). The recursive reapplication of DET+ADJ in a pre- and a postverbal noun phrase is illustrated in below.

17.1.10 COMPLEX NOUN PHRASES WITH ADJECTIVES

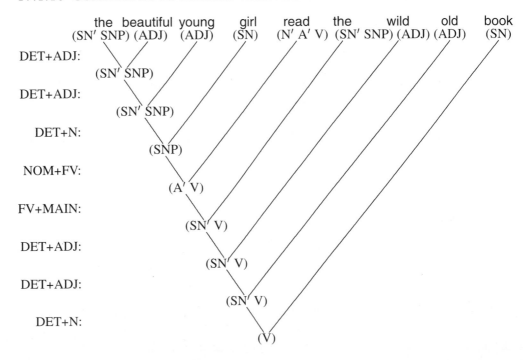

Because the number of adjectives between a determiner and a noun is unrestricted there is no grammatical limit on the length of noun phrases.

In summary, even though complex noun phrases in preverbal position are composed first and then fill the valency position in the verb as a whole, while complex noun phrases in postverbal position first fill the valency position and then are assembled step by step, the categorial patterns of the rules building complex noun phrases, namely DET+N and DET+ADJ, are the same in both instances. Furthermore, the categorial pattern of FV+MAIN accepts the beginning of oblique fillers irrespective of whether they are basic, e.g. it, or complex, e.g. the wild old book.

17.2 English field of referents

Having described the internal agreement restrictions of complex noun phrases in pre- and postverbal position we next analyze the external agreement restrictions between nominal valency fillers and their valency positions in the verb. The set of nominal fillers includes personal pronouns, proper names, and complex noun phrases. Their external agreement restrictions apply to distinctions of *case* (nominative vs. oblique), *number* (singular vs. plural), and *person* (first, second, or third).

From a cognitive point of view, the nominal fillers form an abstract *field of referents* comprising the types of agents and objects which may serve as the arguments of relations. Below, the nominal fillers of English are categorized in a way suitable to specify all external agreement restrictions between nominal fillers and their valency positions in different verb forms.

17.2.1 CATEGORIES OF NOMINAL VALENCY FILLERS IN ENGLISH

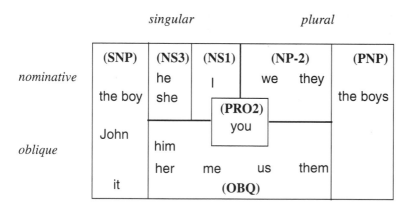

The field of referents is partitioned in a surface compositional way, representing only those properties which are needed for the correct syntactic combinatorics of concrete surfaces. Thus, the distinction between the different sign types of symbol, index, and name (cf. Chapter 6) is not of primary importance here.

For example, the pronoun it is is categorized as an (SNP) because it behaves combinatorially like a proper name or a singular noun phrase. The pronouns I (NS1) and he, she (NS3) are distinguished from singular noun phrases (SNP) because they cannot fill oblique valency positions, in contrast to the latter. For the same reason, we and they of category (NP-2) are distinguished from plural noun phrases (PNP).

Noun phrases of third person singular nominative, i.e. (SNP) and (NS3), are distinguished from other noun phrases of the categories (NS1), (NP-2), and (PNP) because the nominative valency of verb forms like sleeps is restricted to the former, while that of verb forms like sleep is restricted to the latter. The reason for providing I with the special category (NS1) is that the nominative valency position of the auxiliary verb form am is uniquely restricted to it.

The pronouns me, him, her, us, them of category (OBQ) can fill only oblique (i.e. non-nominative) valency positions. For this reason they are distinguished from their nominative counterparts and are unmarked for number or person in their category.

The pronoun you of category (PRO2), finally, is special in that it may fill both nominative and oblique valency positions, in both singular and plural. you is restricted only with respect to *person* in that it may not fill valency positions specified for first (*you am) or third (*you is, *you sleeps) person.

The arrangement of fillers in the field of referents follows their categories. The upper half is occupied by the nominatives (agents). The lower half is occupied by the oblique fillers. The left half contains the singular fillers while the right half contains the plural fillers. All in all, the field of referents is divided into seven different classes.[4]

The vertical peripheries are occupied by the filler categories (SNP) and (PNP), which are unrestricted with respect to case. The category (PRO2), representing the partner in discourse you, is in the center because is is unrestricted with respect to case *and* number. In this way the horizontal sequence S3 S1 SP2 P1 P3 is fixed, with S standing for singular, P for plural, and 1,2,3 for person.

Though the categorization (and associated grouping) of nominal fillers in 17.2.1 arises from grammatical considerations of a strictly surface compositional method, it is also quite telling in terms of cognition. Note that it is not universal, as shown by the analogous analysis 18.2.2 of the German field of referents.

In 17.2.2, the correlation between the nominal filler categories of 17.2.1 and compatible valency positions is shown for finite main verbs of English, using give as the example. Thereby a total of 5 different valency positions must be distinguished, namely the nominative positions N-S3$'$ (in give), N$'$ (in gave), and NS3$'$ (in gives) as well as the oblique positions D$'$ (for dative) and A$'$ (for accusative).

---

[4] On the number seven see G. Miller 1956.

There is in fact one additional class of nominal fillers, namely the pronominal noun phrases mine, yours, his, hers, ours, and theirs. They are restricted in that they may not fill valency positions of first person singular, e.g. *mine am here, *yours am here. Other than that they may fill any nominative, e.g. mine is here, mine are here, and oblique valency positions, e.g. she ate mine, of either number. This agreement pattern may be expressed by the category NP-S1 (noun phrase minus first person singular nominative).

17.2.2 AGREEMENT OF FILLERS AND VALENCY IN MAIN VERBS

(NS1)	I	(SNP)	(SNP)	the boy, John, it
(NP-2)	we, they	(OBQ)	(OBQ)	me, him, her, us, them
(PNP)	the girls	(PNP)	(PNP)	the girls
(PRO2)	you	(PRO2)	(PRO2)	you

[give (N-S3′   V) *]
[gave (N′   D′   A′   V) *]
[gives (NS3′   V) *]

(SNP)	the boy, John, it
(NS3)	he, she

The valency position N-S3′ may be canceled only by the category segments NS1, NP-2, PNP, and PRO3 while the valency position NS3′ may be canceled only by the category segments SNP and NS3. Thus, the possible fillers of N-S3′ and NS3′ are in complementary distribution. The valency position N′ accepts elements from the union of the N-S3′ and NS3′ fillers. The oblique valency positions D′ and A′ accept all filler segments which are not explicitly marked as a nominative, namely SNP, OBQ, PNP, and PRO2.

All main verbs of English have the three finite forms illustrated in 17.2.2, including the irregular ones, and they all share the same pattern of nominative agreement. Also, the agreement restriction between fillers and oblique valency positions is the same for all main verbs. The only variation between different types of main verbs arises in the number of oblique argument positions, which may be zero, e.g. [sleep (N-S3′ V) *], one, e.g. [see (N-S3′ A′ V) *], or two, e.g. [give (N-S3′ D′ A′ V) *].

17.3 Complex verb forms

In addition to the English main verbs there are three auxiliary verbs, do, have, and be, and a larger number of modals, such as can, will, shall, could, would, and should. The modals correspond to the past tense form of main verbs, e.g. [gave (N′ D′ A′ V) *], in that they have no special forms and therefore have no special agreement restrictions regarding their nominative fillers.

The auxiliaries do and have, on the other hand, have three finite forms do, does, did, and have, has, had, respectively, which correspond to those of the main verbs and share their pattern of nominative agreement. The auxiliary be, finally, has the five finite forms am, is, are, was, and were, for which reason it has a special pattern of nominative agreement, described schematically in 17.3.1.

17.3.1 NOMINATIVE AGREEMENT OF THE AUXILIARY be

```
                                                    (PNP)  the girls
                                (NS3)  he, she      (NP-2) we, they
              (NS1) I           (SNP)  the boy, John, it    (PRO2) you
               |                  |                              |
         [am (NS1′ BE′ V) *]   [is (NS3′ BE′ V) *]      [are (N-S13′ BE′ V) *]
                                                        [were (N-S13′ BE′ V) *]


                         (NS3)  he, she
                         (SNP)  the boy, John, it
                         (NS1)  I
                          |
                    [was  (NS13′ BE′ V) *]
```

The categorization of the fillers in 17.3.1 is that of 17.2.1 and has been used already in 17.2.2. Regarding the corresponding valency positions, on the other hand, the additional segments NS1′, NS13′, and N-S13′ are needed in 17.3.1 for handling the special restrictions of am, was, and are/were.

The finite forms of the auxiliaries combine with the nonfinite forms of the main verbs into complex verb forms. In accordance with traditional grammar, the nonfinite forms of English main verbs are the infinitive, e.g. give, the past participle, e.g. given, and the present participle, e.g. giving.

The infinitive has no form of its own, but coincides with the unmarked present tense of the verb.[5] The past participle is marked in some irregular verbs, e.g. given, but usually coincides with the past tense, e.g. worked. The present participle, finally, is always marked, as in giving or working.

The infinitive combines with the finite forms of do into the emphatic, e.g. does give. The past participle combines with the finite forms of have into the present perfect, e.g. has given. The present participle combines with the finite forms of be into the progressive, e.g. is giving.[6]

The finite auxiliary forms all have variants with integrated negation, namely don't, doesn't, didn't, haven't, hasn't, hadn't, ain't, isn't, aren't, wasn't, and weren't. They have the same combinatorial properties as their unnegated counterparts.

The basic categorial structure of combining a finite auxiliary with a nonfinite main verb may be shown schematically as follows.

---

[5] For this reason, one might be tempted to categorize a form like give only once, namely as [give (N-S3′ D′ A′ V) *]. This leads to complications in the syntactic description, however, because it prevents a analogous treatment of, e.g., does give, has given, and is giving as shown in 17.3.2. The potentially adverse effects of assigning two categories, e.g. (N-S3′ D′ A′ V) for the finite form and (D′ A′ DO) for the infinitive, to the base form of all verbs may be largely eliminated by using multicats in the actual implementation (cf. 15.2.4).

[6] In addition there are complex verb forms like is given and has been given (passive) and has been giving (past perfect), which we leave aside. Cf. CoL, p. 100f.

17.3.2 Complex verb forms of English

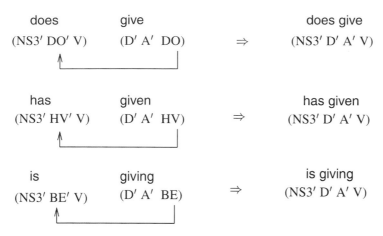

The nominative agrees with the finite auxiliary for which reason its valency position (here NS3′) is located in the auxiliary's category. The oblique valencies, on the other hand, are contributed by the respective nonfinite main verb (here D′ and A′). That the above auxiliaries are finite is marked by the presence of the V segment in their categories. That the main verb forms are nonfinite is shown correspondingly by the absence of a V segment. There is an identity-based agreement between the finite auxiliary and the nonfinite main verb which is expressed in terms of the auxiliary segments DO (for do), HV (for have), and BE (for be), respectively.

The combination of an auxiliary with a nonfinite main verb form, e.g. **has given**, results in a complex verb form which has the same properties in terms of nominative agreement and oblique valency positions as the corresponding finite form of the main verb in question, here **gives**. This holds also in a strictly time-linear derivation, as shown by the following examples.

17.3.3 Comparing basic and complex verb forms of English

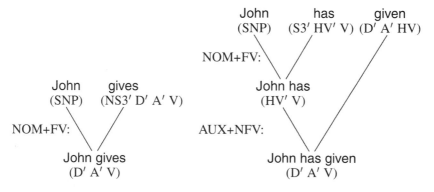

The two partial derivations end in states which may be continued in the same way.

The nonfinite main verb is added by a new rule called AUX+NFV:

17.3.4 AUX+NFV ADDING A NONFINITE VERB

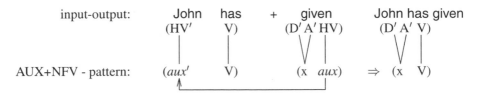

This application of AUX+NFV illustrates the identity-based agreement between auxiliary and main verb as well as the canceling of the auxiliary segment (here HV').

17.4 Finite state backbone of LA-syntax (*LA-E2*)

The handling of (i) complex noun phrases in pre- and postverbal position, (ii) external nominal agreement, and (iii) complex verb forms described intuitively in Sections 17.1–17.3 will now be formalized as the grammar *LA-E2*, which extends *LA-E1* in terms of a larger lexicon, a more detailed variable definition, and the three additional rules DET+ADJ, DET+N, and AUX+NFV.

17.4.1 *LA-E2*: AN ENGLISH LA-SYNTAX WITH COMPLEX NPs

$\text{LX} =_{def}$ {[Julia (SNP) *], [John (SNP) *], [Suzy (SNP) *], [it (SNP) *],
  [boy (SN) *], [boys (PN) *], [girl (SN) *], [girls (PN) *], [book (SN) *],
  [books (PN) *], [a (SN' SNP) *], [every (SN' SNP) *], [the (SN' SNP) *],
  [all (PN' PNP) *], [several (PN' PNP) *], [the (PN' PNP) *]
  [I (NS1) *], [you (PRO2), [he (NS3) *], [she (NS3) *], [it (SNP) *],
  [we (NP-2) *], [they (NP-2) *], [me (OBQ) *], [him (OBQ) *],
  [her (OBQ) *], [us (OBQ) *], [them (OBQ) *]
  [am (NS1' BE' V) *], [is (NS3' BE' V) *], [are (N-S13' BE' V) *]
  [was (NS13' BE' V) *], [were (N-S13' BE' V) *]
  [have (N-S3' HV' V) *], [has (NS3' HV' V) *], [had (N' HV' V) *]
  [do (N-S3' DO' V) *], [does (NS3' DO' V) *], [did (N' DO' V) *]
  [give (N-S3' D' A' V) *], [gives (NS3' D' A' V), [gave (N' D' A' V) *],
  [give (D' A' DO) *], [given (D' A' HV) *], [giving (D A BE) *]
  [like (N-S3' A' V) *], [likes (NS3' A' V), [liked (N' A' V) *]
  [like (A' DO) *], [liked (A' HV) *], [liking (A' BE) *]
  [sleep (N-S3' V) *], [sleeps (NS3' V) *], [slept (N' V) *]
  [sleep (DO) *], [slept (HV) *], [sleeping (BE) *]}

Variable definition:
  np' ε {N', N-S3', NS1', NS3', NS13', N-S13', D', A'}, (valency positions)
  np ε {PRO2, NS1, NS3, NP-2, SNP, PNP, PN, OBQ} (valency fillers), and
    if np = PRO2, then np' ϵ {N', N-S3', N-S13', D', A'},

if np = NS1, then $np' \in$ {N', N-S3', NS1', NS13'},

if np = NS3, then $np' \in$ {NS3', NS13'},

if np = NP-2, then $np' \in$ { N', N-S3'},

if np = SNP, then $np' \in$ { N', NS3', NS13', D', A'},

if np = PNP, then $np' \in$ {N', N-S3', N-S13', D', A'},

if np = OBQ, then $np' \in$ {D', A'},

$n \in$ {SN, PN} and n' correspondingly SN' or PN',

$aux \in$ {DO, HV, BE} and aux' correspondingly DO', HV' or BE'

x, y = .?.?.?.? (arbitrary sequence up to length 4)

$ST_S =_{def}$ { [(x) {1 DET+ADJ, 2 DET+N, 3 NOM+FV}] }

DET+ADJ:	$(n'$ x) (ADJ)	$\Rightarrow$ $(n$ x)	{4 DET+ADJ, 5 DET+N}
DET+N:	$(n'$ x) (n)	$\Rightarrow$ (x)	{6 NOM+FV, 7 FV+MAIN}
NOM+FV:	(np) $(np'$ x V)	$\Rightarrow$ (x V)	{8 FV+MAIN, 9 AUX+NFV}
FV+MAIN:	$(np'$ x V) (y $np)$	$\Rightarrow$ (y x V)	{10 DET+ADJ, 11 DET+N, 12 FV+MAIN}
AUX+NFV:	$(aux'$ V) (x $aux)$	$\Rightarrow$ (x V)	{13 FV+MAIN}

$ST_F =_{def}$ { [(V) rp_{nom+fv}], [(V) $rp_{aux+nfv}$], [(V) $rp_{fv+main}$], [(V) rp_{det+n}]}

The categories of the word forms in LX corresponds to those developed in the previous three sections. The definition of the specified variables np, np', n, and aux characterizes the category segments of different kinds of nominal valency positions, nominal fillers, nouns, and auxiliaries, respectively.

The conditional clauses relating np and np' in the variable definition provide a definition-based characterization of the external agreement restrictions between nominal fillers and their valency positions (cf. Section 17.2).[7] Empirical investigations have shown that a definition of agreement restrictions will use a smaller set of agreeing elements (and will thus be more efficient), if for each filler segment the compatible valency positions are defined, instead of the other way around.

As shown by the LA-grammar 10.2.3 for $a^k b^k c^k$, the relation between rules and their rule packages in LA-grammar defines a *finite state transition network* (cf. 10.2.4). The FSNs of LA-grammars of artificial languages are usually too simple to be of particular interest. The FSNs for LA-grammars for natural languages, on the other hand, may be quite illuminating for the empirical work.

The FSN of *LA-E2* has six different states, namely the start state ST_S (i) and the output states of the five rules DET+ADJ (ii), DET+N (iii), NOM+FV (iv), FV+MAIN (v), and AUX+NFV (vi).

---

[7] The variable definition in this and the following LA-grammars is designed for maximal brevity in a complete description of the complex agreement restrictions of English and German. Alternatively, one could define an equivalent LA-syntax using typed feature structures and unification. It remains to be seen, however, whether translating the definition-based agreement restrictions in 17.4.1 between noun phrases and their verbal positions into typed feature structures would not result in an artificially inflated representation on the level of the lexicon without making the rules any simpler. Furthermore, in order to handle the categorial operations, unification alone would not be sufficient, but would have to be supplemented by operations for canceling and building up valencies, defined in terms of deleting and adding parts of the feature structures.

17.4.2 THE FINITE STATE BACKBONE OF *LA-E2*

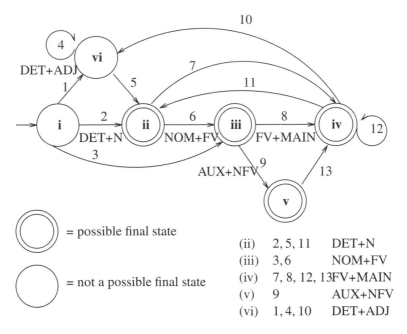

(ii)	2, 5, 11	DET+N
(iii)	3, 6	NOM+FV
(iv)	7, 8, 12, 13	FV+MAIN
(v)	9	AUX+NFV
(vi)	1, 4, 10	DET+ADJ

As in 10.2.4, the states are represented by circles and the transitions by arrows. The additional circles in the states (iii), (iv), (v), and (vi) indicate that they may also serve as final states (cf. ST_F in 17.4.1).

All transitions going into a state correspond to the categorial operation of the *same rule* – though from different preceding states. All transitions leading out of a state, on the other hand, correspond to *different rules* and represent the (content of the) rule package of the rule leading into the state (see also 10.2.4)

A transition corresponds to a successful rule application and thus combines a sentence start with a next word. The possible transitions which a certain rule may perform are defined explicitly by the rule packages which name the rule in question.

The FSNs for natural languages may be quite complex.[8] To make them more transparent it is recommended to number the transitions in the rule packages of the LA-grammar (in 17.4.1 from 1 to 13). These numbers may then be used to name the transition arrows in the FSN. Given that the rules leading into a state are all the same, it is sufficient, and less cluttering, if rule names are added to only *one* transition arrow leading into a state (compare 17.4.2 and 10.2.4).

If each transition of an LA-syntax is given a unique name (e.g. a number), then grammatical derivations within that grammar may be characterized by writing the respective transition names in front of each next word, as in the following examples:

---

[8] Especially in the step by step extension of an LA-grammar for a natural language, the associated FSNs provide a transparent representation of the increase in transitional connections between states. This is shown by the extension of *LA-E1* via *LA-E1.5* (cf. FSNs in 17.5.1) and *LA-E2* (cf. FSN 17.4.2) to *LA-E3* (cf. FSN 17.5.6).

17.4.3 SPECIFYING THE TRANSITION NUMBERS IN THE INPUT

Peter 3 gave 8 Mary 12 a 11 book
the 1 beautiful 4 young 5 girl 6 is 9 reading 13 a 10 wild 4 old 5 book
the 2 boy 6 gave 8 the 11 girl 7 a 11 book
Peter 3 gave 8 Mary 12 Suzy

The numbers in 17.4.3 correspond to those in 17.4.2 and in 17.4.1. Which names are assigned to the transitions in an FSN is essentially arbitrary – instead of numbers one could also use letters or the names of cities. The numbers in 17.4.2 and 17.4.3 are chosen to follow the sequence of rule names in the rule packages of *LA-E2*.

The FSN aspect of left-associative analysis is illustrated in more detail below.

17.4.4 SYNTACTIC ANALYSIS WITH TRANSITION NUMBERS

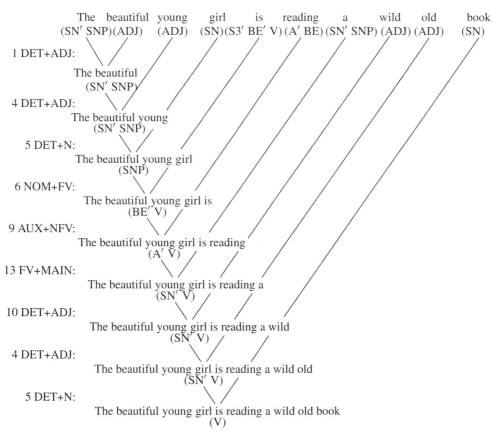

This example has the familiar form of left-associative derivations (e.g. 17.1.2, 17.1.6, 17.1.10) except that the transition numbers of 17.4.2 are specified in front of the rule names. Note that, e.g., DET+ADJ occurs with three different transition numbers, namely 1, 4, and 10, indicating that this rule is called by the rule packages of the start state, of DET+ADJ, and of FV+MAIN. And similarly for DET+N.

The five rules of *LA-E2* fulfill the pattern requirement for C-LAGs (cf. 11.2.5–11.2.7). Furthermore, the five rules have incompatible input conditions (cf. 11.4). Therefore, *LA-E2* is a C1-LAG and parses in linear time.

17.5 Yes/no-interrogatives (*LA-E3*) and grammatical perplexity

The extension of a simple LA-syntax, e.g. *LA-E1*, to a descriptively more powerful system, e.g. *LA-E2*, is facilitated by the fact that each new construction is realized as an additional sequence of transitions within the existing FSN. In the simplest case, a new transition requires no more than adding an existing rule name to an existing rule package. Sometimes the categorial operation of a rule has to be generalized in addition (compare for example FV+MAIN in *LA-E1* and *LA-E2*). At worst, a new rule has to be written and integrated into the network by adding its name to existing rule packages and by writing the names of existing rules into its rule package.

As an example of an extension requiring the definition of new rules consider the addition of complex noun phrases to *LA-E1* (cf. 16.5.2). The finite state backbone of *LA-E1* and its extension into the intermediate system *LA-E1.5* is given below.

17.5.1 EXPANDING *LA-E1* TO *LA-E1.5* HANDLING COMPLEX NPS

LA-E1

LA-E1.5

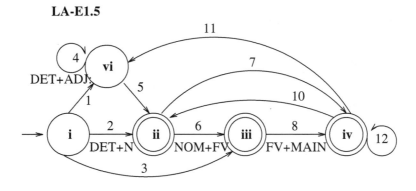

To facilitate comparison, the FSNs of *LA-E1* and *LA-E1.5* use the same names for corresponding states and transitions as the FSN of *LA-E2* (cf. 17.4.2). The grammar *LA-E1.5*, which is omitted here, is like *LA-E2* without the rule AUX+NFV.

To extend *LA-E1* to the handling of complex noun phrases, two locations in the network have to be taken care of. One corresponds to the preverbal position of complex noun phrases in the sentence, the other to the postverbal position(s).

Adding complex noun phrases in preverbal position requires definition of the new rules DET+N and DET+ADJ, resulting in the additional states **ii** and **vi**. Furthermore, the additional transitions 1 and 4 (DET+ADJ), 2 and 5 (DET+N), and 6 (NOM+FV) must be defined. This is done by writing the rule names DET+ADJ and DET+N into the rule packages of START and DET+ADJ, and the rule name NOM+FVERB into the rule package of DET+N.

Adding complex noun phrases in postverbal position re-uses the just added rules DET+N and DET+ADJ without any need for additional state definitions or changes in their categorial operations. All that is required are three additional transitions, namely 7 (FV+MAIN), 10 (DET+ADJ), and 11 (DET+N). These are implemented by writing the rule name FV+MAIN into the rule packages of NOM+FV, and the rule name DET+ADJ and DET+N into the rule package of FV+MAIN.

In the development and extension of LA-grammars, existing rules may be re-used in additional transitions, if the patterns of their categorial operations have been designed in sufficient generality (cf. Sections 17.1 and 17.3). The debugging of rules is greatly facilitated by the fact that LA-syntax is (i) type-transparent and (ii) time-linear. For this reason, errors in the parsing follow directly from errors in the grammar specification. The first occurrence of an error usually causes the time-linear derivation to break off immediately, whereby the cause is displayed explicitly in the category or the rule package of the last sentence start (cf. 10.5.5).

As another example of a possible extension consider adding yes/no-interrogatives to *LA-E2*. In English, this construction is based on inverting the order of the nominative and the finite auxiliary as compared to that of the corresponding declarative.

17.5.2 COMPARING DECLARATIVES AND YES/NO-INTERROGATIVES

Suzy does like the book.	Does Suzy like the book?
Suzy has liked the book.	Has Suzy liked the book?
Suzy is liking the book.	Is Suzy liking the book?

The extension to the interrogative structures should be such that the transitions following the nonfinite main verb remain unchanged. Furthermore, the complex noun phrases serving as nominatives in the additional constructions should be derived by feeding the new transitions back into the preexisting network.

The planned extension requires the definition of two new rules, called AUX+MAIN and IP. AUX+MAIN has the following categorial operation:

17.5.3 CATEGORIAL OPERATION OF AUX+MAIN

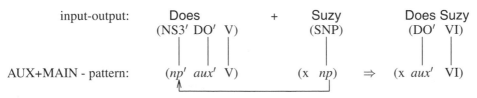

The variable x in the next word pattern (x *np*) anticipates complex nominatives beginning with a determiner, e.g. [every (SN′ SNP) *] as in Does every + child (sleep).
Changing the verb segment from V to VI ensures that the other new rule IP (for interpunctuation, cf. 17.5.4) adds a question mark rather than a full stop.

Next we have to integrate AUX+MAIN into *LA-E2*. Because AUX+MAIN handles
the beginning of interrogative sentences it must be written into the rule package of the
start state ST_S. If the nominative is a complex noun phrase, as in Does the girl or
Does the beautiful young girl, AUX+MAIN should be continued with DET+NOUN
or DET+ADJ, the names of which are therefore written into its rule package.

If the nominative is elementary, on the other hand, AUX+MAIN should be continued with the existing rule AUX+NFV. Because complex nominatives following
AUX+MAIN must also continue with AUX+NFV, the name AUX+NFV must be written not only into the rule package of AUX+MAIN, but also into those of DET+ADJ
and DET+N. Once the already existing rule AUX+MAIN has been reached, the
derivation continues as in *LA-E2*.

The new rule IP adds a question mark to sentence starts of category (VI), and a
question mark or a full stop to sentence starts of category (V). Thus the new system
will handle, e.g., Suzy sleeps., Suzy sleeps?, and Is Suzy sleeping?, while *Is
Suzy sleeping. will be rejected. The categorial operation of IP is illustrated in 17.5.4,
whereby the variable *vt* stands for 'V type.'

17.5.4 CATEGORIAL OPERATION OF IP

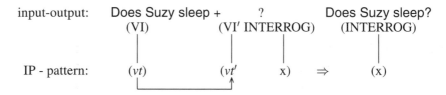

The name of IP must be added to the rule packages of DET+N (e.g. Suzy ate the
apple + .), NOM+FV (e.g. Suzy sleeps + .), FV+MAIN (e.g. Suzy saw Mary + .),
and AUX+NFV (e.g. Suzy is sleeping + . or Is Suzy sleeping + ?).

These extensions are formally captured by the following definition of *LA-E3*.

17.5.5 *LA-E3* FOR ENGLISH YES/NO-INTERROGATIVES

> LX = LX of *LA-E2* plus {[. (V′ decl) *], [? (V′ interrog) *], [? (VI′ interrog)*]}
> Variable definitions = that of *LA-E2* plus $vt \; \varepsilon$ {V, VI},

$ST_S =_{def}$ { [(x) {1 DET+ADJ, 2 DET+N, 3 NOM+FV, 4 AUX+MAIN}] }
DET+ADJ: (*n′* x) (ADJ) ⇒ (*n′* x)   {5 DET+ADJ, 6 DET+N}
DET+N: (*n′* x) (*n*)   ⇒ (x)   {7 NOM+FV, 8 FV+MAIN, 9 AUX+NFV, 10 IP}
NOM+FV: (*np*) (*np′* x V)   ⇒ (x V)   {11 FV+MAIN, 12 AUX+NFV, 13 IP}

FV+MAIN: $(np'$ x V) (y $np)$ $\Rightarrow$ (y x V) {14 DET+ADJ, 15 DET+N, 16 FV+MAIN, 17 IP}
AUX+NFV: $(aux'$ V) (x $aux)$ $\Rightarrow$ (x V)   {18 FV+MAIN, 19 IP}
AUX+MAIN:$(np'$ aux' V) (x $np)$ $\Rightarrow$ (x aux' VI) {20 AUX+NFV, 21 DET+ADJ, 22 DET+N}
IP: (vt) $(vt'$ x) $\Rightarrow$ (x)   {}
$ST_F =_{def}$ { [(decl) rp_{ip}], [(interrog) rp_{ip}]}

The finite state backbone of *LA-E3* has the following structure:

17.5.6 THE FINITE STATE BACKBONE OF *LA-E3*

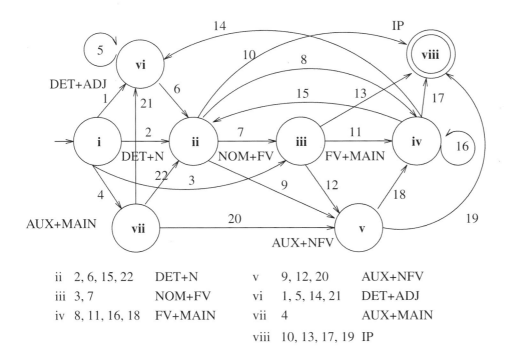

ii	2, 6, 15, 22	DET+N		v	9, 12, 20	AUX+NFV
iii	3, 7	NOM+FV		vi	1, 5, 14, 21	DET+ADJ
iv	8, 11, 16, 18	FV+MAIN		vii	4	AUX+MAIN
				viii	10, 13, 17, 19	IP

The extension of *LA-E2* to *LA-E3* in order to handle yes/no-interrogatives required no revision of the older system, but is based solely on the addition of the rules AUX+MAIN and IP, and the addition of certain rule names to certain rule packages.

The FSN 17.5.6 has 22 transitions, corresponding to the sum total of all rule names in all rule packages of *LA-E3* including the start package. *LA-E3* has seven rules and one start state for which reason the FSN consists of a total of 8 states (i–viii). The new interpunctuation rule IP has the special property that its rule package is empty. Furthermore, the definition of the final states in *LA-E3* treats IP as the only 'completing rule' of the grammar.

The 7 rules of *LA-E3* have incompatible input conditions for which reason *LA-E3* – like its predecessors – is a C1-LAG and parses in linear time. When parsing with

LA-E3, any kind of input is analyzed from left to right, either until the whole input has been analyzed or until the derivation fails.

If the input is parsed to the end and at least one of the last states turns out to fulfill a final state definition of the grammar, the input is classified as a *complete grammatical* sentence. If the input is parsed to the end and none of the last states turns out to fulfill a final state definition, the input is classified as an *incomplete well-formed* expression. If the input is not parsed to the end, on the other hand, the input is classified as *ungrammatical*.

The step by step extension of *LA-E1* to *LA-E3* has resulted in increasingly denser transitions in the FSNs. Nevertheless, the use of rule packages with *selected* rules is still far more efficient than a hypothetical system in which all rule packages blindly list all the rules of the grammar. In the latter case, an LA-grammar with one start state and seven rules (analogous to *LA-E3*) would result in $7 + (7 \cdot 7) = 56$ transitions – in contrast to *LA-E3*, which has only 22 transitions.

The total number of transitions divided by the number of states provides a general measure for the *perplexity* of C1-LAGs. In statistical language recognition, perplexity represents the average number of possible continuations at any point.

> Perplexity is, crudely speaking, a measure of the size of the set of words from which the next word is chosen given that we observe the history of the spoken words.
>
> S. Roukos 1995

In the word sequence **The early bird gets the**, for example, it is statistically highly probable that the next word will be **worm**. Thus, for statistical language recognition the set from which the next word is chosen is very small here and the perplexity therefore very low. Correspondingly, the perplexity of a scientific paper on, e.g. radiology, will be higher than in the above example, but still lower than in general English. An example of general English is the Brown corpus, which has been claimed to have a perplexity of 247.

While statistical language recognition determines perplexity probabilistically with respect to the sequences of word forms observed in a corpus, the new notion of *grammatical* perplexity represents the average number of attempted rule applications in an LA-grammar. For example, in the hypothetical LA-grammar with nonselective rule packages mentioned above, the grammatical perplexity would be 56 : 8, i.e., 7 attempted rule applications per composition. In *LA-E3*, on the other hand, the grammatical perplexity is only 2.75 (= 22 : 8).

The use of selective rule packages does not only result in lower grammatical perplexity and thus in higher efficiency, however. Equally important is the fact that selective rule packages (i) provide a descriptively more adequate linguistic description and (ii) allow simpler definitions of the categorial input patterns of the rules.[9]

---

[9] The second point refers to the fact that the categorial operations need to provide pattern-based *distinctions* only relative to those rules which they share a rule package with. In practice, one will find that specifying the categorial operations for an LA-grammar with an average of 2.75 rules per package is considerably easier than for an equivalent LA-grammar with an average of 7 rules per package.

Exercises

Section 17.1

1. Explain the LA-categorization of English determiners. Does it use a definition-based or an identity-based handling of agreement?
2. Describe the LA-derivations of John gave Mary Fido, the boy gave the girl the dog, and the old boy gave the young girl the big dog, and give detailed explanations of the pattern matching and the categorial operations of the rules involved.
3. Why is the recursive application of DET+ADJ unrestricted? What causes the recursion of FV+MAIN to be restricted to at most two applications?
4. Explain why the time-linear derivation of complex noun phrases in postverbal position creates a problem for rule systems based on hierarchical term *substitution*, but not for rule systems based on possible *continuations* defined in terms of left-associative pattern matching.

Section 17.2

1. What are the seven classes of nominal fillers in English? In what sense do they constitute the field of referents characteristic of the English language?
2. Why is the analysis of nominal fillers in 17.2.1 strictly surface compositional? Why would it be unjustified to transfer this analysis to other languages?
3. Describe the agreement restrictions of nominal fillers and verbal valency positions in English main verbs.
4. What are the agreement restrictions of you, I, he, him, and she?

Section 17.3

1. Which special agreement restrictions are required by the auxiliary to be?
2. What is the categorial structure of English auxiliaries?
3. Describe the nonfinite verb forms of English. To what degree are they marked morphologically, and how are they categorized?
4. Explain the categorial operation of the rule AUX+NFV.

Section 17.4

1. Describe the finite state backbone of *LA-E2* and explain where in 17.4.1 it is defined.
2. Why must the transitions going into a state all have the same rule name, while the transitions going out of a state must all have different rule names?
3. Which components of an LA-grammar determine the exact total number of its transitions?
4. Which component of an LA-grammar determines the choice of transitions at any step of a derivation during analysis?

5. What do LA-grammars have in common with finite automata and which formal aspect of LA-grammars raises their generative power from the class of regular languages to the class of recursive languages?

6. Which complexity-theoretic property makes the C-LAGs especially suitable for the description of natural languages?

Section 17.5

1. What may be required for the extension of an existing LA-syntax to new constructions? Illustrate your explanation with examples.

2. Write an explicit LA-grammar for the finite state backbone of *LA-E1.5* provided in 17.5.1.

3. Explain the categorial operation of AUX+MAIN. Give a formal reason why AUX+MAIN can apply only at the beginning of a sentence?

4. How is the derivation of ungrammatical *Does John sleep. prevented in *LA-E3*? Explain the categorial operation of IP.

5. Expand *LA-E3* to handle sentences like There is a unicorn in the garden.

6. On what grounds is the analysis of an input expression classified as in/complete or un/grammatical? Are the principles of this classification limited to natural languages or do they apply to formal languages as well?

7. Name three reasons why the use of selective rule packages is preferable over nonselective rule packages.

8. Explain the notion of grammatical perplexity as a meaningful measure for the efficiency of C-LAGs. Could it be applied to B-LAGs and A-LAGs as well?

9. Why is it possible to compute the grammatical perplexity of C1-LAGs, but not of C2- or C3-LAGs?

18. LA-syntax for German

The LA-grammar analysis of natural languages should not only provide a formal description of their syntactic structure, but also explain their communicative function. To advance a functional understanding of different types of natural language syntax within a strictly time-linear derivation order, this chapter complements the LA-syntax of an isolating fixed word order language (English, Chapter 17) with the LA-syntax of an inflectional free word order language (German).

Section 18.1 describes the pre- and postverbal derivation of complex noun phrases in German based on a distinctive categorization of determiners. Section 18.2 analyzes the external agreement restrictions of nominal fillers based on a field of referents and formalizes the results in *LA-D2*. Section 18.3 illustrates the far-reaching differences in English and German word order regarding complex verb forms, interrogatives, and subordinate clauses. Section 18.4 presents a detailed categorial analysis of complex verb forms and formalizes the results in *LA-D3*. Section 18.5 expands *LA-D3* to interrogatives and adverbial subclauses, concluding with the formal definition of *LA-D4* and its finite state transition network

18.1 Standard procedure of syntactic analysis

Natural languages are all based on the same time-linear derivation order. The only syntactic differences between individual languages arise in the language specific handling of *agreement* and *word order*, often based on differences in the *valency structure* of certain lexical items. For this reason there is a standard procedure for the grammatical description of any natural language.

To determine the typological properties of a natural language, the first step of its syntactic analysis should be the formal treatment of declarative main clauses with elementary finite verbs and elementary nominal fillers. This is because the position of the finite verb and its nominal fillers, as well as the agreement restrictions between nominal fillers and verbs of differing valency structure, determine the basic word order and agreement properties of the language in principle. This step has been illustrated with *LA-D1* for an inflectional free word order language (German, Section 16.4) and *LA-E1* for an isolating fixed word order language (English, Section 16.5).

The second step is the extension to complex nominal fillers. This requires treatment of the internal and the external agreement restrictions of derived noun phrases, and the time-linear derivation of complex fillers in pre- and postverbal position. Apart from their more complicated internal structure, the addition of derived noun phrases to the initial grammar should result in the same types of sentences as before. This is illustrated by the transition from *LA-E1* (16.5.6) to *LA-E2* (17.4.1).

The third step is the extension to complex verb phrases to treat complex tenses and modalities. This may lead to new variants of word order. Otherwise, the number of different sentence frames handled by the formal grammar may still be rather small, depending on the language type. Yet all aspects of valency, agreement, and word order applicable to these sentence frames should now have their completely detailed and general final treatment. This step is illustrated by the transition from *LA-E2* (17.4.1) to *LA-E3* (17.5.5).

After this initial three-step phase one may well be tempted to explore the syntactic territory further by adding additional constructions. This is not recommended, however. What has to be done first at this point is the definition of a theoretically well-founded semantic and pragmatic interpretation for the syntactic analysis developed so far, which must be demonstrated to be functionally operational.

Only then follows the second phase of expanding the grammar. There are many topics to choose from: (i) the addition of basic and derived modifiers ranging from adverbs over prepositional phrases to subordinate clauses, (ii) the treatment of sentential subjects and objects including infinitive constructions, (iii) the handling of different syntactic moods like interrogative and different verbal moods like passive, and (iv) the treatment of conjunctions including gapping constructions.

Syntactically, the extensions of the second phase should build on the structures of the first phase without any need for revisions in the basic valency and agreement structures. The syntactic analyses of the second phase should be developed directly out of the semantic and pragmatic interpretation, and be provided for both, the speaker and the hearer mode (cf. Chapters 19, 23, 24).

Having illustrated the initial three-step phase of grammar development with the definition of *LA-E1*, *LA-E2*, and *LA-E3* for English, let us turn now to German. As German's free word order in declarative main clauses with elementary fillers has been described in Section 16.4, we are ready to take the second step, i.e. the extension to complex noun phrase fillers in pre- and postverbal position.

Derived noun phrases in German resemble those in English in that their basic structure consists of a determiner, zero or more adjectives, and a noun. They differ from those in English, however, in that (i) they are marked for case and (ii) show a wide variety of inflectional markings resulting in complicated agreement restrictions between the determiner and the noun. Furthermore, (iii) adjectives have inflectional endings which result in agreement restrictions between determiners and adjectives (cf. 15.2.2).

The definite and the indefinite determiners in German have a total of 12 distinct surfaces. Using a definition-based agreement with nouns in the categorization of 14.5.2, an identity-based agreement with adjectives (cf. 18.1.2), and a fusion of cases when possible (e.g. (S3&A) for **das Kind**), a total of 17 different categories for the 12 surfaces is sufficient for a complete treatment of the internal and external agreement restrictions of derived noun phrases in German.

18.1.1 Distinctive categorization of determiners

definite article   *indefinite article*

[der   (E′ MN′ S3)   [ein   (ER′ MN′ S3)
        (EN′ F′ G&D)   (ES′ N-G′ S3&A) INDEF-ART]
        (EN′ P-D′ G)   DEF-ART]   [eines   (EN′ -FG′ G)   INDEF-ART]
[des   (EN′ -FG′ G)   DEF-ART]   [einem   (EN′ -FD′ D)   INDEF-ART]
[dem   (EN′ -FD′ D)   DEF-ART]   [einen   (EN′ M-N′ A)   INDEF-ART]
[den   (EN′ M-N′ A)   [eine   (E′ F′ S3&A)   INDEF-ART]
        (EN′ PD′ D)   DEF-ART]   [einer   (EN′ F′ G&D)   INDEF-ART]
[das   (E′ N-G′ S3&A)   DEF-ART]
[die   (E′ F′ S3&A)
        (EN′ P-D′ P3&A)   DEF-ART]

As explained in Section 15.2, the proper treatment of German adjective agreement is very simple, and consists in matching the ending of the adjective with a corresponding segment of the determiner. This is illustrated by the following examples showing the pattern matching of the rules DET+ADJ and DET+N. They are part of the German LA-grammar *LA-D2* defined in 18.2.5.

18.1.2 Categorial operation of DET+ADJ

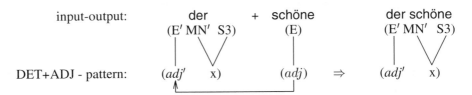

The categorial operation of DET+ADJ leaves the adjective segment in place to allow recursion (compare 17.1.9).

18.1.3 Categorial operation of DET+N

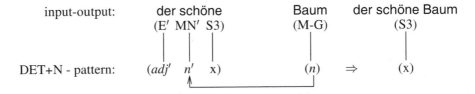

The categorial operation of DET+N removes both, the adjective and the noun segment, from the category because they are no longer needed.

The time-linear derivation of derived noun phrases in pre- and postverbal position is shown in 18.1.4, using the transition numbers of *LA-D2*.

18.1.4 Pre- and postverbal derivation of noun phrases

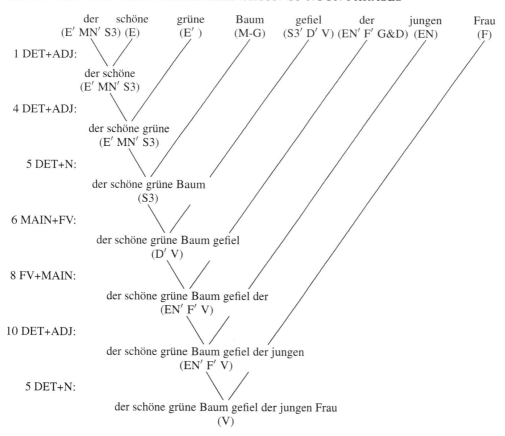

The categorial patterns of DET+ADJ (18.1.2) and DET+N (18.1.3) are suited equally well for the pre- and the postverbal derivation of complex noun phrases. The time-linear derivation is analogous to that in English (cf. 17.1.2–17.1.10).

18.2 German field of referents (*LA-D2*)

Because German is an inflectional language its noun phrases and verbs are tradition-ally represented in the form of exhaustive paradigms such as the following.

18.2.1 Traditional paradigms of German noun phrases

	Masculinum	*Femininum*	*Neutrum*	*Plural*
Nominative	der Mann	die Frau	das Kind	die Männer, etc.
Genitive	des Mannes	der Frau	des Kindes	der Männer, etc.
Dative	dem Mann	der Frau	dem Kind	den Männern, etc.
Accusative	den Mann	die Frau	das Kind	die Männer, etc.

Developed by the grammarians of classic Greek and Latin, exhaustive paradigms characterize all the forms of a word or phrase type in terms of certain parameters such as casus, genus, and numerus in noun prases, person, numerus, tense, and mood in verbs, casus, genus, numerus, and comparation in adjectives, etc.

A paradigm table results from multiplying the relevant parameter values. In 18.2.1, for example, there are 24 slots resulting from multiplying 4 casus, 3 genera, and 2 numeri. Note that only 18 distinct surfaces are distributed over the 24 slots. Surfaces which occur in more than one slot are treated as ambiguous.

Exhaustive paradigms are misleading for the description of German because they disguise the fact that femininum, neutrum, and plural noun phrases of German always have the same surface for the nominative and the accusative. Furthermore, femininum noun phrases of German always have the same surface for the genitive and dative. Thus, a crucial property of German is that the surfaces of certain nominal fillers are *not marked* for grammatical properties that are *relevant* for an unambiguous specification of the associated valency position.

Exhaustive paradigms are misleading also in another respect: they *mark* nominal fillers for properties that are *not relevant* for an unambiguous specification of compatible valency positions. For example, while the distinction of singular and plural in the genitive, dative, and accusative is important semantically, it is combinatorially irrelevant because the oblique valency positions of German verbs – like their English counterparts – are not restricted for number. The same holds for the genus distinction, which is combinatorially irrelevant in all casus.

A surface compositional categorization of nominal fillers in German is shown below. It includes complex noun phrases, personal pronouns, and proper names, and forms an abstract field of referents analogous to that of English (cf. 17.2.1).

18.2.2 DISTINCTIVE CATEGORIES OF NOMINAL FILLERS (GERMAN)

					Singular		**Plural**		
N	du (S2)	ich (S1)	er (S3)	Peter (S3& A& D)	das Kind / es die Frau (S3&A)	sie (S3&P3 &A)	die Männer die Frauen die Kinder (P3&A)	wir (P1)	ihr (P2)
A	dich (A)	mich	ihn den Mann					uns euch (D&A)	
D	dir mir ihm (D) dem Mann			dem Kind / der Frau (G&D)	ihr	ihnen	den Männern den Frauen den Kindern		
G	deiner meiner seiner		des Kindes des Mannes		ihrer (G)		der Männer der Frauen der Kinder	unserer eurer	

Certain surfaces which an exhaustive paradigm would present in several slots are analyzed here as lexical units occurring only once.[1] For example, Peter occurs only once in the field with the category (S3&D&A). And correspondingly for das Kind, es, die Frau (S3&A), sie (S3&P3&A), die Männer, die Frauen, die Kinder (P3&A), uns, euch (D&A), and der Frau (G&D).

Furthermore, certain forms of different words with the same distinctive category are collected into one subfield. Thus, dir, mir, ihm, ihr, dem Mann, ihnen, den Männern, den Frauen, and den Kindern have the same category (D) even though they differ in person or number. This is because these differences do not affect agreement with valency positions specified for oblique cases.

As in the English counterpart 17.2.1, the order of the fields in 18.2.2 is based on the horizontal distinction between number and the vertical distinction between case with the levels N (nominative), A (accusative), D (dative), G (genitive). The center is formed by a nominal filler which happens to be unrestricted with respect to number and combines the nominative and accusative case. This is the pronoun sie with the category (S3&P3&A), serving simultaneously as (i) femininum singular nominative and accusative, (ii) plural nominative and accusative, and (iii) the formal way of addressing the partner(s) of discourse in the nominative and accusative.

To the left and right of the center are placed the other third person singular and plural noun phrases, respectively. Then follow the personal pronouns of second and first person, symmetrical in singular and plural. Different fields are distinguished in terms of person and number on the level N only because only the nominative fillers are restricted in these parameters.

The horizontal levels of case N, A, D, and G are ordered in such a way that fields covering more than one case such as S3&P3&A (e.g. Peter), S3&A (e.g. es, das Kind, die Frau), G&D (e.g. der Frau), and D&A (e.g. uns, euch) can be formed. The vertical columns are arranged symmetrically around the center, forming the horizontal sequence

S2, S1, SP3, P1, P2

in German, as compared to the English sequence

S3, S1, SP2, P1, P3

with S for singular, P for plural and 1,2,3 for person. Accordingly, the center SP2 of English stands for second person singular and plural (you), while the center SP3 of German stands for third person singular and plural (sie).

The distinctions between first person 1 (I, we), second person 2 (you), and third person 3 (he, she, it) may be interpreted as increasing degrees of distance D1, D2, and D3. Combined with the different centers SP2 (you) of the English field of referents and SP3 (sie) of its German counterpart, there result the following correlations.

---

[1] The one exception is ihr, which has two uses differing in person, case, and number. These may be expressed in the common category (P2&D), but not graphically in terms of a single occurrence.

18.2.3 CENTERING AND DISTANCE IN FIELDS OF REFERENCE

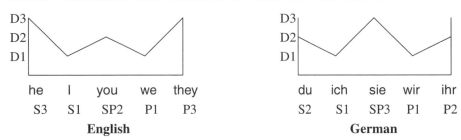

he	I	you	we	they		du	ich	sie	wir	ihr
S3	S1	SP2	P1	P3		S2	S1	SP3	P1	P2
English						**German**				

The center of the English field of referents is focussed onto the medium distance D2, while in German it is focussed onto the maximal distance D3. The peripheries are focussed in English onto the maximal distance D3, and in German onto the medium distance D2. The intermediate positions between center and peripheries are focussed in both language onto the minimal distance D1.

A strictly surface compositional analysis of different natural languages may thus provide insight into their different coding methods – which would be obscured by blindly multiplying out paradigm slots as in an exhaustive analysis like 18.2.1. Even if one prefers to refrain from interpreting the fields of referents 7.2.1 and 18.2.2 cognitively as different language-inherent conceptualizations, there remains the fact that the distinctive categorization is more concrete and combinatorially more efficient[2] than corresponding exhaustive categorizations like 18.2.1.

For the 14 types of nominal fillers there are 11 valency positions in German, represented as S1′, S13′, S2′, S23′, S13′, S2P2′, S3′, P13′, P2′, G′, D′, and A′.

18.2.4 AGREEMENT OF NOMINAL FILLERS AND VERBAL VALENCIES

```
      (S3&D&A)                                       (S3&P3&A)
       (D&A)                                         (S3&D&A)
      (P2&D) (S3&P3&A)                                (S3&A)
      (S1) (D) (A)              (S2)                   (S3)
       |   |   |                 |                      |
[gebe (S1′  D′  A′    V) *]   [gibst (S2′ * V) *]    [gibt (S3′ * V) *]
                             [gabst (S2′ * V) *]
```

```
                                                     (S3&P3&A)
                                                     (S3&D&A)
                                                      (S3&A)
      (P3&A)                   (P2&D)                  (S3)
       (P1)                     (P2)                   (S1)
        |                        |                      |
[geben (P13′ * V)  *]       [gebt  (P2′ * V)  *]    [gab  (S13′ * V) *]
[gaben (P13′ * V)  *]       [gabt  (P2′ * V)  *]
```

---

[2] The distinctive analysis uses only 14 different categories, in contrast to the 72 categories of the exhaustive paradigm analysis. 72 is the product of 3 genders, 2 numbers, 4 cases, and 3 persons.

The paradigms of finite verb forms in German also show certain characteristic 'holes.' They are caused by the fusion of the first and third person plural forms in the present and imperfect tense, represented as P13′, as well as the fusion of first and third person singular forms of the imperfect tense, represented as S13′.

  The internal and external agreement of basic and complex nominal fillers in German is handled explicitly and completely by the variable definition of *LA-D2*.

18.2.5 GERMAN LA-GRAMMAR HANDLING COMPLEX FILLERS (*LA-D2*)

LX = LX of LA-D1 plus the determiners defined in 18.1.1, the nouns defined in 14.5.1, 14.5.2, and the following pronouns

    [ich (S1) *], [du (S2) *], [er (S3) *], [es (S3&A) *], [wir (P1) *],
    [ihr (P2&D) *], [sie (S3&P3&A) *], [deiner (G) *], [uns (D&A) *],
    [euch (D&A) *], [mir (D) *], [dir (D) *], [ihm (D) *], [mich (A) *],
    [dich (A) *], [ihn (A) *]

plus adjectives with comparison

[schöne (E) *]	[schönere (E) *]	[schönste (E) *]
[schönen (EN) *]	[schöneren (EN) *]	[schönsten (EN) *]
[schöner (ER) *]	[schönerer (ER) *]	[schönster (ER) *]
[schönes (ES) *]	[schöneres (ES) *]	[schönstes (ES) *]

plus finite main verb forms of differing valency structures

[gebe (S1′ D′ A′ V) *]	[lese (S1′ A′ V) *]	[schlafe (S1′ V) *]
[gibst (S2′ D′ A′ V) *]	[liest (S23′ A′ V) *]	[schläfst (S2′ V) *]
[gibt (S3′ D′ A′ V) *]	[lesen (P13′ A′ V) *]	[schläft (S3′ V) *]
[geben (P13′ D′ A′ V) *]	[lest (P2′ A′ V) *]	[schlafen (P13′ V) *]
[gebt (P2′ D′ A′ V) *]	[las (S13′ A′ V) *]	[schlaft (P2′ V) *]
[gab (S13′ D′ A′ V) *]	[last (S2P2′ A′ V) *]	[schlief (S13′ V) *]
[gabst (S2′ D′ A′ V) *]	[lasen (P13′ A′ V) *]	[schliefst (S2′ V) *]
[gaben (P13′ D′ A′ V) *]		[schliefen (P13′ V) *]
[gabt (P2′ D′ A′ V) *]		[schlieft (P2′ V) *]

variable definition

$np \in$ {S1, S2, S3, P1, P2, P2&D, G, G&D, D, A, S3&A, S3&D&A, D&A,
    P3&A, S3&P3&A}

$np′ \in$ {S1′, S13′, S2′, S23′, S2P2′, S3′, P13′, P2′, G′, D′, A′}

and if $np \in$ {G, D, A}, then $np′$ is correspondingly G′, D′, or A′

    if np = P1, then $np′$ = P13′
    if np = S1, then $np′ \in$ {S1′, S13′}
    if np = S2, then $np′ \in$ {S2′, S23′}
    if np = S3, then $np′ \in$ {S3′, S23′}
    if np = P3&A, then $np′ \in$ {P13′, A′}
    if np = P2&D, then $np′ \in$ {P2′, D′}

if np = G&D, then np' ϵ {G', D'}

if np = D&A, then np' ϵ {D', A'}

if np = S3&A, then np' ϵ {S3', S23', A'}

if np = S3&D&A, then np' ϵ {S3', S23', D', A'}

if np = S3&P3&A, then np' ϵ {S3', S23', P13', A'}

n ϵ {MN, M-G, M-NP, M-GP, MGP, M-GP-D, F, N-G, -FG, -FD, N-GP, N-GP-D, NDP-D, P, P-D, PD},

n' ϵ {MN', M-N', F', N-G', -FG', -FD', P-D', PD'}, and

if n ϵ {MN, -FG, -FD, F, P-D, PD}, then n' is corresponding

if n =   M-G, then n' ϵ {MN', M-N'}

if n =   M-NP, then n' ϵ {-FG', -FD', P-D', PD' }

if n =   M-GP, then n' ϵ {MN', -FD', M-N', P-D', PD'}

if n =   MGP, then n' ϵ {-FG', P-D', PD'}

if n =   M-GP-D, then n' ϵ {MN', -FD', M-N', P-D'}

if n =   N-G, then n' ϵ {N-G', -FG', -FD'}

if n =   N-GP, then n' ϵ {N-G', -FG', -FD', P-D', PD'}

if n =   N-GP-D, then n' ϵ {N-G', -FG', -FD', P-D'}

if n =   NDP-D, then n' ϵ {-FD', P-D'}

if n =   P, then n' ϵ {P-D', PD'}

adj ϵ {e, en, es, er} and adj' is corresponding

$ST_S =_{def}$ { [(x) {1 DET+ADJ, 2 DET+N, 3 MAIN+FV}] }

DET+ADJ: $(adj'$ x) (adj) $\Rightarrow$ $(adj'$ x) {4 DET+ADJ, 5 DET+N}

DET+N: $(adj'$ n' x) (n) $\Rightarrow$ (x)   {6 MAIN+FV, 7 FV+MAIN}

MAIN+FV: (np) (x np' y V) $\Rightarrow$ (x y V) {8 FV+MAIN}

FV+MAIN: (x np' y V) (z np) $\Rightarrow$ (z x y V){9 FV+MAIN, 10 DET+ADJ, 11 DET+N}

$ST_F =_{def}$ { [(V) $\mathrm{rp}_{MAIN+FV}$], [(V) $\mathrm{rp}_{FV+MAIN}$], [(V) rp_{DET+N}] }

LA-D2 handles the same 18 word order patterns as *LA-D1*, but with the additional capacity for complex nominal fillers. The categorial pattern of the rule FV+MAIN, defined in *LA-D1* as (x np' y V) (np) $\Rightarrow$ (x y V), is generalized in *LA-D2* to (x np' y V) (z np) $\Rightarrow$ (z x y V). In this way FV+MAIN can handle continuations with elementary fillers (e.g. pronouns or proper names) as well as continuations with (the beginning of) complex noun phrases.

18.3 Verbal positions in English and German

The word order of English and German differs in the position of the finite – main or auxiliary – verb in declarative main clauses. English has the verb in *post-nominative* position whereby the nominative is always the first valency filler. German has a *verb-second* structure in declarative main clauses whereby fillers, including the nominative, may be placed freely. This difference is illustrated by the following examples:

18.3.1 FINITE VERB POSITION IN DECLARATIVE MAIN CLAUSES

English: post-nominative	*German*: verb-second
1. Julia *read* a book	Julia *las* ein Buch
2. *a book *read* Julia	Ein Buch *las* Julia
3. Yesterday Julia *read* a book	*Gestern Julia *las* ein Buch
4. *Yesterday *read* Julia a book	Gestern *las* Julia ein Buch
5. Julia yesterday *read* a book	*Julia gestern *las* ein Buch
6. *While Mary slept, *read* Julia a book	Als Maria schlief, *las* Julia ein Buch
7. While Mary slept, Julia *read* a book	*Als Maria schlief, Julia *las* ein Buch

In clauses with a finite main verb and beginning with a nominative, English and German may be alike, as shown by 1. In German, the positions of the nominative and the accusative may be inverted (2), however, leading to a dissimilarity between the two languages described in Sections 16.4 and 16.5. The structure of 2 violates the post-nominative rule of English, but satisfies the verb-second rule of German.

The word order difference in question shows up also in clauses with initial modifiers, such as elementary adverbs (3,4,5) and adverbial clauses (6,7). In English, the preverbal nominative may be preceded (3,7) or followed (5) by an adverbial, which in German leads to a violation of the verb-second rule. In German, adverbials are treated as constituents filling the preverbal position (4,6), which in English leads to a violation of the post-nominative rule.

In other words, adverbials occupy the preverbal position of declarative main clauses *exclusively* in German, but *non-exclusively* in English. Because in German nominal fillers may occur freely in postverbal position there is no reason to squeeze one of them into a preverbal spot already occupied by a modifier. English, in contrast, marks the nominative grammatically by its preverbal position. Rather than precluding modifiers from preverbal position, they may share it with the nominative.

A second difference arises in the positions of finite auxiliaries and nonfinite main verbs. In English, the parts of complex verb forms like is reading or has given are next to each other in *contact position*. In German, they are in *distance position*, with the finite auxiliary in second place and the nonfinite main verb at the end of the clause.

18.3.2 NONFINITE MAIN VERB POSITION IN DECLARATIVE MAIN CLAUSES

English: contact position	*German*: distance position
1. Julia *has slept*	Julia *hat geschlafen*
2. Julia *has read* a book	*Julia *hat gelesen* ein Buch
3. *Julia *has* a book *read*	Julia *hat* ein Buch *gelesen*
4. Yesterday Julia *has read* a book	*Gestern Julia *hat gelesen* ein Buch
5. *Yesterday *has* Julia a book *read*	Gestern *hat* Julia ein Buch *gelesen*
6. Julia *has given* M. a book yesterday	*Julia *hat gegeben* M. ein Buch gestern
7. *Julia *has* M. yesterday a book *given*	Julia *hat* M. gestern ein Buch *gegeben*

In declarative main clauses with a complex one-place verb, English and German may be alike, as shown by 1. In the case of complex two- and three-place verbs, however, the two languages diverge. The word orders in 2, 4, and 6 are grammatical in English, but violate the distance position rule in German, whereby 4 also violates the verb-second rule of German. The word orders 3, 5, and 7 are grammatical in German, but violate the contact position rule of English.

The distance position of complex verb forms in declarative main clauses of German is also called *Satzklammer* (sentential brace).[3] From the viewpoint of communication, the clause final position of the nonfinite main verb serves, first, to mark the end of the clause. Second, it keeps the attention of the hearer, whose interpretation has to wait for the main verb. Third, it gives the speaker more time to select the main verb. As a case in point, consider the following example:

Julia has the offer of the opposing party yesterday afternoon
Julia hat das Angebot der Gegenseite gestern nachmittag *abgelehnt.*   declined
                                                          verworfen.   refused
                                                          kritisiert.   criticized
                                                          zurückgewiesen. rejected

In English, the choice of the main verb must be decided very early, as shown by the corresponding example Julia has *declined/refused/criticized/rejected* the offer of the opposing party yesterday afternoon.

A third word order difference between English and German arises in the position of the – basic or complex – finite verb in subordinate clauses. In English, the position of the verb in subordinate clauses is the same as in declarative main clauses, namely *post-nominative*. In German, on the other hand, the verb is in *clause-final* position, whereby in complex verb forms the nonfinite main verb precedes the finite auxiliary.

18.3.3 VERB POSITION IN SUBORDINATE CLAUSES

English: post-nominative	*German*: clause final
1. before Julia *slept*	bevor Julia *schlief*
2. before Julia *had slept*	*bevor Julia *hatte geschlafen*
3. *before Julia *slept had*	bevor Julia *geschlafen hatte*
4. before Julia *bought* the book	*bevor Julia *kaufte* das Buch
5. *before Julia the book *bought*	bevor Julia das Buch *kaufte*
6. before Julia *had bought* the book	*bevor Julia *hatte gekauft* das Buch
7. *before the book a man *bought*	bevor das Buch ein Mann *kaufte*

In subordinates clauses with a basic one-place verb, English and German may be alike, as shown by 1. In the case of complex verb forms of arbitrary valencies (2,3,6) and basic verbs of more than one valency (4,5,7), the two languages diverge. The word

---

[3] The Satzklammer is one of many structures of natural language syntax which systematically violates the (defunct) principles of constituent structure analysis (cf. Section 8.4).

orders in 2, 4, and 6 are grammatical in English, but violate the clause-final rule for the position of the finitum in subordinate clauses of German. The word orders 3, 5, and 7 are grammatical in German, but violate the post-nominative rule of English, whereby 7 also violates the fixed order of nominative and accusative.

From the viewpoint of communication, the clause-final position of the finite verb in subordinate clauses has a similar function as the clause final position of the nonfinite main verb in declarative main clauses. It serves to mark the end of the clause, keeps the hearer's attention, and gives the speaker maximal time for selecting the main verb.

18.4 Complex verbs and elementary adverbs (LA-D3)

German has two auxiliaries, haben (*have*) and sein (*be*), as well as a number of modals, such as werden (*will*), können (*can*), wollen (*want*), dürfen (*may*), sollen (*should*), and müssen (*must*). Their nominative restrictions in the present and imperfect are the same as those of German main verbs.[4]

18.4.1 LA-PARADIGMS OF GERMAN AUXILIARIES AND MODALS

[bin (S1′ S′ V) *]   [habe (S1′ H′ V) *]   [kann (S13′ M′ V) *]
[bist (S2′ S′ V) *]   [hast (S2′ H′ V) *]   [kannst (S2′ M′ V) *]
[ist (S3′ S′ V) *]   [hat (S3′ H′ V) *]   [können (P13′ M′ V) *]
[sind (P13′ S′ V) *]   [haben (P13′ H′ V) *]   [könnt (P2′ M′ V) *]
[seid (P2′ S′ V) *]   [habt (P2′ H′ V) *]   [konnte (S13′ M′ V) *]
[war (S13′ S′ V) *]   [hatte (S13′ H′ V) *]   [konntest (S2′ M′ V) *]
[warst (S2′ S′ V) *]   [hattest (S2′ H′ V) *]   [konnten (P13′ M′ V) *]
[waren (P13′ S′ V) *]   [hatten (P13′ H′ V) *]   [konntet (P2′ M′ V) *]
[wart (P2′ S′ V) *]   [hattet (P2′ H′ V) *]

The category segments S′ (*sein*), H′ (*haben*), and M′ (*modal*) control the identity-based agreement between the finite auxiliary and a suitable nonfinite form of the main verb. Thereby S and H agree with the past participles of different verbs (e.g. ist begegnet, but hat gegeben), while M agrees with the infinitive (e.g. will sehen).

Whether a past participle agrees with the auxiliary H (*have*) or S (*be*) is a lexical property of the main verb and related to the semantic phenomenon of aspect. Some verbs may combine with either H or S, e.g. ist ausgeschlafen (emphasizing the state of having slept enough) versus hat ausgeschlafen (emphasizing the completion of the process).

The infinitive (e.g. sehen) has the same form as the first and third person plural of the present tense with one exception: the infinitive of the auxiliary sein differs from

[4] Cf. 18.2.5; compare also 17.3.1. For simplicity, the 'Konjunktiv' (subjunctive) is omitted in 18.3.3.

the finite form in question, i.e. **sind**. In analogy to English (cf. 17.3.2), the infinitive is provided with its own category, based on the modal category segment M.

18.4.2 COMPLEX VERB FORMS OF GERMAN

$$
\begin{array}{ccc}
\begin{array}{cc} \text{hat} & \text{gegeben} \\ (S3'\ H'\ V) & (D'\ A'\ H) \end{array} & \Rightarrow & \begin{array}{c} \text{hat gegeben} \\ (S3'\ D'\ A'\ V) \end{array}
\end{array}
$$

$$
\begin{array}{ccc}
\begin{array}{cc} \text{ist} & \text{begegnet} \\ (S3'\ S'\ V) & (D'\ S) \end{array} & \Rightarrow & \begin{array}{c} \text{ist begegnet} \\ (S3'\ D'\ V) \end{array}
\end{array}
$$

$$
\begin{array}{ccc}
\begin{array}{cc} \text{will} & \text{sehen} \\ (S3'\ M'\ V) & (A'\ M) \end{array} & \Rightarrow & \begin{array}{c} \text{will sehen} \\ (S3'\ A'\ V) \end{array}
\end{array}
$$

To explain the special structure of German auxiliary constructions in a strictly time-linear derivation order, let us first consider declarative main clauses with a finite main verb. There, the valency carrier is in second position such that for incoming fillers all the valency positions are available from the beginning. For example, in

   Die Frau *gab* dem Kind den Apfel

or

   Dem Kind *gab* die Frau den Apfel,

the initial nominal filler cancels a compatible position in the adjacent main verb, then the following nominal fillers cancel the remaining positions postverbally.

In corresponding auxiliary constructions, on the other hand, the valency carrier is not present from the beginning such that incoming fillers may be without suitable valency positions for a while. For example, in

   Die Frau *hat* dem Kind den Apfel *gegeben*

the first filler cancels the nominative position in the auxiliary, but the dative and the accusative have no corresponding valency positions until the nonfinite main verb arrives at the end. Furthermore, if the first filler is not a nominative, as in

   Dem Kind *hat* die Frau den Apfel *gegeben*,

no valency position at all can be canceled in the combination of the initial nominal filler and the auxiliary.

Thus, in auxiliary constructions the fillers often have to be remembered in the sentence start category until the valency carrier is added. This requires the following extensions of *LA-D2*: (i) the combination of the first constituent and the auxiliary (MAIN+FV revised as **+FV**), (ii) the combination of the sentence start ending in the auxiliary and the next constituent(s) (FV+MAIN revised as **+MAIN**), and (iii) adding the nonfinite main verb in clause final position whereby the collected fillers must cancel the corresponding positions in the nonfinite main verb (new rule **+NFV**).

In step (i) the following case distinctions must be made.

18.4.3 +FV ALTERNATIVES OF ADDING THE AUXILIARY

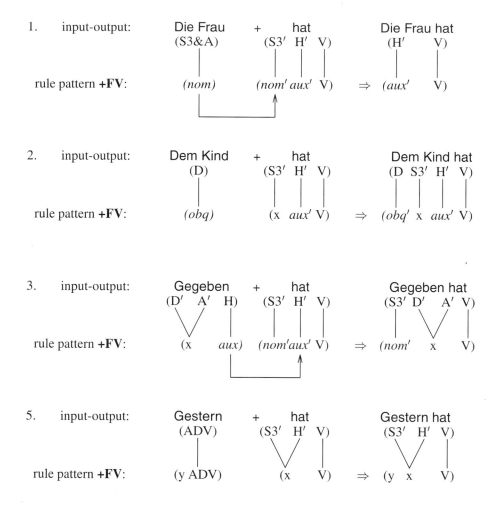

1.   input-output: Die Frau   +   hat   Die Frau hat
                       (S3&A)   (S3′ H′ V)   (H′   V)

rule pattern **+FV**: (nom)   (nom′ aux′ V) ⇒ (aux′ V)

2.   input-output: Dem Kind   +   hat   Dem Kind hat
                       (D)   (S3′ H′ V)   (D S3′ H′ V)

rule pattern **+FV**: (obq)   (x aux′ V) ⇒ (obq′ x aux′ V)

3.   input-output: Gegeben   +   hat   Gegeben hat
                       (D′ A′ H)   (S3′ H′ V)   (S3′ D′ A′ V)

rule pattern **+FV**: (x   aux)   (nom′aux′ V) ⇒ (nom′   x   V)

5.   input-output: Gestern   +   hat   Gestern hat
                       (ADV)   (S3′ H′ V)   (S3′ H′ V)

rule pattern **+FV**: (y ADV)   (x   V) ⇒ (y x   V)

In clause 1, the nominative filler cancels the nominative valency in the finite auxiliary. In clause 2, the presence of the auxiliary segment triggers the oblique filler to be added to the category. In clause 3, a topicalized nonfinite main verb cancels the agreeing auxiliary segment and adds its valency positions, resulting in a category equivalent to a corresponding finite main verb. In clause 5, the combination of an adverb and a finite auxiliary leaves the category of the latter unchanged.

It would be possible to treat each clause in 18.4.3 as a separate rule. However, because these clauses are linguistically related and share the same rule package we introduce a notation for combining several clauses into a complex categorial operation. This notation, illustrated in 18.4.4, is a subtheoretical variant like multicats. It is equivalent to the standard notation of the algebraic definition and does not change the formal status of the grammars with respect to their complexity.

18.4.4 EXTENDING MAIN+FV INTO +FV USING CLAUSES

+FV:
1. (*nom*)(*nom′ aux′* V)   ⇒   (*aux′* V)
2. (*obq*)(x *aux′* V)   ⇒   (*obq* x *aux′* V)
3. (x *aux*)(*nom′ aux′* V)   ⇒   (*nom′* x V)
4. (*np*)(x *np′* y V)   ⇒   (x y V)
5. (y ADV)(x V)   ⇒   (y x V) {+MAIN, +NFV, +FV, +IP}

The input conditions of the clauses in 18.4.4 are applied sequentially to the input. When a clause matches the input, the associated output is derived and the rule application is finished.[5] Clauses 1, 2, 3, and 5 correspond to 18.4.3. Clause 4 handles the standard combination of a nominal filler and a finite main verb inherited from MAIN+FV in *LA-D2*.

Next consider step (ii), i.e., the continuations after the finite auxiliary.

18.4.5 +MAIN ALTERNATIVES AFTER THE AUXILIARY

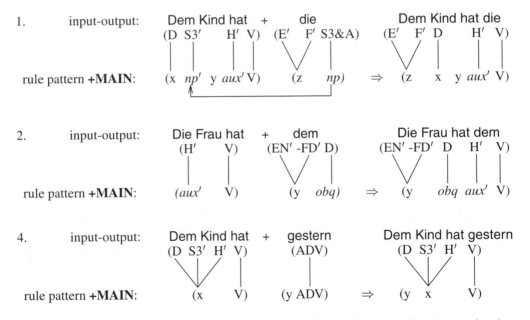

In clause 1, the auxiliary is followed by a nominative which cancels the nominative valency position in the auxiliary. In clause 2, the auxiliary is followed by an oblique filler, whereby the presence of the auxiliary segment triggers the addition of the next word category segments to the sentence start. In clause 4, the continuation with an adverb has no effect on the syntactic category of the sentence start.

These alternatives are joined with the FV+MAIN combination of a finite main verb and a nominal filler (clause 3) into the new rule **+MAIN**.

---

[5] There exists the additional possibility of bracing several clauses to indicate that they apply in parallel, which may result in more than one continuation path.

18.4.6 EXTENDING FV+MAIN INTO +MAIN USING CLAUSES

+MAIN:
$$
\begin{aligned}
&1.\ (x\ nom'\ y\ aux'\ V)(z\ nom) &&\Rightarrow\ (z\ x\ y\ aux'\ V)\\
&2.\ (x\ aux'\ V)(y\ obq) &&\Rightarrow\ (y\ obq\ x\ aux'\ V)\\
&3.\ (x\ np'\ y\ V)(z\ np) &&\Rightarrow\ (z\ x\ y\ V)\\
&4.\ (x\ V)(y\ ADV) &&\Rightarrow\ (y\ x\ V)
\end{aligned}
$$
$$\{+ADJ, +N, +MAIN, +NFV, +FV, +IP\}$$

Clause 2 of **+MAIN** allows the addition of arbitrarily many oblique fillers. The resulting sentence is only grammatical, however, if the nonfinite main verb added at the end happens to provide a corresponding set of valency positions.

Step (iii) consists in adding the nonfinite main verb concluding the clause. It is handled by the new rule **+NFV**, which checks identity-based agreement between the auxiliary and the nonfinite main verb illustrated in 18.4.2. Furthermore, the oblique fillers collected in the sentence start cancel all the oblique valency positions in the nonfinite main verb (the nominative is already canceled at this point). The categorial operation of **+NFV** is illustrated in 18.4.7.

18.4.7 CATEGORIAL OPERATION OF +NFV

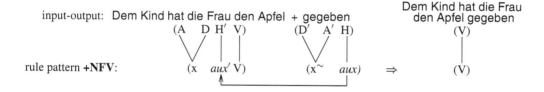

A novel aspect of **+NFV** is the concept of *agreeing lists*. Two lists x and x^ are in linguistic agreement, written as [x = x^], if they are of equal length and for each value in one list there is an agreeing value in the other and vice versa. List agreement will also be used in the handling of German subordinate clauses (cf. Section 18.5).

18.4.8 GERMAN GRAMMAR HANDLING COMPLEX VERB FORMS (*LA-D3*)

LX = LX of *LA-D2* plus auxiliaries defined in 18.4.1, plus
 nonfinite main verb form of 18.4.2, plus adverbials
 [gestern (ADV) *], [hier (ADV) *], [jetzt (ADV) *], plus punctuation signs
 [. (V′ DECL) *], [? (VI′ INTERROG) *], [? (V′ INTERROG) *]
variable definition = variable definition of *LA-D2* plus
 $nom \in np \setminus \{D, A, D\&A\}$ nominative filler[6]
 $nom' \in np \setminus \{D, A\}$ nominative valency positions
 $obq \in \{D, A, D\&A\}$ oblique filler
 $aux \in \{H, B, M\},$ auxiliaries and modals

$vt \, \epsilon \, \{V, VI\},$ mood marker
$sm \, \epsilon \, \{DECL, INTERROG\},$ sentence mood

$\mathrm{ST}_S =_{def} \{ [(x) \{1 \text{ +ADJ, } 2 \text{ +N, } 3 \text{ +FV, } 4 \text{ +NFV}\}] \}$

+ADJ: $(adj' \, x) \, (adj) \quad \Rightarrow \quad (adj \, x) \quad \{5 \text{ +ADJ, } 6 \text{ +N}\}$

+N: $(adj' \, n' \, x) \, (n) \quad \Rightarrow \quad (x) \quad \{7 \text{ +FV, } 8 \text{ +MAIN, } 9 \text{ +NFV, } 10 \text{ +IP}\}$

+FV: $(nom)(nom' \, aux' \, V) \Rightarrow (aux' \, V)$

$\quad\quad\quad (obq)(x \, aux' \, V) \quad \Rightarrow \quad (obq \, x \, aux' \, V)$

$\quad\quad\quad (x \, aux)(nom' \, aux' \, V) \Rightarrow (nom' \, x \, V)$

$\quad\quad\quad (np)(x \, np' \, y \, V) \quad \Rightarrow \quad (x \, y \, V)$

$\quad\quad\quad (ADV)(x \, V) \quad\quad \Rightarrow \quad (x \, V) \quad \{11 \text{ +MAIN, } 12 \text{ +NFV, } 13 \text{ +IP}\}$

+MAIN: $(x \, nom' \, y \, aux' \, V)(z \, nom) \Rightarrow (z \, x \, y \, aux' \, V)$

$\quad\quad\quad (x \, aux' \, V)(y \, obq) \Rightarrow (y \, obq \, x \, aux' \, V)$

$\quad\quad\quad (x \, np' \, y \, V)(z \, np) \Rightarrow (z \, x \, y \, V)$

$\quad\quad\quad (x \, V)(y \, ADV) \quad \Rightarrow \quad (y \, x \, V) \quad \{14 \text{ +ADJ, } 15 \text{ +N, } 16 \text{ +MAIN, } 17 \text{ +NFV,}$
$\quad\quad\quad\quad\quad\quad\quad\quad\quad\quad\quad\quad\quad\quad\quad\quad 18 \text{ +FV, } 19 \text{ +IP}\}$

+NFV: $(x \, aux' \, V)(x^{\sim} \, aux)$

$\quad\quad\quad\quad (x = x^{\sim}) \quad \Rightarrow \quad (V) \quad \{20 \text{ +IP}\}$

+IP: $(vt) \, (vt' \, sm) \quad \Rightarrow \quad (sm) \quad \{\}$

$\mathrm{ST}_F =_{def} \{ [(sm) \, \mathrm{rp}_{+\mathrm{ipt}}] \}$

While **+FV** and **+MAIN** show the use of *clauses*, **+NFV** shows the use of a *subclause*. If the pattern of the main clause of **+NFV** is matched by the input, then the indented subclause checks whether the list agreement between the values of x and the values of $x^{\sim}$ is satisfied. Only if this is the case, an output is derived. Especially in larger applications, subclauses allow a natural structuring of LA-rules into main conditions and subconditions, improving both linguistic transparency and computational efficiency.

The rule **+IP** adds full stops or question marks to declarative main clauses, as in Julia hat gestern ein Buch gelesen? In *LA-D4*, **+IP** will be adopted without change. Other sentence moods are based on *prefinal* derivation steps (cf. Section 18.5).

LA-D3 provides a complete treatment of word order variations and word order restrictions in German declarative main clauses with and without auxiliaries. It handles all internal and external agreement restrictions and valency phenomena arising in 980 types of clauses which differ with respect to the valency structure of the verb, basic vs. complex noun phrases, basic vs. complex verb forms including topicalized nonfinite main verbs, and declarative vs. interrogative mood. If the possible adding of adverbs in various positions is taken into account, the number of constructions handled by *LA-D3* is theoretically infinite.

The application of **+NFV** with the canceling of agreeing lists is illustrated by the following derivation, the transition numbers of which are those of *LA-D3*.

---

[6] If Y is a subset of X, then the notation $X \backslash Y$ (X minus Y) stands for the set of elements of X without the elements of Y. This notation should not be confused with the definition of categories of C-grammar.

18.4.9 DECLARATIVE WITH DATIVE PRECEDING AUXILIARY

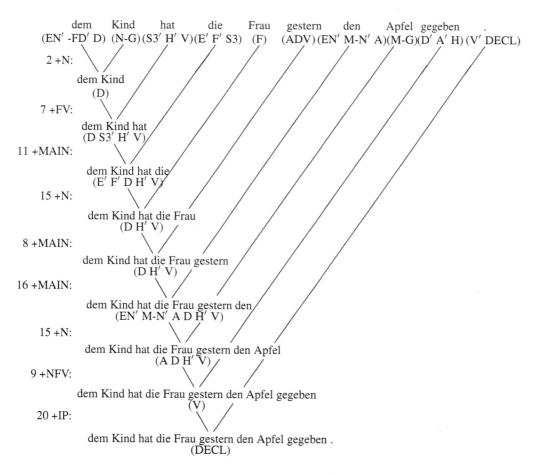

Whether a nominal argument is added to the sentence start category or used to cancel a valency position is controlled by the presence/absence of an *aux* segment, i.e. H, B, or M, in the carrier and by whether the filler is a nominative or not.

18.5 Interrogatives and subordinate clauses (*LA-D4*)

The distance position of auxiliary and main verb in declarative sentences arises in similar form also in interrogatives with auxiliaries and in subordinate clauses. Consider the following examples of interrogatives with the verbal components in italics.

18.5.1 INTERROGATIVE WITH AND WITHOUT AUXILIARY

1. *Hat* die Frau   dem Kind gestern   den Apfel *gegeben* ?
   (*Has the   woman the   child yesterday   the apple given   ?*)

2. *Hat* dem Kind gestern die Frau den Apfel *gegeben*?
3. *Hat* gestern die Frau dem Kind den Apfel *gegeben*?
4. *Gab*   die Frau   dem Kind gestern   den Apfel ?
   (*Gave the woman the   child yesterday   the   apple* ?)
5. *Gab* gestern die Frau dem Kind den Apfel?

A sentential brace is formed in 1, 2, and 3 by the initial auxiliary and the final nonfinite main verb. In 1, the auxiliary is followed by the nominative, in 2 by an oblique filler, and in 3 by an adverb. The examples 4 and 5 begin with a finite main verb, followed by a nominal filler and an adverb, respectively. The initial finite verb is combined with a main component by the new rule **?+MAIN**.

18.5.2 **?+MAIN** STARTING AN INTERROGATIVE MAIN CLAUSE

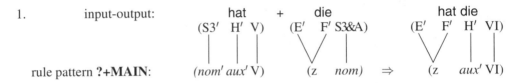

?+MAIN is similar to **+MAIN** except for changing the verb type to VI, thus enforcing the addition of a question mark at the end. By including **?+MAIN**, but not **+MAIN**, in the start state's rule package, a sentence initial finite verb is treated obligatorily as the beginning of an interrogative.

After the application of **?+MAIN**, the derivation of interrogatives is based on existing rules. For example, sentence 1 in 18.5.1 is based on the rule sequence **?+MAIN, +N, +MAIN, +N, +MAIN, +MAIN +N, +NFV, +IP**, while sentence 5 is based on the rule sequence **?+MAIN, +MAIN, +N, +MAIN, +N, +MAIN, +N, +IP**.

Another type of sentential brace is formed by subordinate clauses. As illustrated below, the brace begins with the subordinating conjunction, to the category of which the subsequent nominal fillers are added. The end of the brace is the verb. It requires list agreement to cancel the fillers collected in the sentence start.

18.5.3 SUBORDINATE CLAUSES WITH AND WITHOUT AUXILIARY

1. *Als*   die Frau   dem Kind gestern   den Apfel *gegeben* *hat*
   (*When the woman the   child yesterday the   apple given   has*)
2. *Als* dem Kind gestern die Frau den Apfel *gegeben hat*
3. *Als* gestern die Frau dem Kind den Apfel *gegeben hat*
4. *Als*   die Frau   dem Kind gestern   den Apfel *gab*
   (*When the woman the   child yesterday the apple gave*)
5. *Als* gestern die Frau dem Kind den Apfel *gab*

The beginning of adverbial subclauses is handled by an extension of **+MAIN**:

18.5.4 +MAIN STARTING AN ADVERBIAL SUBCLAUSE

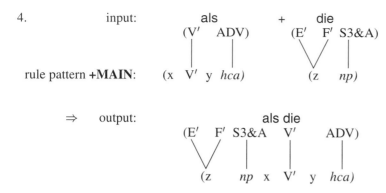

4.   input: als   +   die
                           (V′   ADV)   (E′ F′ S3&A)

rule pattern +MAIN: (x   V′ y *hca*)   (z   *np*)

⇒   output: als die
                       (E′   F′ S3&A   V′   ADV)

                       (z   *np* x   V′ y *hca*)

The variable *hca* (for higher clause attachment) is specified for the values V, VI, and ADV. The addition of nominal fillers to the sentence start category is triggered by the presence of the nonfinal V′, provided here by the subordinating conjunction. The clause final finite main verb is added by an extension of **+FV**.

18.5.5 +FV CONCLUDING SUBCLAUSE WITH FINITE MAIN VERB

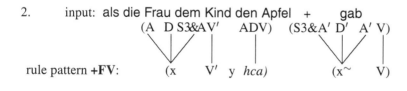

2.   input: als die Frau dem Kind den Apfel +   gab
                (A D S3&AV′   ADV) (S3&A′ D′ A′ V)

rule pattern +FV: (x   V′ y *hca*)   (x~   V)

⇒   output: als die Frau dem Kind den Apfel gab
                          (ADV)

                       (y *hca*)

If the adverbial clause has a preverbal position in the higher sentence, the ADV segment in the category of the subordinating conjunction serves as the result segment (cf. 18.5.5). Once the derivation of the adverbial subclause in initial position is completed it has the category (ADV) and is treated like an elementary adverb (e.g. **gestern** in 18.4.3, 5). If the adverbial clause has a postverbal position in the higher clause, the ADV-segment is deleted in the process of adding the subordinating conjunction (cf. 18.5.6).

Adverbial clauses in any position use the V′-segment of the subordinating conjunction category as the main node for the duration of the subclause derivation. This means that the preverbal nominal fillers are attached to the left of the current V′-segment. The V′-segment of the current subclause is always positioned left-most in the category of the sentence start.

18.5.6 BEGINNING OF AN ADVERBIAL SUBCLAUSE IN POSTVERBAL POSITION

Julia las,	+ als		Julia las, als	+ Maria		Julia las, als Maria
(A′ V)	(V′ ADV)	⇒	(V′ A′ V)	(S3&D&A)	⇒	(S3&D&A V′ A′ V)

The next combination finishes the subordinate clause with the finite main verb. Thereby everything up to and including the first non-final V′ is canceled and the category is the same as before the adverbial subclause was started.

18.5.7 COMPLETION OF AN ADVERBIAL SUBCLAUSE IN POSTVERBAL POSITION

Julia las, als Maria	+ schlief		Julia las, als Maria schlief,
(S3&D&A V′ A′ V)	(S3′ V)	⇒	(A′ V)

In this manner, adverbial subclauses may also be nested, allowing center embedding of arbitrary depth.

18.5.8 NESTING OF ADVERBIAL SUBCLAUSES IN PREVERBAL POSITION

Als Maria, obwohl Julia die Zeitung	+ las		Als Maria, obwohl Julia die Zeitung las,
(A S3&D&A V′ S3&D&A V′ ADV)	(S3′ A′ V)	⇒	(S3&D&A V′ ADV)

The nesting is indicated categorially by the presence of more than one V′ segment in the sentence start category, but only as long as the nested clauses have not been completed.

Subclauses with complex verb phrases are handled in the same way except that the nonfinite main verb and the finite auxiliary are added in two steps.

18.5.9 **+NFV** ADDS NONFINITE MAIN VERB TO SUBCLAUSE

1   input: als dem Kind die Frau den Apfel + gegeben

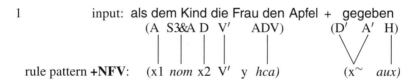

rule pattern **+NFV**:   (x1 *nom* x2 V′   y   *hca)*   (x$^\sim$ *aux)*

⇒   output: als dem Kind die Frau den Apfel gegeben

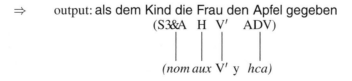

(nom aux V′   y   *hca)*

This example illustrate that the nominative filler may be positioned between the oblique fillers. To achieve here list agreement between the oblique fillers and their valency positions in the nonfinite verb, the sublists x1 and x2 are concatenated, written as (x1 ∘ x2). With this notation, the desired list agreement may be formulated as [(x1 ∘ x2) = x$^\sim$] (cf. **+NFV** in *LA-D4*, defined in 18.5.11).

The second and final step of concluding a subordinate clause with a complex verb is the addition of the finite auxiliary.

18.5.10 +FV CONCLUDES SUBCLAUSE WITH FINITE AUXILIARY

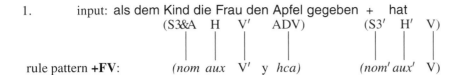

1.   input: als dem Kind die Frau den Apfel gegeben +   hat
                     (S3&A   H   V'   ADV)   (S3'   H'   V)

rule pattern **+FV**:   *(nom aux   V'   y   hca)   (nom' aux'   V)*

⇒   output: als dem Kind die Frau den Apfel gegeben hat
                              (ADV)

                       (y hca)

In 18.5.10, the nominative segments have to agree.

The handling of interrogatives and adverbial subclauses is formalized as the LA-syntax *LA-D4* for German.

18.5.11 LAG HANDLING INTERROGATIVE AND ADVERBIAL CLAUSES (*LA-D4*)

LX = LX of *LA-D3* plus subordinating conjunctions
   [als (V' ADV) *], [nachdem (V' ADV) *], [obwohl (V' ADV) *]
variable definition = variable definition of *LA-D3* plus *hca* ϵ {V, VI, ADV}
$ST_S =_{def}$ { [(x) {1 +ADJ, 2 +N, 3 +FV, 4 +MAIN, 5 ?+MAIN}] }

+N:	*(adj' n' x) (n)*	⇒	*(x)* {6 +FV, 7 +MAIN, 8 +NFV, 9 +IP}
+ADJ:	*(adj' x) (adj)*	⇒	*(adj' x)* {10 +ADJ, 11 +N}

?+MAIN: *(nom' aux' V)(z nom)* ⇒ *(z aux' VI)*
           (nom' aux' V)(y obq) ⇒ *(y obq nom' aux' VI)*
           (x np' y V)(z np) ⇒ *(z x y VI)*
           (x V)(y ADV)   ⇒   *(y x VI)*   {12 +ADJ, 13 +N, 14 +MAIN,
                                          15 +NFV, 16 +IP}

+FV: *(nom aux V' y hca) (nom' aux' V)* ⇒ *(y hca)*
      (x V' y hca)(x~ V)
          [x = x~]   ⇒   *(y hca)*
      (nom)(nom' aux' V) ⇒ *(aux' V)*
      (obq)(x aux' V)   ⇒   *(obq x aux' V)*
      (x aux)(np' aux' V) ⇒ *(x np' V)*
      (np)(x np' y V)   ⇒   *(x y V)*
      (ADV)(x V)   ⇒   *(x V)*   {17 +MAIN, 18 +NFV, 19 +FV, 20 +IP}

+MAIN: *(x nom' y aux' V)(z nom)* ⇒ *(z x y aux' V)*
          (x aux' V)(y obq) ⇒   *(y obq x aux' V)*

$$(x\ np'\ y\ V)(z\ np) \Rightarrow \quad (z\ x\ y\ V)$$
$$(x\ V'\ y\ hca)(z\ np) \Rightarrow (z\ np\ x\ V'\ y\ hca)$$
$$(x\ V)(y\ ADV) \quad \Rightarrow \quad (y\ x\ V) \quad \{21\ +ADJ,\ 22\ +N,\ 23\ +MAIN,\ 24\ +NFV,$$
$$25\ +FV,\ 26\ +IP\}$$

+NFV: $(x1\ nom\ x2\ V'\ y\ hca)(\ x^{\sim}\ aux)$
$$[(x1 \circ x2) = x^{\sim}] \Rightarrow (nom\ aux\ V'\ y\ hca)$$
$$(x\ aux'\ V)(x^{\sim}\ aux) \Rightarrow (V) \qquad \{27\ +FV,\ 28\ +IP\}$$

+IP: $(vt)\ (vt'\ sm) \qquad \Rightarrow \qquad (sm) \qquad \{\}$
$$ST_F =_{def} \{\ [(V)\ rp_{+ipt}],\ [(VI)\ rp_{+ipt}]\}$$

LA-D4 extends *LA-D3* to yes/no-interrogatives and adverbial subclauses, each with simple and complex verb forms, whereby the adverbial subclauses may take various positions, including center embedding.

LA-D4 has the following finite state transition network.

18.5.12 THE FINITE STATE BACKBONE OF *LA-D4*

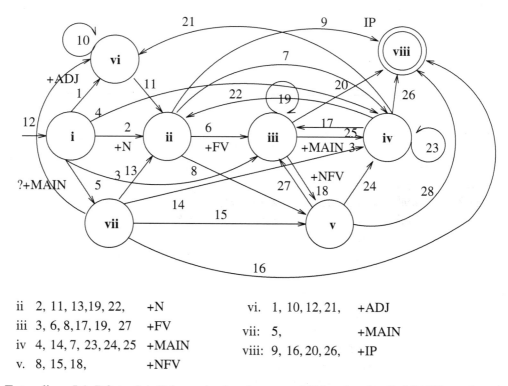

ii	2, 11, 13, 19, 22,	+N		vi.	1, 10, 12, 21,	+ADJ
iii	3, 6, 8, 17, 19, 27	+FV		vii:	5,	+MAIN
iv	4, 14, 7, 23, 24, 25	+MAIN		viii:	9, 16, 20, 26,	+IP
v.	8, 15, 18,	+NFV				

Extending *LA-D3* to *LA-D4* required only one additional rule, **?+MAIN**, and one additional clause in **+NFV**. The extension consisted mostly in adding transitions, implemented by writing additional rule names into rule packages. While *LA-D3* has 6 rules and 19 transitions, *LA-D4* has 7 rules and 28 transitions. The handling of agreement and word order variation of *LA-D3* is preserved in *LA-D4*. Like its predecessor, *LA-D4* is a C1-LAG and parses in linear-time.

At this point the syntactic coverage of English and German could be easily extended further and further. For the methodological reasons explained in Section 18.1, however, the existing grammars should first be *verified* in the following areas.

1. *Syntactic verification*
   The formal grammars for English and German developed so far should be implemented as parsers and tested automatically on increasing sets of positive and negative test sentences.
2. *Morphological and lexical verification*
   The word form recognition of these grammars should be changed from the preliminary full form lexica LX to suitable applications of LA-Morph and be tested on corpus-based word lists in order to provide extensions with sufficient data coverage of the lexicon and the morphology.
3. *Functional verification in communication*
   The formal grammars and parsers for natural languages should be supplemented with an automatic semantic and pragmatic interpretation that is (i) in line with the basis assumptions of the SLIM theory of language and (ii) demonstrated to be functional in automatic applications.

Verification steps 1 and 2 have been provided by the extensive LA-grammars for English and German described in NEWCAT and CoL, from which *LA-E1 – LA-E4*[7] and *LA-D1 – LA-D4* were abstracted. The details of verification step 3, on the other hand, require a closer theoretical description, which will be the topic of Part IV.

Exercises

Section 18.1

1. Explain the three steps and two phases of developing an LA-grammar for the syntactic description of a new language.
2. What is the structure of derived noun phrases in German and how does it differ from English?
3. Describe the agreement of German adjectives.

---

[7] For the definition of *LA-E4* see Section 23.4.

4. Which rule in *LA-D1* must be slightly generalized for the derivation of 18.1.4, and how?

5. Give derivations like 18.1.4 for the sentences der jungen Frau gefiel der schöne grüne Baum, die junge Frau gab dem kleinen Kind einen Apfel, dem kleinen Kind gab die junge Frau einen Apfel, and einen Apfel gab die junge Frau dem kleinen Kind. Explain why the time-linear derivation of complex nominal fillers in pre- and postverbal position can be handled by the same rules.

Section 18.2

1. Explain in what two respects traditional paradigms, e.g. of German noun phrases, are misleading from a syntactic point of view. Why are traditional paradigms not surface compositional?

2. Compare the abstract field of referents of German (18.2.2 and 18.2.3) and English (17.2.1). Explain their surface compositional motivation and relate them to traditional paradigms.

3. Because the genitive case is becoming obsolete in its function as a nominal filler/valency position you may simplify 18.2.2 for the purposes of modern German. How many fields are there left after eliminating the genitive?

4. Describe the agreement between nominal fillers and finite verbs in German.

5. Explain the handling of external and internal agreement of derived noun phrases in *LA-D2*.

6. Provide a finite state backbone – analogous to 17.4.2 – for *LA-D2* using the transition numbers given in 18.2.5.

Section 18.3

1. What is the difference between a post-nominative and a verb-second word order? Give examples from English and German.

2. Where in the word order specification of a language do adverbials play a role?

3. What is the difference between contact and distance positions in complex verb constructions? Give examples from English and German.

4. Explain why the sentential brace (*Satzklammer*) in German systematically violates the principles of constituent structure 8.4.1.

5. Is it possible to motivate distance position in terms of communicative function?

6. Compare the word order of subordinate clauses in English and German.

7. Why does the word order of auxiliary constructions in declarative main clauses and in subordinate clauses of German pose an apparent problem for the time-linear filling of valency positions?

Section 18.4

1. Describe the categorial structure of complex verb forms in German.

2. Explain the five categorial alternatives in adding a finite auxiliary in declarative main clauses of German and describe how they are formalized in the rule +FV.

3. Explain the four categorial alternatives in continuing after a finite auxiliary in declarative main clauses of German and describe how they are formalized in the **+MAIN**.

4. Describe the role of list agreement in the rule **+NFV**.

5. Why does the variable definition of *LA-D3* specify the variables *nom* and *nom'* in addition to *np* and *np'*?

6. Provide a finite state backbone – analogous to 17.5.6 – for *LA-D3* using the transition numbers given in 18.4.8.

Section 18.5

1. Explain why German yes/no-interrogatives with auxiliaries are the maximal form of a *sentential brace*.

2. Why is there a new rule **?+MAIN** for the beginning of yes/no-interrogatives rather than handling its function by existing **+MAIN**?

3. Compare the categorial operations of **?+MAIN** and **+MAIN** in 18.5.11.

4. Why is *LA-D4* a C1-LAG?

5. Determine the grammatical perplexity of *LA-D4*. Does the use of rule clauses and subclauses affect the perplexity value?

6. Extend *LA-D4* to a handling of prepositional phrases in postnominal position as in der Apfel + auf dem Tisch, and in adverbial position as in Auf dem Tisch + lag. Take into account the semantic analysis of prepositional phrases presented in Section 12.5. Adapt the finite state backbone 18.5.12 to your extension.

Semantics and Pragmatics

19. Three system types of semantics

Part I presented the mechanism of natural communication within the SLIM theory of language. Part II described algorithms of generative grammar for automatically analyzing the combinatorics of formal and natural languages. Part III developed detailed morphological and syntactical analyses of natural language surfaces using the time-linear algorithm of LA-grammar. Based on these foundations, we turn in Part IV to the semantic and pragmatic interpretation of natural language.

First different traditional approaches to semantic interpretation will be described in Chapters 19–21, explaining basic notions, goals, methods, and problems. Then a semantic and pragmatic interpretation of LA-grammar within the SLIM theory of language will be developed in Chapters 22–24. The formal interpretation is implemented computationally as an extended database system called word bank.

Section 19.1 explains the structure common to all systems of semantic interpretation. Section 19.2 compares three different types of formal semantics, namely the semantics of logical languages, of programming languages, and of natural languages. Section 19.3 illustrates the functioning of logical semantics with a simple model-theoretic system and explains the underlying theory of Tarski. Section 19.4 shows why the semantics of programming languages is independent of Tarski's hierarchy of metalanguages and which special conditions a logical calculus must fullfil in order to be suitable for implementation as a computer program. Section 19.5 explains why a complete interpretation of natural language is impossible within logical semantics and describes Tarski's argument to this effect based on the Epimenides paradox.

19.1 Basic structure of semantic interpretation

The term semantics is being used in different fields of science. In *linguistics*, semantics is a component of grammar which derives representations of meaning from syntactically analyzed natural surfaces. In *philosophy*, semantics assigns set-theoretic denotations to logical formulas in order to characterize truth and to serve as the basis for certain methods of proof. In *computer science*, semantics consists in executing commands of a programming language automatically as machine operations .

Even though the semantics from these different fields of science differ substantially in their goals, their methods, their applications, and their form, they all share the same basic two-level structure consisting of syntactically analyzed language expressions and associated semantic structures. The two levels are systematically related by means of an assignment algorithm.

19.1.1 THE 2-LEVEL STRUCTURE OF SEMANTIC INTERPRETATION

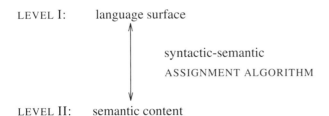

LEVEL I: language surface

                        syntactic-semantic
                        ASSIGNMENT ALGORITHM

LEVEL II: semantic content

For purposes of transmission and storage, semantic content is coded into surfaces of language (representation). When the content is needed, any speaker of the language may reconstruct it by analyzing the surface.[1] The reconstruction consists in (i) syntactic analysis, (ii) retrieval of word meanings from the lexicon, and (iii) deriving the meaning of the whole by assembling the meanings of the parts as specified by the syntactic structure of the surface.

The expressive power of semantically interpreted languages resides in the fact that the inverse procedures of representing and reconstructing content are realized *automatically*: a semantically interpreted language may be used correctly without the user having to be conscious of these procedures, or even having to know or understand their details.

For example, the programming languages summarize frequently used sequences of elementary operations as higher functions, the names of which the user may then combine into complex programs. These programs work in the manner intended even though the user is not aware of – and does not care about – the complex details of the machine or assembler code operations.

The logical language may likewise be used without the user having to go through the full details of the semantic interpretation. One purpose of a logical syntax is to represent the structural possibilities and restrictions of the semantics in such a way that the user can reason truly on the semantic level based solely on the syntactic categories and their combinatorics.

The natural languages are also used by the speaker-hearer without conscious knowledge of the structures and procedures at the level of semantics. In contradistinction to the artificial languages of programming and logic, for which the full details of their

---

[1] For example, it is much easier to handle the surfaces of an expression like 36 · 124 than to execute the corresponding operation of multiplication semantically by using an abacus. Without the language surfaces one would have to slide the counters on the abacus 36 times 124 'semantically' each time this content is to be communicated. This would be tedious, and even if the persons communicating were to fully concentrate on the procedure each time it would be extremely susceptible to error.

semantics are known at least to their designers and other specialists, the exact details of natural language semantics are not known directly even to science.

19.2 Logical, programming, and natural languages

The two level structure common to all genuine systems of semantics allows to control the structures on the semantic level by means of syntactically analyzed surfaces. In theory, different semantic interpretations may be defined for one and the same language, using different assignment algorithms. In practice, however, each type of semantics has its own characteristic syntax in order to achieve optimal control of the semantic level via the combination of language surfaces.

The following three types of semantics comply with the basic structure 19.1.1 of semantic interpretation.

19.2.1 THREE DIFFERENT TYPES OF SEMANTIC SYSTEMS

1. *Logical languages*
   They originated in philosophy.[2] Their complete expressions are called *propositions*. They are designed to determine the truth value of arbitrary propositions relative to arbitrary models. They have a *metalanguage-based semantics* because the correlation between the two levels is based on a metalanguage definition.

2. *Programming languages*
   They are motivated by a practical need to simplify the interaction with computers and the design of software. Their expressions are called *commands*, which may be combined into complex programs. They have a *procedural semantics* because the correlation between the levels of syntax and semantics is based on the principle of execution, i.e., the operational realization of commands on an abstract machine which is usually implemented electronically.

3. *Natural languages*
   They evolve naturally in their speech communities and are the most powerful and least understood of all semantically interpreted systems. Their expressions are called *surfaces*. Linguists analyze the preexisting natural languages syntactically by reconstructing the combinatorics of their surfaces explicitly as generative grammars. The associated semantic representations have to be deduced via the general principles of natural communication because the meanings$_1$ have no concrete manifestation like the surfaces. Natural languages have a *convention-based* semantics because the word surfaces (types) have their meaning$_1$ assigned by means of conventions (cf. de Saussure's first law, Section 6.2).

On the one hand, the semantic interpretations of the three types of languages all share the same two level structure. On the other hand, their respective components differ in

---

[2] An early highlight is the writing of Aristotle, where logical variables are used for the first time.

all possible respects: they use (i) different language expressions, (ii) different assignment algorithms, and (iii) different objects on the semantic level:

19.2.2 THREE TYPES OF SEMANTIC INTERPRETATION

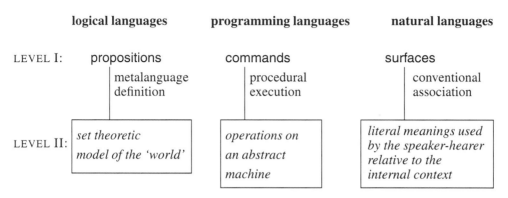

The most important difference in the semantic interpretation of artificial and natural languages consists in the fact that the interpretation of artificial languages is limited to the two-level structure of their semantics whereas the interpretation of natural languages is based on the [2+1] level structure of the SLIM theory of language.

Thus, in logical and programming languages the interpretation is completed on the semantic level. In contrast, the natural languages have an additional interpretation step which is as important as the semantic interpretation. This second step is the pragmatic interpretation in communication. It consists in matching the objects of the semantic level with corresponding objects of an appropriate context of use.

In the abstract, six relations may be established between the three basic types of semantics.

19.2.3 MAPPING RELATIONS BETWEEN THE THREE TYPES OF SEMANTICS

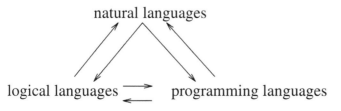

We represent these relations as N→L, N→P, L→N, L→P, P→N, and P→L, whereby N stands for the natural, L for the logical, and P for the programming languages.

In reality, there has evolved a complicated diversity of historical, methodological, and functional interactions between the three systems. These relations may be characterized in terms of the notions *replication, reconstruction, transfer*, and *composition*:

– *Replication*

  The logical languages evolved originally as formal replications of selected natural language phenomena (N→L). Programming languages like LISP and Prolog were designed to replicate selected aspects of logical languages procedurally (L→P). The programming languages also replicate phenomena of natural language directly, such as the concept of 'command' (N→P).

– *Reconstruction*

  When an artificial language has been established for some time and achieved an independent existence of its own, it may be used to reconstruct the language it was originally designed to replicate in part. A case in point is the attempt in theoretical linguistics to reconstruct formal fragments of natural language in terms of logic (L→N). Similarly, computational linguistic aims at reconstructing natural languages by means of programming languages (P→N). One may also imagine a reconstruction of programming concepts in a new logical language (P→L).

– *Transfer*

  The concentrated efforts to transfer methods and results of logical proof theory to programming languages (L→P)[3] have led to important results. A simple, general transfer is not possible, however, because of the differing methods, structures, ontologies, and purposes of these two different types of language.[4] Attempts in philosophy of language to transfer the model-theoretic method to the semantic analysis of natural language (L→N) were also only partially successful.[5]

– *Combination*

  Computational linguistics aims at modeling natural communication with the help of programming languages (P→N), whereby methods and results of the logical languages play a role in both, the construction of programming languages (L→P) and the analysis of natural language (L→N). This requires a functional overall framework for combining the three types of language in a way that utilizes their different properties while avoiding redundancy as well as conflict.

The functioning of the three types of semantics will be explained below in more detail.

19.3 Functioning of logical semantics

In logical semantics, a simple sentence like Julia sleeps is analyzed as a proposition which is either true or false. Which of these two values is denoted by the proposition depends on the state of the world relative to which the proposition is interpreted.

---

[3] D. Scott & C. Strachey 1971.

[4] The transfer of logical proof theory to an automatic theorem prover necessitates that each step – especially those considered 'obvious' – be realized in terms of explicit computer operations (cf. R. Weyhrauch 1980). This requirement has already modified modern approaches to proof theory profoundly (P→L reconstruction).

[5] Cf. Section 19.4.

The state of the world, called the model, is defined in terms of sets and set-theoretic operations.

19.3.1 INTERPRETATION OF A PROPOSITION

LEVEL I   logical language: *sleep (Julia)*

LEVEL II   world (model):

By analyzing the surface **Julia sleeps** formally as *sleep (Julia)* the verb is characterized syntactically as a functor and the name as its argument.

The lexical part of the associated semantic interpretation (word semantics) assigns denotations to the words – here the set of all sleepers to the verb and the individual Julia to the proper name. The compositional part of the semantics derives denotations for complex expressions from the denotations of their parts. In particular, the formal proposition **sleep (Julia)** is assigned the value true (or 1) relative to the model, if the individual denoted by the name is an element of the set denoted by the verb. Otherwise, the proposition denotes the value false (or 0).

A logical language requires definition (1) of a lexicon in which the basic expressions are listed and categorized, and (2) of a syntax which provides the rules for combining the basic expressions into well-formed complex expressions. Its semantic interpretation requires in addition definition (3) of a model, (4) of the possible denotations of syntactic expressions in the model, and (5) of a semantic rule for each syntactic rule.

These five components of model theory are illustrated by the following definition, which in addition to the usual propositional calculus also handles example 19.3.1 in a simple manner.[6]

19.3.2 DEFINITION OF A MINIMAL LOGIC

1. **Lexicon**

   Set of one-place predicates: {sleep, sing}
   Set of names: {Julia, Susanne}

2. **Model**

   A model $\mathcal{M}$ is a two-tuple (A, F), where A is a non-empty set of entities and F a denotation function (see 3).

---

[6] For simplicity, we do not use here a recursive definition of syntactic categories with systematically associated semantic types à la Montague. Cf. CoL, p. 344–349.

3. **Possible Denotations**

  a) If P_1 is a one-place predicate, then a possible denotation of P_1 relative to a model $\mathcal{M}$ is a subset of A. Formally, $F(P_1)\mathcal{M} \subseteq A$.

  b) If α is a name, then the possible denotations of α relative to a model $\mathcal{M}$ are elements of A. Formally, $F(\alpha)\mathcal{M} \in A$.

  c) If ϕ is a sentence, then the possible denotations of ϕ relative to a model $\mathcal{M}$ are the numbers 0 and 1, interpreted as the truth values 'true' and 'false.' Formally, $F(\phi)\mathcal{M} \in \{0,1\}$.

Relative to a model $\mathcal{M}$ a sentence ϕ is a true sentence, if and only if the denotation ϕ in $\mathcal{M}$ is the value 1.

4. **Syntax**

  a) If P_1 is a one-place predicate and α is a name, then $P_1(\alpha)$ is a sentence.

  b) If ϕ is a sentence, then $\neg\phi$ is a sentence.

  c) If ϕ is a sentence and ψ is a sentence, then $\phi \,\&\, \psi$ is a sentence.

  d) If ϕ is a sentence and ψ is a sentence, then $\phi \vee \psi$ is a sentence.

  e) If ϕ is a sentence and ψ is a sentence, the $\phi \rightarrow \psi$ is a sentence.

  f) If ϕ is a sentence and ψ is a sentence, then $\phi = \psi$ is a sentence.

5. **Semantics**

  a) '$P_1(\alpha)$' is a true sentence relative to a model $\mathcal{M}$ if and only if the denotation of α in $\mathcal{M}$ is element of the denotation of P_1 in $\mathcal{M}$.

  b) '$\neg\,\phi$' is a true sentence relative to a model $\mathcal{M}$ if and only if the denotation of ϕ is 0 relative to $\mathcal{M}$.

  c) '$\phi \,\&\, \psi$' is a true sentence relative to a model $\mathcal{M}$ if and only if the denotations of ϕ and of ψ are 1 relative to $\mathcal{M}$.

  d) '$\phi \vee \psi$' is a true sentence relative to a model $\mathcal{M}$ if and only if the denotation of ϕ or ψ is 1 relative to $\mathcal{M}$.

  e) '$\phi \rightarrow \psi$' is a true sentence relative to a model $\mathcal{M}$ if and only if the denotation of ϕ relative to $\mathcal{M}$ is 0 or the denotation of ψ is 1 relative to $\mathcal{M}$.

  f) '$\phi = \psi$' is a true sentence relative to a model $\mathcal{M}$ if- and only if the denotation of ϕ relative to $\mathcal{M}$ equals the denotation of ψ relative to $\mathcal{M}$.

The rules of syntax (4) define the complex expressions of the logical language, those of the semantics (5) specify the circumstances under which these complex expressions are true.

The simple logic system 19.3.2 establishes a semantic relation between the formal language and the world by defining the two levels as well as the relation between them in terms of a metalanguage. The theory behind this method was presented by the Polish-American logician ALFRED TARSKI (1902–1983) in a form still valid today.

In the formal definition of an interpreted language, Tarski 1935 distinguishes between the object language and the metalanguage. The object language is the language to be semantically interpreted (e.g. quoted expressions like 'ϕ & ψ'), while the definitions of the semantic interpretation are formulated in the metalanguage. The metalanguage is presupposed to be known by author and reader at least as well or even better than their mother tongue because there should be no room at all for differing interpretations.

The metalanguage definitions serve to formally interpret the object language. In logical semantics the task of the interpretation is to specify under which circumstances the expressions of the object language are true. Tarski's basic metalanguage schema for characterizing truth is the so-called T-condition. According to Tarski, the T stands mnemonically for *truth*, but it could also be taken for *translation* or *Tarski*.

19.3.3 SCHEMA OF TARSKI'S T-CONDITION

T: x is a true sentence if and only if p.

The T-condition as a whole is a sentence of the metalanguage, which quotes the sentence x of the object language and *translates* it as p. Tarski illustrates this method with the following example:

19.3.4 INSTANTIATION OF TARSKI'S T-CONDITION

'Es schneit' is a true sentence if and only if it snows.

This example is deceptively simple, and has resulted in misunderstandings by many non-insiders.[7] What the provocative simplicity of 19.3.3 and 19.3.4 does not express when viewed in isolation is the exact nature of the *two-level structure* (cf. 19.1.1) which underlies all forms of semantic interpretation and therefore is also exemplified by the particular method proposed by Tarski.

A closer study of Tarski's text shows that the purpose of the T-condition is not a redundant repetition of the object language expression in the metalanguage translation. Rather, the T-condition has a twofold function. One is to construct a systematic connection between the object language and the world in terms of the metalanguage; thus, the metalanguage serves as the means for realizing the *assignment algorithm* in logical semantics. The other is to characterize truth: the truth-value of x in the object language is to be determined via the interpretation of p in the metalanguage.

Both functions require that the metalanguage can refer directly (i) to the object language and (ii) to the correlated state of affairs in the world (model). The connection between the two levels of the object language and the world established by the metalanguage is shown schematically in 19.3.5.

---

[7] A. Tarski 1944 complains about these misunderstandings and devotes the second half of his paper to a detailed critique of his critics.

19.3.5 RELATION BETWEEN OBJECT AND METALANGUAGE

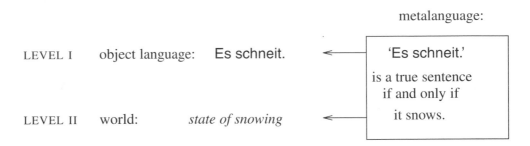

The direct relation of the metalanguage to the world is called *verification*. The verification of T consists in the ability to actually determine whether p holds or not. For example, in order to determine whether Es schneit is true or not, it must be possible to determine whether or not it actually snows. Without ensuring the possibility of verifying p, the T-condition is (i) vacuous for the purpose of characterizing truth (see 19.5.1 below) and (ii) dysfunctional for the purpose of assigning semantic objects.

That Tarski calls the p in the T-condition the 'translation' of the x is misleading because translation in the normal sense of the word is not concerned with truth at all. Instead a translation is adequate, if the speaker meaning₂ of the source and the target language expressions happen to be *equivalent*. For example, translating Die Katze ist auf der Matte as The cat is on the mat is adequate simply because the German source and the English target expression mean the same. There is obviously neither the need nor even the possiblity of verification outside a theory of truth.

Tarski, however, took it to be just as obvious that within a theory of truth the possibility of verification must hold. In contradistinction to 19.3.4, Tarski's scientific examples of semantically interpreted languages do not use some natural language as the metalanguage – because they do not ensure verification with sufficient certainty. Rather, Tarski insisted on carefully *constructing* special metalanguages for certain well-defined scientific domains for which the possibility of verification is guaranteed.

According to Tarski, the construction of the metalanguage requires that (i) all its basic expressions are explicitly listed and that (ii) each expression of the metalanguage *has a clear meaning* (op.cit., p. 172). This conscientious formal approach to the metalanguage is exemplified in Tarski's 1935 analysis of the calculus of classes, which illustrates his method in formal detail. The only expressions used by Tarski in this example are notions like not, and, is contained in, is element of, individual, class, and relation. The meaning of these expressions is immediately obvious insofar as they refer to the most basic mathematical objects and set-theoretic operations. In other words, Tarski limits the construction of his metalanguage to the elementary notions of a fully developed (meta)theory, e.g. a certain area in the foundations of mathematics.

The same holds for the semantic rules in our example 19.3.2, for which reason it constitutes a well-defined Tarskian semantics. The semantic definition of the first rule[8] is shown in 19.3.6 as a T-condition like 19.3.5.

19.3.6 T-CONDITION IN A LOGICAL DEFINITION

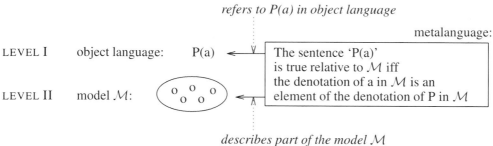

The possibility to verify the T-condition 19.3.6 is guaranteed by no more and no less than the fact that for any given model $\mathcal{M}$ anyone who speaks English and has some elementary notion of set theory can *see* (in the mathematical sense of 'unmittelbare Anschauung' or immediate obviousness) whether the relation specified in the translation part of T holds in $\mathcal{M}$ or not.

The appeal to immediate obviousness has always served as the ultimate justification in the history of mathematics:

> En l'un les principes sont palpables mais éloignés de l'usage commun de sorte qu'on a peine à tourner late tête de ce côte-la, manque d'habitude : mais pour peu qu'on l'y tourne, on voit les principes à peine; et il faudrait avoir tout à fait l'esprit faux pour mal raisonner sur des principes si gros qu'il est presque impossible qu'ils échappent.
> [In [the mathematical mind] the principles are obvious, but remote from ordinary use, such that one has difficulty to turn to them for lack of habit : but as soon as one turns to them, one can see the principles in full; and it would take a thoroughly unsound mind to reason falsely on the basis of principles which are so obvious that they can hardly be missed.]
>
> B. PASCAL (1623 -1662), *Pensées*, 1951:340

In summary, Tarski's theory of truth and logical semantics is clearly limited to those domains which provide sufficiently certain methods of verification, such as mathematics, logic, and natural science.

19.4 Metalanguage-based versus procedural semantics

In contrast to the semantic definitions in 19.3.2, which use only immediately obvious logical notions and are therefore legitimate in the sense of Tarski's method, the following instantiation of the T-conditions violates the precondition of verifiability.

---

[8] Compared to 19.3.5, 19.3.6 is more precise because the interpretation is explicitly restricted to a specific state of affairs, specified formally by the model $\mathcal{M}$. In a world where it snows only at certain times and certain places, on the other hand, 19.3.5 will work only if the interpretation of the sentence is limited – at least implicitly – to an intended location and moment of time.

19.4.1 EXAMPLE OF A VACUOUS T-CONDITION

'A is red' is a true sentence if and only if A is red.

This instantiation of the T-condition is formally correct, but vacuous because it does not relate the meaning of the object language expression red to some verifiable concept of the metalanguage. Instead the expression of the object language is merely repeated in the metalanguage.[9]

Within the boundaries of its set-theoretic foundations, model-theoretic semantics has no way of providing a truth-conditional analysis for content words like red such that its meaning would be characterized adequately in contradistinction to, e.g. blue. There exists, however, the possibility of *extending the metatheory* by calling in additional sciences such as physics.

From such an additional science one may select a small set of new basic notions to serve in the extended metalanguage. The extended metalanguage functions properly, if the meaning of the additional expressions is immediately obvious within the extended metatheory.

In this way we might improve the T-condition 19.4.1 as follows:

19.4.2 IMPROVED T-CONDITION FOR red

'A is red' is a true sentence if and only if A refracts light in the electromagnetic frequency interval between α and β.

Here the metalanguage translation relates the object-language expression red to more elementary notions (i.e., the numbers α and β within an empirically established frequency scale and the notion of refracting light which is well-understood in the domain of physics) and thus succeeds in characterizing the expression in a non-vacuous way which is moreover objectively verifiable.

Examples like 19.4.1 show that the object-language may contain sentences for which there are only vacuous translations in the given metalanguage. This does not mean that a sentence like 'x is red' is not meaningful or has no truth-value. It only means that the metalanguage is not rich enough to provide the basis for an immediately obvious verification of the sentence. This raises the question of how to handle the semantics of the metalanguage, especially with respect to its unanalyzed parts, i.e., the parts which go beyond the elementary notions of its metatheory.

In answer to this question Tarski constructed an infinite hierarchy of metalanguages.

---

[9] Tarski's own example 19.3.4 is only slightly less vacuous. This is because the metalanguage translation in Tarski's example is in a natural language *different* from the object language. The metalanguage translation into another natural language is misleading, however, because it omits the aspect of verification, which is central to a theory of truth. The frequent misunderstandings which Tarski 1944 so eloquently bewails may well have been caused in large part by the 'intuitive' choice of his examples.

19.4.3 HIERARCHY OF METALANGUAGES

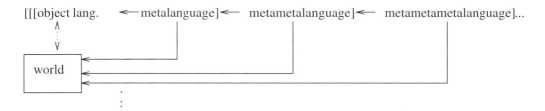

Here some vacuous T-conditions of the metalanguage, e.g. 19.4.1, are repaired by nonvacuous T-conditions, e.g. 19.4.2, in the metametalanguage. Some holes in the metametalanguage are filled in turn by means of additional basic concepts in the metametametalanguage, etc. That an infinite hierarchy of metalanguages makes total access to truth ultimately impossible, at least for mankind, is not regarded as a disadvantage of Tarski's construction – on the contrary, it constitutes a major part of its philosophical charme.

For the semantics of programming and natural languages, however, a hierarchy of metalanguages is not a suitable foundation.[10] Consider for example the rules of basic addition, multiplication, etc. The problem is not to provide an adequate metalanguage definition for these rules, similar to 19.3.2. Rather, the road from such a metalanguage definition to a working calculator is quite long and in the end the calculator will function mechanically – without any reference to these metalanguage definitions and without any need to understand the metalanguage.[11]

This simple fact has been called the *autonomy from the metalanguage*.[12] It is characteristic of all programming languages. Autonomy from the metalanguage does not mean that computers would be limited to uninterpreted, purely syntactic deduction systems, but rather that Tarski's method of semantic interpretation is not the only one possible. Instead of assigning semantic representations to an object language by means of a metalanguage, computers use an operational method in which the notions of the programming language are realized automatically as machine operations.

---

[10] The discussion of Tarski's semantics in CoL, pp. 289-295, 305-310, and 319-323, was aimed at bringing out as many similarities between the semantics of logical, programming, and natural languages as possible. For example, all three types of semantic interpretation were analyzed from the viewpoint of truth: whereas logical semantics *checks* whether a formula is true relative to a model or not, the procedural semantics of a programming language *constructs* machine states which 'make the formula true,' – and similarly in the case of natural semantics. Accordingly, the reconstruction of logical calculi on the computer was euphemistically called 'operationalizing the metalanguage'.

  Further reflection led to the conclusion, however, that emphasizing the similarities was not really justified: because of the differing goals and underlying intuitions of the three types of semantics a general transfer from one system to another is ultimately impossible. For this reason the current analysis first presents what all semantically interpreted systems have in common, namely the basic two-level structure, and then concentrates on bringing out the formal and conceptual differences between the three systems.

[11] See in this connection also 3.4.3.

[12] CoL, p. 307 ff.

Because the semantics of programming languages is procedural (i.e. metalanguage-independent), while the semantics of logical calculi is Tarskian (i.e. metalanguage-dependent), the reconstruction of logical calculi as computer programs is at best difficult.[13] If it works at all, it usually requires profound compromises on the side of the calculus – as illustrated, for example, by the computational realization of predicate calculus in the form of Prolog.

Accordingly, there exist many logical calculi which have not been and never will be realized as computer programs. The reason is that their metalanguage translations contain parts which may be considered immediately obvious by their designers (e.g. quantification over infinite sets of possible worlds in modal logic), but which are nevertheless unsuitable to be realized as empirically meaningful mechanical procedures.

Thus, the preconditions for modeling a logical calculus as a computer program are no different from non-logical theories such as physics or chemistry: the basic notions and operations of the theory must be sufficiently clear and simple to be realized as electronic procedures which are empirically meaningful and can be computed in a matter of minutes or days rather than centuries (cf. 8.2.2).

19.5 Tarski's problem for natural language semantics

Because the practical use of programming languages requires an automatic interpretation in the form of corresponding electronic procedures they cannot be based on a metalanguage-dependent Tarski semantics. But what about using a Tarski semantics for the interpretation of natural languages?

Tarski himself leaves no doubt that a complete analysis of natural languages is in principle impossible within logical semantics.

> The attempt to set up a structural definition of the term 'true sentence' – applicable to colloquial language – is confronted with insuperable difficulties.
>
> A. Tarski 1935, p. 164.

Tarski proves this conclusion on the basis of a classical paradox, called the Epimenides, Eubolides, or liar paradox.

The paradox is based on self-reference. Its original 'weak' version has the following form: if a Cretan says **All Cretans (always) lie** there are two possibilities. Either the Cretan speaks truly, in which case it is false that *all* Cretans lie – since he is a Cretan himself. Or the Cretan lies, which means that there exists at least one other Cretan who does not lie. In both cases the sentence in question is false.[14]

Tarski uses the paradox in the 'strong' version designed by Leśniewski and constructs from it the following proof that a complete analysis of natural language within logical semantics is necessarily impossible.

---

[13] With the notable exception of propositional calculus. See also *transfer* in 19.2.3.

[14] For a detailed analysis of the weak version(s) see C. Thiel, 1995, p. 325–7.

For the sake of greater perspicuity we shall use the symbol 'c' as a typological abbreviation of the expression 'the sentence printed on page 384 , line 4 from the top.' Consider now the following sentence:

c is not a true sentence

Having regard to the meaning of the symbol 'c', we can establish empirically:
(a) 'c is not a true sentence' is identical with c.
For the quotation-mark name of the sentence c we set up an explanation of type (2) [i.e. the T-condition 19.3.3]:
(b) 'c is not a true sentence' is a true sentence if and only if c is not a true sentence.
The premise (a) and (b) together at once give a contradiction:
c is a true sentence if and only if c is not a true sentence.

A. Tarski 1935

In this construction, self-reference is based on two preconditions. First, a sentence located in a certain line on a certain page, i.e. line 4 from the top on page 384 in the current Chapter 19, is abbreviated as 'c'.[15]

Second, the letter 'c' with which the sentence in line 4 from the top on page 384 is abbreviated also occurs in the unabridged version of the sentence in question. This permits to substitute the c in the sentence by the expression which the 'other' c abbreviates. The substitution is schematically described in 19.5.1.

19.5.1 Leśniewski's reconstruction of the Epimenides

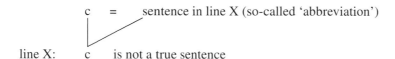

substitution of the c-abbreviation: sentence in line X is not a true sentence

If the sentence in line X is not true, then it holds that it is not the case that c is not a true sentence. It follows from this double negation that c is a true sentence. Thus, it holds both, 'c is not a true sentence' as the original statement and 'c is a true sentence' as obtained via substitution and its interpretation.

To prove that a logical semantics for natural language is impossible, Tarski combines Leśniewski's version of the Epimenides paradox with his T-condition. In this way he turns an isolated paradox into a contradiction of logical semantics.

19.5.2 Inconsistent T-condition using Epimenides paradox

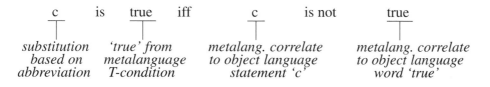

There are three options to avoid this contradiction in logical semantics.

The first option consists in forbidding the abbreviation and the substitution based on it. This possibility is rejected by Tarski because "no rational ground can be given why substitution should be forbidden in general."

The second option consists in distinguishing between the truth predicate $true^m$ of the metalanguage and $true^o$ of the object language.[16] On this approach

c is $true^m$ if and only if c is not $true^o$

is not contradictory because $true^m \neq true^o$. Tarski does not consider this option, presumably because the use of more than one truth predicate runs counter to the most fundamental goal of logical semantics, namely a formal characterization of *the* truth.

The third option, chosen by Tarski, consists in forbidding the use of truth predicates in the object language. For the original goals of logical semantics this third option poses no problem. Characterizing scientific theories like physics as true relations between logical propositions and states of affairs does not require a truth predicate in the object language. The same holds for formal theories like mathematics.

Furthermore, for many mathematical logicians there is no reason to reconstruct the object language as a natural language. On the contrary, the development of semantically interpreted logical calculi is motivated by their desire to avoid the vagueness and contradictions of the natural languages. G. Frege 1896 (1967, p. 221) expresses this sentiment as follows:

> Der Grund, weshalb die Wortsprachen zu diesem Zweck [d.h. Schlüsse nur nach rein logischen Gesetzen zu ziehen] wenig geeignet sind, liegt nicht nur an der vorkommenden Vieldeutigkeit der Ausdrücke, sondern vor allem in dem Mangel fester Formen für das Schließen. Wörter wie >also<, >folglich<, >weil< deuten zwar darauf hin, daß geschlossen wird, sagen aber nichts über das Gesetz, nach dem geschlossen wird, und können ohne Sprachfehler auch gebraucht werden, wo gar kein logisch gerechtfertigter Schluß vorliegt.
> [The reason why the word languages are suited little for this purpose [i.e., draw inferences based on purely logical laws] is not only the existing ambiguity of the expressions, but mainly the lack of clear forms of inference. Even though words like 'therefore,' 'consequently,' 'because' indicate inferencing, they do not specify the rule on which the inference is based and they may be used without violating the well-formedness of the language even if there is no logically justified inference.]

This long-standing, widely held view provides additional support to refrain from any attempt of applying a Tarski semantics to natural language.

If logical semantics is nevertheless applied to natural language, however, the third option does pose a serious problem. This is because the natural languages *must*[17]

---

[15] The page and line numbers have been adjusted from Tarski's original text to fit those of this chapter. This adjustment is crucial in order for self reference to work.

[16] The second option will be explored in Chapter 21, especially Section 21.2, for the semantics of natural language.

[17] This follows from the role of natural languages as the pretheoretical metalanguage of the logical languages. Without the words true and false in the natural languages a logical semantics couldn't be defined in the first place.

contain the words **true** and **false**. Therefore a logical semantic interpretation of a natural (object-)language in its entirety will unavoidably result in a contradiction.

Tarski's student RICHARD MONTAGUE (1930–1970), however, was undaunted by Tarski's proof and insisted on applying logical semantics to natural language.

> I reject the contention that an important theoretical difference exists between formal and natural languages. ... Like Donald Davidson I regard the construction of a theory of truth – or rather the more general notion of truth under an arbitrary interpretation – as the basic goal of serious syntax and semantics.
>
> R. Montague 1970, *"English as a formal language"*[18]

We must assume that Montague knew the Epimenides paradox and Tarski's related work. But in his papers on the semantics of natural languages Montague does not mention this topic at all. Only Davidson, who Montague refers to in the above quotation, is explicit:

> Tarski's ... point is that we should have to reform natural language out of all recognition before we could apply formal semantic methods. If this is true, it is fatal to my project.
>
> D. Davidson 1967

A logical paradox is fatal because it destroys a semantical system. Depending on which part of the contradiction an induction starts with, one can always prove both, a theorem and its negation. And this is not acceptable for a theory of truth.[19]

Without paying much attention to Tarski's argument, Montague, Davidson, and many others insist on using logical semantics for the analysis of natural language. This is motivated by the following parochial prejudices and misunderstandings.

For one, the advocates of logical semantics for natural language have long been convinced that their method is the best-founded form of semantic interpretation. Because they see no convincing alternatives to their metalanguage-dependent method – despite calculators and computers – they apply logical semantics in order to arrive at least at a partial analysis of natural language meaning within formal semantics.

A second reason is the fact that the development of logic began with the description of selected natural language examples. After a long independent evolution of logical systems it is intriguing to apply them once more to natural languages[20] in order to show which aspects of natural language semantics can be easily modeled within logic.

---

[18] R. Montague 1974, p.188.

[19] As a compromise, Davidson suggested to limit the logical semantic analysis of natural language to suitable consistent fragments. This means, however, that the project of a complete logical semantics for natural language is doomed to fail.

Attempts to avoid the Epimenides paradox in logical semantics are S. Kripke 1975, A. Gupta 1982, and H. Herzberger 1982. These systems each define an artificial object language (first order predicate calculus) with truth predicates. That this object language is nevertheless consistent is based on defining the truth predicates as *recursive valuation schemata*.

Recursive valuation schemata are based on a large number of valuations (transfinitely many in the case of Kripke 1975). As a purely technical trick, they miss the point of the Epimendes paradox which is essentially a problem of reference: a symbol may refer on the basis of its meaning and at the same time be a referent on the basis of its form.

[20] As an L→N reconstruction, cf. 19.3.3.

A third reason is that natural languages are often viewed as defective because they can be misunderstood and – in contrast to the logical calculi – implicitly contradictory. Therefore, the logical analysis of natural language has long been motivated by the goal to systematically expose erroneous conclusions in rhetorical arguments in order to arrive at truth. What is usually overlooked, however, is that the interpretation of qnatural languages work quite differently from the metalanguage-dependent logical languages.

Exercises

Section 19.1

1. Which areas of science deal with semantic interpretation?
2. Explain the basic structure common to all systems of semantic interpretation.
3. Name two practical reasons for building semantic structures indirectly via the interpretation of syntactically analyzed surfaces.
4. Explain the inverse procedures of representation and reconstruction.
5. How many kinds of semantic interpretation can be assigned to a given language?
6. Explain in which sense axiomatic systems of deduction are not true systems of semantic interpretation.

Section 19.2

1. Describe three different types of formal semantics.
2. What purpose is served in programming languages by the level of syntactically analyzed surfaces?
3. On which principle is the semantics of programming languages based and how does it differ from that of logical languages?
4. What is the basic difference between the semantics of natural languages, on the one hand, and the semantics of logical and programming languages, on the other?
5. Why is a syntactical analysis presupposed by a formal semantic analysis?
6. Discuss six possible relations between different types of semantics.
7. What kind of difficulties arise in the replication of logical proof theory as computer programs for automatic theorem provers?

Section 19.3

1. Name the components of a model-theoretically interpreted logic and explain their function.
2. What are the goals of logical semantics?
3. What is Tarski's T-condition and what is its purpose for semantic interpretation?
4. Why is verification a central part of Tarski's theory of truth?
5. What is the role of translation in Tarski's T-condition?
6. Why does Tarski *construct* the metalanguage in his example of the calculus of classes? Which notions does he use in this construction?
7. What does immediate obviousness do for verification in mathematical logic?

Section 19.4

1. Explain a vacuous T-condition with an example.
2. What is the potential role of non-mathematical sciences in Tarski's theory of truth?
3. For what purpose does Tarski construct an infinite hierarchy of metalanguages?
4. Why is the method of metalanguages unsuitable for the semantic interpretation of programming languages?
5. What is the precondition for realizing a logical calculus as a computer program?
6. In what sense does Tarski's requirement that only immediately obvious notions may be used in the metalanguage have a counterpart in the procedural semantics of the programming languages?

Section 19.5

1. How does Tarski view the application of logical semantics to the analysis of natural languages?
2. Explain the Epimenides paradox.
3. Explain the three options to avoid the inconsistency of logical semantics caused by the Epimenides paradox.
4. What is the difference between a false logical proposition like 'A & ¬A' and a logical inconsistency caused by a paradox?
5. What difference does Montague see between the artificial and the natural languages, and what is his goal in the analysis of natural languages?
6. Name three reasons for applying logical semantics to natural languages. Are they valid?

20. Truth, meaning, and ontology

This chapter explains how a theory of truth and a theory of meaning are related in the logical semantics of natural language. It presents four basic types of ontologies for theories of semantics, and shows that the ontology presumed plays an important role for the analysis of intensional contexts, propositional attitudes, and vagueness.

Section 20.1 describes natural language phenomena which have motivated extensions of logical semantics, in particular Frege's distinction between sense and reference. Section 20.2 explains Carnap's reconstruction of Frege's sense as formal functions called intensions. Section 20.3 shows why the phenomenon of propositional attitudes is incompatible with the ontological assumptions of logical truth conditions. Section 20.4 explains why there are in principle four basic types of ontology for theories of semantic interpretation. Section 20.5 describes the basic concepts of many-valued logics and shows how the choice of a particular ontology can greatly influence the outcome of a formal analysis, using the phenomenon of vagueness as an example.

20.1 Analysis of meaning in logical semantics

The application of logical semantics to natural language is intended to characterize meaning. It is not obvious, however, how a theory of truth is supposed to double as a theory of meaning. To indirectly motivate the logical semantics of natural language as a theory of meaning, the following principle is used.

20.1.1 THE MEANING PRINCIPLE OF LOGICAL SEMANTICS

> If a speaker-hearer knows the meaning of a sentence, (s)he can say for any
> state of affairs whether the sentence is true or false with respect to it.

This principle equates the intuitive knowledge of a natural sentence meaning with the knowledge of its truth conditions. In logical semantics, the meanings of natural sentences are characterized by describing the sentences' truth conditions.

However, if the meanings of the natural (meta-)language are used to define the truth conditions of logical semantics, and the truth conditions of logical semantics are used to define the meanings of the natural (object) language, then there arises the danger of a vicious circle. Such a circle can only be avoided, if the meaning analysis within logical semantics is limited to *reducing* complex natural language meanings to elementary logical notions.

For this, the logical notions must be restricted explicitly to a small, well-defined arsenal of elementary meanings which are presupposed to be immediately obvious – in accordance with Tarski's requirement for the metalanguage (cf. Section 19.4). The natural language fragment (object language) is analyzed semantically by translating it into the logical language with the goal of bringing out the truth-conditional aspects of the natural language as much as possible.[1]

The logical analysis of natural language has repeatedly brought to light properties which seemed puzzling or paradoxical from a logical point of view. This has evoked ambivalent reactions among the logicians involved.

On the one hand, they felt confirmed in their view that natural languages are imprecise, misleading, and contradictory.[2] On the other hand, some of these phenomena were taken as a challenge to expand logical systems so that they could handle selected natural language puzzles and explain them in terms of their extended logical structure (N→L reconstruction, cf. Section 19.2).

The discrepancies between the intuitive assumptions of logical semantics and the meaning structures of natural language as well as attempts at overcoming them at least in part are illustrated by a classical example from the history of logic. It is based on two rules of inference developed from Aristotelian logic by medieval scholasticists, namely *existential generalization* and *substitutivity of identicals*.

According to the rule of existential generalization it follows from the truth of a proposition F(a,b) that a exists and that b exists. For example, the sentence Julia kissed Richard is analyzed semantically as a *kiss*-relation between the entities Julia and Richard. If Julia kissed Richard is true, then it must be true that Julia exists and Richard exists.

The rule of substitutivity of identicals says that, given the premises F(b) and b = c, F(b) implies F(c). For example, if Richard = Prince of Burgundy, then the truth of the sentence Julia kissed Richard implies the truth of the sentence Julia kissed the Prince of Burgundy. This substitutivity of Richard and Prince of Burgundy *salva veritate*, i.e. preserving the truth-value, is based on the fact that these two different expressions denote the same object.

But what about the following pairs of sentences?

---

[1] Thereby it was initially considered sufficient to assign logical translations informally to suitably selected examples, based on intuition. W.v.O. Quine 1960, for example, presented the formula $\wedge x[raven(x) \rightarrow black(x)]$ as the semantic representation of the sentence All ravens are black, using his native speaker understanding of English and his knowledge of logic (informal translation).

A rigorously formal method of analyzing the truth-conditional aspect of natural languages was pioneered by Richard Montague. In Montague grammar, syntactically analyzed expressions of natural language are mapped systematically into equivalent logical formulas by means of a well-defined algorithm (formal translation).

[2] It would be a worthwhile topic in history of science to collect and classify statements to this effect in the writings of Frege, Russell, Tarski, Quine, etc.

20.1.2 VALID AND INVALID INSTANCES OF EXISTENTIAL GENERALIZATION

1) Julia finds a unicorn. > A unicorn exists.
2) Julia seeks a unicorn. ≯ A unicorn exists.

The symbols > and ≯ represent *implies* and *doesn't imply*, respectively. The premises in these two examples have exactly the same syntactic structure, namely F(a,b). The only difference consists in the choice of the verb. Yet in (1) the truth of the premise implies the truth of the consequent, in accordance with the rule of existential generalization, while in (2) this implication does not hold.

Example (2) raises the question of how a relation can be established between a subject and an object if the object does not exist. How can Julia seeks a unicorn be grammatically well-formed, meaningful, and even true under realistic circumstances? Part of the solution consisted in specifying certain environments in natural sentences in which the rule of existential generalization does not apply, e.g., in the scope of a verb like seek. These environments are called the *uneven* (Frege 1892), *opaque* (Quine 1960), or *intensional* (Montague 1974) *contexts*.

This solves only part of the puzzle, however. If the meaning of a unicorn is not an object existing in reality, what else could it be? And how should the difference in the meaning of different expressions for non-existing objects, such as square circle, unicorn, and Pegasus, be explained within the logical framework?

The necessity of distinguishing between these meanings follows from the second inference rule, the substitutivity of identicals. For example, if we were to use the empty set as the referent of square circle, unicorn, and Pegasus in order to express that no real objects correspond to these terms, then the truth of Julia seeks a unicorn would imply the truth of Julia seeks Pegasus and Julia seeks the square circle because of the substitutivity of identicals. Such a substitution would violate the intuitive fact that the resulting sentences have clearly non-equivalent meanings – in contrast to the earlier the Richard/Prince of Burgundy examples.

The non-equivalence of Julia seeks a unicorn and Julia seeks a square circle leads to the conclusion that in addition to the real objects in the world there also exist natural language meanings which are independent of their referents. This was recognized in Frege's 1892 distinction between *sense* (Sinn) and *reference* (Bedeutung).

Frege concluded from examples like 20.1.2 that all expressions of language have a meaning (sense), even square circle and Pegasus. Based on their meaning some of these expressions, e.g. Julia, refer to real objects (referents), whereas others, e.g. a unicorn, happen to have no referent.[3] In this way a sentence like Julia seeks a unicorn can be properly explained: the proposition establishes a relation between the real individual denoted by Julia and the meaning (sense) of a unicorn.

Frege's move to distinguish the meaning of language expressions (sense) and the object referred to (referents) is correct from the viewpoint of natural language analysis,

---

[3] G. Frege proposed to use the empty set as the referent for non-referring expressions – to make referring a total function. This was declared *plainly artificial* by B. Russell 1905.

but dangerous for the main concern of analytic philosophy, i.e. the characterization of truth. Whereas (i) the signs of language and (ii) the objects in the world are real, concrete, and objective, this does not hold in the same way for the entities called meaning (sense). As long as it is not completely clear in what form these meanings exist they pose an ontological problem.

From the ontological problem there follows a methodological one: how can a logical theory of truth arrive at reliable results if it operates with concepts (senses) the nature of which has not been clearly established? Doesn't the everyday use of language meanings show again and again that utterances may be understood in many different ways? This 'arbitrariness' of different speaker-hearer meanings threatens to compromise the main concern of philosophical logic.

Frege's way out was to attribute a similar form of existence to the meanings of natural language as to the numbers and their laws in mathematical realism. Mathematical realism proceeds on the assumption that the laws of mathematics exist even if no one knows about them; mathematicians *discover* laws which have extemporal validity.[4] Frege supposed the meanings of natural language to exist in the same way, i.e., independently of whether there are speakers-hearers who have discovered them and use them more or less correctly.

20.2 Intension and extension

Frege's proposal for making language meaning 'real' was successful insofar as it rekindled interest in solving puzzles of natural language. On the one hand, there were attempts to handle the phenomena treated by Frege without postulating a sense (B. Russell 1905, W.v.O. Quine 1960). On the other hand, there were attempts to reconstruct the notion of sense logically.

A highlight among the efforts of the second group is Carnap's reconstruction of Frege's sense as formal intensions. R. Carnap 1947 proceeds on the assumption that the meaning of **sleep**, for example, is the set of sleepers (as in 19.3.1). In addition, Carnap uses the fact that the elements of this set may vary in time: two individuals awake, another one falls asleep, etc. Also, the set of sleepers may vary from one place to another.

Using the set I of different points in time and the set J of different places, Carnap constructs an index (i,j), with $i \in I$ and $j \in J$. Then he defines the *intension* of a word or expression as a function from $I \times J$ into 2^A, i.e. the power set over the set of entities A. Carnap calls the value of an intension at an index the *extension*. The set-theoretic type of an extension depends on whether the expression is a sentence, a proper name, or a predicate.

---

[4] The counterposition to mathematical realism is constructivism according to which a mathematical law begins to exist only at the point when it has been constructed from formal building blocks and their inherent logic by mathematicians.

20.2.1 EXAMPLES OF CARNAP'S *Intensions*

intension

proposition: $I \times J \rightarrow \{0,1\}$

extension

intension

proper name: $I \times J \rightarrow a \in A$

extension

intension

1-pl. predicate: $I \times J \rightarrow \{a1, a2, ..\} \subseteq A$

extension

The notion intension refers to the function as a whole. Its domain is $I \times J$ and its range is called extension. The extensions of propositions are the truth values defined as elements of $\{0,1\}$, the extensions of proper names are objects defined as elements of A, and the extensions of predicates are relations defined as subsets of A.

Carnap's systematic variation of denotations relative to different indices required an extended definition of the logical language (as compared to, e.g., 19.3.2). To this purpose the model $\mathcal{M}$ is expanded into a so-called model structure $\mathcal{MS}$. A model structure consists of a set of models such that each index is associated with a model specifying the extension of all words of the logical language at that index.

In order to determine the truth-value of a sentence relative to a model structure and an index, the model structure must be explicitly defined. Even for a very small model structure this definition is extremely cumbersome and complex. Moreover, the formal interpretation of a sentence relative to such a definition produces nothing that isn't already known to begin with: the truth value of the sentence is an exact reflection of what the logician defined in the model structure at the index in question.

That model structures have no practical purpose is not regarded as a drawback by model-theoretic semanticists, however. Their goal is not a complete model of the world in order to determine the actual truth-value of sentences, but rather to design a logical semantics in which the truth-value of arbitrary well-formed sentences *could in principle* be determined formally if the model structure were defined.

From this viewpoint, Carnap's formal intensions fulfill many desiderata of Frege's sense. Intensions characterize the meaning (sense) of an expression insofar as anyone who knows the metalanguage definition of the model structure and the metalanguage definition of the logical language can determine the extension of the expression at any index. Furthermore, intensions serve as denotations in intensional contexts: the sentence Julia seeks an apple, for example, may be analyzed as a *seek*-relation between the extension of Julia and the intension of an apple, irrespective of whether or not there exist any apples at the index in question.

In addition, Carnap uses the index parameters I and J for the definition of modal and temporal operators. Based on the temporal parameter J, the past operator H is defined as follows: a proposition H(p) is true relative to a model structure $\mathcal{MS}$ and an index (i,j) if there exists an index j', j' < j, and p is true relative to $\mathcal{MS}$ and (i,j'). And accordingly for future operator W (temporal logic).

Furthermore, by generalizing the location parameter I to range over 'possible worlds,' Carnap defines the modal operators for necessity $\square$ and possibility $\diamond$. In modal logic, two expressions are treated as necessarily equivalent, iff their extensions are the same in *all* possible worlds; two expressions are possibly equivalent, iff there is *at least one* possible world in which they have the same extension.

By defining intensions in this way, Carnap managed to avoid some of the problems with substitution *salva veritate*. Assume, for example, that neither apples nor pears exist at an index (i,j), such that the extensions of the notions **apple** and **pear** are the same at (i,j), namely the empty set. Because the distribution of apples and pears is not necessarily (i.e., not in all possible worlds) the same, they have different intensions. Therefore the truth of **Julia seeks an apple** relative to (i,j) does not imply the truth of **Julia seeks a pear** relative to (i,j) – provided that substitutivity of identicals applies here at the level of intensions, as it should in the intensional context created by **seek**.

This type of analysis has been extended to prevent expressions for fictitious objects such as **unicorn** or **Pegasus** to have the same meaning. While admitting that neither unicorns nor Pegasus exist in the real world, it is argued that their non-existence is not *necessary* (i.e., does not hold in *all* possible worlds).

Thus, given that the expressions in question have meaning it may be assumed that there is at least one 'non-actual' possible world in which unicorns and Pegasus exist and have different extensions, ensuring different intensions. It is this use of possible worlds which was especially offensive to, e.g., W.v.O. Quine 1960 on ontological grounds.

Even though Carnap's formal notion of intension[5] exhibits properties with regard to substitutivity of identicals and existential generalization which are similar to Frege's notion of sense, the two underlying theories are profoundly different, both in terms of structure and content. This difference is characterized schematically in 20.2.2 using the binary feature [±sense].

20.2.2 TWO APPROACHES TO MEANING

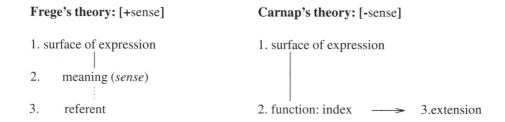

Frege's theory: [+sense]	Carnap's theory: [-sense]
1. surface of expression	1. surface of expression
2. meaning (*sense*)	
3. referent	2. function: index ⟶ 3.extension

Frege's approach is [+sense] because it uses a separate level for the meaning of language. The surface and the meaning of expressions form a fixed unit which faces the level of referents.[6] In this respect, Frege's approach and the [2+1] level structure of the SLIM theory of language are similar.

Carnap's approach is [–sense] because expressions refer directly to the 'world.' Apart from the definition of a few additional operators, the only difference between Carnap's intensional logic and the extensional system defined in 19.3.2 consists in the fact that for Carnap the world is represented not just by a single model $\mathcal{M}$ but rather by a model structure $\mathcal{MS}$. The model structure represents different states of the world as a multitude of models which have different indices. The indices provide the formal domain for functions which Carnap calls intensions.

Though the relation between intensions and the intended referents is treated in Carnap's system, it is by metalanguage definition only. The meaning (intension) refers to a referent at a given index by using the index as the argument of the intension function, thus rendering the associated extension as the value. Because Carnap defines extensions solely in terms of set-theoretic structures, he does not characterize the meaning of expressions like **square** (cf. 4.2.2), **triangle**, **red**, or **blue** any more than Frege.

20.3 Propositional attitudes

Despite the successful treatment of intensional contexts, temporal and modal operators, and a number of other puzzles in logical semantics there remain two basic problems which in principle cannot be solved within this framework, namely

– the Epimenides paradox (cf. Section 19.5) and
– the problem of propositional attitudes.

Propositional attitudes are expressed by sentences which describe the relation between a cognitive agent and a propositional content. For example, the sentence
  Suzanne believes that Cicero denounced Catiline.
expresses the propositional attitude of *belief* as a relation between Suzanne and the

---

[5] Carnap's formal notion of intensions plays a central role in Montague grammar. There, analyzed language expressions are formally translated into a typed lambda calculus. This *intensional logic* was designed by Montague to accommodate many traditional and new puzzles of natural language in a clean formal fashion within logical semantics.

   For example, in his "Proper treatment of quantification in ordinary English" (PTQ), Montague presents and solves the new puzzle **The temperature is 30 and rising**: the extension of temperature equals the extension of a certain number, here 30; because a number like 30 cannot increase, the predicate **rise** is applied not to the extension of **temperature**, but to the intension!

[6] According to Frege, expressions of language refer by means of their sense. Frege never specified how the meaning (sense) of an expression should exactly be defined and how reference to the intended object should exactly be guided by the properties of the meaning. But the main point is that his general approach provides a certain structural correlation between the surface, the meaning, and the referent.

proposition Cicero denounced Catiline. What are the truth conditions of propositional attitudes?

According to the intuitions of modal logic, a proper name denotes the same individual in all possible worlds (rigid designator).[7] For example, because Cicero and Tullius are names for one and the same person it holds necessarily (i.e, in all possible worlds) that Cicero = Tullius. Therefore, it follows necessarily from the truth of Cicero denounced Catiline that Tullius denounced Catiline.

However, if one of these sentences is embedded under a predicate of propositional attitude, e.g. believe, the substitution *salva veritate* is not valid even for proper names. Thus, according to intuition, Suzanne believes that Cicero denounced Catiline does not imply that Suzanne believes that Tullius denounced Catiline. Even though the referent of Cicero is necessarily identical with the referent of Tullius, it could be that Suzanne is not aware of this. Accordingly, a valid substitution *salva veritate* would require in addition the truth of Suzanne believes that Cicero is Tullius.

Because different human beings may have different ideas about the external reality, a treatment of propositional attitudes in the manner of Carnap and Montague would have to model not only the realities of natural science, but also the belief structures of all the individual speakers-hearers.[8] Such an attempt at handling this particular natural language phenomenon by yet another extension of model-theoretic semantics, however, would violate the basic assumptions of a theory of truth.

As shown in connection with Tarski's semantics in Section 19.3, a theory of truth is not only concerned with a formal definition of implications, but just as much with the verification of the premises of those implications. Only if the second aspect is fulfilled, is it possible to establish a true relation between a language and the world.

In order to determine what an individual believes, however, one is dependent on what the individual chooses to report whereby it cannot be checked objectively whether this is true or not. For this reason, individual 'belief-worlds' have justly been regarded as a prime example of what lies outside any scientific approach to truth.[9]

The phenomenon of propositional attitudes raises once more the fundamental question of logical semantics for natural language (cf. Section 20.1), namely

---

[7] Compare S. Kripke's 1972 model-theoretic treatment of proper names with their SLIM-theoretic treatment in terms of internal name markers (Sections 6.1, 6.4, and 6.5).

[8] In purely formal terms one could define a 'believe-operator' $\mathcal{B}$ as follows:

$\mathcal{B}(x, p)^{\mathcal{MS},i,j,g}$ is 1 iff $p^{\mathcal{MS},b,j,g}$ is 1, whereby b is a belief-world of x at index i,j.

However, one should not be fooled by this seemingly exacting notation which imitates Montague's PTQ. This T-condition is just as vacuous as 19.4.1 as long as it is not clear how the metalanguage definition should be verified relative to belief-worlds.

[9] In logical semantics, an ontological problem similar to individual belief-worlds is created by individual sensations, like a tooth ache, which do not exist in the same way as real objects in the world. The so-called *double aspect theory* attempts to make such sensations 'real' to the outside observer by means of measuring brain waves. By associating the phenomenon *pain* with both, (i) the individual sensation and (ii) the corresponding measurement, this phenomenon is supposed to obtain an ontological foundation acceptable to logical semantics. A transfer of this approach to the truth-conditional analysis of belief would require infallible lie detectors.

'Definition of truth (conditions) by means of meaning or
definition of meaning in terms of truth (conditions)?'

though in a different form relating to ontology:

20.3.1 THE BASIC ONTOLOGICAL PROBLEM OF MODEL THEORY

Is the speaker-hearer part of the model structure or
is the model structure part of the speaker-hearer?

If the goal of logical semantics is to characterize truth, then one may use only logical meanings which are presupposed to be immediately obvious and eternal. On this
approach the speaker-hearer must be part of the model structure, just like all the other
objects. Thereby, the relation of truth between expressions and states of affairs exists
independently of whether it is discovered by this or that speaker-hearer or not.[10]

If the goal of logical semantics is to analyze language meaning, then the system, developed originally for the characterization of truth based on logical meanings, is used
for a new purpose, namely the description of language meanings in the form of truth
conditions. This new purpose of bringing out the logical aspect of natural language
meanings brings about a profound change in the original ontological assumptions.

In order for the meanings of language to be *used in communication* they must be
part of the speaker-hearer's cognition. The cognitive (re-)interpretation of the model
as part of the speaker-hearer, however, is incompatible with the goals and methods
of traditional theories of truth. Conversely, the 'realistic' interpretation of the model
within a theory of truth is incompatible with an analysis of natural language meanings
which are used in communication.[11]

There may be examples in mathematics in which a formal theory happens to allow
more than one interpretation, e.g. geometry. This does not mean, however, that any
formal theory may in general be used for any new interpretation desired. A case in
point is logical semantics, whose formalism cannot be interpreted simultaneously as
a general description of truth and a general description of natural language meaning –
as shown by the phenomenon of propositional attitudes.

The alternative stated in 20.3.1 is characterized schematically in 20.3.2 using the
binary feature [±constructive].

---

[10] The 'received view' in philosophical logic and theory of science does not require that a representation of scientific truth include the speaker-hearer as part of the model. The only reason why a
speaker-hearer is added by some to a model-theoretic logic is the treatment of special phenomena
characteristic of natural language, especially the interpretation of indexical pronouns like I and you.
Thereby the speaker-hearer is in principle part of the model structure – making it impossible to provide an adequate truth-conditional treatment of propositional attitudes for the reasons given above.
A detailed critique of the outmoded 'received view', as well as its alternatives, may be found in F.
Suppe 1977.

[11] Early examples of the misguided attempt to reinterpret the model structure as something cognitive
inside the speaker-hearer are Hausser 1978, 1979c, 1981, 1983b, and SCG.

20.3.2 TWO ONTOLOGICAL INTERPRETATIONS OF MODEL THEORY

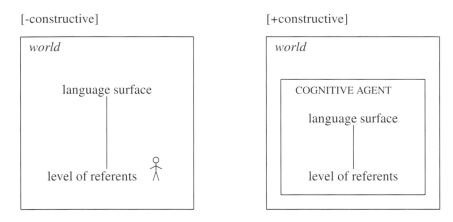

The [-constructive] interpretation establishes the relation between the language surfaces and the level of referents outside the cognitive agent out there in the real world. The agent is itself an object at the level of referents, and may at best observe this somehow god-given, direct relation between language and the objects of the world.

The [+constructive] interpretation, on the other hand, establishes the relation between the language surfaces and the level of referents inside the cognitive agent. What the agent does not perceive in the world or does not know plays no role in his reference procedures, though what (s)he feels, wishes, plans, etc., does.

The most fundamental difference between the two ontologies consists in the fact that they require different types of semantics:

– Any system based on a [–constructive] ontology must have a metalanguage-based semantics. This is because scientific statements believed to be eternally true and independent of any speaker-hearer cannot be meaningfully operationalized. The only option remaining for establishing the external relation between the expression and the state of affairs is in terms of metalanguage definitions.
– Any system based on a [+constructive] ontology must have a procedural semantics. This is because such systems would be simply useless without a procedural semantics. Neither a computer nor a cognitive agent could practically function on the basis of a metalanguage-based semantics.

For different phenomena of logical analysis, the distinction between a [-constructive] and a [+constructive] ontology has the different implications. In propositional calculus, the difference between the two ontologies happens to make no difference. In first-order predicate calculus, quantification over infinite sets (universal quantifier) will differ within the two ontologies. In intensional logic, the treatment of opaque contexts and modality is incommensurable with their alternative treatment in a [+constructive] cognitive model. A treatment of propositional attitudes, finally, is incompatible with a [-constructive] ontology, but poses no problem in a [+constructive] system.

20.4 Four basic ontologies

The features [± sense] (cf. 20.2.2) and [± constructive] (cf. 20.3.2) are independent of each other and can therefore be combined. This results in four types of semantic interpretation based on four different ontologies, namely [–sense, –constructive], [+sense,–constructive], [–sense,+constructive], and [+sense,+constructive].

20.4.1 ONTOLOGIES OF SEMANTIC INTERPRETATION

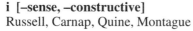

i [–sense, –constructive]
Russell, Carnap, Quine, Montague

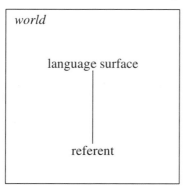

ii [+sense, –constructive]
Frege

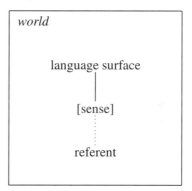

iii [–sense, +constructive]
Newell & Simon, Winograd, Shank

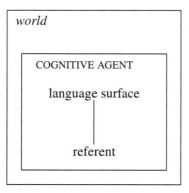

iv [+sense, +constructive]
Anderson, CURIOUS, SLIM-machine

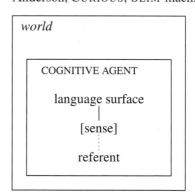

As indicated by the authors' names, these different ontologies have been adopted by different approaches to semantic interpretation.

The [–sense,–constructive] ontology (**i**) is the basis of logical semantics. Concerned with a solid foundation for truth, logical semantics uses only referents which are considered to be ontologically real. In nominalism, these are the concrete signs of language and the states of affairs built up from concrete objects. In mathematical realism, the ontology is extended to include abstract objects like sets and numbers. Both versions have in common that the semantics is defined in a metalanguage as a direct, external relation between language and the world.

This type of semantics has been adopted by the main stream of analytic philosophy, from Russell via the early Wittgenstein, Carnap, Montague, to Putnam. Given its ontological foundations, logical semantics is in principle unsuitably for a complete analysis of natural language meaning. This has resulted in a rather ambivalent view of natural language in philosophical logic.

The [+sense,–constructive] ontology (**ii**) was used by Frege in his attempt to analyze uneven (opaque, intensional) readings in natural language. For modeling the mechanics of natural language communication, this type of semantics is only half a step in the right direction. As a theory of truth, any [–constructive], metalanguage-based semantics is necessarily incompatible with representing cognitive states.[12]

The [–sense,+constructive] ontology (**iii**) underlies the semantics of programming languages. The user puts commands (surfaces of a programming language) into the computer which turns them directly into corresponding electronic procedures. When a result has been computed it is communicated to the user by displaying language expressions on the screen. In this traditional use, a computer is still a far cry from a cognitive agent. However, there is already the important distinction between the *task environment* in the 'world' (cf. 3.1.3) and the computer internal *problem space*, whereby the semantic interpretation is located in the latter.

Because of their origin as conventional programs on conventional computers (cf. 1.1.3) most systems of artificial intelligence are based – subconsciously, so to speak – on a [–sense,+constructive] ontology. This holds, for example, for SHRDLU (T. Winograd 1972), HEARSAY (R. Reddy et al. 1973), and SAM (R. Schank & R. Abelson 1977). In cognitive psychology this ontology has been used as well, for example in the *mental models* by P. Johnson-Laird 1983.

Within artificial intelligence, A. Newell & H. Simon 1972, p. 66, have argued explicitly against an intermediate level of sense – for purely ontological reasons. In their view, a distinction between language meanings (sense) and the computer internal referents would result "in an unnecessary and unparsimonious multiplication of hypothetical entities that has no evidential support."

A direct, fixed connection between language expressions and their referents, however, prevents any autonomous classification of new objects in principle. Therefore, [–sense, +constructive] systems are limited to closed toy worlds created by the programmer. Examples are the chess board (Newell & Simon, Reddy et al.), the blocks world (Winograd), or the restaurant script (Schank & Abelson). It is by no means accidental that these systems have no components of artificial perception: because

---

[12] Frege argued explicitly against misinterpreting his system as representing cognitive states, which would be what he called 'psychologistic'. Recently, situation semantics (J. Barwise & J. Perry 1983) and discourse semantics (H. Kamp & U. Reyle 1993) have attempted to revive the [+sense, –constructive] type of semantics. Their inherently anti-cognitive point of view is clearly depicted in Barwise & Perry 1983, p. 226, in the form of diagrams.

they lack the intermediate level of concepts (sense) they could not utilize perception (e.g. artificial vision) to classify and to automatically integrate new objects into their domain (cf. Section 4.3).

The [+sense,+constructive] ontology (**iv**), finally, underlies the SLIM theory of language. SLIM bases its [+sense] property structurally on the matching of meaning$_1$ and the context of use within its [2+1] level structure, while its [+constructive] property is based on the fact that this matching occurs inside the cognitive agent.

In cognitive psychology, this type of semantics has been used by J.R. Anderson & G.H. Bower 1973 and 1980. They present a general psychological model of natural language understanding which may be interpreted as an internal matching of language concepts onto a context structure and insofar resembles the functioning of CURIOUS as described in Chapter 4.

The theoretical relation between the four ontologies may be analyzed by either emphasizing their differences or their similarities. In the latter case one will present one's semantics as a purely formal structure which may be assigned different interpretations without affecting the formal essence. Different ontologies may be related by analyzing one as a simplifying generalization of the other.

For example, the difference between a [+sense] and a [-sense] ontology may be minimized by interpreting the latter as a simplification of the former. Assume that (i) the world is closed such that objects can neither appear nor disappear, (ii) relations between language expressions and their referents are fixed once and for all, and (iii) there is no spontaneous use of language by the speaker-hearer. Then there is no reason for postulating a level of sense, thus leading to a [-sense] system as a special case of the [+sense] system.

Because of this simplification one might view the [-sense] system as more valid or more essential than the [+sense] system. One should not forget, however, that there are empirical phenomena which simply cannot be handled within a [-sense] ontology, such as the reference to new objects of a known type.

The difference between a [+constructive] and a [-constructive] ontology may also be minimized in terms of a simplification. Assume that the cognitive agent has perfect recognition such that the distinction between the external objects (i.e. language expressions and referents) and their internal cognitive representations may be neglected. Then there is no reason to distinguish between the external reality and its internal cognitive representation, thus leading to a [-constructive] ontology as a special case of a [+constructive] ontology.

Because of this simplification one might view the [-constructive] system as more valid and more essential than the [+constructive] system. One should not forget, however, that there are empirical phenomena which simply cannot be handled within a [-constructive] ontology, such as propositional attitudes.

The choice between the four different types of ontologies depends on the intended application. Therefore, when (i) *expanding* a given semantics to a new application or when (ii) *transferring* partial analyses from one application to another, one should be as well-informed about the basic structural differences between the four ontologies as about possible equivalences obtained at the price of simplifying generalizations.

20.5 Sorites paradox and the treatment of vagueness

The importance of ontology for the empirical analysis of semantic phenomena may be illustrated with the example of vagueness. In logical semantics, the treatment of vagueness takes a classical paradox from antiquity as its starting point, namely the Sorites paradox or paradox of the heap.

> One grain of sand does not make a heap. Adding an additional grain still doesn't make a heap. If n grains do not form a heap, then adding another single grain will not make a heap either. However, if this process of adding a grain is continued long enough, there will eventually result a genuine heap.

The Sorites paradox has been carried over to the logical semantics of natural language by arguing as follows: consider the process of, e.g., a slowly closing door. Doesn't it raise the question at which point the sentence **The door is open** is still true and at which point it is false? Then one goes one step further by asking to which *degree* the sentence is true in the various stages of the door closing.

> Sensitive students of language, especially psychologists and linguistic philosophers, have long been attuned to the fact that natural language concepts have vague boundaries and fuzzy edges and that, consequently, natural-language sentences will very often be neither true, nor false, nor nonsensical, but rather true to a certain extent and false to a certain extent, true in certain respects and false in other respects.
>
> G. Lakoff 1972, p. 183

Another situation which as been presented as an example of truth-conditional vagueness is the classification of colors. If an object is classified as *red* in context a, but as *non-red* in context b, doesn't it follow that the natural language concept *red* must be vague? If the predicate **x is red** is applied to the transition from red to orange in a color spectrum, the situation resembles the slowly closing door.

If these assumptions are accepted, then the traditional two-valued (bivalent) logic does not suffice and must be extended into a many-valued (non-bivalent) logic.[13] Such systems evolved originally in connection with the question of whether a proposition must always be either true or false.

> Throughout the orthodox mainstream of the development of logic in the West, the prevailing view was that every proposition is either true or else false - although which of these is the case may well neither be *necessary* as regards the matter itself nor *determinable* as regards our knowledge of it. This thesis, now commonly called the "Law of Excluded Middle", was, however, already questioned in antiquity. In Chap. 9 of his treatise *On Interpretation (de interpretatione)*, Aristotle discussed the truth status of alternatives regarding "future-contingent" matters, whose occurrence – like that of the sea battle tomorrow – is not yet determinable by us and may indeed actually be undetermined.
>
> N. Rescher, 1969, p. 1

Non-bivalent systems exceed the founding assumptions of logic, as shown by the remarkable term 'future-contingent.' While standard logic takes the viewpoint of an all-knowing being and is concerned with the interpretation of absolute propositions, future-contingent propositions are concerned with more mundane situations. For example, the question of which truth-value the sentence Tomorrow's sea battle will be lost by the Persians has *today* is obviously not asked by an all-knowing being. Furthermore, the content of this proposition is neither mathematical nor scientific in nature, and therefore cannot be verified.

In modern times, non-bivalent logics began with the Scotsman Hugh MacColl (1837 – 1909), the American Charles Sanders Peirce (1839 – 1914), and the Russian Nikolai A. Vasil'ev (1880 – 1940). The non-bivalent logics may be divided into two basic groups, namely the *three-valued* logics, in which a proposition can be true (1), false (0), or undetermined (#), and the *many-valued* logics, in which truth-values are identified with the real numbers between 0 and 1, e.g. 0.615.

The three-values logics and the many-valued logics all suffer from the same basic problem:

> Which truth-value should be assigned to complex propositions based on component propositions with non-bivalent truth-values?

Thus, a three-valued system raises the question: What should be the value of, e.g., A&B if A has the value 1 and B has the value #? Similarly in a many-valued system: if the component proposition A has the truth-value 0.615 and B has the value 0.423, what value should be assigned to A&B?

There is an uncomfortable wealth of possible answers to these questions. N. Rescher 1969 describes 51 different systems of non-bivalent logics proposed in the literature up to that date. Of those, only a few will be briefly described here.

J. Łukasiewicz 1935 assigns the following truth-values to logical conjunction: if both A and B are 1, then A&B is 1; if one of the conjuncts is 0, then A&B is 0; but if one of the conjuncts is 1 and the other is #, then A&B is #. This assignment reflects the following intuitions: if it turns out that one conjunct of A& B is 0, then the other conjunct needn't even be looked at because the whole conjunction will be 0 anyhow. But if one conjunct is 1 and the other is #, then the whole is indeterminate and assigned the value #.

The same reasoning is used by Łukasiewicz regarding logical disjunction: if one of the two disjuncts is 1, then A∨B is 1; if both disjuncts are 0, then A∨B is 0; but if one disjunct is 0 and the other is #, then the conjunction A∨B is #.

A different value assignment is proposed by D.A. Bochvar 1939, who proceeds on the assumption that a complex proposition is always # if one of its components is #. Thus, if A is 0 and B is #, then Bochvar does not assign 0 to A&B (as in the system of

---

[13] Besides vagueness, non-bivalent logics have been motivated by semantic presuppositions (see for example Hausser 1973 and 1976).

Łukasiewicz), but rather #. And similarly for A∨B. Bochvar justifies this assignment by interpreting # as *senseless* rather than unknown.

A third method of assigning non-bivalent truth-values is *supervaluations* by B. van Fraassen 1966, 1968, 1969. His concern is to maintain the classic tautologies and contradictions in a three-valued system. According to van Fraassen, it should not matter for, e.g., A∨~A (tautology) or A&~A (contradiction), whether A has the value 1, 0, or #, because according to the classic view tautologies *always* have the value 1 and contradictions *always* have the value 0.

Supervaluations are basically a complicated motivational structure which allows assigning bivalent truth-values to tautologies and contradictions while treating contingent propositions similar to Łukasiewicz's system. The motivation is based on the ontologically remarkable assumption that the truth-value of an elementary proposition may be checked repeatedly (as in scientific measuring).[14]

The problem of justifying the choice of a certain a value assignment is even worse in the case of multi-valued systems. For example, if the component proposition A has the truth-value 0.615 and B has the truth-value 0.423, what should be the value of A&B? 0.615? 0.423? the average 0.519? 1? 0? And similarly for A∨B. For each of these different assignments one may find a suitable application. In the majority of the remaining cases, however, the principle chosen will turn out to be counterintuitive or at best artificial.

From a history of science point of view such a multitude of more than 50 different alternatives is a clear case of an *embarrassment of riches* in combination with *descriptive aporia* (cf. Section 22.2). These two syndromes are an infallible sign that there is something seriously wrong with the basic premises of the approach in question.

In multivalued logic systems, the mistake resides in the premise formulated in the above quotation from Lakoff that propositions may obviously have non-bivalent truth-values. Once this premise is accepted one is stuck in a futile search for adequate value assignments to complex propositions, e.g. the question of which truth-value should be assigned to A&B if A and B have the truth-values 0.615 and 0.423, respectively.

Instead of accepting the premise we should ask instead how such peculiar truth-values come about in the first place. And with this question we come back to the issue of the underlying ontology. More precisely: what impact has the structural difference between the [-sense,-constructive] and the [+sense,+constructive] ontology on the formal analysis of vagueness?

Consider the sentence The door is open and red. In logical semantics it is analyzed as the proposition A&B, where A = [The door is open] and B = [The door is red]. If A and B have the truth-values 0.615 and 0.423, respectively, then the analy-

---

[14] The appeal of van Fraassen's construction resides in the fact that multiple value assignments are essentially a hidden procedural component. This procedural component is coached within the overall metalanguage-based approach and partial in that it is limited to elementary propositions. The use of a procedural component is ironic insofar as van Fraassen aims at saving the bivalence of logical tautologies and contradictions as compared to, e.g., J. Łukasiewicz 1935.

sis of this conjunction within the appropriate [-sense,-constructive] ontology has the following structure:

20.5.1 VAGUENESS IN [-SENSE,-CONSTRUCTIVE] SEMANTICS

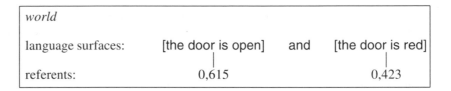

Assigning the truth-values 0.615 and 0.423 to the propositions A and B may be motivated by giving a long story about how the door is slightly non-open and how its color is slightly non-red. Such stories, though not part of logical theory, are intended to make us accept the peculiar, artificial values assigned to A and B. Once these assignments have been accepted, attention is focussed on the question of which truth-value should be assigned to the complex proposition A&B.

A completely different analysis results if The door is open and red is analyzed as an *utterance* within the [+sense, +constructive] ontology. This alternative framework provides four different positions which may be possible sources of vagueness. They are marked in 20.5.2 by the letters *a–d*.

20.5.2 VAGUENESS IN [+SENSE,+CONSTRUCTIVE] SEMANTICS

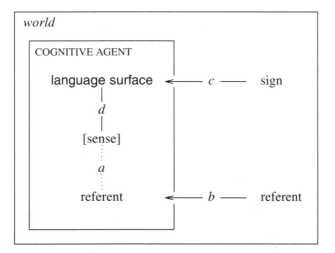

Position *d* corresponds to the vertical denotation lines in 20.5.1 because there the meanings are assigned to the language surfaces. The additional places *a, b,* and *c*

come about in the transition to a [+sense,+constructive] ontology. In contradistinction to position d, they have in common that they are interfaces based on matching.

Within a [+sense,+constructive] ontology the most natural place for handling vagueness is a. This is because there a language meaning, e.g. the M-concept of red, is matched with a restricted set of potential contextual referents (I-concepts$_{loc}$).

For example, the word red may be used to refer to a pale pink stone provided that all the other objects in the subcontext are, e.g., grey. As soon as a bright red stone is added, however, the candidate for best match changes and the pale pink stone will now be counted among the non-red objects. This is not due to a special 'vagueness'-property of the color concept *red*, but rather to a change in the context of use (see also the handling of metaphoric reference in 5.2.1).

Other places in which vagueness may arise naturally are b and c. In the case of b, vagueness is caused by imprecise perception of the task environment. In the case of c, it is caused by imprecise hearing or pronouncing of spoken language expressions, and accordingly for written language. In either case, vagueness originates in the interaction of the cognitive agent with the external environment and may influence communication by affecting the matching procedure a.

Thus, within a [+sense,+constructive] ontology the alleged vagueness of the color words does not arise in their semantics at all. Instead it is a normal instance of matching an M-concept and a contextual referent (I-concept$_{loc}$) in the pragmatics, a procedure based on the principle of best match relative to a restricted context of use. As shown in Chapters 3 and 4, this analysis of the semantics and pragmatics of the color terms can easily be realized operationally within the construction of CURIOUS by defining M-concepts like red as intervals of electromagnetic frequency.

Exercises

Section 20.1

1. Which principle is used to apply the logical characterization of truth indirectly to the analysis of natural language meanings?
2. How can a *circulus vitiosus* be avoided, if truth is defined in terms of meanings and meanings are defined in terms of truth?
3. Describe the rules of existential generalization and substitutivity of identicals, and explain them with examples.

4. Describe the properties of intensional contexts. By what other names have they been called?
5. Why would it not suffice to let terms for which no referents exist, like unicorn and **Pegasus**, denote the empty set?
6. Read Frege 1892 (pp. 143–162 in Frege 1967) and explain the distinction between sense and reference using his morning/evening star example.
7. What does Frege accomplish with this distinction, why is it ontologically problematic, and how does mathematical realism help him in overcoming this problem?

Section 20.2

1. Explain Carnap's formal reconstruction of Frege's notion of sense.
2. What is an intension? Specify the domain/range structure of intensions for expressions of different categories.
3. What is the relation between a logical model and a model structure?
4. Define a formal model structure for 6 verbs, 5 nouns, and 4 adjectives using the definition of intensional logic in SCG, p. 81 f. The sets A, I, and J should each contain 10 elements. For a simplified example see 22.2.1.
5. Name two modal operators and explain their formal definition within model theory.
6. What is the formal basis for defining the tense operators H and W?
7. What are possible worlds used for in model theory?

Section 20.3

1. Explain in what sense the treatments of intensional contexts by Frege and Carnap achieve similar results, and in what sense they differ ontologically.
2. What is Montague's main accomplishment in the logical analysis of natural language?
3. Name two problems which are in principle unsolvable for a logical semantics of natural language,
4. What is a propositional attitude?
5. Explain in detail why a formal treatment of propositional attitudes in logical semantics would be incompatible with the goals of a theory of truth.
6. Why is it impossible to operationalize a semantics based on a [-constructive] ontology?
7. Why would a semantics based on a [+constructive] ontology be pointless if it is not operationalized?

Section 20.4

1. Describe four different kinds of ontology for systems of semantics.
2. Explain which types of ontology the semantics of logical, programming, and natural language are based on.

3. Explain how a [-sense,-constructive] ontology may be viewed as both, a special case of a [+sense,+constructive] ontology and a higher form of abstraction.
4. Name a phenomenon which can be handled only in systems with a [+sense] ontology.
5. Name a phenomenon which can be handled only in systems with a [+constructive] ontology.
6. Which ontological property limits systems of classic AI, such as SHRDLU, in principle to toy worlds?

Section 20.5

1. Name the Sorites Paradox? How does it relate to the alleged vagueness of natural language?
2. Explain the notion of semantic presuppositions (cf. CoL, p. 333–344). How may they be used to motivate non-bivalent logic?
3. What is the *law of excluded middle*?
4. What are future-contingent propositions?
5. Name two different interpretations of the third truth-value #.
6. Look up the three-valued system of Kleene in N. Rescher 1969 and compare it with those of Łukasiewicz and Bochvar.
7. How are the truth-values of complex propositions computed from the component propositions in a many-valued logic? Use the example $A \vee B$, where $A = 0.37$ and $B = 0.48$.
8. How many different systems are described in N. Rescher's 1969 investigation of non-bivalent logics? Does this number reflect positively or negatively on non-bivalent approaches to logic? What could be the reason behind the invention of so many alternative solutions?
9. Explain how the treatment of vagueness differs in systems with a [-sense,-constructive] and a [+sense,+constructive] ontology.
10. What is the role of pragmatics in the analysis of vagueness?
11. Why is it desirable from the viewpoint of classic logic to handle vagueness without a many-valued logic?

21. Absolute and contingent propositions

The ontology of a semantic theory can influence empirical analysis profoundly. This has been demonstrated with the phenomenon of vagueness which leads to an embarrassment of riches within the [–sense,–constructive] ontology of logical semantics, but falls into the spectrum of normal pragmatic interpretation within the [+sense,+constructive] ontology of the SLIM theory of language.

It is therefore promising to reanalyze other classical problems of logical semantics within the [+sense,+constructive] ontology in order to disarm them. Of special interest is the Epimenides paradox, which was used by A. Tarski 1935, 1994 to prove that a complete logical semantics of natural language is impossible. Will a [+sense,+constructive] reanalysis of the Epimenides enable object languages to contain the words true and false without making their semantics inconsistent?

Section 21.1 develops a distinction between the logical truth values 1 and 0 and the natural truth values $true^c$ and $false^c$ based on a comparison of absolute and contingent propositions. Section 21.2 reconstructs the Epimenides paradox in a [+sense,+constructive] system. Section 21.3 strengthens Montague's notion of a homomorphic semantic interpretation into a formal version of surface compositionality. Section 21.4 shows that the strictly time-linear derivation of LA-grammar is not in conflict with a homomorphic semantic interpretation. Section 21.5 explains why the mathematical complexity of a system may be greatly increased by a semantic interpretation and how this is to be avoided.

21.1 Absolute and contingent truth

In logic, the term proposition has acquired a specialized use, representing sentences which do not require knowledge of the utterance situation for their semantic interpretation. Such propositions are usually obtained by translating selected natural language examples into formulas of logic. Compared to the natural language examples, the formal translations are partly simplified and partly supplemented, depending on what the logical language provides and requires for a proper logical proposition.

From the viewpoint of the SLIM theory of language, this specialized use of the term proposition[1] (German 'Aussage') is problematic because it constitutes a hybrid

---

[1] See Section 3.4 for a description of propositions proper.

between an *utterance* (i.e. a pragmatically interpreted or interpretable token) and an *expression* (i.e. a pragmatically uninterpreted type). The suppressed distinction between utterance token and expression type may be brought to light by investigating the difference between absolute and contingent propositions.

Absolute propositions express scientific or mathematical contents. These contents are special in that they make the interpretation largely independent from the usual role of the speaker. For example, in the proposition

> In a right-angled triangle, it holds for the hypotenuse A and the cathetes B and C that $A^2 = B^2 + C^2$

the circumstances of the utterance have no influence on the interpretation and the truth value of the sentence in question, for which reason they are ignored.

The special properties of absolute propositions are reflected in logical truth. This notion is formally expressed by the metalanguage words false and true referring to the abstract set-theoretic objects $\emptyset$ (empty set) und $\{\emptyset\}$ (set of empty set), respectively, of the model structure.[2]

Thereby, logical truth is based on the system of truth conditions of the language at hand. The referential objects $\emptyset$ und $\{\emptyset\}$ serve merely as model-theoretic fix points into which the denotations of propositions are mapped by the metalanguage rules of interpretation (e.g. 19.3.2).

Contingent propositions, on the other hand, are based on sentences with everyday contents such as

> Your dog is doing well.

Contingent propositions can only be interpreted – and thereby evaluated with respect to their truth value – if the relevant circumstances of the utterance situation are known and systematically entered into the interpretation.

This requires that the parameters of the STAR point be known, i.e., the spatial location S, the time T, the author A, and the intended recipient R (cf. Section 5.3). The characteristic properties of contingent propositions correspond to a natural notion of truth, represented by the contingent truth values $true^c$ and $false^c$. Intuitively, a proposition such as

> The Persians have lost the battle

may be regarded as $true^c$, if the speaker is an eye witness who is able to correctly judge and communicate the facts, or if there exists a properly functioning chain of communication between the speaker and a reliable eye witness.

Within the SLIM theory of language, the natural truth values $true^c$ and $false^c$ have a procedural definition. A proposition – or rather a statement – uttered by, e.g., a robot is evaluated as $true^c$, if all procedures contributing to communication work correctly. Otherwise it is evaluated as $false^c$.[3]

---

[2] Instead of $\emptyset$ and $\{\emptyset\}$, other notations use 0 and 1, $\bot$ and $\top$, no and yes, etc.

[3] A limiting case of this definition is the possibility that two errors accidentally cancel each other such that a statement happens to be true despite faulty processing. The crucial fact here is not that an isolated statement turns out to be true by accident, but rather the combination of errors. Statistically,

What follows from this distinction between logical and natural truth? Using two different notions of truth for absolute and contingent sentences would clearly be suboptimal from both, the viewpoint of philosophical logic and the SLIM theory of language. Instead, the goal is an overall system with a uniform semantics which can correctly interpret any utterance of the form C is true no matter whether C happens to be a contingent or an absolute sentence.

A straightforward way of unifying the semantics of absolute and contingent statements is treating one type as a special case of the other. For a [−constructive,−sense] approach it would thus be desirable, if logical semantics would allow a complete treatment of contingent statements as a special case of absolute statements. Conversely, for a [+constructive,+sense] approach it would be desirable, if natural semantics would allow a treatment of absolute statements as a special case of contingent statements. Given the choice between the two possibilities, the one applicable in greater generality is to be preferred.

In logical semantics, the handling of absolute statements may be extended to contingent statements in many instances – as shown by Montague's model-theoretic analysis of English.[4] The phenomenon of propositional attitudes (Section 20.3) has shown, however, that a proper semantic interpretation – that is an ontologically justified assignment of the values 1 or 0 – is *not always* possible. Furthermore, according to Tarski, sentences of the form C is (not) true are forbidden in the object language. For these two reasons a general treatment of contingent statements as a special case of absolute statements is excluded.

In natural semantics, on the other hand, absolute statements may always be treated as a special case of contingent statements. For the SLIM theory of language, absolute statements are special only insofar as (i) they can be interpreted independently of their STAR-point and (ii) the cognitive responsibility for their content is transferred from the individual speaker to society and its historically grown view of the world.

Thus, an absolute statement like The chemical formula of water is H_2O is true^c, if there exists a correctly functioning chain of communication between the speaker and the responsible experts.[5] The true sentences of absolute scientific and logically-mathematical systems are thus reconstructed contingently by interpreting them as cognitive accomplishments of the associated human – and thus fallible – society.

---

this special case will occur very rarely such that a doubly faulty cognition of a robot will reveal itself in the obvious defects of the vast majority of its utterances.

[4] See the sample analyses at the end of PTQ (Chapter 8 in R. Montague 1974).

[5] The notion of a 'causal chain' from one speaker to the next has been emphasized by S. Kripke 1972, especially with regards to proper names and natural classes. The central role of 'experts' in the scientific specification of certain meanings in the language community – e.g. analyzing water as H_2O – was stressed by H. Putnam 1975a, but with the absurd conclusion *that meanings just ain't in the head* (op.cit., p. 227). These authors investigate meaning and reference as a precondition for the foundation of truth, but they fail to make the necessary distinctions between the semantics of logical and natural languages, between a [−sense,−constructive] and a [+sense,+constructive] ontology, and between absolute and contingent truth.

From this anthropological point of view it is quite normal that an absolute statement may be considered true^c at certain times – due to the majority opinion of the experts – yet later turn out to be false^c. Such mishaps happened and still happen quite frequently in the history of science, as shown by statements like Fire is based on the material substance of phlogiston or, closer to home, The surface is determined by repeated application of certain formal operations called "grammatical transformations" to [base phrase markers].[6]

The differences in the truth predicates of natural and logical semantics derive directly from structural difference between their respective [–sense,–constructive] and [+sense, +constructive] ontologies.

21.1.1 ONTOLOGICAL FOUNDATION OF NATURAL AND LOGICAL TRUTH

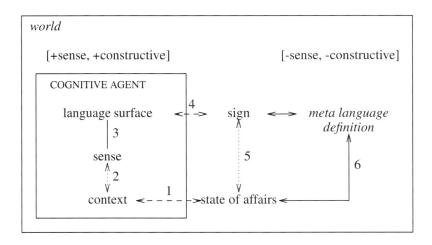

Both systems treat truth as relation 5 between the external sign and the external state of affairs. But they use different methods and concepts to realize this relation.

The [–sense,–constructive] system defines relation 5 directly by means of a suitable metalanguage definition 6. The analysis is done by the logician, who – in concord with the ontology presumed – concentrates on the truth relation between the expression and the state of affairs, abstracting from all interpersonal aspects of communication.[7] The logical model and the rule-based interpretation of the expression are designed to realize formally what is assumed as obvious to begin with. The purpose of the logical system is the explicit derivation of truth values.

In a [+sense, +constructive] system, on the other hand, a real task environment is presupposed. It must be analyzed automatically by the cognitive agent in certain relevant aspects whereby a corresponding context representation is constructed internally.

---

[6] See 4.5.2 and 8.5.4.

[7] For the logician, the state of affairs is not given by nature, but must be defined as a specific formal model $\mathcal{M}$. Strictly speaking, logic does not define relation 5 in the world, but only in a formal system designed to resemble the world.

Relation 5 between the external sign and the external state of affairs is thus established solely in terms of cognitive procedures based on the components 1 (contextual cognition/action), 2 (pragmatic interpretation), 3 (semantic interpretation), and 4 (language-based cognition/action). The purpose of the system is communicating contextual contents by means of natural language to other cognitive agents.

In summary, the contingent truth values $true^c$ and $false^c$ concentrate on the cognitive functions of concrete speaker-hearers in the evaluation of concrete utterances. The logical truth values 1 and 0, on the other hand, leave these aspects aside, taking the view of an omniscient being, who evaluates the relation between expression types and states of affairs independently of the existence of concrete speaker-hearers.

21.2 Epimenides in a [+sense,+constructive] system

Based on the contingent truth values $true^c$ and $false^c$, let us reanalyze the Epimenides paradox. In contradistinction to Tarski's analysis (cf. Section 19.5), the following [+sense, +constructive] analysis permits an object-language to contain the words true and false without causing its semantics to be inconsistent.

In preparation of this reanalysis, consider a benign use of the expression C is not a true sentence. This expression, used by Tarski to derive the Epimenides paradox, consists of a language-based abbreviation, C, and a negative truth statement. Its legitimate use within a [+sense,+constructive] system is based on the following structure.

21.2.1 BENIGN CASE OF A LANGUAGE-BASED ABBREVIATION

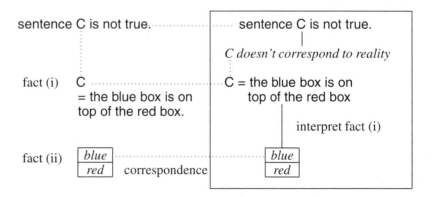

C abbreviates the expression The blue box is on the red box. The abbreviation is shown in the external task environment as fact (i). In addition, the task environment contains the state of affairs described by the sentence abbreviated as C, shown as fact (ii).

When the expression C is not (a) true (sentence) is processed by a [+sense, +constructive] system, e.g. CURIOUS, the semantics assigns to the surface a meaning$_1$ which may be paraphrased as *C doesn't correspond to reality*. For this semantic representation the pragmatics attempts to supply a matching contextual structure.

Thereby it turns out that C is defined as an abbreviation of the expression The blue box is on the red box according to fact (i). The remaining part of the input expression, i.e. is not a true sentence, is processed by the pragmatic component by checking whether the content of the long version of C corresponds to reality. The meaning$_1$ of The blue box is on the red box is matched with the corresponding subcontext, namely fact (ii), whereby it turns out that they *do* correspond. Thus the original input C is not a true sentence is evaluated as false^c.

This result may cause a suitably developed robot to react in various different ways. If it is in a righteous mood, it may protest and argue that C is in fact true. If it is forbearing, it might quietly register that the speaker was joking, cheating, or plain wrong. If it is cooperative, it will discuss the facts with the speaker to discover where their respective interpretations diverge in order to arrive at an agreement.

There are many language-based abbreviations in combination with natural truth statements which are as benign as they are normal.[8] For example, the position of the boxes in fact (ii) of 21.2.1 may be inverted, in which case the input sentence would be evaluated as true^c. Or fact (ii) may be removed from the task environment of CURIOUS, in which case the robot could not check the truth of the input sentence on its own. Whether the robot will use this unchecked information should depend on whether the speaker has earned the status of a reliable partner or not.

A special case of a language-based abbreviation is the Epimenides paradox. Its [+sense,+constructive] reanalysis has the following structure.

21.2.2 RECONSTRUCTION OF THE EPIMENIDES PARADOX

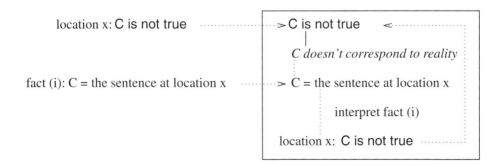

---

[8] This type of example also includes language-based abbreviations without the words true or false, for example C consists of eight words or C consists of seven words. Relative to the situation 21.2.1 these sentences would be evaluated as true^c and false^c, respectively. See also W.v.O. Quine 1960.

In a clearly marked location x, the robot reads C is not (a) true (sentence) and assigns to it the meaning$_1$ *C doesn't correspond to reality*. As in 21.2.1, the pragmatics attempts to supply a subcontext corresponding to this meaning$_1$.

Taking into account fact (i), it turns out that C is defined as an abbreviation of The sentence at location x. The remaining part of the input sentence, i.e. is not true, is processed by the pragmatics by checking whether the content of what C abbreviates corresponds to reality. For this, the meaning$_1$ of The sentence at location x is matched with a corresponding subcontext. In contrast to 21.2.1, where the meaning$_1$ of The blue box is on the red box is matched with the contextual (non-verbal) fact (ii), the meaning$_1$ of The sentence at location x in 21.2.2 leads to the language-based referent (sign) C is not true.

At this point, the pragmatics may treat the referential object C is not true as an uninterpreted or as an interpreted sign. Treating it as an uninterpreted sign would make sense in combination with, e.g., *is printed in sans serif*. In 21.2.2, however, treatment as an uninterpreted sign would make no sense. Rather, the most natural action would seem to interpret the sign – which starts the semantic-pragmatic interpretation procedure all over again.

Thus, if the external circumstances bring a [+sense,+constructive] system into the special situation of the Epimenides paradox, it will get into a blind cycle and – without additional assumptions – will remain there. As shown schematically in 21.2.2, the C in C is not true will be replaced again and again with the corresponding sentence at location x.

Our ontologically-based reanalysis of the Epimenides paradox does not result in its resolution, but rather in its transformation. What appears as a logical contradiction on the level of the semantics in Tarski's [–sense,–constructive] system (cf. Section 19.5) reappears as an infinite recursion of the semantic-pragmatic interpretation. The [+constructive,+sense] reanalysis disarms the Epimenides paradox, both on the level of the semantics and the theory of communication, because

– the words true^c and false^c may be part of the object language without causing a logical contradiction in its semantics, and
– the recursion caused by the Epimenides paradox can be recognized in the pragmatics and taken care of as a familiar[9] type of failing interpretation without adversely affecting the communicative functioning of the system.

The reanalysis avoids Tarski's contradiction in the semantics because the metalanguage distinguishes between (i) the logical truth values 1 and 0 from the T-condition, (ii) the natural truth values true^c and false^c from the object language, and (iii)

---

[9] It holds in general of pragmatic interpretation that a continuous repetition in the analysis of one and the same contextual object should be avoided, for example by means of counters. In this way the recursion caused by the Epimenides paradox may be recognized and stopped. Discontinuing a particular interpretation attempt in order to choose an alternative scheme of interpretation or to ask for clarification is a normal part of pragmatics.

their procedural metalanguage correlates *does (not) correspond to reality*. If we were to assume for the sake of the argument that the semantic component of CURIOUS were a logical semantics like Montague grammar, then the [+sense,+constructive] reanalysis of the Epimenides paradox would not result in Tarski's contradiction

$a.$ C is 1 if and only if C is not 1

but rather in the contingent statement

$b.$ C is 1 if and only if C does not correspond to reality.

In contrast to version a, version b does not contain a logical contradiction.

For the SLIM theory of language, the reanalysis of the Epimenides paradox is important. The reanalysis opens the way to define a *complete* semantics of natural language because it avoids Tarski's contradiction even if the language to be modeled contains the words true and false.

For the logical semantics of natural language, on the other hand, the reanalysis is of little help. This is because the procedural notion of contingent truth – essential for avoiding Tarski's contradiction – can be neither motivated nor implemented outside a [+constructive,+sense] ontology.

21.3 Frege's principle as homomorphism

In artificial languages, the form of the syntax and the associated semantics is decided by the language designers. Their job is to *construct* the artificial language as best as possible for a given task. The natural languages, on the other hand, are given in all their variety as historically grown conventions of their speech communities.

Computational linguistics has the task to functionally *reconstruct* the mechanics of natural language communication as realistically as possible (reverse engineering). The starting point of this reconstruction is the natural surfaces because they are manifested in the acoustical or visual medium as concrete signs.

The meanings, on the other hand, are of a cognitive nature. They lie in the dark for the outside observer (cf. 4.3.2), and can only be deduced from (i) the lexical and syntactic properties of the surfaces and (ii) their use in different contexts of interpretation.

According to Frege's principle (cf. 4.4.1), the meaning of a complex expression results from the meaning of the parts and the mode of their composition. The communicative function of natural syntax is the composition of semantic representations by means of composing the associated surfaces.

The build up of complex meanings$_1$ via syntactic composition is achieved by defining (i) for each word form a semantic counterpart and (ii) for each syntactic operation a simultaneous semantic operation. Montague formalized this structural correlation between syntax and semantics mathematically as a *homomorphism*.[10]

---

[10] The formal definitions may be found in Montague's paper *Universal Grammar*, R. Montague 1974, especially pp. 232,3.

The notion of a homomorphism captures the intuitive concept of a *structural simi-larity* between two complex objects. A structural object so is homomorphic to another structural object SO, if for each basic element of so there is a (not necessarily basic) counterpart in SO, and for each relation between elements in so there is a corresponding relation between corresponding elements in SO.

To express the structural similarity of the semantic level to the level of the surface, Montague defined a homomorphism formally as a relation between two (uninterpreted) languages.

21.3.1 FORMAL DEFINITION OF A HOMOMORPHISM

Language-2 is homomorphic to language-1 if there is a function T which

- assigns to each word of category a in language-1 a corresponding expression of category A in language-2, and
- assigns to each n-place composition f in language-1 a corresponding n-place composition F in language-2, such that
- T(f(a,b)) = F((T(a))(T(b)))

According to this definition, it is equivalent whether a and b are first combined in language-1 via f(a,b) after which the result a_b is translated by T into A_B of language-2, or whether a and b are first translated via T(a) and T(b) into A and B, respectively, and then combined in language-2 via F(A,B) into A-B.

$$
\begin{array}{lll}
\text{language-1:} & f\,(a,\,b) \longrightarrow & a\_b \\
& T \quad\ |\ \ | & | \\
\text{language-2:} & F\,(A,B) \longrightarrow & A\text{-}B
\end{array}
$$

A grammar in which the semantics (i.e. language-2) is homomorphic to the syntax (i.e. language-1) satisfies the so-called homomorphism condition. In such a system (i) each word form (basic element) in the syntax must be assigned a semantic counterpart (meaning$_1$) and (ii) each syntactic composition of word forms and/or expressions must be assigned a corresponding composition on the semantic level.

21.3.2 SYNTACTIC COMPOSITION WITH HOMOMORPHIC SEMANTICS

$$
\begin{array}{lll}
\text{analyzed surfaces:} & a \circ b \longrightarrow & ab \\
& |\quad | & | \\
\text{meanings}_1\text{:} & A \circ B \longrightarrow & AB
\end{array}
$$

The homomorphism condition by itself, however, is not sufficient as a formalization of Frege's principle insofar as it is defined for *analyzed* surfaces (cf. Section 4.4), whereas natural language communication is based on *unanalyzed* surfaces.

The problem is that the transition from unanalyzed to analyzed surfaces (interpretation) and vice versa (production) has been misused to enrich the levels of the analyzed surface and/or the meaning$_1$ by means of zero elements or identity mappings. From the viewpoint of the SLIM theory of language, this is strictly illegal because it violates the methodological principle of surface compositionality.

The use of zero elements and identity mappings is illustrated schematically in 21.3.3 and 21.3.4, respectively, where the zero elements are marked by #.

21.3.3 USE OF A ZERO ELEMENT (illegal)

1. Smuggling in during interpretation (↓) – Filtering out during production (↑)

$$
\begin{array}{lccc}
\text{unanalyzed surfaces:} & a' & & a' \\
& | & & | \\
\text{analyzed surfaces:} & a \circ b\# & \Rrightarrow & ab\# \\
& | \quad | & & | \\
\text{meanings}_1: & A \circ B & \Rrightarrow & AB
\end{array}
$$

2. Filtering out during interpretation (↓) – Smuggling in during production (↑)

$$
\begin{array}{lcccc}
\text{unanalyzed surfaces:} & a' & b' & & ab' \\
& | & | & & | \\
\text{analyzed surfaces:} & a \circ b\# & & \Rrightarrow & ab\# \\
& | & & & | \\
\text{meanings}_1: & A & & \Rrightarrow & A
\end{array}
$$

Zero elements of type 1 are postulated whenever the unanalyzed surface does not contain what the grammar theory at hand would like to find. Examples are the postulation of a 'zero determiner' in

Peter drank DET# wine

or a 'zero subject' in the imperative

YOU# help me!

Zero elements of type 2 are postulated whenever the surface contains something which the grammar theory at hand would not like to find. Examples are the conjunctions that in sentential objects, e.g.

Peter believes THAT# Jim is tired.

and to in infinitives, which have been regarded as superfluous syntactic sugar.

The two types of zero elements are also combined, as in passive

DET# wine WAS# ordered BY# Peter

or in infinitives

Peter promised Jim TO# Peter# sleep

Peter persuaded Jim TO# Peter# sleep.

Being at the core of nativism's linguistic generalizations, zero elements continue to be popular in C- and PS-grammar. They are usually supported with elaborate linguistic argumentations, as for example by Chomsky, who calls his zero elements 'traces.'

Because zero elements are marked neither in the unanalyzed surface nor in the meaning$_1$ they must be inferred by the parser. This is done by (i) adding them hypothetically into all possible positions and (ii) testing each case in terms of a derivation attempt.

As soon as a formal theory of grammar admits a single zero element, any given unanalyzed surface or meaning$_1$ raises the question of where and how often this zero element should be postulated. For this reason the use of zero elements pushes the complexity of such systems sky high, making them either $\mathcal{NP}$-complete or undecidable (cf. Chapters 8 and 12).

Equivalent to the problem caused by zero elements is the one caused by identity mappings of the form $x \circ y \to x$. This is shown by the schematic examples in 21.3.4 which use no zero elements in the input to the composition rules yet have the same outputs as in 21.3.3. The reason is that the rule of composition suppresses the contribution of b in the surface (type 1) or in the meaning (type 2).

21.3.4 USE OF AN IDENTITY MAPPING (illegal)

1. Filtering out during production (↑) – Smuggling in during interpretation (↓)

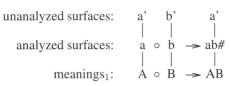

2. Smuggling in during production (↑) – Filtering out during interpretation (↓)

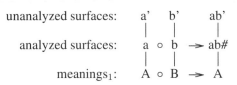

It has been argued that zero elements and identity mappings do not violate the homomorphism definition 21.3.1.[11] This requires, however, a choice as to whether the element marked by # is to be regarded as part of the homomorphism (as required by the respective type 1 structures in 21.3.3 and 21.3.4) or to be ignored (as required by the respective type 2 structures) – a clear violation of mathematical method.

In short, zero elements and identity mappings alike (i) destroy the systematic correlation between syntax and semantics, (ii) have a devastating effect on mathematical complexity, and (iii) fail to maintain the minimal methodological standard of concreteness. To ensure a proper functioning of the homomorphism condition and to prevent the use of zero elements and identity mappings, we present a formally oriented variant of the SC-I principle (cf. 4.4.2).

---

[11] Even Montague used quasi-transformational derivations of anaphoric pronouns and a syncategorematic treatment of logical operators, thus violating the spirit of Frege's principle and the homomorphism condition. A detailed account may be found in SCG.

21.3.5 SURFACE COMPOSITIONALITY II (SC-II PRINCIPLE)

A semantically interpreted grammar is surface compositional if and only if

- the syntax is restricted to the composition of concrete word forms (i.e. no zero elements and no identity mappings),
- the semantics is homomorphic to the syntax (in the sense of 21.3.1), and
- objects and operations on the level of semantics which correspond to the syntax in accordance with the homomorphism condition may not be realized by zero elements or identity mappings.

The SC-I principle applies Frege's principle to concrete surfaces of language in order to arrive at (i) a strictly compositional syntax and (ii) a clear separation of semantics and pragmatics. The SC-II principle makes this goal more precise by defining surface compositionality as a formal strengthening of the homomorphism condition.

21.4 Time-linear syntax with homomorphic semantics

The structures of semantic interpretation are hierarchical, while the surfaces of natural language are linear. In order to supply the time-linear derivation with a *homomorphic* semantics, the functor-argument structure must be built in a manner that is orthogonal to the resulting semantic hierarchy.[12] This comparatively new method[13] consists of two steps which correspond to the conditions of the homomorphism condition:

21.4.1 TIME-LINEAR BUILD-UP OF SEMANTIC HIERARCHIES

- Step 1: *Translation of word forms into component hierarchies*
  Each word form is mapped into a semantic component hierarchy (tree). The structure of the tree is determined by the syntactic category of the word form.
- Step 2: *Left-associative combination of component hierarchies*
  For each combination of the left-associative syntax there is defined a corresponding combination of component hierarchies on the level of the semantics.

  Step 1 is illustrated below with the analyzed word forms **the** and **man**.

---

[12] This is different from the traditional method of building semantic hierarchies by means of possible substitutions. For example, phrase structure grammar derives semantically motivated hierarchies (constituent structures of the deep structure) by substituting elementary nodes with more complex structures (top-down branching). Categorial grammar derives such hierarchies by substituting complex structures with elementary nodes (bottom-up amalgamating).

  In other words, the formalisms of PS- and C-grammar in their respective context-free forms are alike in that their linguistic applications are (i) conceptually based on constituent structure and (ii) achieve the build-up of the constituent structure hierarchies by using syntactic derivations which directly reflect the structure of the underlying semantic intuitions. This method is not compatible with a time-linear derivation order, however, and therefore unsuitable for semantically interpreting an LA-grammar.

[13] An informal description was first presented in CoL, p. 42 f., and illustrated with an LA-parser for a semantically interpreted fragment of English (op.cit., p. 345–402).

21.4.2 DERIVATION OF COMPONENT HIERARCHIES FROM WORD FORMS

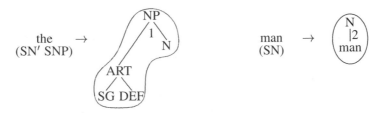

The two subtrees are derived automatically from the categories (SN' SNP) and (SN), respectively. The subtree of content words contains the associated M-concept, represented by the base form of the surface (here *mam*). Next consider step 2.

21.4.3 TIME-LINEAR COMPOSITION WITH HOMOMORPHIC SEMANTICS

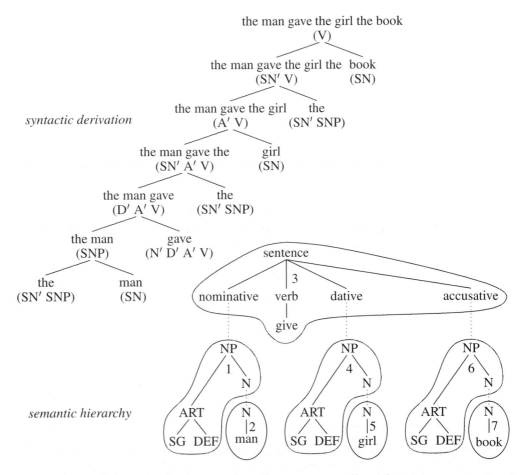

For each word form in the syntax there is a corresponding elementary component tree and for each left-associative composition in the syntax there is a composition of component trees in the semantics. To indicate the strictly compositional nature of the semantic hierarchy, the elementary component trees are outlined graphically and each is marked with the position number of the word form from which it was derived.

For example, component tree 1 (for the in position 1) is a functor, which takes the component tree 2 (for man in position 2) as its argument. The resulting (complex) component tree 1+2 (i.e., the NP representing the man) is a sentence start which serves as the argument for the (elementary) component tree 3. Component tree 3 is derived from the third word form gave, whereby the syntactic category (N′ D′ A′ V) determines the form of the tree. The resulting complex tree 1+2+3 is then combined with the elementary component tree 4, which serves as an argument derived from the determiner the. The complex component tree 1+2+3+4 in turn takes the elementary component tree 5 derived from the noun woman as argument, etc.

The category-based derivation of elementary subtrees from the word forms is handled by the automatic word form recognition (LA-Morph). The time-linear combination of semantic component trees is handled by semantic clauses in the combination rules of LA-syntax.

The simultaneous syntactic-semantic derivation 21.4.3 shows in principle[14] how a time-linear LA-syntax may be supplied with a surface compositional, homomorphic semantic interpretation. Moreover, because zero-elements or identity mappings are used neither in the syntax nor in the semantics, it is strictly surface compositional in the sense of SC-II.

The semantic hierarchy in 21.4.3 expresses that the verb give forms a relation between the actants man, woman, and book, whereby their roles are characterized by different cases. At first glance, this may seem to resemble the constituent structures of PS-grammar (cf. Sections 8.4–9.5). From the viewpoint of constituent structure analysis, however, the hierarchy in 21.4.3 satisfies neither its intuitive assumptions nor its formal definition 8.4.1.

More specifically, a constituent structure analysis would proceed on the assumption that gave is semantically closer to the woman and the book than to the man. These assumptions, supported by movement (cf. 8.4.8) and substitution (cf. 8.4.7) tests, result in a tree structure in which the subject and the verb phrase are treated as sister nodes (in accordance with the rewriting rule S → NP VP), a structure not accommodated by the hierarchy in 21.4.3.

Thus, all constituent structures are by definition semantic hierarchies, but not all semantic hierarchies are constituent structures. A semantic hierarchy which is not a constituent structure is illustrated in 21.4.3. It is motivated linguistically in terms of two general principles, namely (i) the functor-argument structure and (ii) the time-linear derivation order of natural language.

---

[14] The analysis of 21.4.3 represents the state of development characteristic of CoL, where this type of semantic interpretation has been implemented as a program and tested on a sizable fragment of English. What is still missing there, however, is a *declarative* presentation of the semantic rules. These will be presented in Chapter 23 for an advanced form of LA-semantics.

21.5 Complexity of natural language semantics

According to the CoNSyx hypothesis 12.5.7, the natural languages fit into the class of C1-languages and parse in linear time. The benefits of an efficient syntax are wasted, however, if the associated semantic interpretation has mathematical properties which push the overall system into a complexity class higher than that of the syntax alone. For this reason a formal semantic interpretation of an LA-syntax for natural language is empirically suitable only if the semantic interpretation does not increase the complexity of the resulting overall system as compared to the syntax alone.

That the low complexity of a syntactic system may easily be pushed sky high by the semantic interpretation is illustrated by the following examples from mathematics:

(a) π (b)   1:3
        | |
        3.14159265...   1' :' 3' = 0.333...

Both examples satisfy the homomorphism condition. In example (a), the word form π (*Pi*), standing for the ratio of the circumference of a circle to its diameter, denotes an infinitely long transcendental number. In example (b), a simple syntactic structure 1:3 denotes an infinitely long periodic number.

These examples show that an elementary word form or a very simple composition can denote infinitely long numbers in mathematics or non-terminating procedures in computer programs. Thereby, the semantic interpretation pushes the complexity from linear in the original syntax to undecidable in the overall system.

How can natural semantics retain low complexity, if it includes mathematical objects which are infinite and thus of high complexity? The crucial structural basis for this is the principled distinction between (i) the meaning$_1$ of language expressions, (ii) the internal subcontext providing the contextual referents, and (iii) the external counterparts of the contextual referents.

According to the SLIM-theoretic analysis of natural communication, meaning$_1$ functions basically as a *key* to information which is stored in the intended 'files' of the internal subcontext. To perform this function, minimal meanings$_1$ suffice. They must be differentiated only to the degree that the correct contextual referents and relations can be matched. This process is supported by preselecting and restricting the relevant subcontext using the pragmatic circumstances of the interpretation (especially the STAR-point).

The basic accessing function of natural language meaning shows up in the example **Suzanne is writing a thesis on the Trakhtenbrod Theorem**. We have no trouble understanding this sentence even if we have no idea of the mathematical content referred to with the noun phrase **Trakhtenbrod Theorem**. The relation between the key, the context file, and the mathematical content may be represented schematically as follows.

21.5.1 INTERPRETATION OF 'TRAKHTENBROD THEOREM'

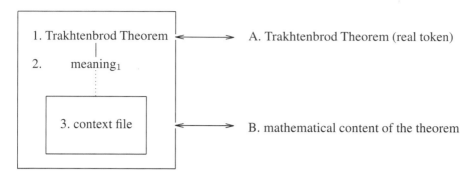

The left hand side shows the speaker-hearer's internal processing of language while the right hand side shows the external situation with (A) the sign **Trakhtenbrod Theorem** and (B) the corresponding mathematical content. Analogous to 20.5.2 and 21.1.1, the internal analysis is based on the [+sense,+constructive] ontology of the SLIM theory of language, while the associated external analysis represents the [−sense, −constructive] ontology of mathematical realism.

In accordance with mathematical realism, the content of the theorem is located in the external real world, outside the cognitive agent.[15] The meaning$_1$ of the expression **Trakhtenbrod Theorem** in natural semantics, on the other hand, is a cognitive, speaker-hearer-internal concept. It may be paraphrased roughly as *mathematical theorem discovered by a person named Trakhtenbrod*. This minimal semantic representation is sufficient as the key to contextual files the content of which may vary widely from one speaker-hearer to another, depending on their knowledge of mathematics in general and Trakhtenbrod's Theorem in particular.

For example, if a speaker-hearer encounters the expression **Trakhtenbrod Theorem** for the first time, a new contextual file is opened which contains no more than the expression and the circumstances of the utterance situation. An average speaker-hearer may nevertheless be said to understand the expression in such a situation – as shown by the fact the (s)he would be able to, e.g., procure literature on the theorem.

A more demanding task would be to recognize the theorem solely on the basis of its content, as when chosing it from a collection of unnamed theorems. For this, the relevant contextual file of the speaker-hearer would have to contain specialized mathematical knowledge. The acquisition of this knowledge does not affect the literal

---

[15] This example shows clearly that the difficulties of logical semantics with the Epimenides paradox (Section 19.5), the analysis of propositional attitudes (Section 20.3), and the treatment of vagueness (Section 20.5) do not argue against mathematical realism and its [−sense,−constructive] ontology at all. Rather, the described difficulties of logic result solely from the misguided attempt to transfer a [−sense, −constructive] semantics designed for the treatment of mathematical and natural science to the meaning analysis of natural language.

meaning of the expression Trakhtenbrod Theorem, however, but only the associated contextual file of an individual speaker-hearer.[16]

Structures of high mathematical complexity have no place in the semantic component (2) of natural language. Like the vastness of the universe, the laws of physics, or real beer mugs, they exist outside of the cognitive agent in position B and in a secondary way as contextual structures in position 3 of 21.5.1. Even contextual referents which do not originate in the external reality, like the acute individual tooth ache[17] of a cognitive agent, should not be treated in position 2 but located instead in the relevant internal subcontext 3.

The functioning of language meanings as keys for accessing the potentially complex – though always finite – contents of the internal context does not require that the semantics be of any higher complexity than the associated syntax. This conclusion is summarized in 21.5.2 as the **C**omplexity of **N**atural language **Sem**antics hypothesis, or CoNSem hypothesis for short.

21.5.2 CoNSem hypothesis
(Complexity of Natural language Semantics)

> The interpretation of a natural language syntax within the C-LAGs is empirically adequate only if there is a finite constant k such that
>
> - it holds for each elementary word form in the syntax that the associated semantic representation consists of at most k elements, and
> - it holds for each elementary composition in the syntax that the associated semantic composition increases the number of elements introduced by the two semantic input expressions by maximally k elements in the output.
>
> This means that the semantic interpretation of syntactically analyzed input of length n consists of maximally $(2n - 1) \cdot$ k elements.

A semantic interpretation which complies with the CoNSem hypothesis will increase the complexity of the overall system – as compared to the syntax alone – by only a constant. For example, if k = 5, then the semantic interpretation of a syntactically analyzed input of length 3 will consist of maximally $(2 \cdot 3 - 1) \cdot 5 = 25$ elements:

$$
\begin{array}{ccccccc}
\text{a} & \cdot & \text{b} & \longrightarrow & \text{ab} & \cdot & \text{c} & \longrightarrow & \text{abc} \\
| & & | & & | & & | & & | \\
[1\text{--}5] & \& & [6\text{--}10] & & [1\text{--}15] & \& & [16\text{--}20] & & [1\text{--}25]
\end{array}
$$

In other words, CoNSem systems are in the same overall complexity class as their C-LAG syntax alone.

---

[16] A similar analysis holds for the expression π. From the viewpoint of mathematical realism, the referent of π may be regarded as an infinitely long number in position B of 21.5.1. Its internal, cognitive counterpart in position 3, on the other hand, is a finite approximation, contained in a context file which may be accessed by a minimal meaning₁ (position 2).

[17] Cf. footnote 9 in Section 20.3.

The CoNSem hypothesis is applicable to the whole class of C-LAGs. This is because (i) C-LAGs limit the complexity of syntactic composition by a finite constant and (ii) different degrees of complexity within the C-LAGs are caused solely by different degrees of ambiguity.[18] For the analysis of natural languages, however, the CoNSem hypothesis is of special interest in connection with the CoNSyx hypothesis 12.5.7, which puts the syntax of natural languages into the linear class of C1-LAGs.

CoNSyx and CoNSem are empirical hypotheses and as such they can be neither logically proven nor refuted. Instead they serve as formal constraints. Their joint fulfillment guarantees any empirical analysis of natural language to be of linear complexity.

To refute the two hypotheses one would have to present constructions of natural language which clearly cannot be analyzed within the boundaries of CoNSyx and CoNSem. Conversely, to support the two hypotheses it must be shown in the long run that maintaining them does not create unsurmountable difficulties – in contradistinction to, e.g., the historical precedent of constituent structure.[19]

Whether or not the empirical analysis of a problematic construction will satisfy CoNSyx and CoNSem depends not only on the skill of the syntactic-semantic analysis, but also on the question of whether the problem in question is really of a syntactic-semantic nature (cf. 12.5.5 and 12.5.6). Therefore the possibility of maintaining the CoNSyx and CoNSem hypotheses can only be properly evaluated relative to a functioning overall model of natural communication.

Exercises

Section 21.1

1. Explain the notion of a logical proposition in comparison to the notions utterance and expression.
2. Describe the difference between absolute and contingent propositions using examples.
3. How are the logical truth values represented formally?
4. What is the difference between the notions of logical and natural truth?

---

[18] See Sections 11.4 and 11.5.

[19] The principle of constituent structure turned out to be incompatible with the empirical facts because of the discovery – or rather the belated recognition – of discontinuous structures. See Sections 8.3 and 8.4.

5. What are the preconditions for the interpretation of contingent propositions, and why is it that absolute propositions seem to be free from these preconditions?

6. Why is it impossible for logical semantics to treat contingent propositions generally as a special case of absolute propositions?

7. How is the status of truth affected by the reanalysis of absolute propositions as special cases of contingent propositions?

8. Which role is played by the 'experts' in determining the natural truth value of absolute propositions?

9. Describe how [+sense, +constructive] and [−sense,−constructive] systems differ in their respective methods of establishing a relation between language expressions and states of affairs.

10. What is meant by the words **true** and **false** in everyday communication?

Section 21.2

1. Explain the interpretation of the sentence **C is not true** within the framework of a [+sense,+constructive] system using the analysis of a benign example.

2. Explain the reanalysis of the Epimenides paradox in a [+sense,+constructive] system.

3. Why does the reanalysis of the Epimenides paradox require a change of Tarksi's ontology?

4. What is the difference between treating a given language expression as an uninterpreted vs. an interpreted sign? Explain the difference using examples from W.v.O. Quine 1960.

5. Is the Epimenides paradox based on a contingent or an absolute proposition?

6. In what sense is the Epimenides paradox preserved in the [+sense,+constructive] reanalysis and in what sense is it disarmed?

7. Why is the introduction of natural truth values crucial for the definition of an object language which can contain the words **true** and **false** without making the overall semantics inconsistent?

8. To what degree is a transfer of logical semantic analyses to the semantics of natural language possible?

9. Why is the reanalysis of the Epimenides paradox of little help to the logical semantics of natural language?

10. What kind of errors in the cognitive processing of a [+sense, +constructive] system result in false statements? When does the system speak truly?

Section 21.3

1. How do the artificial and the natural languages differ from the viewpoint of a language designer?

2. What is the connection between Frege's principle, surface compositionality, and Montague's use of a homomorphism to relate syntax and semantics?

3. Explain the formal definition of a homomorphism and illustrate it with a formal example.
4. Explain why the use of zero elements and identity mappings makes the homomorphism condition vacuous.
5. Which mathematically dubious assumption is needed for arguing that zero elements and identity mappings do not formally violate the homomorphism condition?
6. Why does the use of zero elements increase the mathematical complexity of a system?
7. Why does the use of zero elements violate the most minimal methodological standard of concreteness?

Section 21.4

1. Explain the SC-II principle in comparison with SC-I.
2. Why is the fixed surface compositional connection between the semantics and the syntax functionally necessary in natural communication?
3. Describe three different methods of building up a semantic hierarchy.
4. Why is a time-linear build up of semantic hierarchies compatible with maintaining a homomorphic semantics?
5. What is meant by the 'functor-argument structure' of a semantics?

Section 21.5

1. What is the complexity class of an LA-syntax for natural language?
2. Show with examples why a semantic interpretation can increase the complexity of a syntactic system.
3. Explain the functioning of natural language meanings using the example Trakhtenbrod Theorem.
4. Summarize why the underlying theory of language has a direct impact on the complexity of syntax and semantics, using the example of PP-attachment (Section 12.5) for syntax and the example π for semantics.
5. Summarize why the underlying theory of language has a direct impact on the way in which relevant phenomena are analyzed, using the examples Epimenides paradox, propositional attitudes, vagueness, intensional contexts, and semantic presuppositions.
6. Why is a [−sense,−constructive] ontology appropriate for mathematical realism, but not for the analysis of natural language?
7. What is the CoNSem hypothesis and what would be required to refute it?

22. Database semantics

The construction of a cognitive machine capable of communicating in natural language requires the explicit definition of the context of use. For this, a special data structure is needed which is suitable for reading language-based information in and out in a time-linear fashion.

Section 22.1 illustrates the basic mechanism of natural communication with a simple example. Section 22.2 explains why logical models and frame-theoretic knowledge bases lead to problems of the type 'descriptive aporia' and 'embarrassment of riches' when reinterpreted as a context of use. Section 22.3 presents the new data-structure of bidirectional, co-indexed feature structures, called proplets, in which the functor-argument structure of elementary propositions and their extrapropositional relations are coded. Section 22.4 analyzes the context of use as a set of proplets in a network database called word bank. Section 22.5 illustrates the functioning of proplets with an example of a word bank.

22.1 Database metaphor of natural communication

The representation of individual knowledge relative to which natural language is interpreted is called the context. Representing the context as a speaker-hearer-internal database provides a familiar computational framework which allows to characterize the basic differences between a user's interaction with a database (DB interaction), on the one hand, and natural communication (NL communication), on the other.

22.1.1 INTERACTION WITH A CONVENTIONAL DATABASE

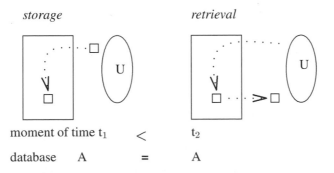

The big boxes represent the computer containing the database. The small boxes represent the language signs serving as input and output. The ovals represent the user controlling the in- and output.

22.1.2 INTERACTION BETWEEN SPEAKER AND HEARER

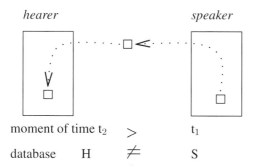

Here the big boxes represent cognitive agents which may be natural or artificial. The differences between DB interaction and NL communication may be summarized as follows.

22.1.3 DB INTERACTION AND NL COMMUNICATION

- ENTITIES INVOLVED
  Database interaction:
  The interaction takes place between two different entities, the user and the database.
  NL communication:
  The interaction takes place between two similar and equal cognitive agents, the speaker and the hearer.
- ORIGIN OF CONTROL
  Database interaction:
  The database operations of input and output are controlled by the user.
  NL communication:
  There is no user. Instead, the cognitive agents control each other by alternating in the speaker- and the hearer-mode (*turn taking*).
- METHOD OF CONTROL
  Database interaction:
  The user controls the operations of the database with a programming language the commands of which are executed as electronic procedures.
  NL communication:
  The speaker controls language production as an autonomous agent, coding the parameters of the utterance situation into the output expressions (cf. 5.4.2). The hearer's interpretation is controlled by the incoming language expression (cf. 5.4.1).

– TEMPORAL ORDER

Database interaction:

The output (database as 'speaker') occurs necessarily *after* the input (database as 'hearer').

NL communication:

Language production (output procedure of the speaker) occurs necessarily *before* language interpretation (input procedure of the hearer).

From the fact that the speaker and the hearer constitute two different[1] databases follows the notion of successful natural communication. Speaker and the hearer understand each other if the contextual substructure to be communicated by the speaker is reconstructed *analogously* by the hearer – in terms of a correct embedding of the copy of the speaker's subcontext at a corresponding location in the hearer's database (cf. 4.5.5 as well as Section 23.5). For example, the post card in 5.3.1 does not refer to an arbitrary dog in an arbitrary kitchen. Rather, communication between the author of the post card and the addressee can only be called successful, if both refer to corresponding subcontexts containing corresponding referents.

The basic mechanism of natural communication is illustrated below with the interpretation of the sentence **Fido likes Zach** relative to a simple context of use.[2]

22.1.4 SKETCH OF A SIMPLE SUBCONTEXT

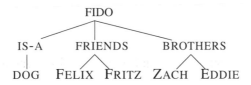

The subcontext is depicted as a semantic hierarchy in a preliminary conventional form familiar from knowledge representations in artificial intelligence. According to this representation, Fido is a dog, Felix and Fritz are his friends, and Zach and Eddie are his brothers.

In the hearer mode, the pragmatic interpretation of the expression **Fido likes Zach** consists in *embedding* the expression's meaning$_1$ into the context structure 22.1.4. This means that the context is extended to contain the additional relation **like** between **Fido** and **Zach**, as illustrated below.

---

[1] Except when talking to oneself. This, however, is more a verbalization of thought without an addressee than access to certain substructures of ones context controlled by language. See also footnote 4 in Section 5.3.

[2] See also CoL, p. 28f.

22.1.5 PRAGMATIC INTERPRETATION OF 22.1.1

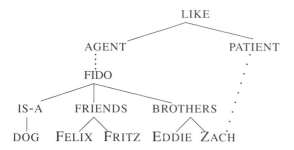

In the speaker mode the pragmatic interpretation consists in *extracting* the meaning[1] of **Fido likes Zach** from the subcontext. This means that the speaker copies a relevant part of his contextual substructure and maps it into natural language.

The task of the pragmatics is to describe the embedding (cf. 5.4.1) and extraction (cf. 5.4.2) between the semantic representation and the context in terms of explicit, programmable rules. This requires the definition of (i) a grammar including syntax and semantics, of (ii) an internal context of use, and of (iii) a pragmatics defined as a matching procedure between the semantic representation and the context of use.

22.2 Descriptive aporia and embarrassment of riches

In order to facilitate implementation of the matching procedure, the representation of (i) the language meaning and (ii) the context of use should be defined in terms of the *same* formalism. Which formalism would be suitable for this purpose? We begin by investigating two well established systems from different traditions, namely (a) formal logic and (b) frame theory.

The logical analysis of natural language is exemplified by Montague grammar, which is widely admired for its high standard of formal explicitness and differentiation of content. Within this framework, the information contained in 22.1.4 may be represented equivalently as follows.

22.2.1 MODEL-THEORETIC DEFINITION OF A CONTEXT

Let $\mathcal{MS}$ be a model structure $(A, I, J, \leq, F)$, where A, I, J are sets, $\leq$ is a simple ordering on J, and F is a denotation function.

A, I, J, and F have the following definition:

$A = \{a_0, a_1, a_2, a_3, a_4\}$
$I = \{i_1\}$
$J = \{j_1\}$
$F(\text{fido'})(i_1, j_1) = a_0$
$F(\text{felix'})(i_1, j_1) = a_1$

$$F(fritz')(i_1, j_1) = a_2$$
$$F(zach')(i_1, j_1) = a_3$$
$$F(eddie')(i_1, j_1) = a_4$$
$$F(dog')(i_1, j_1) = \{a_0\}$$
$$F(fido\text{-}friends')(i_1, j_1) = \{a_1, a_2\}$$
$$F(fido\text{-}brothers')(i_1, j_1) = \{a_3, a_4\}$$

At the index (i_1,j_1) the proper names fido, felix, zach, and eddie denote the model-theoretic individuals a_0, a_1, a_2, a_3, and a_4, respectively, while the properties fido-friends and fido-brothers denote the sets $\{a_1, a_2\}$ and $\{a_3, a_4\}$, respectively.

The original purpose of such a definition is to serve in the explicit interpretation of logical propositions (cf. Section 19.3). For example, the formal interpretation of fido-friend'(felix') relative to 22.2.1 would render the truth value 1 whereas fido-friend'(zach') would be evaluated as 0. This metalanguage-based derivation of truth-values relative to a formal model presupposes a [–sense, –constructive] ontology and treats the formal model as a representation of the external real world.

It is possible, however, to reinterpret the formal model structure 22.2.1 as a specification of the internal subcontext in a [+sense,+constructive] system.[3] This instantiates the strategy – used frequently in science – of specifying a new concept (here: internal context of use) in terms of a known, well-defined formalism (here: model theory).

The reinterpretation uses truth conditions to *build* models which make specific propositions true. These models are used to represent (i) the meaning of language and (ii) the context of use. The goal is to realize internal matching pragmatics by embedding meaning models into context model structures and by extracting meaning models from context model structures in a rhetorically meaningful way.

For example, extending the hearer context to the meaning of a new sentence such as Fido likes Zach would require automatically adding the formula

$$F(like)(i_1, j_1) = \{(a_0, a_3)\}$$

to 22.2.1. In this way, the additional relation like would be defined to hold between the individuals a_0 (the denotation of Fido) and a_3 (the denotation of Zach) at index (i_1,j_1) – analogous to 22.1.5. Formalizing this procedure requires specification of the intended referents and the correct index, which is difficult to do (descriptive aporia).[4]

Another method to formally define a context of use like 22.1.4 is offered by the programming languages. Especially suitable for this purpose, at least at first glance, are the so-called *frames*, which have been widely used in artificial intelligence and its favorite programming language LISP.

A frame is an abstract data type consisting of a frame name, an arbitrary number of slots, and for each slot an arbitrary number of fillers. A larger collection of frames is called a knowledge base. Frame systems provide a simple method for adding new information into – and to retrieve specific data from – the knowledge base.

---

[3] This approach was explored in SCG.
[4] For this reason volume II of SCG was never written.

A new frame is created with the command[5] `(make-frame FRAME (SLOT (value FILLER ...) ...)`. The structure of 22.1.4, for example, can be defined equivalently as a frame by the following command.

22.2.2 CREATING A *frame*

```
(make-frame
  fido
    (is-a (value dog))
    (friends (value felix fritz))
    (brothers (value zach eddie))
)
```

The result of this operation is stored in the computer as follows.

22.2.3 DEFINITION OF 22.4.2 AS A *frame*

```
(fido
  (is-a (value dog))
  (friends (value felix fritz))
  (brothers (value zach eddie))
)
```

The frame name is **fido**. It has the slots **is-a**, **friends**, and **brothers**. The slot **is-a** has the value **dog**, the slot **friends** has the values **felix** and **fritz**, and the slot **brothers** has the values **zach** and **eddie** as fillers.

The original purpose of such a definition is storing information which may later be retrieved using commands like `(get-values FRAME SLOT)`. For example, the command

```
(get-values 'FIDO 'FRIENDS)
```

would retrieve the values

```
(FELIX FRITZ)
```

assuming that the frame FIDO is defined as in 22.2.3. This retrieval of slot values can be surprisingly useful in larger knowledge bases.[6]

For the purpose of internal matching pragmatics one may reinterpret a frame definition like 22.2.3 as a specification of an internal subcontext.[7] This instantiates once more the strategy of specifying a new concept (here again: internal context of use) in terms of a known, implemented formalism (here: frame theory).

The reinterpretation uses frame definitions to represent (i) the meaning of natural language and (ii) the context of use. The goal is to realize internal matching pragmatics by automatically embedding language frames into the context knowledge base

---

[5] The names of the commands vary between the numerous different implementations of frames. As an introduction see for example P.H. Winston & B.K. Horn 1984, p. 311 ff.

[6] Frame systems usually offer a wealth of additional operations and options, such as removing information, inheritance, defaults, demons, and views. Despite of these additional structural possibilities, or perhaps because of them, frame systems typically suffer from uncontrolled growth combined with a lack of transparency and difficulties in checking consistency.

[7] This approach was explored in CoL.

and by extracting language frames from the context knowledge base as the basis of rhetorically meaningful utterances.

For example, extending the hearer context to the meaning of a new sentence such as **Fido likes Zach** would require representing this information as a frame, e.g.

```
(fido
 (like (value Zach)
)
```

and to automatically add the part

```
(like (value Zach)
```

as a new slot into 22.2.3.

The problem is that a given piece of information, e.g. a set of propositions, may be represented in many alternative ways in frame theory. For example, instead of making **fido** the frame name in in 22.2.3, we could have arranged the information alternatively by using **dog, friends,** or **brothers** as frame names. Each of these alternatives requires that the commands for adding and retrieving information are formulated accordingly.

The original design of a knowledge base expects the *user* to maintain a systematic structuring of the data by operating directly on the knowledge base with the commands of a programming language. Such a structuring is necessary to ensure proper storage and retrieval of the data.

The problem with reinterpreting frame theory for internal matching is that is impossible to (i) restrict the format as required for the purpose of storage and retrieval, and at the same time (ii) accommodate the structures occurring in the interpretation and production of natural language. Instead, effective storage and retrieval is prevented by the structural variety of natural language which leads to many alternative ways of structuring the data (embarrassment of riches).[8]

The problems encountered with the reinterpretation of model theory and frame theory, respectively, lead to the conclusion that using the same formalism for defining language meaning and interpretation context is a necessary, but not a sufficient condition for successfully realizing internal matching pragmatics. Furthermore, the attempts at reinterpreting these traditional formalisms fail ultimately for the same reason.

Because both formalisms are originally based on a [–sense] ontology, their reinterpretation in the [+sense, +constructive] environment of the SLIM theory of language results in the juxtaposition of two levels never designed to interact with each other. The holistic presentation of the two levels as separate, closed entities may be in line with the intuitive explanation of internal matching pragmatics (cf. Chapters 3–6), but fails to suggest an abstract algorithm for embedding meaning$_1$ into the context (hearer mode) and for extracting meaning$_1$ out of the context (speaker mode).

The underlying difficulty manifested itself as descriptive aporia in the model-theoretic approach and as embarrassment of riches in the frame-theoretic approach.

---

[8] For this reason the frame-theoretic approach of CoL was abandoned.

These problem types are symptomatic for situations in which a given formal method is inherently unsuitable for the description of the empirical phenomena at hand.

Descriptive aporia arises in situations in which the adopted grammar system does not suffice for the analysis of the phenomena at hand: the alternatives all seem to be equally bad. Embarrassment of riches arises in situations in which the phenomena at hand may be analyzed in several different ways within the adopted grammar system: the alternatives all seem to be equally good. Descriptive aporia and embarrassment of riches may also occur simultaneously (cf. Sections 9.5 and 20.5).

22.3 Propositions as sets of coindexed proplets

In order to achieve an efficient time-linear embedding and extraction of meaning$_1$ in internal matching pragmatics, the new approach of database semantics[9] was developed. Its data structure consists of concatenated elementary propositions at both, the level of language and the level of context..

According to the classic view described in Section 3.4, elementary propositions consist of the basic building blocks *functor*, *argument*, and *modifier*. The relation between these basic building blocks within elementary propositions as well as the relation between two elementary propositions was illustrated in 3.4.2 with graphical means. A database semantics turns this graphical representation into a format which is suitable for databases in general and a context database in particular.

Thereby, the graphical representation is coded alternatively as a *set* of bidirectionally co-indexed proplets. A proplet is a basic element of a proposition, defined as a feature structure.

22.3.1 PROPOSITION 3.4.2 AS A SET OF PROPLETS (preliminary format)

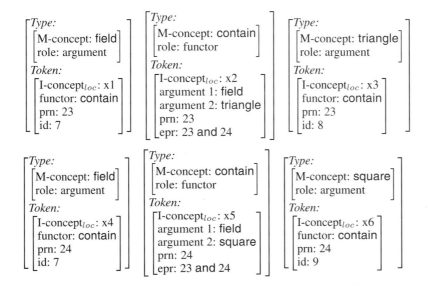

Like 3.4.2, 22.3.1 represents a sequence of contextual recognitions of the robot CU-RIOUS which may be paraphrased as The field contains a triangle and a square. For the moment, the proplets consist of two substructures, the *type* and the *token*.

The *type* contains the associated M-concept, i.e. the abstract concept with which parameter constellations in contextual cognition are classified (recognition) or realized (action) as I-concepts$_{loc}$ (cf. 3.3.5). An example of such an M-concept is 3.3.2, which is called square in English. Because M-concepts have a complex structure they are represented in 22.3.1 by the associated English words. In addition, the type specifies which functor-argument role the associated I-concepts$_{loc}$ play within elementary propositions.

The *token* contains the I-concept$_{loc}$, i.e. an individual recognition or action. An example of such an I-concept$_{loc}$ is 3.3.1. Because I-concepts$_{loc}$ have a complex structure their representation is simplified in 22.3.1 as x1–x6. These symbols are used as names to refer to particular proplets. In addition, a token feature structure contains one or more *intra*propositional continuation predicates (functor, argument 1, argument 2), the proposition number prn, and an open number of *extra*propositional continuation predicates (epr or id).

An elementary proposition is defined as a set of proplets with the same proposition number prn. For example, the first three proplets in 22.3.1 constitute a proposition because they have the same prn 23, while the remaining three proplets constitute proposition 24.

The functor-argument structure of an elementary proposition is coded into the intrapropositional continuation features of its proplets. Thereby, nominal proplets specify the related functor and verbal proplets specify the related arguments.[10] For example, the verbal proplet x5 in 22.3.1 has the continuation features [argument 1: field] and [argument 2: square]. Correspondingly, the nominal proplets x4 and x6 each contain the continuation feature [functor: contain].

In this way, the grammatical structure of the associated proposition may be reconstructed for any proplet in the database. For example, the proplet x4 in 22.3.1 specifies that it is (i) of the type field, (ii) functions as an argument, (iii) belongs to proposition 24, and (iv) that the associated functor is a proplet with the M-concept contain and the prn 24. With these informations, the continuation proplet x5 may be found in the database. This proplet confirms that it (a) serves as functor and that (b) its argument 1 is of the type field. Furthermore, it provides (c) another continuation proplet, namely [argument 2: square].

The concatenation of elementary propositions is defined in terms of the proplets' *extra*propositional continuation features. In nominal proplets, this is the identity number id, which specifies coreference or non-coreference with other nominal proplets.

---

[9] First published in Hausser 1996.

[10] From the simplified viewpoint of contextual propositions, argument 1 functions here intuitively as the grammatical-semantic subject and argument 2 as the object.

For example, that the triangle and the square are contained in the *same* field is expressed in 22.3.1 by the field proplets x1 and x4 having the same id Wert (here 7).

In verbal proplets, extrapropositional continuations are specified by their epr feature. For example, the epr feature of proplet x5 specifies that the preceding proposition has the prn number 23 and that the conjunction of this concatenation is and. When proposition 24 has been traversed intrapropositionally, the epr feature of x5 may be used to navigate extrapropositionally to proposition 23.

In summary, the relations between the different parts of a proposition and between different propositions, expressed graphically in 3.4.2, are coded equivalently in 22.3.1 in terms of features in individual proplets. Theses relations are realized bidirectionally. Proplets which are arguments specify the associated functor. Proplets which are functors specify the associated arguments. And similarly for modifier and modified.

22.4 Proplets in a classic database

In the area of databases, classic and nonclassic databases are distinguished.[11] Classic databases are based on fixed structures called *records*. For many decades, classic databases have served as stable and powerful software tools in practical applications of often gigantic size.

The nonclassic databases are called knowledge bases. They permit structures of varying size and form based on the principle of slot and filler.[12] Especially in large applications, nonclassic database systems are not as widely used as classic ones.

Among the classic databases there have evolved three basic types, namely the *relational, hierarchical*, and *network* databases. Of these the relational databases are the most common. Moreover, the characteristics of hierarchical and network databases may be simulated within relational databases. Thus, the type of a relational database is the most powerful as compared to the other two types of classic databases.

Because the proplets in a context database correspond in structure to a small set of fixed patterns,[13] a context database may be realized as a record-based, classic database. The distributed, bidirectional structure defined by the proplets of a context database is based on the following relations:

22.4.1 RELATIONS BETWEEN PROPLET FEATURES

$$
\begin{aligned}
\text{type} &\leftrightarrow \text{token} \\
\text{token} &\leftrightarrow \text{prn} \\
\text{prn} &\leftrightarrow \text{epr} \\
\text{token} &\leftrightarrow \text{id} \\
\text{functor} &\leftrightarrow \text{argument} \\
\text{modifier} &\leftrightarrow \text{modified}
\end{aligned}
$$

---

[11] See R. Elmasri & S.B. Navathe 1989.

Based on these relations a context database may be realized as a *relational* database.

This is very useful, especially in the initial phase of evolving a SLIM-theoretical context database, because one may rely on one of several existing commercial industrial grade software products capable of storing and accessing billions of context tokens. At the same time, however, one should be clearly aware of the different goals of a commercial relational database and a SLIM-theoretic context database.

A relational database is designed to efficiently store arbitrary data whereby the operations of the database are controlled by the user's commands in a programming language (called SQL or structured query language). In contrast, a context database is designed to autonomously and automatically turn natural language utterances into corresponding propositions, to store them correctly, to extract propositions in a meaningful way, and to turn them into rhetorically suitable utterances.

For a relational database, the preliminary presentation 22.3.1 as an *unordered* set of proplets is no problem. This is because the sorting and processing of the data is controlled by abstract software principles such that the individual proplets exist only virtually based on the relations defined within the database.

For humans, on the other hand, an unordered set provides no structural support for finding legitimate continuation proplets for a given proplet. We are therefore looking for a structural principle to order proplets in such a way that a corresponding presentation of the database contents would enable humans to find continuation proplets systematically, without the help of the computer software.

A suitable structural principle for this is ordering the proplets as a *word bank*.

22.4.2 PROPOSITIONS 3.4.2 AS A WORD BANK

TYPES   SIMPLIFIED PROPLETS

$$
\begin{bmatrix} \text{M-concept: contain} \\ \text{role: functor} \end{bmatrix}
\quad
\begin{bmatrix} \text{I-concept}_{loc}: \text{x2} \\ \text{argument 1:field} \\ \text{argument 2:triangle} \\ \text{prn: 23} \\ \text{epr: 23 and 24} \end{bmatrix}
\begin{bmatrix} \text{I-concept}_{loc}: \text{x5} \\ \text{argument 1:field} \\ \text{argument 2:square} \\ \text{prn: 24} \\ \text{epr: 23 and 24} \end{bmatrix}
$$

$$
\begin{bmatrix} \text{M-concept: field} \\ \text{role: argument} \end{bmatrix}
\quad
\begin{bmatrix} \text{I-concept}_{loc}: \text{x1} \\ \text{functor: contain} \\ \text{prn: 23} \\ \text{id: 7} \end{bmatrix}
\begin{bmatrix} \text{I-concept}_{loc}: \text{x4} \\ \text{functor: contain} \\ \text{prn: 24} \\ \text{id:7} \end{bmatrix}
$$

$$
\begin{bmatrix} \text{M-concept: square} \\ \text{role: argument} \end{bmatrix}
\quad
\begin{bmatrix} \text{I-concept}_{loc}: \text{x6} \\ \text{functor: contain} \\ \text{prn: 24} \\ \text{id: 9} \end{bmatrix}
$$

---

[12] Non-classic databases were considered in Section 22.2 in connection with the attempt to define a frame-based semantics for natural language.

[13] Namely the patterns for functors, arguments, and modifiers, each with two or three sub-patterns.

$$\begin{bmatrix} \text{M-concept: triangle} \\ \text{role: argument} \end{bmatrix} \qquad \begin{bmatrix} \text{I-concept}_{loc}\text{: x3} \\ \text{functor: contain} \\ \text{prn: 23} \\ \text{id: 8} \end{bmatrix}$$

Compared to the preliminary format 22.3.1, the proplets are reduced to the feature structures representing their token aspect. The feature structures representing the types are each used only once and ordered alphabetically. The tokens (simplified proplets) are ordered behind their types. The horizontal lines of a word bank, consisting of one type and a sequence of associated proplets, are called *token lines*.

In contradistinction to the unordered representation 22.3.1, the word bank format illustrated in 22.4.2 allows to systematically find possible continuation proplets – without the help of the data base software, solely on the basis of how the data are ordered. For example, proplet x4 specifies that it is of the type field, that its functor (i) is a proplet of the type contain and (ii) that its proposition number is 24. Because the types are arranged alphabetically it is easy to find the token line of contain. By going through this token line searching for the proposition number 24, the correct continuation proplet x2 is found.

From a linguistic viewpoint, a word bank has the lexical structure of alphabetically ordered lemmata. Especially in language-based propositions, each lemma (token line) consists of the type of a content word and a sequence of proplets representing associated word forms used in concrete utterances. The type describes the general properties of the word, e.g. the grammatical category (role) and the M-concept. A particular *use* of the type (proplet), on the other hand, contains the proposition number, the continuation predicates within that proposition (functors, arguments, modifiers, modified), as well as a specification of the extrapropositional relations.

From computational viewpoint, the structure of a word bank instantiates the classic type of a network database. A network database defines a 1:n relation between two kinds of records, the owner records and the member records. In 22.4.3, for example, the different departments of a university are treated as owners and their respective students as members:

22.4.3 EXAMPLE OF A NETWORK DATABASE

owner records	*member records*			
Comp.Sci.	Riedle	Schmidt	Stoll	...
Mathematics	Müller	Barth	Jacobs	...
Physics	Weber	Meier	Miele	...

In this simplified example, the different records are represented by names. In reality, the owner record type 'department' would specify attributes like name, address, phone number, etc. while the member record type 'student' would specify attributes to characterize each person.

The number of member records for a given owner record is variable in a network database. Maintaining the 1:n relation between an owner and its member records requires, however, that any given member record is assigned to a unique owner. For example, to ensure that for any member record there exists exactly one owner, no student in example 22.4.3 may have more than one major.

In a word bank, the types function as owner records and the associated proplets as member records. Just as in 22.4.3 each student is assigned to exactly one department via his or her major, each proplet is assigned to exactly one type. The owner records of a language-based word bank may be derived directly from the lexicon of the natural language in question (initializing). The member records, on the other hand, are read into the word bank automatically by means of an LA-grammar.

As a network database, a word bank corresponds to the especially simple variant with single member sets because the member records for a given owner are always of the same record structure. Moreover, a word bank corresponds to a non-recursive network database because each owner type always differs from the associated member types.

At the same time a word bank goes beyond a classic network database because it defines *possible continuations* within its record-based structure. These form the basis for a kind of operation which conventional databases do not provide, namely the autonomous linear navigation through the database which is independent from conventional, user-controlled methods.

The navigation-friendly implementation of the functor-argument structure of concatenated propositions in a word bank is only a first step, however. The next question is how a word bank should be interpreted in accordance with the SLIM theory of language and its internal matching pragmatics. This will be the topic of the next two chapters.

22.5 Example of a word bank

To strengthen the intuitive grasp of the word bank concept, let us translate the subcontext example 22.1.4 into an equivalent word bank. For this, the hierarchical structure of 22.1.4 must be represented alternatively as a sequence of elementary propositions.

22.5.1 PROPOSITIONAL PRESENTATION OF SUBCONTEXT 22.1.4

1. Fido is a dog.
2. Fido has friends.
3. The friends are Zach and Eddie.
4. Fido has brothers.
5. The brothers are Felix and Fritz.
6. Fido likes Zach.

The propositions 1–5 in 22.5.1 correspond to the content of the initial subcontext 22.1.4, while proposition 6 corresponds to the subcontext extension illustrated in 22.1.5. Using the graphical style of 3.4.2, the above sequence of elementary propositions may be represented as follows.

22.5.2 GRAPHICAL PRESENTATION OF THE PROPOSITIONS IN 22.5.1

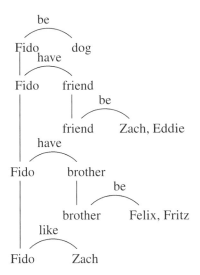

This graphical representation of the internal context as a sequence of concatenated elementary proposition shows that the word bank approach is conceptually completely different from the intuitive hierarchy (cf. 22.1.4) and its model-theoretic (cf. 22.2.1) or frame-theoretic (cf. 22.2.3) realizations.

Compared with the intuitive hierarchy 22.1.4, the graphical representation 22.5.2 consists of complete propositions whereby the verbs establish intrapropositional relations between subjects and objects. The vertical lines indicate coreference between nouns and illustrate the id-type of extrapropositional concatenation.

From the graphical presentation it is only a small step to the corresponding word bank. While 22.5.2 expresses intrapropositional relations in terms of the verbal connecting lines, 22.5.3 joins proplets into propositions by means of common prn values. Furthermore, the extrapropositional sequencing of 22.5.1 is expressed in 22.5.3 in terms of suitable epr values.

22.5.3 SUBCONTEXT 22.1.1 AS A WORD BANK

TYPES   PROPLETS

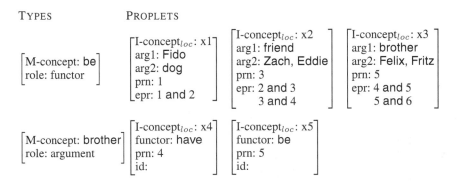

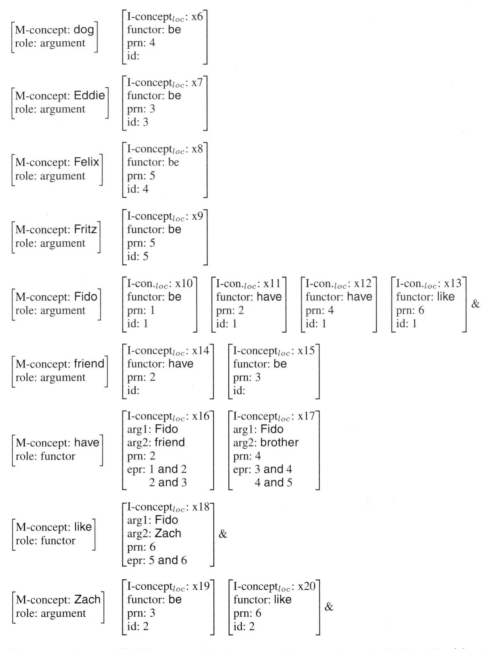

The prn values in 22.5.3 agree with the proposition numbers in 22.5.1. The id values of proper names correspond to their order of appearance. In other nominal proplets, the id values are left unspecified: they may either be added later in accordance with the content of the text, or reconstructed in terms of inferences. The ordering of the propositions is expressed in terms of the epr values of verbal proplets, e.g. epr: 5 and 6. In order to illustrate the reading-in of a new proposition (interpretation), the proplets derived from proposition 6 (cf. 22.5.1) have been marked in 22.5.3 with '&'.

Consider now the reading-in of proposition 6 into a state of the above word bank without the marked proplets. The semantic representation of proposition 6 is the following small word bank. It is derived automatically by a semantically interpreted LA-grammar of English (cf. Chapter 23).

22.5.4 SEMANTIC REPRESENTATION OF PROPOSITION 6

TYPES   PROPLETS

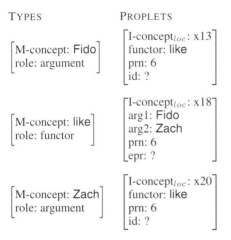

$$
\begin{bmatrix} \text{M-concept: Fido} \\ \text{role: argument} \end{bmatrix}
\quad
\begin{bmatrix} \text{I-concept}_{loc}: x13 \\ \text{functor: like} \\ \text{prn: 6} \\ \text{id: ?} \end{bmatrix}
$$

$$
\begin{bmatrix} \text{M-concept: like} \\ \text{role: functor} \end{bmatrix}
\quad
\begin{bmatrix} \text{I-concept}_{loc}: x18 \\ \text{arg1: Fido} \\ \text{arg2: Zach} \\ \text{prn: 6} \\ \text{epr: ?} \end{bmatrix}
$$

$$
\begin{bmatrix} \text{M-concept: Zach} \\ \text{role: argument} \end{bmatrix}
\quad
\begin{bmatrix} \text{I-concept}_{loc}: x20 \\ \text{functor: like} \\ \text{prn: 6} \\ \text{id: ?} \end{bmatrix}
$$

The pragmatic interpretation of this semantic representation consists in sorting the proplets of the small word bank into the large word bank representing the context of use. This embedding procedure of the internal matching pragmatics requires (i) adding the new proplets in the corresponding token line (as shown in 22.5.3) and (ii) assigning the correct epr and id values by means of counters and inferences.

Reading a propositional content out of a word bank is as simple as the reading it in. It is based on navigating through the proplets of the word bank whereby a special type of LA-grammar computes the possible continuations (see Chapter 24). The proplets traversed are copied into a buffer. They form a sequence which corresponds to the underlying navigation. The elements of this sequence are realized as word forms.

Exercises

Section 22.1

1. Why may the internal context of use be viewed as a database?

2. How does the interaction between a user and a database differ from the interaction between a speaker and a hearer?
3. Why is there no turn taking in the user's interaction with a database?
4. Explain the elementary procedure underlying natural language communication according to the SLIM theory of language using a simple example of a context.
5. How does this procedure differ in the speaker and the hearer?

Section 22.2

1. Is it possible to reinterpret model-theoretic semantics as a description of the context of use? What would be required for this, and in what way would such a reinterpretation modify the original goal of logical semantics?
2. Expand the model structure defined in 22.2.1 by adding one more index and the predicate **sleep** such that the falling asleep of Felix and Fritz and the waking up of Zach and Eddie is being modeled.
3. Describe the basic principles of frame theory and compare it with model theory.
4. Is it possible to reinterpret frame-theoretic semantics as a description of the context of use? In what way would such a reinterpretation modify the original purpose of frame-theoretic semantics?
5. Why does the use of frame-theoretic semantics within the SLIM theory of language necessitate a change of the original ontology?
6. Is the use of a uniform formalism for representing the two levels of meaning$_1$ and context a necessary or a sufficient condition for the successful definition of an internal matching pragmatics for natural language?
7. Is it a good idea to construct a [+sense] system by combining two instantiations of a [-sense] system? What kinds of problems are likely to occur?

Section 22.3

1. What is a proplet?
2. Explain how the graphical representation of a proposition may be expressed equivalently as a set of bidirectionally related proplets.
3. By means of which feature are the proplets of a proposition held together?
4. What is the difference between an intra- and extrapropositional continuation?
5. How the principle of possible continuations formally coded into the proplets' feature structure?

Section 22.4

1. What is the difference between classic and non-classic databases?
2. Name three different types of classic databases.
3. What does SQL stand for?
4. Explain the formal structure of a network database.
5. In what sense can a word bank be treated as a network database?
6. Which record types function as owner and which as member in a word bank?

7. What is a token line?
8. Explain the difference between a word bank and an unordered set of proplets.
9. In which respect does a word bank go beyond the structure of a conventional network database?
10. What is the function of the proposition number in a word bank?

Section 22.5

1. Explain the conceptual difference between the graphical representations 22.1.4 and 22.5.2.
2. Why does the graphical representation 22.1.4 contain implicit instances of repeated reference, and how are they treated explicitly in a word bank?
3. How does a word bank handle the embedding of internal matching pragmatics?
4. How does a word bank handle the extraction of internal matching pragmatics?
5. Why is language interpretation and language production relative to a word bank time-linear?

23. SLIM machine in the hearer mode

The format of a word bank is so general that the functor-argument structure and the concatenation of arbitrary propositions may be represented in it. This is the foundation for modeling natural communication as a SLIM machine.

Section 23.1 describes its external connections and motor algorithms, and reconstructs the distinction between meaning$_1$ and context. Section 23.2 presents the ten SLIM states of cognition. Section 23.3 describes the feature structures and elementary operations of the natural language semantics. Section 23.4 illustrates the semantic interpretation with a sample derivation. Section 23.5 provides a definition of meaning$_1$ and describes the transition from semantic to pragmatic interpretation.

23.1 External connections and motor algorithms

The mechanism of natural communication requires corelating the levels of language and context. In a SLIM machine, this is realized structurally by arranging two word banks on top of each other, the upper one for storing language-based propositions, and the lower one for storing propositions which represent the context.

23.1.1 STATIC STRUCTURES OF THE SLIM MACHINE

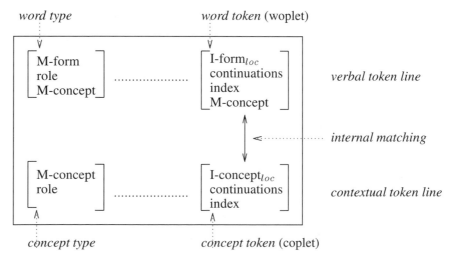

For each token line of the upper level there is a corresponding token line on the lower level. To distinguish the proplets of the language level and the context level, the word tokens are called *woplets* and the concept tokens are called *coplets*. If no such distinction is needed because we are dealing with propositions in general, *proplet* will be used as the generic term.

The relation between a word type and a corresponding concept type is based on their having the same M-concept and the same role. Word types differ from corresponding concept types only in that the word types have a language surface, represented by the M-form, which is absent in the concept types.

The relation between a woplet and a corresponding coplet is based on the type/token relation (cf. 4.2.2): woplets contain M-concepts where coplets usually contain I-concepts$_{loc}$. Woplets differ from corresponding coplets in that the woplets contain a surface, represented by the I-form$_{loc}$, which is absent in the coplets.

A SLIM Machine is based on altogether three type-token relations. They hold (i) between a word type and its woplets in the token lines of the upper level (horizontal), (ii) between the M-concept of a woplet and the I-concept$_{loc}$ of a corresponding coplet (vertical), and (iii) between a concept type and its coplets in the token lines of the lower level (horizontal).

In order to turn 23.1.1 into an active, autonomous, cognitive machine, the static structure must be complemented with

– *external connections* to the perception and action parameters, and
– *motor algorithms* which power the cognitive operations.

23.1.2 DYNAMIC PROCEDURES OF THE SLIM MACHINE

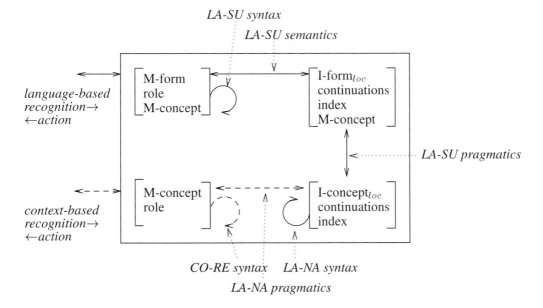

A SLIM machine is powered by three motor algorithms. The (i) LA-SU syntax (*Left-Associative SUrface syntax*) analyzes incoming expressions, maps them into corresponding sets of woplets, and powers language interpretation. The (ii) LA-NA syntax (*Left-Associative NAvigation syntax*) powers the autonomous navigation through the contextual database and is the basis of verbal and non-verbal action. The (iii) CO-RE syntax (*COntextual REcognition syntax*) analyzes incoming contextual perceptions, maps them into corresponding coplets, and powers contextual interpretation.

A secondary, passive function of the LA-SU and the CO-RE syntax is to help realize the output in language- and context-based action, respectively. During language- and context-based recognition, the LA-NA syntax is put into gear to passively control the handling of incoming coplets.

While contextual recognition and action may regarded as being outside linguistic analysis in the narrow sense, a modeling of immediate reference in language-based control (cf. 23.2.6 and 23.2.8) and context-based commenting (cf. 23.2.7 and 23.2.9) would be impossible without it. Therefore, mapping contextual recognition into propositional representations, and mapping propositions into contextual actions must be included in a systematic overall analysis.

23.2 Ten SLIM states of cognition

The distinctions between language-based and contextual recognition and action result in ten states of activation which are characteristic of the cognitive machine 23.1.2. They are represented schematically as specific machine constellations, called SLIM 1 to SLIM 10. The respective activation point of each state is indicated by ❋.

23.2.1 SLIM 1: RECOGNITION (contextual)

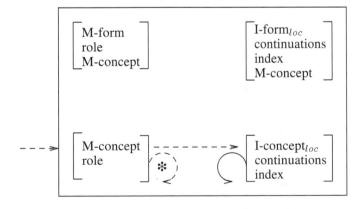

SLIM 1 is powered by the CO-RE Syntax. Contextual parameter values are matched with M-concepts, resulting in coplets with corresponding I-concepts$_{loc}$, which are combined into concatenated elementary propositions (cf. Sections 3.2–3.4). When the coplets are read into the contextual word bank, the LA-NA syntax is put into gear, passively controlling the operation.

23.2.2 SLIM 2: ACTION (contextual)

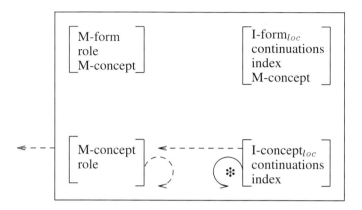

SLIM 2 is powered by the LA-NA syntax, i.e., the autonomous navigation through the contextual word bank. During this navigation, some of the propositions traversed are interpreted as actions (non-verbal pragmatic interpretation of the LA-NA syntax). Thereby the CO-RE syntax is put into gear to help realizing the contextual actions.

23.2.3 SLIM 3: INFERENCE (contextual)

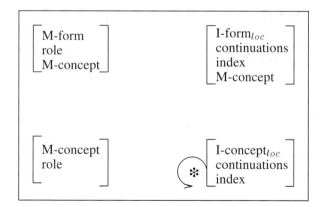

SLIM 3 consists solely in the autonomous navigation through the contextual word bank. The external connections are not active. As in SLIM 2, the origin of activation is the LA-NA syntax.[1] SLIM 1, SLIM 2 and SLIM 3 function also in cognitive agents

---

[1] When viewed in isolation, the concatenated propositions of the contextual word bank are like a rail road system while the LA-NA syntax is like a locomotive navigating through the propositions.

which have not developed the language level, as for example a dog. Thereby SLIM 3 is the cognitive state corresponding to, e.g., a dreaming animal.

23.2.4 SLIM 4: INTERPRETATION OF LANGUAGE (mediated reference)

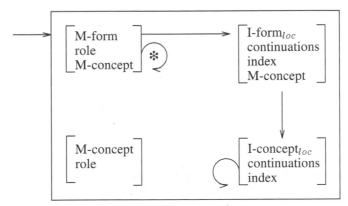

SLIM 4 is powered by language recognition[2] and the LA-SU syntax. The sequence of word types is mapped into language-based I-propositions which are read into the upper, linguistic word bank as bidirectionally related woplets (cf. Section 23.4). These are interpreted pragmatically relative to the lower, contual word bank (cf. Section 23.5). When the resulting coplets are read into the contextual word bank, the LA-NA syntax is put into gear, passively controlling the operation.

23.2.5 SLIM 5: PRODUCTION OF LANGUAGE (mediated reference)

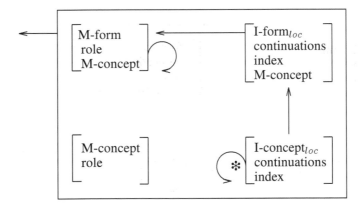

SLIM 5 is powered by the autonomous navigation through the propositions of the contextual word bank based on the LA-NA syntax.[3] The coplets traversed are assigned

---

[2] The upper left arrow represents a $d \Rightarrow i$ transfer, e.g. speech recognition (cf. Section 1.4). As described in Section 4.1, language recognition is based on a matching between M-forms and language parameter values, resulting in access to a corresponding lexical word type.

[3] As in SLIM 2 and SLIM 3, see also 5.4.2. The upper left arrow represents an $i \Rightarrow d$ transfer, e.g. speech synthesis (cf. Section 1.4).

corresponding woplets by finding for each I-concept$_{loc}$ (lower level) the associated M-concept (upper level) and thus the word type. The I-forms$_{loc}$ of the woplets are realized as language surfaces with the help of the M-forms of the corresponding word types. Thereby the LA-SU syntax is put in gear, passively controlling well-formedness of the output.

23.2.6 SLIM 6: LANGUAGE-CONTROLLED ACTION (immediate reference)

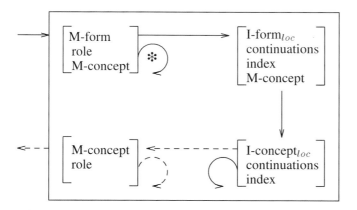

SLIM 6 is powered by language input which controls actions in the external task environment. An example would be an instruction like **Peel the potatoes and put them into boiling water.** SLIM 6 may be analyzed as a combination of language-based SLIM 4 and contextual SLIM 2. Note that SLIM 4 and SLIM 5 are instances of mediated reference, while SLIM 6 to SLIM 9 are instances of immediate reference.

23.2.7 SLIM 7: COMMENTED RECOGNITION (immediate reference)

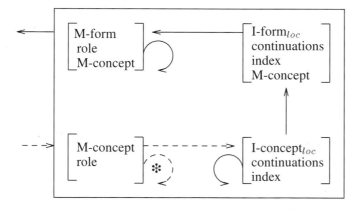

SLIM 7 is powered by contextual recognition which is put into words. An example is direct reporting like **Now a man with a violin case shows up**. SLIM 7 may be analyzed as a combination of contextual SLIM 1 and language-based SLIM 5.

23.2.8 SLIM 8: LANGUAGE-CONTROLLED RECOGNITION (immediate reference)

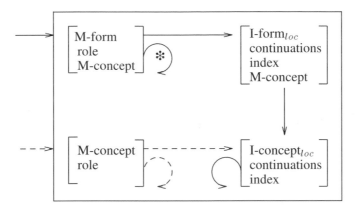

SLIM 8 is powered by language input which controls contextual recognition. Thereby both language-based and context-based propositions are read into the lower, contextual word bank. An example is an instruction like **In the upper right drawer you find a silver key**. SLIM 8 may be analyzed as a combination of language-based SLIM 4 and contextual SLIM 1.

23.2.9 SLIM 9: COMMENTED ACTION (immediate reference)

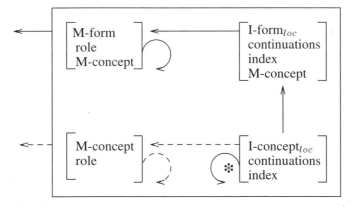

SLIM 9 is powered by the autonomous navigation through the propositions of the contextual word bank. These are simultaneously put into contextual action and into words. An example is a comment like **I operate the lever of the door lock and open it**. SLIM 9 may be analyzed as a combination of language-based SLIM 5 and contextual SLIM 2.

23.2.10 SLIM 10: COGNITIVE STILLSTAND

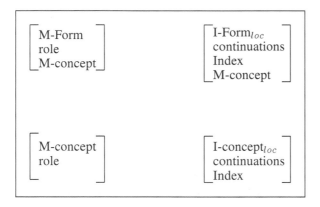

While in natural cognitive agents there is always a minimum of cognitive activity (SLIM 3) as long as they are alive and well, in artificial agents the time-linear navigation may be halted without jeopardizing the database contents. For this reason, SLIM 10 is included as a legitimate cognitive state representing the limiting case.

The typology of SLIM 1 to SLIM 10 is firstly complete because it is structurally impossible to define additional cognitive procedures of this kind, due to its systematic nature. Secondly, it is real because each of the ten procedures is supported by natural examples. Thirdly, it is useful because it allows to define important theoretical notions as simple variations of the same well-motivated structure.

In particular, *context-based* cognition is represented by SLIM 1 to SLIM 3, *language-based* cognition is represented by SLIM 4 and SLIM 5, while simultaneous context- and language-based cognition is represented by SLIM 6 to SLIM 9. Context-based cognition distinguishes between *recognition* (SLIM 1), *action* (SLIM 2), and *inferencing* (SLIM 3). Language-based cognition distinguishes between the *hearer mode* (SLIM 4, SLIM 6, SLIM 8), and the *speaker mode* (SLIM 5, SLIM 7, SLIM 9). *Immediate* reference (SLIM 4, SLIM 5) is distinguished from *mediated* reference (SLIM 6 to SLIM 9). In immediate reference, language-based *control* (SLIM 6, SLIM 8) is distinguished from context-based *commenting* (SLIM 7, SLIM 9).

23.3 Semantic interpretation of LA-SU syntax

The formal rules of semantic interpretation will now be illustrated in the context of SLIM 4, i.e. the interpretation of language with mediated reference. Aa shown in Section 21.4, a homomorphic interpretation of the LA-SU syntax requires a semantic interpretation of (i) word forms and of (ii) syntactic composition.

For word form recognition and production, a component of automatic morphology (cf. Chapters 13–15) is used by the SLIM machine. This results in an extension of the word types insofar as inflectional variants are integrated into their structure as follows.

23.3.1 REPRESENTING INFLECTIONAL VARIANTS IN A WORD TYPE

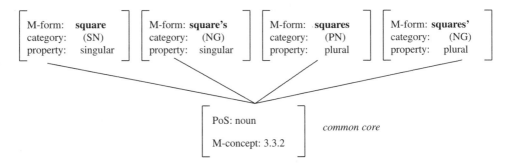

This structure combines (i) the common core, i.e. properties of the *word*, and (ii) the inflectional variants, i.e. properties characteristic of specific *word forms*. The common core serves as the owner record of the word bank while the inflectional variants are used by (i) word form recognition, (ii) syntactic analysis, (iii) syntactic generation, and (iv) word form production.

The first step of the syntactic-semantic interpretation consists in deriving woplets from the recognized word forms. This task is performed by the automatic word form recognition in combination with lexical lookup.

23.3.2 WORD FORM RECOGNITION AND DERIVATION OF A WOPLET

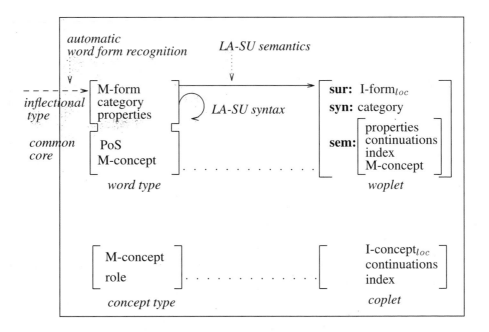

All woplets have the features sur, syn, and sem (for surface, syntax, and semantics, respectively). Woplets of different word types differ solely in their sem-features.

23.3.3 NOMINAL, VERBAL, AND ADJECTIVAL WOPLET STRUCTURES

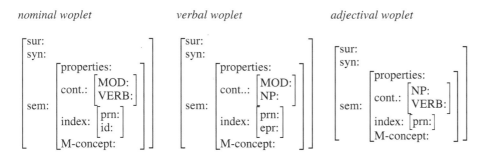

nominal woplet   *verbal woplet*   *adjectival woplet*

The **sem** feature of nominal woplets specifies the continuation attributes **MOD** for adjectives and **VERB** for the functor of the elementary proposition of which it is a part – as well as characteristic properties such as number. The **index** feature contains the attributes **prn** and **id**. In elementary propositions consisting of several, possibly equal woplets, these may be distinguished in terms of different **id** values.

The **sem** feature of verbal woplets specifies the continuation attributes **MOD** for adverbs and **NP** for the valency fillers – as well as properties such as tense and mood. The **index** feature contains the attributes **prn** and **epr**. The **epr** attribute specifies extrapropositional relations based on conjunctions, e.g. [**epr:** 2 then 3] (cf. 23.4.8).

The **sem** feature of adjectival woplets specifies the continuation attributes **NP** for adjectival use and **VERB** for adverbial use – as well as properties such as comparison. The **index** feature contains neither an **id** nor an **epr** attribute.[4]

The second step of the syntactic-semantic interpretation consists in reconstructing the functor-argument structure by filling in the woplets' continuation features as well as assigning their index values. This task is performed by semantic clauses which complement the concatenation rules the LA-SU syntax.

23.3.4 SCHEMA OF SEMANTICALLY INTERPRETED LA-SU RULE

rule:

   syn: ⟨ss-pattern⟩   ⟨nw-pattern⟩ ⟹ ⟨ss'-pattern⟩
   sem: semantic operations

input: output:

$$
\begin{bmatrix} \text{sur:} \\ \text{syn: } \langle a\rangle \\ \text{sem:} b \end{bmatrix}_1 \quad \dots \quad \begin{bmatrix} \text{sur: } m \\ \text{syn: } \langle c\rangle \\ \text{sem: } d \end{bmatrix}_i \quad + \quad \begin{bmatrix} \text{sur: } n \\ \text{syn: } \langle e\rangle \\ \text{sem: } f \end{bmatrix}_{i+1} \qquad \begin{bmatrix} \text{sur:} \\ \text{syn: } \langle a\rangle \\ \text{sem: } b \end{bmatrix}_1 \quad \dots \quad \begin{bmatrix} \text{sur: } m+n \\ \text{syn: } \langle g\rangle \\ \text{sem: } h \end{bmatrix}_{i+1}
$$

---

[4] For reasons of space, the following analyses of propositions will concentrate on nominal and verbal woplets, leaving the treatment of adjectives and adverbials aside.

The uninterpreted LA-syntax for natural language described in Chapters 16–18 continues to operate unchanged[5] based on the syn values (categories) of the woplets. However, while the uninterpreted syntax operates with a sentence start defined as an ordered triple, the semantically interpreted syntax operates with a *set* of woplets of which only one is *syntactically active*. It is the one with a non-NIL sur value (here [sur: *m*]). After application of the rule to the sentence start and the next word *n*, the syntactically active woplet in the resulting sentence start has the sur value *m+n*.

In the beginning of an LA-grammatical derivation, the sentence start consists of a one-element set containing the woplet of the first word. This woplet is the result of lexical lookup, for which reason it has a non-NIL sur value.[6] Subsequently, the syntactically active woplet of the sentence start is determined by the syntactical rules.

While the LA-syntactic analysis ignores woplets with empty sur values, the LA-semantic interpretation may take all input woplets into account. The semantic interpretation of an LA-SU syntax is based on the following six operations.

23.3.5 THE SIX BASIC OPERATIONS OF THE LA-SU SEMANTICS

1. copy_{ss}: include the woplets of the sentence start in the result.
2. copy_{nw}: include the woplet of the next word in the result.
3. $n_1.x \overset{\boxed{a}}{\longmapsto} n_2.y$: copy the values of the source feature x in n_1 additively into the goal feature y in n_2, whereby n_1 and n_2 may be the woplets of the sentence start or the next word.
4. $n_1.x \overset{\boxed{e}}{\longmapsto} n_2.y$: copy the values of the source feature x in n_1 exclusively into the goal feature y in n_2, whereby the value of y must be NIL (empty value).
5. $n_1.x \overset{\boxed{r}}{\longmapsto} n_2.①$: substitute all occurrences of the variable ① in n_2 simultaneously with the value of the source feature x in n_1.
6. $n.x \overset{\boxed{m}}{\longmapsto} n.x$: mark the first value of the source feature x in n, whereby the value of x must be a list.

Additive and exclusive copying are illustrated schematically in 23.3.6, whereby ⋆ indicates the position(s) in the resulting feature structures which have been modified.

23.3.6 COMPARISON OF ADDITIVE AND EXCLUSIVE COPYING

Additive:

$\text{nw.y} \overset{\boxed{a}}{\longmapsto} \text{ss.x}$
copy_{ss}

$\left[\text{x: } a\right]_1 \left[\text{x:}\right]_2 + \left[\text{y: } b\right]_3 \implies \left[\star\text{ x: } a\, b\right]_1 \left[\star\text{ x: } b\right]_2$

Exclusive:

$\text{nw.y} \overset{\boxed{e}}{\longmapsto} \text{ss.x}$
copy_{ss}

$\left[\text{x: } a\right]_1 \left[\text{x:}\right]_2 + \left[\text{y: } b\right]_3 \implies \left[\text{x: } a\right]_1 \left[\star\text{ x: } b\right]_2$

---

[5] A purely formal adaptation is the use of the list brackets ⟨...⟩ instead of (...) for feature structures.
[6] The value NIL in feature structures is represented here as the 'empty' value.

In these simplified examples, the sentence starts each consist of two woplets. In woplet$_1$ the attribute x has the value a, while in woplet$_2$ the value of x is empty. The next word, represented as woplet$_3$, has the attribute y with the value b.

The additive copying of the first example adds the y value of the next word to all x attributes of the sentence start woplets. The exclusive copying of the second example adds the y value of the next word only to those x attributes of sentence start woplets which have the value NIL.

If a semantic operation specifies copying of a value which happens to be not yet defined (i.e. NIL) in the source attribute, a variable is written into the source attribute and the target attribute(s) of the operation (cf. 23.4.6). When a later semantic operation supplies a defined value for an attribute containing this variable, all instances of the variable are simultaneously substituted by the defined value (cf. 23.4.7).

In exclusive and additive copying, values of the feature structure **sem** are restricted to target attributes of woplets which have the same proposition number as the source woplet(s). Copying values into target attributes of woplets with another proposition number is only permitted in attributes of the feature structure **inx** because there the purpose is the definition of *extra*propositional relations (cf. 23.4.8).

23.4 Example of syntactic-semantic derivation (*LA-E4*)

The semantic interpretation of a LA-SU syntax will now be illustrated with the syntactic-semantic derivation of the following sentence.

23.4.1 The man gave Mary a flower because he loves her.

Each derivation step will be represented explicitly in the format of 23.3.4.

Syntactically, the derivation is based on an extension of *LA-E3* (cf. 17.5.5), called *LA-E4*. The extension is necessary because sentence 23.4.1 contains an adverbial sub-clause, a construction not handled by *LA-E3*. The definition of *LA-E4* uses the format of rule alternatives as presented in 18.4.4 following.

23.4.2 *LA-E4* FOR ADVERBIAL SUBCLAUSES OF ENGLISH

LX = LX of *LA-E3* plus {(slowly (ADP) *), (because (# ADP) *)}
Variable definitions = those of *LA-E3* plus $mn \in \{np \cup \{V, VI\}\}$

$ST_S =_{def}$ { [(x) { 1 DET+ADJ, 2 DET+N, 3 NOM+FV, 4 AUX+MAIN, 5 STRT-SBCL}] }
DET+ADJ: (n x) (ADJ)   ⇒ (n x)   { 6 DET+ADJ, 7 DET+N }
DET+N: (n x) (n)   ⇒ (x)   { 8 NOM+FV, 9 FV+MAIN, 10 AUX+NFV,
                                            11 ADD-ADP, 12 IP }
NOM+FV: (np # x) (np' y V) ⇒ (y # x)

$$(np)\ (np'\ \text{x V}) \quad\quad \Rightarrow\ (\text{x V})\ \{13\ \text{FV+MAIN},\ 14\ \text{AUX+NFV},\ 15\ \text{ADD-ADP},$$
$$16\ \text{IP}\}$$

FV+MAIN: $(np'\ \#\text{x})\ (\text{y}\ np) \quad \Rightarrow\ (\text{y x})$

$(np'\ \text{x}\ \#\text{y})\ (\text{z}\ np) \quad \Rightarrow\ (\text{z x}\ \#\text{y})$

$(np'\ \text{x V})\ (\text{y}\ np) \quad \Rightarrow\ (\text{y x V})\ \{17\ \text{DET+ADJ},\ 18\ \text{DET+N},\ 19\ \text{FV+MAIN},\ 20\ \text{IP}\}$

AUX+NFV: $(aux\ \#\text{x V})\ (aux) \quad \Rightarrow\ (\text{x V})$

$(aux\ \#\text{x V})\ (\text{y}\ aux) \quad \Rightarrow\ (\text{y}\ \#\text{x V})$

$(aux\ \text{V})\ (\text{x}\ aux) \quad \Rightarrow\ (\text{x V})\quad \{21\ \text{FV+MAIN},\ 22\ \text{IP}\}$

AUX+MAIN:$(np\ aux\ \text{V})\ (\text{x}\ np') \quad \Rightarrow\ (\text{x}\ aux\ \text{VI})\ \{23\ \text{AUX+NFV},\ 24\ \text{DET+ADJ},\ 25\ \text{DET+N}\}$

ADD-ADP: $(\text{x ADP})\ (mn\ \text{y}) \quad \Rightarrow\ (\text{x}\ mn\ \text{y})$

$(mn\ \text{y})\ (\text{x ADP}) \quad \Rightarrow\ (\text{x}\ mn\ \text{y})\ \{26\ \text{STRT-SBCL},\ 27\ \text{NOM+FV},\ 28\ \text{FV+MAIN}\}$

STRT-SBCL:$(\#\ \text{x})\ (\text{y}\ np) \quad \Rightarrow\ (\text{y}\ np\ \#\text{x})\ \{29\ \text{DET+ADJ},\ 30\ \text{DET+N},\ 31\ \text{NOM+FV},$
$$32\ \text{ADD-ADP}\}$$

IP: $(vt)\ (vt\ \text{x}) \quad\quad \Rightarrow\ (\text{x}) \quad\quad \{\}$

$ST_F =_{def} \{\ [(\text{V})\ \text{rp}_{ip}],\ [(\text{VI})\ \text{rp}_{ip}]\}$

The feature structures of the next words will be based on automatic conversion of lexical entries into woplets (cf. 23.3.3). For reasons of space, the features **property**, **cont.**(inuation), **index**, and **M-concept** are abbreviated as P, C, I, and M, respectively.

The derivation of example 23.4.1 begins with an application of DET+N:

23.4.3 APPLYING DET+N TO *the + man*

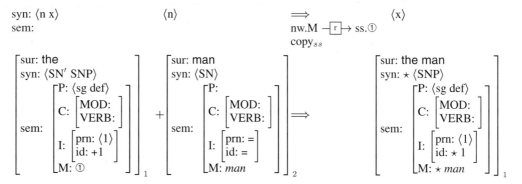

The semantic operation of DET+N replaces the variable ① in the ss by the value of the feature M in the nw (written as **nw.M** $-\boxed{r}\!\!\to$ **ss.**①). Then the woplet of the ss, but not of the nw, is included in the resulting ss (written as **copy**$_{ss}$). The proposition number is incremented by the control structure at the beginning of a new sentence (**prn:** $\langle 1 \rangle$). The incrementation of the **id** value in the result woplet is lexically-based (**id:** +1).

The result of 23.4.3 represents a definite noun phrase with the M-concept *man*. Thus, noun phrases consisting of a function word and a content word like **the man** are built into one woplet by the semantics. This is done by copying the relevant values of the content word into the woplet of the function word. The same holds for complex verb phrases, e.g. **has seen** – with **has** as the function word (cf. 17.3.2). In contrast, the phrasal parts of a complex noun or verb phrase, such as relative or adverbial clauses, are analyzed as elementary propositions which are concatenated with the higher sentence via their id or epr value.

The next composition is based on the rule NOM+FV.

23.4.4 APPLYING NOM+FV TO *the man + gave*

syn: $\langle np \rangle$ $\langle np'\ x\ V \rangle$ $\Longrightarrow$ $\langle x\ V \rangle$
sem:

nw.M $-\boxed{e}\mapsto$ ss.VERB
ss.M $-\boxed{a}\mapsto$ nw.NP
copy$_{ss}$ copy$_{nw}$

$$
\begin{bmatrix}
\text{sur: the man} \\
\text{syn: } \langle SNP \rangle \\
\text{sem: } \begin{bmatrix}
\text{P: } \langle sg\ def \rangle \\
\text{C: } \begin{bmatrix} \text{MOD:} \\ \text{VERB:} \end{bmatrix} \\
\text{I: } \begin{bmatrix} \text{prn: } \langle 1 \rangle \\ \text{id: 1} \end{bmatrix} \\
\text{M: } man
\end{bmatrix}
\end{bmatrix}_1
+
\begin{bmatrix}
\text{sur: gave} \\
\text{syn: } \langle N'\ D'\ A'\ V \rangle \\
\text{sem: } \begin{bmatrix}
\text{P: past tense} \\
\text{C: } \begin{bmatrix} \text{MOD:} \\ \text{NP:} \end{bmatrix} \\
\text{I: } \begin{bmatrix} \text{prn: =} \\ \text{epr:} \end{bmatrix} \\
\text{M: } give
\end{bmatrix}
\end{bmatrix}_3
$$

$$
\Longrightarrow
\begin{bmatrix}
\text{sur:} \\
\text{syn: } \langle SNP \rangle \\
\text{sem: } \begin{bmatrix}
\text{P: } \langle sg\ def \rangle \\
\text{C: } \begin{bmatrix} \text{MOD:} \\ \text{VERB: } \star\ give \end{bmatrix} \\
\text{I: } \begin{bmatrix} \text{prn: } \langle 1 \rangle \\ \text{id: 1} \end{bmatrix} \\
\text{M: } man
\end{bmatrix}
\end{bmatrix}_1
\begin{bmatrix}
\text{sur: the man gave} \\
\text{syn: } \star\ \langle D'\ A'\ V \rangle \\
\text{sem: } \begin{bmatrix}
\text{P: } \langle past\ tense \rangle \\
\text{C: } \begin{bmatrix} \text{MOD:} \\ \text{NP: } \star\ \langle man \rangle \end{bmatrix} \\
\text{I: } \begin{bmatrix} \text{prn: } \star\ \langle 1 \rangle \\ \text{epr:} \end{bmatrix} \\
\text{M: } give
\end{bmatrix}
\end{bmatrix}_3
$$

The semantic operations specify that the M-concept of the ss is to be copied into the continuation attribute of the nw and vice versa (written as ss.M $-\boxed{a}\mapsto$ nw.NP and nw.M $-\boxed{e}\mapsto$ ss.VERB, respectively). The woplets of the ss and the nw are all included in the result (written as copy$_{ss}$ and copy$_{nw}$, respectively). Note that only one woplet in the resulting ss has a non-NIL sur value.

The next composition is based on the rule FV+MAIN.

23.4.5 APPLYING FV+MAIN TO *the man gave + Mary*

syn: $\langle np'\ x\ V \rangle$ $\langle y\ np \rangle$ $\Longrightarrow$ $\langle y\ x\ V \rangle$
sem:

nw.M $-\boxed{a}\mapsto$ ss.NP
ss.M $-\boxed{e}\mapsto$ nw.VERB
copy$_{ss}$ copy$_{nw}$

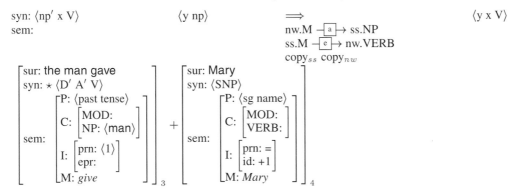

$$
\Longrightarrow
\begin{bmatrix}
\text{sur: the man gave Mary} \\
\text{syn: } \star \langle A'\ V \rangle \\
\text{sem: }
\begin{bmatrix}
\text{C: }
\begin{bmatrix}
\text{P: } \langle \text{past tense} \rangle \\
\text{MOD:} \\
\text{NP: } \langle \text{man}, \star \text{Mary} \rangle
\end{bmatrix} \\
\text{I: }
\begin{bmatrix}
\text{prn: } \langle 1 \rangle \\
\text{epr:}
\end{bmatrix} \\
\text{M: } give
\end{bmatrix}
\end{bmatrix}_3
\begin{bmatrix}
\text{sur:} \\
\text{syn: } \langle \text{SNP} \rangle \\
\text{sem: }
\begin{bmatrix}
\text{C: }
\begin{bmatrix}
\text{P: } \langle \text{sg name} \rangle \\
\text{MOD:} \\
\text{VERB: } \star give
\end{bmatrix} \\
\text{I: }
\begin{bmatrix}
\text{prn: } \star \langle 1 \rangle \\
\text{id: } \star 2
\end{bmatrix} \\
\text{M: } Mary
\end{bmatrix}
\end{bmatrix}_4
$$

The semantic operation nw.M $-\boxed{a}\!\!\rightarrow$ ss.NP in 23.4.5 illustrates additive copying of a value: even though the NP-attribute of *give* already has a value (namely man), Mary is added as an additional filler.

For reasons of space, the following rule applications will not explicitly list woplets which are not modified any further (here the woplet of the man). These woplets are presupposed implicitly, however, because they are needed in the pragmatic embedding of the semantic representation into the contextual word bank (cf. 23.5.3).

The next composition is based on a reapplication of the rule FV+MAIN. Even though the nw is a proper name in 23.4.5 and a determiner in 23.4.6, the rule application is based in both instances on the same syntactic and semantic patterns.

23.4.6 APPLYING FV+MAIN TO *the man gave Mary + a*

$$
\begin{array}{llll}
\text{syn: } \langle np'\ x\ V \rangle & \langle y\ np \rangle & \Longrightarrow & \langle y\ x\ V \rangle \\
\text{sem:} & & \text{nw.M } -\boxed{a}\!\!\rightarrow \text{ ss.NP} & \\
& & \text{ss.M } -\boxed{e}\!\!\rightarrow \text{ nw.VERB} & \\
& & \text{copy}_{ss}\ \text{copy}_{nw} &
\end{array}
$$

$$
\begin{bmatrix}
\text{sur: the man gave Mary} \\
\text{syn: } \langle A'\ V \rangle \\
\text{sem: }
\begin{bmatrix}
\text{C: }
\begin{bmatrix}
\text{P: } \langle \text{past tense} \rangle \\
\text{MOD:} \\
\text{NP: } \langle \text{man, Mary} \rangle
\end{bmatrix} \\
\text{I: }
\begin{bmatrix}
\text{prn: } \langle 1 \rangle \\
\text{epr:}
\end{bmatrix} \\
\text{M: } give
\end{bmatrix}
\end{bmatrix}_3
+
\begin{bmatrix}
\text{sur: a} \\
\text{syn: } \langle \text{SN}'\ \text{SNP} \rangle \\
\text{sem: }
\begin{bmatrix}
\text{C: }
\begin{bmatrix}
\text{P: } \langle \text{sg indef} \rangle \\
\text{MOD:} \\
\text{VERB:}
\end{bmatrix} \\
\text{I: }
\begin{bmatrix}
\text{prn: } = \\
\text{id: } +1
\end{bmatrix} \\
\text{M: } ①
\end{bmatrix}
\end{bmatrix}_5
$$

$$
\Longrightarrow
\begin{bmatrix}
\text{sur: the man gave Mary a} \\
\text{syn: } \star \langle \text{SN}'\ V \rangle \\
\text{sem: }
\begin{bmatrix}
\text{C: }
\begin{bmatrix}
\text{P: } \langle \text{past tense} \rangle \\
\text{MOD:} \\
\text{NP: } \langle \text{man, Mary}, \star ① \rangle
\end{bmatrix} \\
\text{I: }
\begin{bmatrix}
\text{prn: } \langle 1 \rangle \\
\text{epr:}
\end{bmatrix} \\
\text{M: } give
\end{bmatrix}
\end{bmatrix}_3
\begin{bmatrix}
\text{sur:} \\
\text{syn: } \langle \text{SN}'\ \text{SNP} \rangle \\
\text{sem: }
\begin{bmatrix}
\text{C: }
\begin{bmatrix}
\text{P: } \langle \text{sg indef} \rangle \\
\text{MOD:} \\
\text{VERB: } \star give
\end{bmatrix} \\
\text{I: }
\begin{bmatrix}
\text{prn: } \langle 1 \rangle \\
\text{id: } \star 3
\end{bmatrix} \\
\text{M: } ①
\end{bmatrix}
\end{bmatrix}_5
$$

Because the nw is a determiner, its M-attribute has the variable ① as value. This value is copied as a further valency filler into the NP-attribute of *give*.

The next composition illustrates a post-verbal application of DET+N (cf. 23.4.3).

23.4.7 APPLYING DET+N TO *The man gave Mary a + flower*

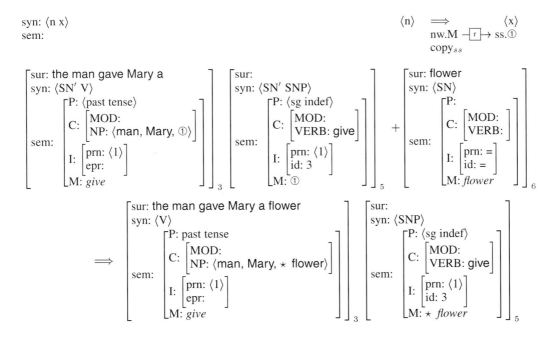

The operation nw.M —[r]→ ss.① replaces all occurrences of the variable ① – placed in the previous rule application – simultaneously by flower. Again, the pre- and postverbal application of this rule is based on the same syntactic and semantic patterns.

23.4.8 APPLYING ADD-ADP TO *The man gave Mary a flower + because*

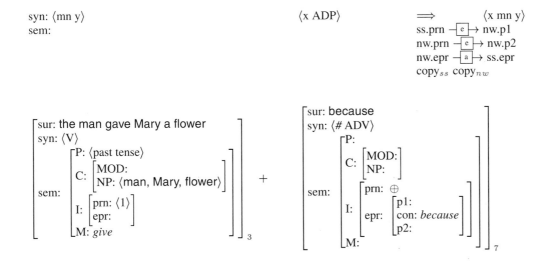

$$\Longrightarrow \quad \begin{bmatrix} \text{sur:} \\ \text{syn: } \langle V \rangle \\ \text{sem:} \begin{bmatrix} \text{P: } \langle \text{past tense} \rangle \\ \text{C: } \begin{bmatrix} \text{MOD:} \\ \text{NP: } \langle \text{man, Mary, flower} \rangle \end{bmatrix} \\ \text{I: } \begin{bmatrix} \text{prn: } \langle 1 \rangle \\ \text{epr: } \star \begin{bmatrix} \text{p1: 1} \\ \text{con: } because \\ \text{p2: 2} \end{bmatrix} \end{bmatrix} \\ \text{M: } give \end{bmatrix} \end{bmatrix}_3 \quad \begin{bmatrix} \text{sur: the ... flower because} \\ \text{syn: } \star \langle \# V \rangle \\ \text{sem:} \begin{bmatrix} \text{P:} \\ \text{C: } \begin{bmatrix} \text{MOD:} \\ \text{NP:} \end{bmatrix} \\ \text{I: } \begin{bmatrix} \text{prn: } \star \langle 2, 1 \rangle \\ \text{epr: } \begin{bmatrix} \text{p1: } \star 1 \\ \text{con: } because \\ \text{p2: } \star 2 \end{bmatrix} \end{bmatrix} \\ \text{M:} \end{bmatrix} \end{bmatrix}_7$$

Controlled by the lexical analysis of **because**, the proposition number of the sub-clause is incremented ([prn: $\oplus$]) and added in front of the previous proposition number (prn: $\langle 2, 1 \rangle$). Also, at the beginning of a new elementary proposition, the nominal identity number is automatically set back to 0 by the control structure.

The conjunction **because** introduces the following **epr** feature structure.

$$\begin{bmatrix} \text{epr: } \begin{bmatrix} \text{p1:} \\ \text{con: } because \\ \text{p2:} \end{bmatrix} \end{bmatrix}$$

Intuitively this may be read as **p1 because p2**, whereby the attributes **p1** and **p2** take proposition numbers as their values. The semantic operations in 23.4.8 first provide values for p1 and p2 (**ss.prn** $-\boxed{e}\!\!\mapsto$ **nw.p1** and **nw.prn** $-\boxed{e}\!\!\mapsto$ **nw.p2**) and then copy the completed **epr** attribute into the verb-woplet of the main clause (**nw.epr** $-\boxed{a}\!\!\mapsto$ **ss.epr**). For simplicity, this complex **epr** value will be written as [epr: 1 bec 2].

23.4.9 APPLYING START-SUBCL TO *The man gave Mary a flower because + he*

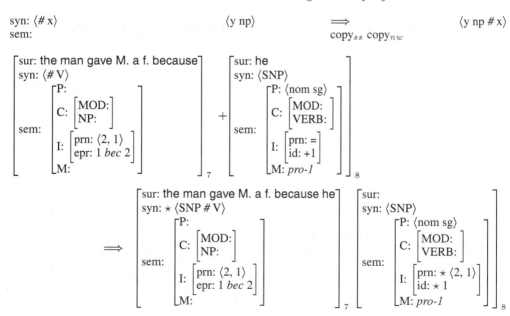

Here the semantic operations consist solely in retaining the ss- and nw-woplets in the resulting ss. Controlled by the lexical analysis of he, the identity number of the first subclause filler is incremented to '1'.

23.4.10 APPLICATION OF NOM+FV TO *The man g. M. a f. because he + loves*

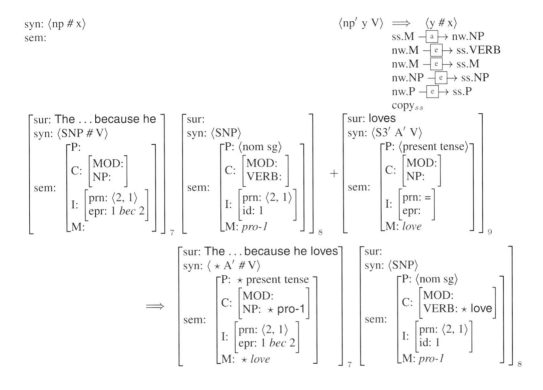

The semantic operations copy the initial noun phrase of the subclause (Pro-1) into the NP-attribute of the verb (ss.M —a⟩→ nw.NP) and the M-concept of the nw into the VERB-attribute of the clause initial noun phrase (nw.M —e⟩→ ss.VERB). Then the relevant values of the nw are copied into the woplet of the conjunction nw.M —e⟩→ ss.M, etc.). Finally, only the complemented woplets of the ss are taken into the resulting ss (copy$_{ss}$).

The last combination in the derivation of 23.3.2 adds the object noun phrase of the subordinate clause using the first rule alternative of FV+MAIN (cf. 23.4.2).

23.4.11 APPLICATION OF FV+MAIN TO *The m. g. M. a f. because he loves + her*

syn: ⟨np′ # x⟩
sem:

⟨y np⟩ ⟹ ⟨y x⟩
nw.np —a⟩→ ss.NP
ss.verb —e⟩→ nw.VERB
ss.prn —m⟩→ ss.prn
copy$_{ss}$ copy$_{nw}$

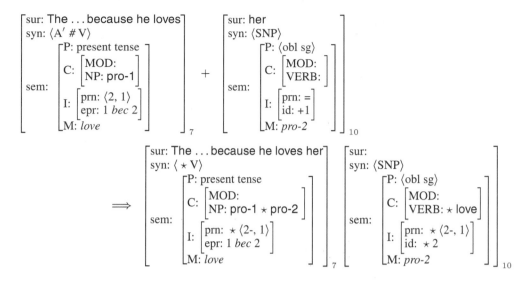

In addition to the usual cross-copying between the verb and the new noun phrase, the semantic operations of this rule contain ss.prn —[m]↦ ss.prn, which serves to deactivate the proposition number of the embedded clause after its completion. The purpose of this operation is especially apparent in embedded subclauses.

23.4.12 PROPOSITION NUMBER OF EMBEDDED SUBCLAUSE

the man,   gave her a flower.
  prn: $\langle 1 \rangle$ because he loves Mary   prn: $\langle 2\text{-}, 1 \rangle$
             prn: $\langle 2, 1 \rangle$

The woplets of the subclause require the proposition number 2, yet after the completion of the subclause the woplets of the main clause remainder must have the earlier proposition number 1. To this purpose the number of the new, embedded proposition is written before the number of the old proposition (thus remembering rather than overwriting the old proposition number). When the embedded proposition is completed, the semantic operation ss.prn —[m]↦ ss.prn deactivates the first unmarked element of the prn value (here $\langle 2\text{-}, 1 \rangle$). In subsequent rule applications, the first non-deactivated (i.e. unmarked) proposition number is used. This method based on a simple stack is suitable to provide the correct proposition number in embedding constructions of arbitrary depth (e.g. 9.2.1).

23.5 From SLIM semantics to SLIM pragmatics

The left-associative syntactic-semantic derivation of example 23.3.2 has resulted in an unordered set of coindexed, bidirectional woplets which is summarized below.

23.5.1 SLIM SEMANTIC REPRESENTATION OF EXAMPLE 23.4.1

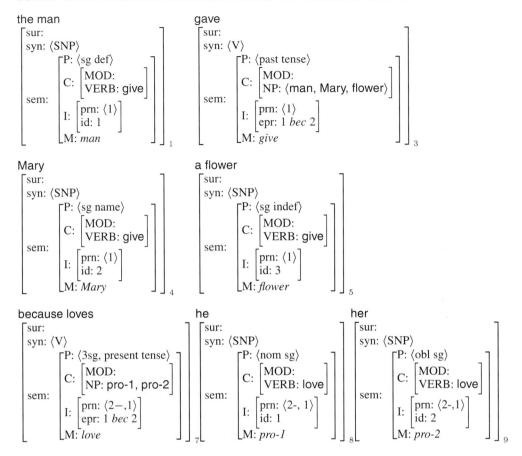

This meaning$_1$ representation of a complex expression illustrates that the semantic representation of natural language consists of the following parts.

23.5.2 COMPONENTS OF MEANING$_1$

– Compositional semantics (sentence semantics)
  1. Decomposition of input into elementary propositions.
  2. Functor-argument structure within an elementary proposition.
  3. Extrapropositional relations among elementary propositions.
– Lexical semantics (word semantics)
  1. Properties and M-concepts of woplets.
  2. Extrapropositional relations between word types by means of *absolute propositions*.

Examples of absolute propositions are a flower is a plant, Mary is a human, etc. They are defined on the contextual level and specify meaning relations between the contextual types by defining the usual *is-a*, *has-a*, *is-part-of*, etc., hierarchies (cf. 24.5.7).

Absolute propositions may be read into the word bank using natural language and differ from episodic propositions only because of the open *loc* values of their I-concepts$_{loc}$. The time-linear navigation through absolute propositions serves SLIM-theoretic inferencing (cf. Section 24.2) in (i) the pragmatic interpretation of the hearer, (ii) the pragmatic generation of the speaker, as well as (iii) non-verbally.

The pragmatic interpretation of the meaning$_1$ representation 23.5.1 begins with embedding the woplets into the contextual word bank.

23.5.3 EMBEDDING 23.5.1 INTO THE CONTEXTUAL WORD BANK

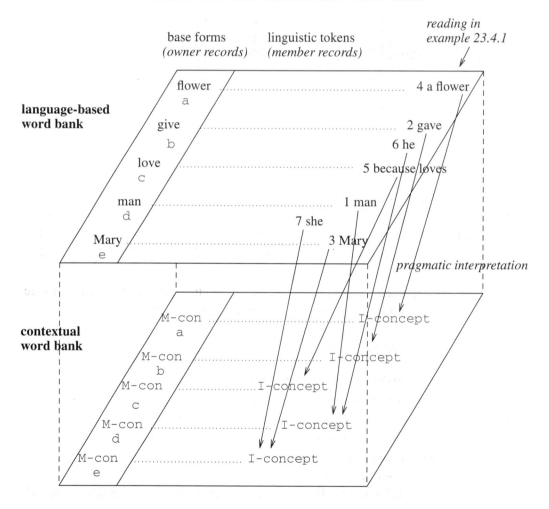

The two horizontal levels correspond to those of CURIOUS in 4.1.3. The upper level contains the linguistic woplets, the lower level the contextual coplets – each in accordance with the structure of the owner and member records of a network database.

Thereby the contextual token lines are arranged directly below the corresponding linguistic token lines. The association between a contextual token line and its linguistic counterpart is based on the M-concept which their respective types have in common (cf. 23.1.1).

For example, in 23.5.3 the contextual token line of *flower* is located directly beneath the linguistic token line of **flower**. In this way the literal use is treated as the default of reference: the search for a suitable contextual referent during pragmatic interpretation begins in the contextual token line located directly underneath the linguistic woplet to be interpreted.[7]

In the pragmatic interpretation of the hearer, the linguistic woplets may refer either (i) to structures already present in the context or (ii) to structures not yet introduced. In the first case, the hearer is reminded of something already known, in the second case the hearer is told something new. A mixture of the two cases is illustrated in 23.5.4 using example 23.3.2.

23.5.4 CONTEXTUAL RECONSTRUCTION OF LANGUAGE INFORMATION

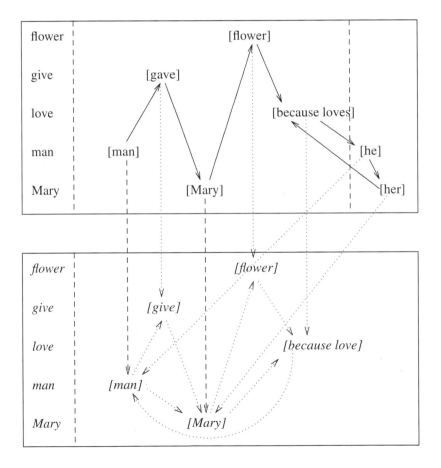

The two boxes represent the linguistic and the contextual level of a SLIM machine whereby contextual features are indicated by italics. The pronouns he and her are stored in a special area for index words.

The meaning$_1$ of the sentence, derived in detail in Section 23.4 and summarized in 23.5.1, is indicated in 23.5.4 as a navigation through the upper, linguistic word bank which powers and controls a corresponding navigation through the lower contextual word bank. For pragmatic interpretation, all linguistic woplets with their intra- and extrapropositional relations are transformed into corresponding contextual coplets and sent to the lower level.

This process, indicated in 23.5.4 by the vertical arrows, is formally facilitated by the fact that woplets and coplets are similar in structure. Turning a linguistic woplet into a contextual coplet requires only deleting the sur feature of the woplet and instantiating the M-concept of the woplet as an I-concept$_{loc}$. For the latter, a pro forma instantiation may be used. In the case of flower, for example, details about how it looks and what its exact *loc* values are may be left unspecified.

The meaning$_2$ understanding of propositions newly embedded into the lower level consists in their pragmatic *anchoring*. This means that the language-based content is related to (i) the time-space parameters and (ii) the content already present in the contextual database. In the case of nominal woplets it requires that (i) their *loc* values specify the place and (ii) their id values specify identity or non-identity with certain other nominal woplets. In the case of verbal woplets it requires that (i) their *loc* values specify the time and (ii) their epr values specify extrapropositional relations to certain other propositions.

Assuming, for example, that the hearer is already familiar with the referents of *the man* and *Mary*, then these referents may exist in the hearer's contextual word bank as a multitude of coplets which all have the id values, e.g. 325 and 627, respectively. In this case, a correct pragmatic anchoring will require that the formerly linguistic proplets of the man and Mary are likewise assigned the id values 325 and 627, respectively.

---

[7] The M-concepts of the language level are fixed to the surfaces of a specific natural language – via definitions which model conventions (cf. 6.2.2). These surfaces belong to an *external* language which has evolved naturally in a language community.

The M-concepts and I-concepts$_{loc}$ of the context level, on the other hand, are the elements of an *internal* system (cf. 3.4.3). This means that they do not have conventionally established connections to particular surfaces. Instead, they function in contextual recognition and action as well as in the representation, storing, and deriving of contextual propositions. Thereby, the contextual M-concepts and I-concepts$_{loc}$ may be used by the system either directly or via some system internal abbreviations, e.g., numbers.

Because the system-internal representations of the context level are not easily read by humans one may refer to them by using the corresponding words of a familiar natural language. In this sense we have used the word *flower* to refer to the corresponding contextual I-concept$_{loc}$. This practice should not be mistaken, however, as if the contextual LA-NA syntax were using the surfaces of some natural language.

Rather, it must be kept in mind that the feature structures of the lower level and the LA-NA syntax operating on them are in principle independent of any particular natural language. This is crucial for theoretical analyses, such as language acquisition, and practical applications, such as machine translation.

Regarding specification of their respective *loc* values, on the other hand, a pro forma instantiation will suffice.

Let us assume that the hearer has not previously known about the giving of the flower and the reason for it, and that the propositions represented in 23.5.1 have been read into the contextual database in terms of the embedding described. To properly anchor this propositional complex pragmatically, it will be sufficient to identify the coplets derived from **the man** and **Mary** with the known entities 325 and 627. This is because the content of the whole propositional complex 23.5.1 is internally connected by intra- and extrapropositional relations.

Additional anchorings may be provided by specifying the time in the *loc* values of the proplets *give* and *because love*. Furthermore, the pragmatic interpretation of the pronouns *he* and *her* requires to assign the correct id values, as with any noun. Though the M-concepts of pronouns have variables as values, their woplets have to be sorted into the token lines of the associated referents, here *man* and *Mary*.

In summary, the pragmatic anchoring of language-based propositions is a gradual procedure. The better a proposition, or a concatenation of propositions, is anchored in (i) the space-time parameters and (ii) the existing contents of the contextual database, the better the meaning$_2$ is understood. Language-based contents read into the contextual word bank of a SLIM machine as pragmatically anchored propositions constitute an extension of the 'rail road system' which may subsequently be traversed by the autonomous time-linear navigation.

Because the coding of concatenated propositions as bidirectionally related, coindexed proplets has been developed as general representation for any content, it is suitable as an interlingua – a language between languages – for machine translation (cf. Section 2.5). Because this interlingua is contextual (i.e., not language-based, cf. 3.4.3) it allows to represent the contents of any natural language. It is thus possible to build a SLIM machine as a multilingual robot.

Exercises

Section 23.1

1. Describe the static structure of a SLIM machine.
2. Describe the motor algorithms of a SLIM machine.
3. Describe the external connections of a SLIM machine.
4. Compare the difference between word types and concept types.

5. Compare the difference between woplets and coplets.
6. Explain the relation between word types and woplets.
7. Explain the relation between concept types and coplets.
8. What is the theoretical principle for relating a woplet to a corresponding coplet?
9. Describe the type-token relations used in the design of the SLIM machine.
10. Compare the external connections of the SLIM machine with the structure 4.1.3 of CURIOUS.
11. What is the central processing unit of a SLIM machine and what are its tasks?
12. How is the [2+1] level structure of the SLIM theory of language realized by the SLIM machine?

Section 23.2

1. Describe the mechanism of natural communication as a system of formal mappings.
2. What is the difference between language-based (verbal) and context-based (non-verbal) cognition in a SLIM machine?
3. What is the difference between immediate and mediated reference in a SLIM machine?
4. How do the speaker and the hearer mode differ in a SLIM machine?
5. Which formal condition controls the choice of woplets for a given sequence of coplets in the speaker mode?
6. Which formal condition controls the choice of coplets for a given sequence of woplets in the hearer mode?
7. Why does a linguistic analysis of immediate reference require a formal treatment of contextual recognition and action in the speaker and the hearer?
8. Explain how immediate reference can be viewed as a special case of mediated reference.

Section 23.3

1. What is the role of morphology in a SLIM machine?
2. Explain the differences between the woplets of nouns, verbs, and adjectives. How are these differences motivated?
3. How is a woplet derived from an analyzed word form during language recognition?
4. Explain the schema of a semantically interpreted LA-SU rule. Why is the LA-SU syntax unaffected by the semantic interpretation?
5. How is the syntactically active woplet of a sentence start formally marked?
6. Explain the semantic operations of a semantically interpreted LA-SU syntax.
7. Explain the difference between additive and exclusive copying. Show examples in which this distinction is needed.
8. Explain why the semantic interpretation of the LA-SU syntax (i) satisfies the homomorphism condition and (ii) is strictly surface compositional.

9. Disassemble the sentences The man who gave Mary a flower loves her and The man bought a flower and gave it to Mary into elementary proposition and depict them in the style of 22.5.2.

Section 23.4

1. Compare the time-linear derivation of a complex noun phrase in pre- and postverbal position from the viewpoint of semantics (cf. 23.4.3 and 23.4.7). Does this require different rules? Are there differences in the derivation?
2. Explain the woplets of function words like the, a, or because. What are the values of their M-attributes?
3. In which way is the extrapropositional relation between the two elementary propositions of 23.3.2 formally realized? Why is it a bidirectional relation?
4. Why is the woplet of the next word not included in the result in 23.4.10?
5. Extend the semantic interpretation of *LA-E4* to a handling of adjectives and adverbs. Explain you extension with the derivation of The man *yesterday* gave Mary a *red* flower.

Section 23.5

1. What is the complexity class of semantically interpreted *LA-E4*?
2. What is the meaning$_1$? Which components does it consist of?
3. Which aspects of meaning$_1$ are captured in the semantic representation 23.5.1?
4. How are woplets turned into corresponding coplets?
5. What does the pragmatic anchoring of a once language-based proposition consist of? What is the difference between the anchoring of nominal and verbal proplets?
6. What is required by the pragmatic interpretation of the pronouns in 23.5.4?
7. How are the meaning$_1$ of expressions and the meaning$_2$ of utterances represented in a SLIM machine?
8. Why is pragmatic interpretation a gradual procedure?
9. The SLIM theory of language comprises the algorithm of LA-grammar, the principles of pragmatics PoP-1 to PoP-7 (including the distinction between meaning$_1$ and meaning$_2$, the STAR-point, and the sign-theoretic analysis of symbols, indices, and names), the principle of surface compositionality, the network database structure of a word bank, the two level structure of a SLIM machine, and the cognitive states of SLIM 1 to SLIM 10 (which define language-based vs. contextual cognition, action, recognition, and inferencing, speaker mode vs hearer mode, mediated vs. immediate reference, and language-based control vs. context-based commenting). Explain how these different components of the theory presuppose and complement each other.

24. SLIM machine in the speaker mode

Modeling the mechanism of natural communication results in an artificial system, though one that resembles its natural prototype as much as possible. This resemblance is needed theoretically for verification of the underlying theory of language and practically for man-machine communication which is maximally user-friendly.

As in all artificial systems, however, the first priority is overall functioning. This requires the reconstruction of procedures the details of which are hidden in the natural model. A case in point is the speaker mode, which requires a modeling of extracting content and mapping it into concrete surfaces.

Section 24.1 illustrates the formal notion of a subcontext in a SLIM machine with an example of nine concatenated propositions. Section 24.2 defines the universal LA-NA syntax for autonomous navigation through subcontexts, and the tracking principles which ensure that the free navigation does not get tangled in loops. Section 24.3 describes how the same navigation may be realized as different language-dependent word order types, namely VSO, SVO, and SOV. Section 24.4 expands the LA-NA syntax to the hypotactic embedding of elementary propositions into each other, which serves as the universal basis of adverbial and relative subclauses. Section 24.5 describes search and inference as additional forms of LA-navigation which are formally treated by means of suitable LA-NA grammars.

24.1 Subcontext as concatenated propositions

According to the SLIM theory of language, the speaker's conceptualization is based on the autonomous navigation through the concatenated propositions of the contextual word bank. Thereby the coherence[1] of language follows immediately from the coherence of the subcontext traversed because during production the navigation is realized immediately in language.

Depending on whether the propositions of a SLIM machine have been read in via direct recognition or via indirect (i.e., language-based, film-based, etc.) representations,

---

[1] Coherence must be distinguished from cohesion (see M.A.K. Halliday & R. Hasan 1976 and R.-A. Beaugrande & W.U. Dressler 1981). Coherence refers to a content making sense and applies to the conceptualization in language production (*what to say*). Cohesion refers to the form in which a content is represented in language and applies to the correct placement of pronouns, the correct theme-rheme structure etc. (*how to say it*). As shown by example 6.1.2, a text may be coherent even if its cohesion is deficient.

we distinguish between immediate and mediated subcontexts. In immediate subcontexts, the coherence of the content follows directly from the coherence of the external world which they reflect, i.e., the temporal and spatial sequence of events, the part-whole relations of objects, etc.

For example, the representation of a swimmer standing at the pool side, diving into the water, and disappearing with a splash is coherent. In contrast, a representation in which a pair of feet appears in the foaming water and a swimmer flies feet first into the air landing on the pool side, would be incoherent – unless it is specified in addition that the representation happens to be, e.g., a backward running movie.

Correspondingly, a representation of people talking with each other would be coherent. In contrast, a similar representation of a deer conversing with a skunk in English would be incoherent – unless it is specified in addition that the representation happens to be fictional.

Mediated subcontexts thus have the special property that the elements familiar from direct recognition may be reordered and reconnected by the author at will. Specification of the author in the STAR point (cf. Sections 5.3 – 5.5) is thus not only needed for the correct pragmatic anchoring of language-based content, but also in order to judge the reliability, seriousness, and purpose of the information.

Mediated subcontexts may also reflect the coherence of the external world, however. In language, the coherence of the world is reflected whenever the speaker navigates through an immediate subcontext and puts its contents automatically into words. This simplest case of language-based coherent subcontexts comes about in terms of the following sequence of mappings:

$$\text{world} \rightarrow \text{speaker context} \rightarrow \text{language} \rightarrow \text{hearer context} \rightarrow \text{world}$$

Here coherence of the immediate speaker context and the mediated hearer context follows from their representing external reality directly and indirectly, respectively.[2]

The autonomous navigation through a coherent subcontext and its representation in language require the definition of a subcontext in the word bank format. In contradistinction to the earlier examples 22.4.2, 22.5.3, and 23.5.4, the following subcontext describes a sequence of events. To facilitate the intuitive orientation, the subcontext is presented first as a sequence of simple sentences.

24.1.1 A SEQUENCE OF PROPOSITIONS FORMING A SUBCONTEXT

1. Peter leaves the house. 2. Peter crosses the street. 3. Peter enters a restaurant. 4. Peter orders a salad. 5. Peter eats the salad. 6. Peter pays the salad. 7. Peter leaves the restaurant. 8. Peter crosses the street. 9. Peter enters the house.

This language-based representation consists of 11 types and 27 tokens of content words and corresponds to the following word bank.

---

[2] The hearer can often check the coherence of a mediated context by comparing it with the corresponding part of the world.

24.1.2 EQUIVALENT REPRESENTATION OF 24.1.1 AS A WORD BANK

CONCEPT TYPES: COPLETS:

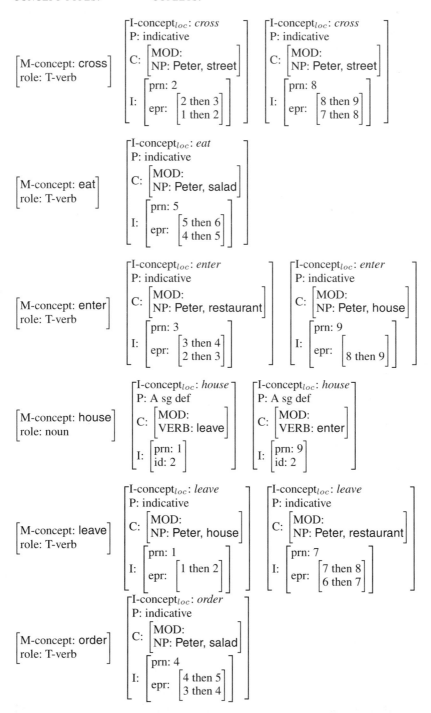

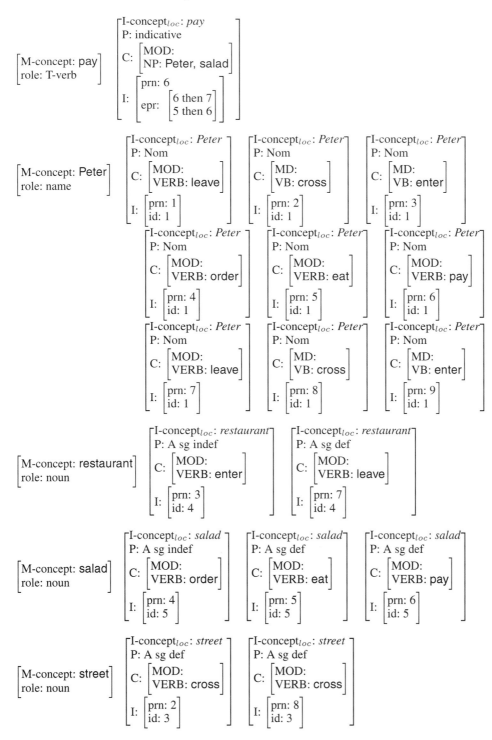

The proplets have no **sur**, **syn**, and **M-concept** features (compare 23.5.1) because this example represents the contextual level. The place of the **sur** features is taken instead by I-concepts$_{loc}$ represented by corresponding base forms.

Like the linguistic woplets, the contextual coplets code *intra*propositional relations in terms of continuation features and the proposition number. Furthermore, *extra*propositional relations are coded in the epr-features of verbal and the id-features of nominal coplets, respectively.

Verbal coplets at the beginning or end of an epr-concatenation have an epr-attribute with only one value. All the other verbal coplets in an extrapropositional chain have epr-attributes with two values, one for the predecessor and one for the successor proposition. Thus, the epr-attribute of verbal coplets results in *bidirectional* relations between propositions.[3]

The extrapropositional relations coded by the epr- and id-values are firstly motivated semantically because they express important aspects of content. Secondly, they realize the notion of subcontext needed to restrict the set of possible referential candidates (cf. Section 5.2). Thirdly, they constitute the 'rail road system' for the most simple (i.e., non-inferring) form of navigation, which in turn is the basis for generating certain syntactic structures of natural language.

In particular, epr-concatenations of verbal coplets provide the basis for a navigation type which natural languages realize universally as the conjunction of main clauses (Section 24.3) and the embedding of adverbial clauses (Section 24.4). Furthermore, id-concatenations of nominal coplets provide the basis for a navigation type which natural languages realize universally as relative clauses (Section 24.4).

24.2 Tracking principles of LA-navigation

The most general form of navigation is the accidental, aimless motion of the focus point through the contents of a word bank analogous to *free association* in humans. Giving thought free rein is the best starting point for a general treatment of language production because it requires the automatic realization of arbitrary thought paths as a sequence of grammatically and rhetorically adequate natural language expressions.[4]

The navigation through coherent contexts ensures the coherence of the associated language expressions and provides an essential aspect of cohesion, namely the serialization of the language expressions. In this way, the navigation-based conceptual-

---

[3] While the different verbal coplets of an epr-concatenation are usually located in different token lines, e.g. leave-prn:1 then cross-prn:2, the nominal coplets with identical id-values occur usually in the same token line, e.g. Peter-prn:1-id:1, Peter-prn:2-id:1, Peter-prn:3-id:1, etc. If there are several coplets with different id-values for a given type, e.g. ... restaurant-prn:7-id:4 ... restaurant-prn:15-id:11 ... restaurant-prn:18-id:4 (assuming a suitable extension of 24.1.2), then for a given coplet the counterparts with the same id-value can be easily found by searching linearly through the token line.

Should the need arise to express more complex identity relations between the elements of different nominal token lines, a general id-index may be defined to provide for any id-value all corresponding nominal coplets in the database.

[4] Free association may be overruled by external or internal demands which activate specific navigation patterns appropriate to the situation at hand.

ization of the Slim theory of language minimizes the traditional distinction between content (*what to say*) and form (*how to say it*).

The dynamic process of navigation through a word bank is powered by the LA-NA syntax (motor algorithm). Its rules have the task of moving the mental focus point from a given coplet to a possible continuation coplet. A successful LA-NA rule application is illustrated schematically in 24.2.1, 24.2.2 and 24.2.3 as a three step procedure, using the example of an intrapropositional navigation.

The first step consists in matching the start pattern of the rule onto a START coplet.

24.2.1 Step 1 of a LA-NA rule application

$$
\text{rule}_{1+2\Rightarrow2}: \quad
\overset{\text{START}}{\begin{bmatrix} m1: a \\ M2: b \\ prn: c \end{bmatrix}}
\overset{\text{NEXT}}{\begin{bmatrix} m2: b \\ M1: x\,a\,y \\ prn: c \end{bmatrix}} \Longrightarrow
\overset{\text{NEW START}}{\begin{bmatrix} m2: b \\ \\ \end{bmatrix}} \text{rule package}_{1+2\Rightarrow2}
$$

$$
\begin{matrix} \text{coplets} \\ \text{of the} \\ \text{word bank} \end{matrix} \quad
\begin{bmatrix} \ldots \\ m1: c1 \\ \ldots \\ M2: c2 \\ \ldots \\ prn: c3 \\ \ldots \end{bmatrix}_1
$$

This operation is formally based on those attributes which the rule pattern and the start coplet have in common – here m1, M2, and prn. If the start pattern contains attributes not matched by the coplet, the rule is not applicable.

By matching the start pattern onto a word bank coplet, the variables in the pattern are assigned values. For example, the variable a in 24.2.1 is assigned the value $c1$, the variable b the value $c2$, and the variable c the value $c3$. These assignments hold for the duration of the rule application.

The second step has the purpose of *finding* the NEXT.

24.2.2 Step 2 of an LA-NA rule application

$$
\text{rule}_{1+2\Rightarrow2}: \quad
\overset{\text{START}}{\begin{bmatrix} m1: a \\ M2: b \\ prn: c \end{bmatrix}}
\overset{\text{NEXT}}{\begin{bmatrix} m2: b \\ M1: x\,a\,y \\ prn: c \end{bmatrix}} \Longrightarrow
\overset{\text{NEW START}}{\begin{bmatrix} m2: b \\ \\ \end{bmatrix}} \text{rule package}_{1+2\Rightarrow2}
$$

$$
\begin{matrix} \text{coplets} \\ \text{of the} \\ \text{word bank} \end{matrix} \quad
\begin{bmatrix} \ldots \\ m1: c1 \\ \ldots \\ M2: c2 \\ \ldots \\ prn: c3 \\ \ldots \end{bmatrix}_1
\;+\;
\begin{bmatrix} \ldots \\ m2: c2 \\ \ldots \\ M1: ...c1.. \\ \ldots \\ prn: c3 \\ \ldots \end{bmatrix}_2
$$

This step is structurally based on that each coplet in a word bank explicitly specifies its continuation predicates.[5] For example, the START pattern in 24.2.1 contains the continuation feature [M2: b], the variable of which is repeated in the feature [m2: b] of the NEXT pattern. Because the variables have been assigned values in the initial step of the rule application, the coplet matching the NEXT pattern is uniquely determined.[6]

The third step of the LA-NA rule application illustrated here consists in outputting the coplet matching the NEXT as the result (value of the NEW START).

24.2.3 STEP 3 OF A LA-NA RULE APPLICATION

$$
\text{rule}_{1+2\Rightarrow 2}: \quad
\overset{\text{START}}{\begin{bmatrix} \text{m1: } a \\ \text{M2: } b \\ \text{prn: } c \end{bmatrix}}
\quad
\overset{\text{NEXT}}{\begin{bmatrix} \text{m2: } b \\ \text{M1: } x\,a\,y \\ \text{prn: } c \end{bmatrix}}
\Longrightarrow
\overset{\text{NEW START}}{\begin{bmatrix} \text{m2: } b \\ \\ \\ \end{bmatrix}}
\text{rule package}_{1+2\Rightarrow 2}
$$

coplets of the word bank

$$
\begin{bmatrix} \dots \\ \text{m1: } c1 \\ \dots \\ \text{M2: } c2 \\ \dots \\ \text{prn: } c3 \\ \dots \end{bmatrix}_1
+
\begin{bmatrix} \dots \\ \text{m2: } c2 \\ \dots \\ \text{M1: } ..c1.. \\ \dots \\ \text{prn: } c3 \\ \dots \end{bmatrix}_2
\Longrightarrow
\begin{bmatrix} \dots \\ \text{m2: } c2 \\ \dots \\ \text{M1: } ..c1.. \\ \dots \\ \text{prn: } c3 \\ \dots \end{bmatrix}_2
$$

The next LA-NA rule takes the coplet of the new START as the sentence start and uses it to look for another next coplet. In this way, a suitable LA-NA syntax may traverse all the intra- and extrapropositional relations of a subcontext.

During free navigation, two problems may arise which require a general solution, namely relapse and split. A relapse arises if an LA-navigation continues to traverse the same coplet sequence in a loop. A split arises if a given START coplet is accepted by more than one LA-NA rule in an active rule package.

Relapse and split are avoided by general tracking principles which control free LA-navigation. Their formal basis are counters which indicate for each coplet when it was traversed. Adding traversal counters gives the static nature of a word bank a dynamic aspect insofar as they make the course of the navigation visible as a track.

24.2.4 TRACKING PRINCIPLES OF LA-NAVIGATION

1. *Completeness*
   Within an elementary proposition those coplets are preferred which have not yet been traversed during the current navigation.
2. *Uniqueness*
   If several START or NEXT coplets are available, no more than one of each are

---

[5] The attributes of continuation predicates are formally marked by upper case.

[6] In intrapropositional continuations, a variable in the START pattern is useful only if it is repeated in the NEXT pattern. An exception is sequence variables (e.g. x and y in [M1:xay]) used for a flexible handling of coplet patterns.

selected whereby the choice may be at random or – if activated – based on a specific navigation pattern.

3. *Recency*

In extrapropositional navigations, propositions which have been least recently traversed are preferred.

4. *Frequency*

When entering a new subcontext, the navigation prefers paths most frequently traversed in previous navigations.

The first principle prevents intrapropositional relapses and premature extrapropositional continuations. The second principle prevents an LA-navigation from traversing several thought paths simultaneously. The third principle prevents extrapropositional relapses and ensures that a subcontext is traversed as completely as possible in the course of a free navigation. According to the fourth principle, the free navigation of a new subcontext first traverses the propositions which have been traversed most frequently in the past – whereby it is assumed that under the normal circumstances of everyday life the most important navigation patterns are activated most often.[7]

Viewed in isolation, a LA-NA syntax simply motors the navigation through a word bank whereby the focus point is represented by the current 'next coplet' of the current LA-NA rule application. The following example of a LA-NA syntax consists of four rules which control the navigation through (i) elementary propositions (V+NP1, V+NP2) and (ii) extrapropositional relations (V+epr, NP+id).

24.2.5 DEFINITION OF UNIVERSAL LA-NA SYNTAX

ST_S: $\{([\text{M-np: } a] \{1 \text{ V+NP1}, 2 \text{ V+NP2}\})\}$

V+NP1: $\begin{bmatrix} \text{M-verb: } a \\ \text{NP: } x\,b\,y \\ \text{prn: } m \end{bmatrix} \begin{bmatrix} \text{M-np: } b \\ \text{VERB: } a \\ \text{prn: } m \end{bmatrix} \implies \begin{bmatrix} \text{M-verb: } a \\ \\ \end{bmatrix}$ $\{3 \text{ V+NP1}, 4 \text{ V+NP2}, 5 \text{ V+epr}\}$

V+NP2: $\begin{bmatrix} \text{M-verb: } a \\ \text{NP: } x\,b\,y \\ \text{prn: } m \end{bmatrix} \begin{bmatrix} \text{M-np: } b \\ \text{VERB: } a \\ \text{prn: } m \end{bmatrix} \implies \begin{bmatrix} \text{M-np: } b \\ \\ \end{bmatrix}$ $\{6 \text{ NP+id}\}$

V+epr: $\begin{bmatrix} \text{M-verb: } a \\ \text{NP: } x \\ \text{prn: } m \\ \text{epr: } m\,C\,n \end{bmatrix} \begin{bmatrix} \text{M-verb: } b \\ \text{NP: } y \\ \text{prn: } n \\ \text{epr: } m\,C\,n \end{bmatrix} \implies \begin{bmatrix} \text{M-verb: } b \\ \\ \end{bmatrix}$ $\{7 \text{ V+NP1}, 8 \text{ V+NP2}\}$

NP+id: $\begin{bmatrix} \text{M-np: } a \\ \text{VERB: } b \\ \text{prn: } k \\ \text{id: } m \end{bmatrix} \begin{bmatrix} \text{M-np: } a \\ \text{VERB: } c \\ \text{prn: } l \\ \text{id: } m \end{bmatrix} \implies \begin{bmatrix} \text{M-verb: } c \\ \text{NP: } x\,a\,y \\ \text{prn: } l \end{bmatrix}$ $\{9 \text{ V+NP1 } 10 \text{ V+NP2}\}$

ST_F: $\{ ([\text{M-verb: x}] \text{ rp }_{\text{V+NP1}})\}$

---

[7] Cf. E. de Bono 1969.

The attributes M-verb and M-np take verbal and nominal M-concepts, respectively, as values, which match the corresponding I-concepts$_{loc}$ of the input coplets. The variables in 24.2.5 have the following restrictions:

a, b, c stand for individual M-concepts;

k, l, m, n stand for numbers;

x, y, z stand for arbitrary sequences of zero, one, or more M-concepts;

C stands for conjunctions like and, then, because, etc.

The pattern [epr: $m\ C\ n$] in VERB+epr can be interpreted forward (from proposition n to m) and backward (from proposition m to n, cf. 24.4.9).

V+NP1 and V+NP2 have the same input conditions, but render the verb and the noun as output, respectively. The rule V+NP1 serves (i) intrapropositional navigation and (ii) extrapropositional epr-continuation because its output is the verb. The rule V+NP2 is leads directly into an extrapropositional id-continuation because its output is the noun.

V+NP1 and V+NP2 are applied in parallel (cf. rule packages rp$_{V+NP1}$, rp$_{V+epr}$, and rp$_{NP+id}$), which may result in a split between an epr- and an id-continuation. If several continuations are possible simultaneously, one must be selected according to the second tracking principle (uniqueness).

The navigation through an elementary three-place proposition with a subsequent extrapropositional epr-navigation has the following continuation structure.

24.2.6 EXTRAPROPOSITIONAL epr-NAVIGATION

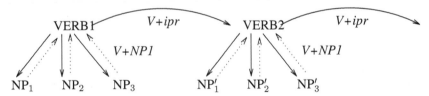

After the initial navigation from a three-place verb to a nominal coplet (V+NP1), the focus is set back to the verb. After the third application of V+NP1 the proposition has been traversed completely. Now the verb of another proposition may be accessed using V+epr, if the current verbal coplet provides an epr-relation.

An extrapropositional navigation via the id-feature from one coreferential nominal coplet to the next has the following continuation structure.

24.2.7 EXTRAPROPOSITIONAL id-NAVIGATION

In contradistinction to 24.2.6, the focus is not set back to the verb after traversing the last NP of the first proposition, but NP₃ is used instead as the input to the rule NP+id. This rule takes two coreferential nominal coplets as input and renders the verbal coplet associated with the second nominal coplet as output.

To illustrate the working of LA-NA we select an arbitrary verb in the word bank 24.1.2, for example the coplet *eat* with the prn-value 5. The start state ST_S of LA-NA provides the rules V+NP1 and V+NP2. Because they take the same input, they apply simultaneously.[8]

In accordance with the tracking principle of uniqueness we choose the coplet *salad* as the continuation.[9] This constitutes the beginning of a navigation within proposition 5 which may be realized in English as **The salad was eaten (by Peter)**.

24.2.8 FIRST APPLICATION OF V+NP1 IN THE WORD BANK 24.1.2

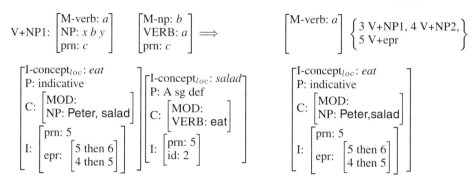

The output of V+NP1 is the coplet **eat**. Of the rules activated by the rule package rp_{V+NP1}, V+NP1, and V+NP2 are equally applicable, while V+epr is ignored in accordance with the tracking principle of completeness.

24.2.9 SECOND APPLICATION OF V+NP1 IN THE WORD BANK 24.1.2

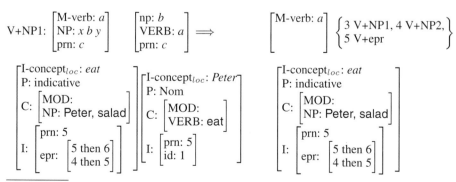

---

[8] This does not violate the second tracking principle (uniqueness) because only *one input* is provided for the composition. The path established by V-NP2 is discontinued at the next composition step because *eat* is a two-place verb (tracking principle of completeness).

[9] As a universal navigation syntax, LA-NA is independent of specific natural languages. Therefore, the 'deep' cases (cf. C. Fillmore 1968, 1977) of nominal coplets in elementary propositions may be traversed in any order.

The application of V+NP2 would be similar except that it would render the coplet *Peter* as output. At this point, all the coplets of proposition 5 have been traversed.

After the simultaneous application of V+NP1 and V+NP2, two START coplets are available, namely *eat* and *Peter*. The first may be continued with V+epr, the second with NP+id. In line with the tracking principle of uniqueness one of them must be chosen (here V+epr).

24.2.10 APPLICATION OF V+EPR IN THE WORD BANK 24.1.2

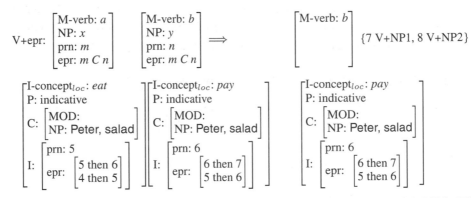

Next the navigation continues as in 24.2.8 either using V+NP1 or V+NP2. The navigation described would be realized in English as The salad was eaten by Peter. Then he paid for it.

24.3 Interpreting autonomous LA-navigation with language

The rules of the LA-NA syntax define a finite state backbone, just like the LA-SU grammars for natural language syntax (see for example 17.5.6 or 18.5.9),

24.3.1 THE FINITE STATE BACK BONE OF LA-NA

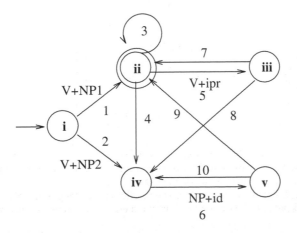

The language specific LA-SU syntax and the universal LA-NA syntax constitute the two end points of natural language interpretation and production. They are connected via LA-SU semantics↔ and LA-SU pragmatics ↕.

24.3.2 UNIVERSALITY AND LANGUAGE SPECIFICITY IN A SLIM MACHINE

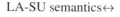

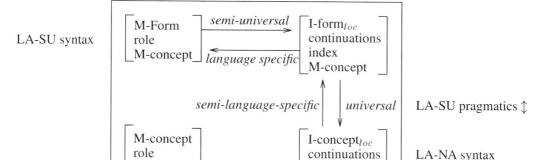

In the hearer mode, LA-SU syntax, LA-SU semantics→ , LA-SU pragmatics↓, and LA-NA syntax are switched together (cf. SLIM 4, 6, and 8 in Section 23.2). The input is analyzed by the language specific LA-SU syntax (motor algorithm) and mapped by the semi-universal LA-SU semantics→ into a set of woplets. Compared to the sequential input to the LA-SU syntax, this set is semi-universal because language specific properties such as word order, inflectional markings, function words, etc., have been eliminated (cf. 23.5.1). The result of the LA-SU pragmatics↓ is universal because the woplets of the semantic interpretation are modified into pragmatically anchored coplets which correspond to a non-language-based representation.

In the speaker mode, LA-NA syntax, LA-SU pragmatics↑, LA-SU semantics←, and LA-SU syntax are switched together (cf. SLIM 5, 7, and 9 in Section 23.2). The input to the LA-SU pragmatics↑ is a sequence of coplets provided by the universal LA-NA syntax (motor algorithm). The LA-SU pragmatics↑ is semi-language-specific because word order properties of the natural language to be generated are being handled at this stage.[10] The LA-SU semantics← is language specific because it transfers the woplet sequence produced by the LA-SU pragmatics↑ into natural language surfaces by providing the function words and the proper inflectional forms.

The order of the proplet sequence generated by the LA-SU pragmatics↑ depends firstly on the underlying navigation and secondly on the natural language. Accord-

---

[10] In certain respects, the LA-SU pragmatics↑ may be regarded as a *semantic* interpretation of the LA-NA syntax because it provides the syntactic navigation procedure with a second level (cf. 19.1.1), i.e. the linguistic woplets. The choice of the term LA-SU pragmatics↑ instead of LA-NA semantics is motivated by the fact that the coplets traversed by the LA-NA syntax and copied as input to the LA-NA pragmatics are strictly speaking not a language (cf. 3.4.3).

ing to J. Greenberg's 1963 language typology there are three basic word orders, namely VSO (verb-subject-object), SVO (subject-verb-object), and SOV (subject-object-verb).

Different LA-SU pragmatics↑ generate these word order types by using a buffer into which the coplets traversed by the time-linear LA-NA syntax are copied. By incrementally delaying the realization vis-à-vis the copying, different LA-SU pragmatics↑ may generate different word order types using the same LA-NA navigation.

For example, assuming the LA-NA navigation traverses the coplets a, b, and c, then a is copied first into the buffer by LA-SU pragmatics↑ – but not yet realized. When the next coplet b is copied into the buffer, there is a choice as to whether a or b should be realized. If b is realized and c is copied into the buffer, the buffer content is < a b* c >, whereby * marks the coplet already realized. Again, there is a choice as to whether a or c should be realized. If the latter is chosen, the realization will result in the surface order b c a, even though the navigation has the order < a b c >.

In this way, the basic word orders VSO, SVO, and SOV may be realized as follows.

24.3.3 REALIZATION PRINCIPLES OF THE BASIC WORD ORDERS

VSO languages

```
                                     realization
                                          ↑
V+NP
buffer: [Verb] + [NP₁]  ⟹  [Verb] [NP₁]
```

$$\text{V+NP} \quad \text{realization} \atop \uparrow$$

$$\text{buffer: [Verb]} + [NP_1] \Longrightarrow \text{[Verb] } [NP_1]$$

$$\text{V+NP}$$
$$\text{buffer: *[Verb] } [NP_1] + [NP_2] \Longrightarrow \text{*[Verb] } [NP_1] [NP_2]$$
realization ↑

$$\text{V+NP}$$
$$\text{buffer: *[Verb] *}[NP_1] [NP_2] + [NP_3] \Longrightarrow \text{*[Verb] *}[NP_1] [NP_2] [NP_3]$$
realization 1↑ 2↑

SVO languages

$$\text{V+NP}$$
$$\text{buffer: [Verb]} + [NP_1] \Longrightarrow \text{[Verb] } [NP_1]$$
realization ↑

$$\text{V+NP}$$
$$\text{buffer: *}[NP_1] \text{ [Verb]} + [NP_2] \Longrightarrow \text{*}[NP_1] \text{ [Verb] } [NP_2]$$
realization ↑

$$\text{V+NP}$$
$$\text{buffer: *}[NP_1] \text{ *[Verb] } [NP_2] + [NP_3] \Longrightarrow \text{*}[NP_1] \text{ *[Verb] } [NP_2] [NP_3]$$
realization 1↑ 2↑

SOV languages

$$\text{V+NP}$$
$$\text{buffer: [Verb]} + [NP_1] \Longrightarrow \text{[Verb] } [NP_1]$$
realization ↑

$$\text{V+NP}$$
$$\text{buffer: *}[NP_1] \text{ [Verb]} + [NP_2] \Longrightarrow \text{*}[NP_1] \text{ [Verb] } [NP_2]$$
realization ↑

$$\text{V+NP}$$
realization 2↑ 1↑

buffer: *[NP$_1$] *[NP$_2$] [Verb] + [NP$_3$] $\Longrightarrow$ *[NP$_1$] *[NP$_2$] [Verb][NP$_3$]

The different realization types are based on the same navigation [Verb] [NP$_1$] [NP$_2$] [NP$_3$].[11] The copying of the current next coplet into the buffer is indicated by '+'. The content of the buffer after application of a LA-NA rule is shown to the right of the arrow. The coplets which have already been realized are marked by *.

The SVO and the SOV languages show that an incremental delay of one coplet (regarding the pragmatic realization vis-à-vis the LA-NA syntax 24.2.5) is structurally necessary. On the one hand, the verb is needed first for specifying the nominal continuation values at the level of the LA-NA syntax. On the other hand, the verb must follow one or more nominals at the level of the surface.

24.4 Subordinating navigation

The id- and epr-continuations permit a paratactic traversal of elementary propositions which is realized in language as a sequence of main clauses. For example, the two concatenated elementary propositions 1 and 2 of the subcontext 24.1.2 allow a temporal and an antitemporal epr-navigation corresponding to the following sentences.

24.4.1 EPR-CONCATENATION

> Peter leaves the house. Then he crosses the street.
> Peter crosses the street. Before that he leaves the house.

The subcontext 24.1.2 permits also paratactic id-navigations such as the following.

24.4.2 ID-CONCATENATION

> Peter orders a salad. The salad is eaten by Peter.

Besides paratactic concatenation of elementary propositions, the natural languages permit hypotactic *embedding*. One universal prototype is *adverbial clauses*. It is based on an epr-continuation.

24.4.3 EPR-SUBORDINATION (ADVERBIAL CLAUSES)

> Before Peter crosses the street, he leaves the house.
> Peter, before he crosses the street, leaves the house.
> Peter leaves, before he crosses the street, the house.
> Peter leaves the house, before he crosses the street.
> After Peter leaves the house, he crosses the street.
> Peter, after he leaves the house, crosses the street.
> Peter crosses, after he leaves the house, the street.
> Peter crosses the street, after he leaves the house.

The other universal prototype is *relative clauses*. It is based on an **id**-continuation.

24.4.4 ID-SUBORDINATION (RELATIVE CLAUSE)

Peter, who leaves the house, crosses the street.

The surfaces in 24.4.1, 24.4.3, and 24.4.4 are based on the same propositional content. The differences in the surfaces are direct reflections of different navigations.[12]

In the embedding constructions 24.4.3 and 24.4.4, traversal of a new proposition is initiated before the current elementary proposition has been completely traversed. The formal handling of these embedding navigations is based on a slight modification of the tracking principles, leaving the LA-NA syntax 24.2.5 unchanged. The modification consists in placing return markers ▼ in the word bank. The return marker indicates where the current proposition has been left for an embedded one.

Each time a return marker has been placed, the tracking principle of completeness is transferred from the current proposition to the embedded proposition. This procedure is recursive such that each embedding may be followed by another one. As soon, however, as an embedded proposition has been traversed completely, the navigation returns automatically to the return marker placed last and removes it. Thereby the tracking principle of completeness is automatically reactivated for the next higher proposition. As soon as that proposition has been traversed completely, the procedure is repeated until the highest elementary proposition is reached again.

Consider, for example, an **id**-navigation underlying the beginning of a relative clause.

24.4.5 APPLYING NP+ID IN THE WORD BANK 24.1.2

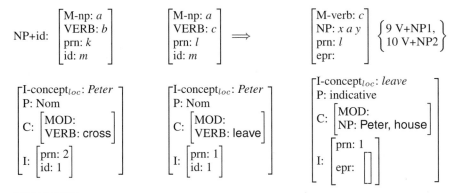

---

[11] Intrapropositional navigation uses only the rule V+NP. The distinction between V+NP1 and V+NP2 applies to *extra*propositional **epr**- and **id**-continuations, for which reason it is omitted here.

[12] Within the SLIM theory of language the phenomena of topic-comment (or theme-rheme) structure in combination with word order, choice of verbal mood, of pronomina, of hypo- versus parataxis, etc., are all treated uniformly as linguistic reflexes of particular types of navigation for particular communicative purposes.

A corresponding navigation traversing the propositions 2 and 1 of the word bank 24.1.2 is shown in 24.4.6, which indicates prn- and id-values of the coplets, the return marker ▼, and its removal – represented as ▲.

24.4.6 Adnominal embedding navigation (preverbal)

Peter, who leaves the house, crosses the street.

▼*cross*	*Peter*	NP+id:		*street*
prn:2	prn:2	*leave*	*house* ▲	prn: 2
	id: 1	prn:1	prn:1	id: 2
			id:3	

The return marker is placed by the rule V+NP2 at the verbal coplet because there the nominal continuations – needed after the return into the current clause – are specified. V+NP2 calls the rule NP+id. The output of NP+id is the verbal coplet specified by the second nominal input coplet. In this way the nominal embedding may be continued immediately with the rule V+NP.

The language specific word order corresponding to this type of universal LA-NA syntax is controlled by the respective semi-language-specific LA-SU pragmatics↑ (cf. 24.3.3).[13] For example, the clause-final position of the verb in subordinate clauses of German is realized as follows.

24.4.7 Word order of adnominal embedding in German

Peter, der das Haus verlassen hat, überquert die Straße.
Peter, who the house left-has, crosses the street.

V+NP2
$$\text{buffer: } [\text{Verb}] + [\text{NP}_1] \implies \overset{\text{realization}}{\overset{\uparrow}{[\text{Verb}]}} \, ▼ \, [\text{NP}_1]$$

NP+id
$$\text{buffer: } {}^*[\text{NP}_1] \, [\text{Verb}] \, ▼ + [\text{NP1}'] \implies {}^*[\text{NP}_1] \, \text{PRO} \, \overset{\text{realization}}{\overset{\uparrow}{[\text{Verb}]}} \, ▼ \, [\text{Verb}']$$

V+NP1
$$\text{buffer: } {}^*[\text{NP}_1] \, {}^*[\text{PRO}] \, [\text{Verb}] \, ▼ \, [\text{Verb}'] + [\text{NP2}'] \implies {}^*[\text{NP}_1] \, {}^*[\text{PRO}] \, [\text{Verb}] \, ▲ \, [\text{Verb}'] \, [\text{NP2}']$$

with realization 2↑ 1↑

V+NP1
$$\text{buffer: } {}^*[\text{NP}_1] \, {}^*[\text{PRO}] \, {}^*[\text{NP2}'] \, [\text{Verb}'] \, [\text{Verb}] ▲ + [\text{NP}_2]$$
$$\implies {}^*[\text{NP}_1] \, {}^*[\text{PRO}] \, {}^*[\text{NP2}'] \, [\text{Verb}'] \, [\text{Verb}] ▲ \, [\text{NP}_2]$$

with realization 1↑ 2↑

---

[13] The incremental delay and the control of different word orders using different LA-SU pragmatics↑ could be avoided only at the price of postulating different LA-NA syntaxes for different language types. This, however, would be in conflict with the assumption of one *universal* LA-NA syntax for language- and context-based recognition and action.

Also, natural languages exhibit word order variants in addition to their basic word order, such as the main- and subclause order in German. These may be handled in a much simpler and more transparent manner in terms of language specific LA-SU pragmatics↑ than via the complicating assumption of many different LA-NA syntaxes.

The rule NP+id goes from two coreferential input nominals directly to the verb[14] of the second nominal. Therefore, the relative pronoun PRO is contributed to the buffer not by the universal LA-NA syntax, but rather by the LA-SU pragmatics↑ of German.

As an example of an adverbial embedding consider the epr-navigation underlying the beginning of an adverbial sentence.

24.4.8 APPLICATION OF V+EPR IN THE WORD BANK 24.1.2

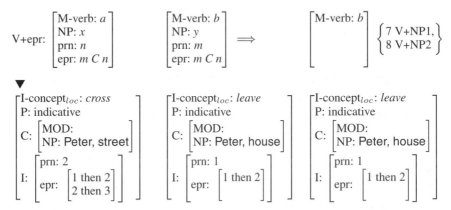

This rule application navigates from the verb *cross* (START) extrapropositionally to the verb *leave* (NEXT) which results as the NEW START. The return marker ▼ is added to the first verb coplet.

In epr-navigations, the conjunction (in its universal, language-independent form) is contained in the epr-feature of the verb coplet. In natural languages it has different realizations depending on whether the extrapropositional navagation is forward or backward, and on whether it is coordinating (paratactic) or subordinating (hypotactic).

24.4.9 DIFFERENT REALIZATIONS OF CONJUNCTIONS

	temporal	causal	modal
coordinating forward:	P1. Then P2.	P1. Therefore P2.	P1. Thus P2.
coordinating backward:	P2. Earlier P1.		
subordinating forward:	p1, before P2, p1.	p1, for which reason P2, p1.	p1, as P2, p1
subordinating backward:	p2, after P1, p2.	p2, because P1, p2.	

In 24.4.8, the conjunction is realized by the Englisch LA-SU pragmatics↑ as **after**, due to (i) the direction and (ii) the hypotactic nature of the navigation.

Once the embedded proposition has been traversed completely using V+NP1, the navigation returns into the next higher proposition to traverse the rest of its coplets. This is shown schematically in 24.4.10, analogous to 24.4.6.

---

[14] Note that the SOV language Korean does not have relative pronouns. There the above sentence transliterates as *The house leaves Peter the street crosses* or *The street the house leaves Peter crosses*.

24.4.10 Adverbial embedding navigation

Peter crossed, after he left the house, the street.

▼*cross*	*Peter*	V+epr			*street*
prn:2	prn:2	*leave*	*Peter house* ▲		prn: 2
(2 then 3)	id: 1	prn: 1	prn:1 prn:1		id: 3
(1 then 2)		(1 then 2)	id:1 id:2		

In German, this universal type of LA-NA navigation is realized as follows.

24.4.11 Word order of adverbial embedding in German

Peter überquert, nachdem er das Haus verlassen hat, die Straße.
(*Peter crosses, after he the house left-has, the street.*)

V+NP1   realization
                                                  ↑
buffer: [Verb] + [NP$_1$] $\Longrightarrow$ [Verb] [NP$_1$]

                                                      realization
V+epr   1↑   2↑
buffer: *[NP$_1$] [Verb] + [Verb$'$] $\Longrightarrow$ *[NP$_1$] [Verb]▼ [CNJ] [Verb$'$]

                                                                              realization
V+NP1   1↑
buffer: *[NP$_1$] *[Verb]▼ *[CNJ] [Verb$'$] + [NP1$'$] $\Longrightarrow$ *[NP$_1$] *[Verb]▼ *[CNJ] [Verb$'$] [NP1$'$]

V+NP1   realization
buffer: *[NP$_1$] *[Verb]▼ *[CNJ] *[NP1$'$] [Verb$'$] + [NP2$'$]   2↑   1↑
    $\Longrightarrow$ *[NP$_1$] *[Verb]▲ *[CNJ] *[NP1$'$] [Verb$'$] [NP2$'$]

V+NP1   realization
buffer: *[NP$_1$] *[Verb]▲ *[CNJ] *[NP1$'$] *[NP2$'$] *[Verb$'$] + [NP$_2$]   1↑
    $\Longrightarrow$ *[NP$_1$] *[Verb]▲ *[CNJ] *[NP1$'$] *[NP2$'$] *[Verb$'$] [NP$_2$]

The LA-SU pragmatic↑ interpretation of V+epr extracts the conjunction CNJ from the verb and produces a language-specific surface in accordance with the distinctions presented in 24.4.9.

Depending on the direction in which proposition 1 and 2 are traversed in the word bank 24.1.2 there are numerous additional types of adverbial embedding navigation. Their English surface reflexes are listed in 24.4.3.

As shown by the following example, marking the beginning of elementary propositions by V+NP2 and V+epr leaving the return marker ▼ is sufficient even in multiple embeddings.

24.4.12 Multiple center embeddings in German

Peter, der den Salat, den er gegessen hatte, bezahlt hatte, verließ das Restaurant.
(*Peter, who the salad, which he paid-had, eaten-had, left the restaurant.*)

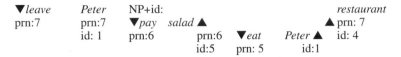

Each new embedded proposition allows additional embeddings. As soon, however, as an embedded proposition has been traversed completely, it is the turn of the remaining coplets of the next higher proposition to be traversed. The moment when a proposition has been traversed completely is determined by the word bank's traversal counters.

24.5 LA-search and LA-inference

Special cases of free LA-navigation are (i) LA-*search* and (ii) LA-*inference*. They are handled in terms of special LA-grammars which operate as purely syntactic algorithms in a word bank.

An LA-search is initiated by a query specifying the desired answer. In natural language there are two general query types called (i) Wh-questions and (ii) yes/no-questions.

24.5.1 BASIC TYPES OF QUESTIONS IN NATURAL LANGUAGE

Wh-question   **Yes/no-question**
Who entered the restaurant?   Did Peter enter the restaurant?

Formally, the hearer's interpretation of a Wh-question results in a verbal coplet in which the value occupied by the Wh-word is represented by the variable σ-1 and the value of prn is represented by σ-2. The hearer's interpretation of a yes/no-question, on the other hand, results in a verbal coplet in which only the prn-attribute contains a variable as value.

24.5.2 SEARCH COPLETS OF THE TWO BASIC TYPES OF QUERIES

Wh-question   **Yes/no-question**

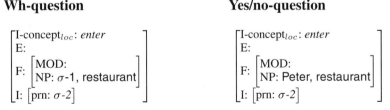

A query is answered by attempting to match the search coplet with the last (most recent) coplet of the corresponding token line (here the line of *enter* in 24.1.2).

If the continuation values of this last coplet match the search coplet, the answer has been found and there is no need for further search. Otherwise, the token line is systematically searched proceeding backwards from the last coplet.

In the case of WH-questions this search is based on the following LA-grammar.

24.5.3 LA-Q1 (WH-QUESTIONS)

ST_S: $\{([a]\{1\ r_1,\ 2\ r_2\})\}$

$$r_1: \begin{bmatrix} \text{M-verb: } a \\ \neg\text{NP: } y\ \sigma\ z \\ \text{prn: } m \end{bmatrix} \begin{bmatrix} \text{M-verb: } a \\ \neg\text{NP: } y\ \sigma\ z \\ \text{prn: } m-1 \end{bmatrix} \implies \begin{bmatrix} \text{M-verb: } a \\ \neg\text{NP: } y\ \sigma\ z \\ \text{prn: } m-1 \end{bmatrix} \{3\ r_1\ 4\ r_2\}$$

$$r_2: \begin{bmatrix} \text{M-verb: } a \\ \neg\text{NP: } y\ \sigma\ z \\ \text{prn: } m \end{bmatrix} \begin{bmatrix} \text{M-verb: } a \\ \text{NP: } y\ \sigma\ z \\ \text{prn: } m-1 \end{bmatrix} \implies \begin{bmatrix} \text{M-verb: } a \\ \text{NP: } y\ \sigma\ z \\ \text{prn: } m-1 \end{bmatrix} \{5\ r_3\}$$

$$r_3: \begin{bmatrix} \text{M-verb: } a \\ \text{NP: } y\ \sigma\ z \\ \text{prn: } n \end{bmatrix} \begin{bmatrix} \text{M-np: } \sigma \\ \text{VERB: } a \\ \text{prn: } n \end{bmatrix} \implies \begin{bmatrix} \text{M-np: } \sigma \\ \text{VERB: } a \\ \text{prn: } n \end{bmatrix} \{\ \}$$

ST_F: $\{([\text{M-np: } \sigma]\ \text{rp}_3)\}$

The proposition numbers in the patterns of LA-Q1 follow the convention that $m - 1$ (m minus one) stands for the proposition number of the coplet immediately preceding the current coplet m in the token line.

The first two LA-Q1 rules apply, if their START-pattern does *not* match the continuation predicates of the input coplet. If the next coplet does likewise not match, then the rule r_1 fires and the search is continued with the next coplet as the new START. If the next coplet matches, on the other hand, rule r_2 fires and the variable σ-1 is bound to the value searched for. Finally, rule r_3 navigates to the coplet of the queried continuation value and returns this coplet as the desired answer.

For example, applying the pattern of the Wh-question in 24.5.2 to the last coplet of the token line **enter** in the word bank 24.1.2 will result in failure because the continuation values [Peter, house] do not match the search pattern [σ-1, restaurant]. Thus, the START-conditions of the rules r_1 and r_2 of LA-Q1 are satisfied.

When applying r_1 and r_2 to the next preceding coplet (here with **prn** 3), r_2 happens to be successful. Thereby the variable σ-1 is bound to the answer of the query (i.e. **Peter**). Rule r_3 navigates to this value (i.e., the coplet **Peter** of proposition 3) and returns it as the answer.[15]

A query based on a yes/no-question is handled in a similar manner, using the following LA-grammar.

24.5.4 LA-Q2 (YES/NO-QUESTIONS)

ST_S: $\{([a]\{1\ r_1,\ 2\ r_2\})\}$

$$r_1: \begin{bmatrix} \text{M-verb: } a \\ \neg\text{NP: } x \\ \text{prn: } m \end{bmatrix} \begin{bmatrix} \text{M-verb: } a \\ \neg\text{NP: } x \\ \text{prn: } m-1 \end{bmatrix} \implies \begin{bmatrix} \text{M-verb: } a \\ \neg\text{NP: } x \\ \text{prn: } m-1 \end{bmatrix} \{3\ r_1\ 4\ r_2\}$$

---

[15] At this point a suitable control system of the cognitive agent could pass the result of LA-Q1 on to LA-NA and initiate a more detailed answer based on navigating through the subcontextual surroundings of the answer coplet.

$$r_2: \begin{bmatrix} \text{M-verb: } a \\ \neg\text{NP: } x \\ \text{prn: } m \end{bmatrix} \begin{bmatrix} \text{M-verb: } a \\ \text{NP: } x \\ \text{prn: } m-1 \end{bmatrix} \implies \begin{bmatrix} \text{M-verb: } a \\ \text{NP: } x \\ \text{prn: } m-1 \end{bmatrix} \{\ \}$$

ST_F: {([verb: a] rp$_1$) ([verb: a] rp$_2$)}

Here the answers are based on the final states ST_F, whereby the first final state represents the answer **no** and the second the answer **yes**.

For example, in the attempt to answer the yes/no-question **Did Peter cross the restaurant?** relative to the word bank 24.1.2, the token line **cross** will be searched in vain. When there are no preceding coplets left in the token line, LA-Q2 terminates in the final state ([verb: a] rp$_1$), which is realized linguistically as the answer **no**. The yes/no-question **Did Peter enter the restaurant?**, on the other hand, results in the final state ([verb: a] rp$_2$), which is realized as the answer **yes**.

LA-navigation and LA-search have in common that they do not change the set of propositions contained in a word bank. LA-inference, on the other hand, has the task of deriving new propositions from the information stored. As a consequence, the word bank content may be modified.[16]

For SLIM-theoretic language generation and interpretation, inferences are needed in many tasks. For example, the linguistic realization of a navigation requires that temporal and modal forms be correctly inferred by the speaker and coded into the language surface. The hearer in turn must infer the correct temporal and modal parameter values from those surfaces. Furthermore, the coreference between nominal coplets must be coded into the language surfaces by the speaker using pronouns or definite descriptions, which in turn must be decoded by the hearer. Also, in the answering of questions a hierarchy of hyper- and hyponyms must be used to infer instantiations which are not contained directly in the word bank, etc.

Inferences of this kind have been studied in detail in logical systems from antiquity to contemporary AI (cf. W. Bibel 1993). This raises the question of whether and how the inferences of classical as well as modern systems should be recreated within the formalism of a word bank. As a simple example consider some classical inferences of propositional calculus.

24.5.5 INFERENCE SCHEMATA OF PROPOSITIONAL CALCULUS

1. $\dfrac{A, B}{\vdash A\&B}$ 2. $\dfrac{A \lor B, \neg A}{\vdash B}$ 3. $\dfrac{A \to B, A}{\vdash B}$ 4. $\dfrac{A \to B, \neg B}{\vdash \neg A}$

5. $\dfrac{A\&B}{\vdash A}$ 6. $\dfrac{A}{\vdash A \lor B}$ 7. $\dfrac{\neg A}{\vdash A \to B}$ 8. $\dfrac{\neg\neg A}{\vdash A}$

---

[16] LA-inference and LA-navigation have in common that they may apply freely, without the need for external influences. The SLIM-theoretic interpretation of language in general and LA-search in particular, on the other hand, exemplify an external control which may interrupt or guide the otherwise autonomous LA-navigation and LA-inference.

In propositional logic the first inference is called *conjunction*: the truth of two arbitrary propositions A and of B implies the truth of the complex proposition A&B.

If this inference is transferred into database semantics, it amounts to an operation which establishes new extrapropositional relations based on the conjunction and. This operation may be realized by means of the following LA-grammar rule.

24.5.6 LA-RULE FOR THE PROPOSITIONAL INFERENCE OF CONJUNCTION

$$
\text{inf1:} \quad
\begin{bmatrix} \text{M-verb: } a \\ \text{prn: } m \end{bmatrix}
\begin{bmatrix} \text{M-verb: } b \\ \text{prn: } n \end{bmatrix}
\implies
\begin{bmatrix} \text{M-verb: } a \\ \text{prn: } m \\ \text{epr: } m \text{ and } n \end{bmatrix}
\begin{bmatrix} \text{M-verb: } b \\ \text{prn: } n \\ \text{epr: } m \text{ and } n \end{bmatrix}
$$

The rule inf1 produces the new extrapropositional relation [epr: *m and n*] between two arbitrary propositions *m* and *n*, enabling navigation from any proposition to any other proposition asserted in a word bank.

In its logical interpretation, the inference in question expresses a conjunction of truth and is as such intuitively obvious. The S<small>LIM</small>-theoretic interpretation, on the other hand, raises the question of *why* two – up to now unconnected – propositions (data base assertions) should be concatenated with *and*. Even though such a concatenation would not result in a falsehood, an uncontrolled application of inf1 would destroy the cognitive coherence of a word bank.

Next consider a type of inference which models the *is-a*, *is-part-of*, and *has-a* hierarchies of classic AI. In a word bank these conceptual hierarchies are represented in terms of *absolute* propositions (cf. Section 21.1) A natural language correlate of an absolute proposition is, e.g., A dog is an animal.

Absolute propositions express general knowledge, in contrast to the *episodic* propositions considered up to now, which express individual knowledge about specific events and objects, such as Peter crosses the street. Formally absolute propositions differ from episodic propositions only in terms of the open *loc* values of their I-concepts$_{loc}$. Other than that absolute propositions form normal subcontexts which can be traversed both intra- and extrapropositionally.

Absolute propositions are an important component of a word bank and serve in part to specify the literal meaning of words (cf. 23.5.2). In the token lines of a word bank the coplets of absolute propositions are positioned between the respective base form (type) and the episodic coplets.[17] As an example of an absolute proposition consider 24.5.7.

24.5.7 COPLETS OF AN ABSOLUTE PROPOSITION

$$
\begin{bmatrix} \text{I-concept}_{loc}\text{: } be \\ \text{NP: dog, animal} \\ \text{prn: abs327} \end{bmatrix}
\begin{bmatrix} \text{I-concept}_{loc}\text{: } dog \\ \text{VERB: be} \\ \text{prn: abs327} \end{bmatrix}
\begin{bmatrix} \text{I-concept}_{loc}\text{: } animal \\ \text{VERB: be} \\ \text{prn: abs327} \end{bmatrix}
$$

---

[17] Cf. Hausser 1996.

These coplets express the absolute proposition **A dog is an animal.** This way of formalizing absolute propositions is suitable to express the contents of any of the usual conceptual hierarchies in a word bank. At the same time such absolute propositions serve as the basis for various kinds of inferences.

Assume, for example, that the word bank contains the following coplet as part of the episodic proposition **Peter saw a dog.**

24.5.8 COPLET OF AN EPISODIC PROPOSITION

$$
\begin{bmatrix}
\text{I-concept}_{loc}\text{: } see \\
\text{NP: Peter, dog} \\
\text{prn: 969}
\end{bmatrix}
$$

Without further provision, the question **Did Peter see an animal?** would be answered with **no** by the word bank. However, given the absolute proposition 24.5.7 and the following rule of inference **inf2**, the word bank could infer that Peter saw indeed an animal when he saw the dog, and answer the question correctly with **yes.**

24.5.9 INFERENCE RULE INF2 FOR ABSOLUTE PROPOSITIONS

$$
\text{inf2: }
\begin{bmatrix}
\text{M-verb: } a \\
\text{NP: } x\ b\ y \\
\text{prn: } n
\end{bmatrix}
\begin{bmatrix}
\text{M-verb: } be \\
\text{NP: } b\ c \\
\text{prn: } abs
\end{bmatrix}
\implies
\begin{bmatrix}
\text{M-verb: } a \\
\text{NP: } x\ c\ y \\
\text{prn: } n
\end{bmatrix}
$$

In our example, the START-pattern of rule **inf2** is matched onto the episodic coplet 24.5.8 whereby the variable b is assigned the value **dog.** This enables the rule **inf2** to navigate to the verbal coplet of the absolute proposition 24.5.9 as the NEXT and to derive the new episodic proposition **Peter saw an animal** as a variant of the original proposition. Based on this inference the question **Did Peter see an animal?** could be answered correctly even though the word bank originally contained only the proposition **Peter saw a dog.**

The rule **inf2** defined in 24.5.9 is applicable to verbal coplets of all absolute propositions with the verb **be**, i.e., to all the elements of the *is-a* hierarchy. Similar rules may be defined for the other hierarchies.

In an analogous manner the distinction between intensional and extensional contexts (cf. Section 20.2) may be handled. Instead of treating, e.g., **seek** and **find** by assigning ontologically (Frege) or formally (Carnap) different objects as denotations, the ultimately desired differences in inference are simply written as absolute propositions concerning these respective verbs.[18] The same applies to inchoative verbs like

---

[18] The absolute propositions in question may be represented intuitively as follows:
   If x seeks y, then possible y does not exist.
   If x finds y, then y exists.
In order to formalize these absolute propositions, the feature structures of the word bank must be extended to the representation of modality. Furthermore, associated rules of inference (like 24.5.9) must be defined such that arbitrary coplets of **seek** and **find** will allow derivation of the correct implications as additional propositions.

fall asleep or **awake**, which imply that the subject was not sleeping or awake before, respectively.[19]

The work of correctly analyzing the various inferences of human cognition in the context of a word bank is as extensive as the work of designing a complete syntax and semantics for a natural language. The format of concatenated elementary propositions is general and flexible, however. It allows to model propositional inferences as well as conceptual hierarchies and word specific implications in a uniform, simple, and descriptively powerful manner. For this reason there are good prospects to finally overcome the well-known problems of contemporary inference systems, which have been clearly identified in the scientific literature (e.g. S.J. Russell & P. Norvig 1995).

The modeling of S<small>LIM</small>-theoretic inferences is to be approached as follows. First, traditional inferences are translated into equivalent LA-inferences and investigated with respect to their modified empirical content.[20] Second, new types of search and inference are developed.[21] Third, these cognitive tools are applied to automatically maintain consistency of the word bank contents.[22] Fourth, the autonomous internal navigation is extended into a control module which guides the agents interaction with the external task environment.[23]

These tasks lead beyond traditional, theoretical, and computational linguistics into neighboring sciences such as robotics, AI, psychology, philosophy of language, and mathematical logic. The goal of man-machine-communication in unrestricted natural language, however, is worthy of a fresh, broadly-based effort. Ninetysix years after the first motorized flight and thirty years after landing on the moon, its successful realization will soon change everyday life more profoundly than any of the previous achievements of science.

---

[19] In order to formalize the related absolute propositions, the feature structures of the word bank must be extended to handle temporal relations.

[20] As illustrated with example 24.5.6.

[21] As illustrated with examples 24.5.3, 24.5.4, 24.5.9.

[22] Thereby a choice is given between a realistic model of human cognition with the possibility of inconsistencies remaining and an artificial system with no tolerance for inconsistencies. The realistic model is based on inferential navigation such that determining and repairing inconsistencies (i) requires varying degrees of mental work and (ii) is locally restricted.

[23] This comprises problem solving, turn taking in dialog, and other appropriate behavior strategies for adjusting to varying situations.

Exercises

Section 24.1

1. How is a subcontext formally realized in a word bank?
2. From what does the cognitive coherence of contextual propositions and their concatenation ultimately derive?
3. What are the two main types of extrapropositional relations?
4. Which formal property characterizes the beginning and the end of a propositional concatenation in a word bank?
5. Name three reasons for defining extrapropositional relations in a word bank.

Section 24.2

1. Why does a general linguistic treatment of language production, including spontaneous speech, require a formal representation of thought?
2. How does the formalism of LA-grammar function in the navigation through a word bank?
3. What may cause a split in LA-navigation, and why should it be avoided?
4. What is a relapse in LA-navigation?
5. Explain the tracking principles of LA-navigation.
6. When does LA-navigation return to the verb after traversing a nominal coplet, and what are the alternatives?

Section 24.3

1. How is the relation between language specificity and the universality contextual propositions handled in a SLIM machine?
2. Explain the LA-SU pragmatic↑ interpretation of a LA-NA navigation in its simplest form.
3. Why does the LA-SU pragmatics↑ generate *sequences* of proplets?
4. What are the three main types of language typology?
5. Show with the example of SOV languages that realization requires an incremental delay.

Section 24.4

1. Give derivations like 24.4.10 for the embedding navigations in 24.4.3.
2. Which piece of information in the verbal coplet is needed first in the derivation of an adverbial subclause?
3. Why is the clause final position of the finite verb in German subclauses not in conflict with the need to begin the derivation of adverbial subclauses with the traversal of the subclause verb?
4. How is the return of a navigation into the next higher proposition formally handled after traversal of an embedded clause?

Section 24.5

1. What is the connection between a question of natural language and a search in a word bank?
2. Describe two forms of questions and the formal procedures by which they are realized in a word bank.
3. Explain the inference schemata of classical propositional calculus. Can they be transferred to LA-inference in a word bank?
4. What are absolute propositions, how are they represented in a word bank, and how do they differ from episodic propositions?
5. How are the *is-a*, *is-part-of*, and *has-a* hierarchies represented in a word bank?
6. Explain why the realization of inferences in a word bank is a special form of LA-navigation?
7. Does a word bank have problems with propositional attitutes (cf. Section 20.2)?
8. How are the notions true and false (cf. Section 21.1) formally implemented in a word bank?

Bibliography

Aho, A.V. & J.D. Ullman (1972) *The Theory of Parsing, Translation, and Compiling*, Vol. I: Parsing, Prentice-Hall, Englewood Cliffs, New Jersey.

Aho, A.V. & J.D. Ullman (1977) *Principles of Compiler Design*, Addison-Wesley, Reading, Massachusetts.

Aho, A.V., J.E. Hopcroft, & J.D. Ullman (1983) *Data Structures and Algorithms*, Addison-Wesley, Reading, Massachusetts.

Aho, A.V., B.W. Kerninghan, & P. Weinberger (1988) *The AWK Programming Language*, Addison-Wesley, Reading, Massachusetts.

Ajdukiewicz, K. (1935) "Die syntaktische Konnexität," *Studia Philosophica*, 1:1–27.

Anderson, J.R. (1990) *Cognitive Psychology and its Implications*, 3rd ed., W.H. Freeman and Company, New York.

Anderson, J.R. & G.H. Bower (1973) *Human Associative Memory*, V.H. Winston, Washington, D.C.

Anderson, J.R. & G.H. Bower (1980) *Human Associative Memory: A Brief Edition*, Lawrence Erlbaum Associates, Hillsdale, New Jersey.

Austin, J.L. (1962) *How to do Things with Words*, Clarendon Press, Oxford, England.

Bach, E. (1962) "The Order of Elements in a Transformational Grammar of German," *Language* 38:263–269.

Bar-Hillel, Y. (1964) *Language and Information. Selected Essays on their Theory and Application*, Addison-Wesley, Reading, Massachusetts.

Barcan-Marcus, R. (1960) "Extensionality," *Mind* 69:55–62, reprinted in L. Linsky (ed.) 1971.

Barthes, R. (1986) *The Rustle of Language*, Hill and Wang, New York.

Barton, G., R.C. Berwick, & E.S. Ristad (1987) *Computational Complexity and Natural Language*, M.I.T. Press, Cambridge, Massachusetts.

Barwise, J. & J. Perry (1983) *Situations and Attitudes*, M.I.T. Press, Cambridge, Massachusetts.

Beaugrande, R.-A. & W.U. Dressler (1981) *Einführung in die Textlinguistik*, Max Niemeyer Verlag, Tübingen.

Behagel, O. (1923–1932) *Deutsche Syntax*, Band 4: Wortstellung, Periodenbau. Carl Winter, Heidelberg.

Bergenholtz, H. (1976) "Zur Morphologie deutscher Substantive, Verben und Adjektive. Probleme der Morphe, Morpheme und ihrer Beziehungen zu den Wortarten," in A. Hoppe (ed.): *Beihefte zur kommunikativen Grammatik*, Bonn.

Bergenholtz, H. (1989) "Korpusproblematik in der Computerlinguistik. Konstruktionsprinzipien und Repräsentativität," in H. Steger (ed.): *Handbücher zur Sprach- und Kommunikationswissenschaft* (Vol.IV). Berlin, New York.

Berkeley, G. (1710) *A Treatise Concerning the Principles of Human Knowledge*, reprinted in Taylor 1974.

Berwick, R.C. & A.S. Weinberg (1984) *The Grammatical Basis of Linguistic Performance: Language Use and Acquisition*, M.I.T. Press, Cambridge, Massachusetts.

Beutel, B. (1997) *Malaga 4.0*, CLUE-Manual, CL lab, Friedrich Alexander Universität Erlangen Nürnberg.

Bibel, W. (1993) *Wissensrepräsentation und Inferenz*, Vieweg, Braunschweig-Wiesbaden.

Biber, D. (1993) "Representativeness in Corpus Design," *Literary and Linguistic Computing*, Vol. 8.4.

Bierwisch, M. (1963) *Grammatik des Deutschen Verbs*, Studia Grammatica II, Akademie-Verlag, Berlin.

Billmeier, G. (1969) *Worthäufigkeitsverteilung vom Zipfschen Typ, überprüft an deutschem Textmaterial*, IPK-Forschungsbericht 69-6, Bonn.

Blair, D.C. & M.E. Maron (1985) "An Evaluation of Retrieval Effectiveness for a full-Text Document-Retrieval System," *Comm. ACM*.

Bloomfield, L. (1933) *Language*, Holt, Rinehart, and Winston, New York.

Bochvar, D. A. (1939) "Ob odnom tréhznačnom isčislénii i égo priménénii k analizu paradoksov klassičéskogo rasšsirennogo funkcional'nogo isčisléniá" (On a 3-valued logical calculus and its application to the analysis of contradictions), *Matématibrevecéskij sbornik*, Vol. 4:287–308. [Reviewed by Alonzo Church, The Journal of Symbolic Logic, Vol.4:98–99, and Vol 5:119.]

Boeken, H. (1992) *Methoden und Ergebnisse einer Tagging-Analyse des Deutschen*, CLUE-betreute Diplomarbeit der Informatik, Friedrich Alexander Universität Erlangen Nürnberg.

Book, R.V. (ed.) (1980) *Formal Language Theory: Perspectives and Open Problems*, Academic Press, New York.

Bono, E. de (1969) *The Mechanism of Mind*, Jonathan Cape, London.

Bresnan, J. (ed.) (1982) *The Mental Representation of Grammatical Relation*, M.I.T. Press, Cambridge, Massachusetts.

Briandais, R. de la (1959) Proc. Western Joint Computer Conf. 15, 295-298.

Bröker, N. (1991) *Optimierung von LA-Morph im Rahmen einer Anwendung auf das Deutsche*, CLUE-betreute Diplomarbeit der Informatik, Friedrich Alexander Universität Erlangen Nürnberg.

Brooks, R.A. (1986) "A Robust Layered Control System for a Mobile Robot," *IEEE Journal of Robotics and Automation* 2.1:14-23.

Brooks, R.A. (1990) "Elephants Don't Play Chess," *Robotics and autonomous systems*, Vol. 6, Numbers 1 and 2, June 1990. Reprinted in P. Maes (ed.) 1990, 3–15.

Brown, P., S. Della Pietra, V. Della Pietra, & R. Mercer (1991) "Word Sense Disambiguation Using Statistical Methods," Proceedings of the 29th Annual Meeting of the Association for Computational Linguistics, Berkeley, CA, June 1991, 264-270.

Brown, P., V. Della Pietra et al. (1992) *'Class-based n-gram Models of Natural Language,'* paper read at the Pisa Conference on European Corpus Resources.

Bühler, K. (1934) *Sprachtheorie: die Darstellungsfunktion der Sprache*, Fischer, Stuttgart.

Burnard, L. & C.M. Sperberg-McQueen (1995) *TEI Lite: An Introduction to Text Encoding Interchange*, http://www-tei.uic.edu/orgs/tei/intros/teiu5.tei.

Burnard, L. (ed.) (1995) *Users Reference Guide British National Corpus Version 1.0*, Oxford University Computing Services, Oxford, England.

Bybee, J.L. (1985) *Morphology: A Study of the Relation between Meaning and Form*, John Benjamins, Amsterdam/Philadelphia.

Carnap, R. (1928) *Der logische Aufbau der Welt*, 2nd ed., Felix Meiner, Hamburg 1961.

Carnap, R. (1947) *Meaning and Necessity*, The University of Chicago Press, Chicago. Illinois.

Chafe, W. (1970) *Meaning and the Structure of Language*, The University of Chicago Press, Chicago. Illinois.

Chafe, W. (1979) "The Flow of Thought and the Flow of Language," in T. Givon (ed.), *Syntax and Semantics*, Vol. 12, Academic Press, New York.

Chomsky, N. (1957) *Syntactic Structures*, Mouton, The Hague.

Chomsky, N. (1965) *Aspects of the Theory of Syntax*, M.I.T. Press, Cambridge, Massachusetts.

Chomsky, N. (1981) *Lectures on Government and Binding*, Foris, Dordrecht.

Church, A. (1951) "The Need for Abstract Entities in Semantic Analysis," *Contributions to the Analysis and Synthesis of Knowledge*, Proceedings of the American Academy of Arts and Sciences, 80, No. 1, 100-112.

Church, A. (1956) *Introduction to Mathematical Logic, Vol. I*, Princeton University Press, Princeton, New Jersey.

Church, K. & R.L. Mercer (1933) "Introduction to the Special Issue on Computational Linguistics Using Large Corpora," *Computational Linguistics*, Vol. 19.1:1–24.

CoL = *Computation of Language*, Hausser 1989a.

Cole, R. (ed.) (1998) *Survey of the State of the Art in Human Language Technology*, Edinburgh University Press, Edinburgh, Scotland.

Collins, A.M. & E.F. Loftus (1975) "A Spreading-Activation Theory of Semantic Processing," *Psychological Review* 82.6:407–428.

Condon, E.U. (1928) "Statistics of Vocabulary," *Science*, 68.1733:300.

Conrad, R. (ed.) (1981) *Kleines Wörterbuch sprachwissenschaftlicher Termini*, Leipzig.

Courant, M. (1899) "Supplément à la Bibliographie Coréenne," Jusquen 1899, Paris: Imprimerie Nationale, p. 70–71. Chapitre II, Bouddisme 3,738.

Date, C.J. (1990) *An Introduction to Database Systems*, 5th ed., Addison-Wesley, Reading, Massachusetts.

Davidson, D. (1967) "Truth and Meaning," *Synthese*, VII:304–323.

Davidson, D. & G. Harman (eds.) (1972) *Semantics of Natural Language*, D.Reidel, Dordrecht.

DeRose, S. (1988) "Grammatical Category Disambiguation by Statistical Optimization," *Computational Linguistics*, 14.1:31–39.

Dretske, F. (1981) *Knowledge and the Flow of Information*, Bradford Books/M.I.T. Press, Cambridge, Massachusetts.

Dreyfus, H. (1981) "From Micro-Worlds to Knowledge Representation: AI at an Impasse," in J. Haugeland (ed.).

Drosdowski, G. (ed.) (1984) *Band 4: Grammatik der Deutschen Gegenwartssprache*, Herausgegeben und bearbeitet von Günther Drosdowski, Dudenverlag Mannheim.

Earley, J. (1970) "An Efficient Context-Free Parsing Algorithm," *Comm. ACM* 13.2:94–102, reprinted in B. Grosz, K. Sparck Jones, & B.L. Webber (eds.) 1986.

Eco, U. (1975) *A Theory of Semiotics*, Indiana University Press, Bloomington, Indiana.

Eisenberg, P. (1986) *Grundriß der deutschen Grammatik*, J.B. Metzlersche Verlagsbuchhandlung, Stuttgart.

Elmasri, R. & S.B. Navathe (1989) *Fundamentals of Database Systems*, Benjamin-Cummings, Redwood City, CA.

Engel, U. (1991) *Deutsche Grammatik*, 2nd ed., Julius Groos Verlag, Heidelberg.

Erben, J. (1968) *Deutsche Grammatik*, Fischer Taschenbuch Verlag, Frankfurt a. M.

Estoup, J.B. (1916) *Gammes Sténographiques*, 4th ed., Paris.

Fillmore, C. (1968) "The Case for Case," in E. Bach & R. Harms (eds.), *Universals of Syntactic Theory*, Holt, Rinehard, and Winston, New York.

Fillmore, C. (1977) "The Case for Case reopened," in P. Cole & J. Sadock (eds.) *Syntax and Semantics*, Vol. 8, Grammatical Relations, Academic Press, New York.

Floyd, R.W. (1967) "Assigning Meaning to Programs," in J.T. Schwartz (ed.), 19–32.

Fraassen, B. van (1966) "Singular Terms, Truth-Value Gaps, and Free Logic," *The Journal of Philosophy* 63:481–495.

Fraassen, B. van (1968) "Presupposition, Implication, and Self-Reference," *The Journal of Philosophy* 65:136–152.

Fraassen, B. van (1969) "Presuppositions, Supervaluations, and Free Logic," in K. Lambert (ed.), *The Logical Way of Doing Things*, Yale University Press, New Haven.

Francis, W.N. (1980) "A tagged corpus: Problems and Prospects," in S. Greenbaum, G. Leech, & J. Svartvik (eds.), 192–209.

Francis, W.N. & H. Kučera (1982) *Frequency Analysis of English Usage: Lexicon and Grammar*, Houghton Mifflin, Boston.

Fredkin, E. (1960) "Trie Memory," *Comm. ACM* 3.9:490–499.

Frege, G. (1967) *Kleine Schriften*, hrsg. von Ignacio Angelelli, Wiss. Buchgesellschaft, Darmstadt.

Fruminka, R.M. (1970) "Über das sogenannte 'Zipfsche Gesetz'," Übersetzt von Arne Schubert, in: Forschungsberichte des Instituts für deutsche Sprache, Vol. 4:117–131.

Gaifman, C. (1961) *Dependency Systems and Phrase Structure Systems*, P-2315, Rand Corporation, Santa Monica, Ca.

Garey, M.R. & D.S. Johnson (1979) *Computers and Intractability: A Guide to the Theory of NP-Completeness*, W. H. Freeman, San Francisco.

Garside, R., G. Leech, & G. Sampson (1987) *The Computational Analysis of English*, Longman, London.

Gazdar, G. (1981) "Unbounded Dependencies and Coordinate Structure," *Linguistic Inquiry* 12.2:155–184.

Gazdar, G., E. Klein, G. Pullum, & I. Sag (1985) *Generalized Phrase Structure Grammar*, Harvard University Press, Cambridge, Massachusetts, and Blackwell, Oxford, England.

Geach, P. (1972) "A Program for Syntax," in D. Davidson & G. Harman (eds.), 483–497.

Ginsburg, S. (1980) "Formal Language Theory: Methods for Specifying Formal Languages – Past, Present, Future," in R.V. Book (ed.), 1–22.

Girard, R. (1972) *La violence et le sacré*, Bernard Grasset, Paris.

Givón, T. (1985) "Iconicity, Isomorphism, and Non-Arbitrary Coding in Syntax," in J. Haiman (ed.) 1985a, *Iconicity in Syntax*.

Goldfarb, C. F. (1990) *The SGML handbook*, Clarendon Press, Oxford, England.

Görz, G. (1988) *Strukturanalyse natürlicher Sprache*, Addison-Wesley, Bonn-Reading.

Greenbaum, S., G. Leech, & J. Svartvik (eds.) (1980) *Studies in English Linguistics: For Randolph Quirk*, Longman, London.

Greenbaum, S. & N. Yibin (1994) "Tagging the British ICE Corpus: English word classes," in N. Oostdijk & P. de Haan (eds.).

Greenberg, J. (1963) "Some Universals of Grammar with Particular Reference to the Order of Meaningful Elements," in J. Greenberg (ed.).

Greenberg, J. (ed.) (1963) *Universals of Language*, M.I.T. Press, Cambridge, Massachusetts.

Greibach, S. (1973) "The hardest context-free language," *SIAM J. Comput.* 2:304–310.

Grice, P. (1957) "Meaning," *Philosophical Review*, 66:377–388.

Grice, P. (1965) "Utterer's Meaning, Sentence Meaning, and Word Meaning," *Foundations of Language*, 4:1–18.

Griffith, T. & Petrik, S. (1965) "On the Relative Efficiencies of Context-Free Grammar Recognizers," *Comm. ACM* 8:289-300.

Grosz, B. & C. Sidner (1986) "Attention, Intensions, and the Structure of Discourse," *Computational Linguistics*, Vol. 12.3:175–204.

Grosz, B., K. Sparck Jones, & B.L. Webber (eds.) (1986) *Readings in Natural Language Processing*, Morgan Kaufmann, Los Altos, CA.

Groenendijk, J.A.G., T.M.V. Janssen, & M.B.J. Stokhof (eds) (1980) *Formal Methods in the Study of Language*, Mathematical Center Tracts 135, University of Amsterdam.

Gupta, A. (1982) "Truth and Paradox," *Journal of Philosophical Logic*, 11:1–60.

Haiman, J. (ed.) (1985a) *Iconicity in Syntax*, Typological Studies in Language Vol. 6, John Benjamins, Amsterdam/Philadelphia.

Haiman, J. (1985b) *Natural Syntax, Iconicity, and Erosion*, Cambridge University Press, Cambridge, England.

Halliday, M.A.K. (1985) *An Introduction to Functional Grammar*, Edward Arnold, London.

Halliday, M.A.K. & R. Hasan (1976) *Cohesion in English*, Longman, London.

Hanrieder, G. (1996) *Inkrementelles Parsing gesprochener Sprache mit einer linksassoziativen Unifikationsgrammatik*, Inaugural Dissertation, CL lab, Friedrich Alexander Universität Erlangen Nürnberg (CLUE).

Harman, G. (1963) "Generative Grammar without Transformational Rules: a Defense of Phrase Structure," *Language* 39:597–616.

Harrison, M. (1978) *Introduction to Formal Language Theory* Addison-Wesley, Reading, Massachusetts.

Haugeland, J. (ed.) (1981) *Mind Design*, M.I.T. Press, Cambridge, Massachusetts.

Hausser, R. (1973) "Presuppositions and Quantifiers," *Papers from the Ninth Regional Conference*, Chicago Linguistics Society, Chicago,. Illinois.

Hausser, R. (1976) "Presuppositions in Montague Grammar," *Theoretical Linguistics* 3:245–279.

Hausser, R. (1978) "Surface Compositionality and the Semantics of Mood," reprinted in J. Searle, F. Kiefer, and M. Bierwisch (eds.), *Speech Act Theory and Pragmatics*, 1980, Reidel, Dordrecht.

Hausser, R. (1979a) "How do Pronouns Denote?" in H. Schnelle et al. (eds.), *Syntax and Semantics*, Vol. 10, Academic Press, New York.

Hausser, R. (1979b) "A Constructive Approach to Intensional Contexts," *Language Research*, Vol. 18.2, Seoul, Korea, 1982.

Hausser, R. (1979c) "A New Treatment of Context in Model-Theory," in *Sull' Anaphora, atti del seminario Accademia della Crusca*, Florence, Italy, 1981.

Hausser, R. (1981) "The Place of Pragmatics in Model-Theory," in J.A.G. Groenndijk et al. (eds.).

Hausser, R. (1982) "A Surface Compositional Categorial Grammar," *Linguistic Journal of Korea,* Vol 7.2.

Hausser, R. (1983a) "Vagueness and Truth," *Conceptus, Zeitschrift für Philosophie,* Jahrgang XVII, Nr. 40/41.

Hausser, R. (1983b) "On Vagueness," *Journal of Semantics* 2:273–302.

Hausser, R. (1984a) *Surface Compositional Grammar*, Wilhelm Fink Verlag, Munich.

Hausser, R. (1984b) " The Epimenides Paradox," presented at the Fourth Colloquium on Formal Grammar, Amsterdam.

Hausser, R. (1985) "Left-Associative Grammar and the Parser NEWCAT," Center for the Study of Language and Information, Stanford University, IN-CSLI-85-5.

Hausser, R. (1986) *NEWCAT: Parsing Natural Language Using Left-Associative Grammar*, Lecture Notes in Computer Science 231, Springer-Verlag, Berlin-New York.

Hausser, R. (1987a) "Left-Associative Grammar: Theory and Implementation," Center for Machine Translation, Carnegie Mellon University, CMU-CMT-87-104.

Hausser, R. (1987b) "Modelltheorie, Künstliche Intelligenz und die Analyse der Wahrheit," in L. Puntel (ed.) *Der Wahrheitsbegriff*, Wissenschaftliche Buchgesellschaft, Darmstadt.

Hausser, R. (1988a) "Left-Associative Grammar, an Informal Outline," *Computers and Translation*, Vol. 3.1:23–67, Kluwer, Dordrecht.

Hausser, R. (1988b) "Algebraic Definitions of Left-Associative Grammar," *Computers and Translation*, Vol. 3.2, Kluwer, Dordrecht.

Hausser, R. (1989a) *Computation of Language, An Essay on Syntax, Semantics and Pragmatics in Natural Man-Machine Communication*, Symbolic Computation: Artificial Intelligence, Springer-Verlag, Berlin-New York.

Hausser, R. (1989b) "Principles of Computational Morphology," Laboratory of Computational Linguistics, Carnegie Mellon University.

Hausser, R. (1992) "Complexity in Left-Associative Grammar," *Theoretical Computer Science*, Vol. 106.2, Elsevier.

Hausser, R. (1996) "A Database Interpretation of Natural Language," *Korean Journal of Linguistics*, Vol. 21.1,2:29–55.

Hausser, R. (ed.) (1996) *Linguistische Verifikation. Dokumentation zur Ersten Morpholympics*, Max Niemeyer Verlag, Tübingen.

Hayashi, T. (1973) "On Derivation Trees of Indexed Grammars—an Extension of the uvwxy Theorem," *Publications of the Research Institute for Mathematical Sciences* 9.1:61–92.

Heidolph, K.E., W. Flämig, & W. Motsch (1981) *Grundzüge einer deutschen Grammatik*, Akademie-Verlag, Berlin.

Herwijnen, E. van (1990) *Practical SGML*, 2nd ed. 1994, Kluwer, Dordrecht.

Herzberger, H. (1982) "Notes on Naive Semantics," *Journal of Philosophical Logic*, 11:61–102.

Hess, K., J. Brustkern, & W. Lenders (1983) *Maschinenlesbare deutsche Wörterbücher*, Max Niemeyer Verlag, Tübingen.

Hoare, C.A.R. (1985) *Communicating Sequential Processes*, Prentice-Hall. Englewood Cliffs, New Jersey.

Hockett, C.F. (1958) *A Course in Modern Linguistics*, Macmillan, New York.

Hofland, K. & S. Johansson (1980) *Word Frequencies in British and American English*, Longman, London.

Hooper, J. (1976) *An Introduction to Natural Generative Phonology*, Academic Press, New York.

Hopcroft, J.E. & Ullman, J.D. (1979) *Introduction to Automata Theory, Languages, and Computation*, Addison-Wesley, Reading, Massachusetts.

Hovy, E.H. (1987) "Some Pragmatic Decision Criteria in Generation," in G. Kempen (ed.), 3–17.

Hubel, D.H. & T.N. Wiesel (1962) "Receptive Fields, Binocular Interaction, and Functional Architecture in the Cat's Visual Cortex," *Journal of Physiology*, 160:106–154.

Hume, D. (1748) *An Enquiry Concerning Human Understanding*, reprinted in Taylor 1974.

Hutchins, W.J. (1986) *Machine Translation: Past, Present, Future*, Ellis Horwood Lmt., Chichester.

Ickler, T. (1997) *Die Disziplinierung der Sprache, Fachsprachen in unserer Zeit*, Gunter Narr Verlag, Tübingen.

Illingworth et al. (ed.) (1990) *Dictionary of Computing*, Oxford University Press, Oxford, England.

Jackendoff, R. (1972) *Semantic Interpretation in Generative Grammar*, M.I.T. Press, Cambridge, Massachusetts.

Jespersen, O. (1921) *Language. Its Nature, Development, and Origin*, reprinted in the Norton Library 1964, W.W. Norton and Company, New York.

Jespersen, O. (1933) *Essentials of English Grammar*, reprinted by the University of Alabama Press 1964.

Johnson-Laird, P.N. (1983) *Mental Models*, Harvard University Press, Cambridge, Massachusetts.

Joos, M. (1936) "Review of G.K.Zipf: The psycho-biology of language," *Language* 12:3.

Joshi, A.K., L.S. Levy, & M. Takahashi (1975) "Tree Adjunct Grammars," *Journal of Computer and Systems Sciences* 10.

Joshi, A.K., K. Vijay-Shanker, & D.J. Weir (1991) "The Convergence of Mildly Context-Sensitive Grammar Formalisms," in P. Sells, S. Shieber, & T. Wasow (eds.), *Foundational Issues in Natural Language Processing*, M.I.T. Press, Cambridge, Massachusetts.

Kaeding, W. (1897/8) *Häufigkeitswörterbuch der deutschen Sprache*, Steglitz.

Kamp, J.A.W. (1981) "A Theory of Truth and Semantic Representation," in J.A.G. Groenendijk et al. (eds.).

Kamp, J.A.W. & U. Reyle (1993) *From Discourse to Logic*, Part 1 & 2, Kluwer, Dordrecht.

Kasami, T. (1965) "A Syntax Analysis Procedure for Unambiguous Context-Free Grammars," *J. ACM* 16:3, 423–431.

Kasami, T. & Torii, K. (1969) "A Syntax-Analysis Procedure for Unambiguous Context-Free Grammar Recognizers," *J. ACM* 16:289–300.

Katz, J. & P. Postal (1964) *An Integrated Theory of Linguistic Descriptions*, M.I.T. Press, Cambridge, Massachusetts.

Kay, M. (1980) *"Algorithmic Schemata and Data Structures in Syntactic Processing"*, reprinted in Grosz, Sparck Jones, Webber (eds.) 1986.

Kempen, G. (ed.) *Natural Language Generation*, Martinus Nijhoff Publishers, Dordrecht.

Kim, Doo-Chong (1981) *Hankwuk-ko-insway-kiswul-sa*, Tham-ku-tang [A History of Old Typography in Korea].

Kleene, S.C. (1952) *Introduction to Metamathematics*, Amsterdam.

Knorr, O. (1997) *Entwicklung einer JAVA-Schnittstelle für Malaga*, CLUE-betreute Studienarbeit der Informatik, Friedrich Alexander Universität Erlangen Nürnberg.

Knuth, D.E. (1973) *The Art of Computer Programming: Vol 3. Sorting and Searching*, Addison-Wesley, Reading, Massachusetts.

Knuth, D.E. (1984) The TEXbook, Addison-Wesley, Reading, Massachusetts.

Koskenniemi, K. (1983) *Two-Level Morphology*, University of Helsinki Publications No. 11.

Kripke, S. (1972) "Naming and Necessity," in D. Davidson & G. Harmann (eds.), 253–355.

Kripke, S. (1975) "Outline of a Theory of Truth," *The Journal of Philosophy* 72:690–715.

Kuhn, T. (1970) *The Structure of Scientific Revolutions*, 2nd ed., The University Press of Chicago, Chicago, Illinois.

Kučera, H. & W.N. Francis (1967) *Computational analysis of present-day English*, Brown University Press, Providence, Rhode Island.

Lakoff, G. (1968) "Pronouns and Reference," University of Indiana Linguistics Club, Bloomington, Indiana.

Lakoff, G. (1972) "Linguistics and Natural Logic," in D. Davidson & G. Harman (eds.), 545–665.

Lakoff, G. (1972) "Hedges: a Study in Meaning Criteria and the Logic of Fuzzy Concepts," in P.M. Peranteau, J.N. Levi, & G.C. Phares (eds.): *Papers from the Eights Regional Meeting of the Chicago Linguistic Society*, 183–228. Chicago, Illinois.

Lambek, J. (1958) "The Mathematics of Sentence Structure," *American Mathematical Monthly* 65:154–170.

Lamport, L. (1986) *LTEX, A Document Preparation System*, Addison-Wesley, Reading, Massachusetts.

Langacker, R. (1969) "Pronominalization and the Chain of Command," in D.A. Reibel & S. A. Shane (eds.), 160–186.

Lawson, V. (1989) "Machine Translation," in C. Picken (ed.), 203–213.

Lee, K.-Y. (1994) "Hangul, the Korean Writing System, and its Computational Treatment," *LDV-Forum*, Band 11.2:26–43.

Lee, K.-Y. (1995) "Recursion Problems in Concatenation: A Case of Korean Morphology," *Proceedings of PACLIC 10, the 10th Pacific-Asian Conference on Language, Information, and Computation*, 215-224.

Lee, K.-Y. (1999), *Computational Morphology*, Korea University Press, Seoul, Korea. [in Korean]

Lee, M.-H. (1993) "Parsing of Korean using a Left-associative Grammar," in Proceedings of HCI '93, p. 21-29. [in Korean]

Lee, M.-H. (1995) "On the Categorization in a Left-associative Grammar," Linguistic Journal of Korea, Vol. 19.2, p. 489-512. [in Korean]

Leech, G., R. Garside, & E. Atwell (1983) "The Automatic Grammatical Tagging of the LOB Corpus," *ICAME Journal* 7, 13–33.

Leech, G. (1991) "The State of Art in Corpus Linguistics," in K. Aijmer & B. Altenberg (eds.), *English Corpus Linguistics, Studies in honour of Jan Svartvik*. New York.

Leech, G. (1995) "A Brief User's Guide to the Grammatical Tagging of the British National Corpus," web site.

Lees, R.B. (1960) "The Grammar of English Nominalizations," *International Journal of American Linguistics*, 26.3, Part 2.

Leidner, J. (1998) *Linksassoziative morphologische Analyse des Englischen mit stochastischer Disambiguierung*, Magisterarbeit, CL lab, Friedrich Alexander Universität Erlangen Nürnberg (CLUE).

Lenders, W. & G. Willee (1986) *Linguistische Datenverarbeitung*, Westdeutscher Verlag, Opladen.

Leśniewski, S. (1929) *"Grundzüge eines neuen Systems der Grundlagen der Mathematik,"* Fundamenta Mathematicae 14:1–81, Warsaw.

Lewis, D. (1972) "General Semantics," in D. Davidson & G. Harman (eds.), 169–218.

Levison, S.C. (1983) *Pragmatics*, Cambridge University Press, Cambridge, England.

Lieb, H.-H. (1992) *"The Case for a new Structuralism,"* in H.-H. Lieb (ed.), Prospects for a New Structuralism. John Benjamin, Amsterdam-Philadelpia.

Linsky, L. (ed.) (1971) *Reference and Modality*, Oxford, England.

Locke, J. (1690) *An Essay Concerning Human Understanding*. In four books. Printed by Eliz. Holt for Thomas Basset, London. Reprinted in Taylor 1974.

Lorenz, O. & G. Schüller (1994) "Präsentation von LA-MORPH," *LDV-Forum*, Band 11.1:39–51. Reprinted in Hausser 1996.

Lorenz, O. (1996) *Automatische Wortformerkennung für das Deutsche im Rahmen von Malaga*, Magisterarbeit, CL lab, Friedrich Alexander Universität Erlangen Nürnberg (CLUE).

Łukasiewicz, J. (1935) *"Zur vollen dreiwertigen Aussagenlogik,"* Erkenntnis 5:176.

Lyons, J. (1968) *Introduction to Theoretical Linguistics*, Cambridge University Press, Cambridge, England.

MacWhinney, B. (1978) *The Acquisition of Morphophonology*, Monographs of the Society for Research in Child Development, No. 174, Vol. 43.

Maes, P. (ed.) (1990) *Designing Autonomous Agents*, M.I.T./Elsevier.

Mandelbrot, B., Apostel, L., Morf, A. (1957) "Linguistique Statistique Macroscopique," in J. Piaget (ed.), *Logique, Langage et Théorie de L'information*, Paris.

Marciszewski, W. (ed.) (1981) *Dictionary of Logic as applied in the study of language*, Nijhoff, The Hague.

Marr, D. (1982) *Vision*, W.H. Freeman and Company, New York.

Marshall, I. (1983) "Choice of Grammatical Word-Class without global Syntactic Analysis: Tagging Words in the LOB Corpus', *Computers and the Humanities*, Vol. 17:139–150.

Marshall, I. (1987) "Tag Selection Using Probabilistic Methods," in Garside et al. (eds.).

Matthews, P.H. (1972) *Inflectional Morphology. A Theoretical Study Based on Aspects of Latin Verb Conjugation*, Cambridge University Press, Cambridge, England.

Matthews, P.H. (1974) *Morphology. An Introduction to the Theory of Word Structure*, Cambridge Textbooks in Linguistics, Cambridge University Press, Cambridge, England.

McCawley, J.D. (1982) *Thirty Million Theories of Grammar*, The University of Chicago Press, Chicago, Illinois.

McClelland, D. (1991) "OCR: Teaching Your Mac to Read," *MACWORLD*, November 1991, 169–175.

Meier, H. (1964) *Deutsche Sprachstatistik*, Erster Band, Hildesheim.

Miller, G. (1956) "The Magical Number Seven, plus or minus two: some Limits on our Capacity for Processing Information," *Psychological Review*, Vol. 63.2:81–97.

Miller, G. & N. Chomsky (1963) "Finitary Models of Language Users," in *Handbook of Mathematical Psychology*, Vol. 2, R. Luce, R. Bush, & E. Galanter (eds.), John Wiley, New York.

Montague, R. (1974) *Formal Philosophy*, Yale University Press, New Haven.

Neisser, U. (1967) *Cognitive Psychology*, Appleton-Century-Crofts, New York.

NEWCAT = *NEWCAT: Parsing Natural Language Using Left-Associative Grammar*, Hausser 1986.

Newell, A. & H.A. Simon (1972) *Human Problem Solving*, Prentice-Hall, Englewood Cliffs, New Jersey.

Newell, A. & H.A. Simon (1975) "Computer Science as Empirical Inquiry: Symbols and Search," in J. Haugeland (ed.).

Norman, D.A. & D.E. Rumelhart (eds.) (1975) *Explorations in Cognition*, W.H. Freeman and Company, San Francisco.

Ogden, C.K. & I.A. Richards (1923) *The Meaning of Meaning*, Routledge and Kegan Paul LTD, London.

Oostdijk, N. (1988) "A Corpus Linguistic Approach to Linguistic Variation," in G. Dixon (ed.): *Literary and Linguistic Computing*, Vol. 3.1.

Oostdijk, N. & P. de Haan (1994) *Corpus-based Research into Language*, Editions Rodopi B.V., Amsterdam-Atlanta, GA.

Palmer, S. (1975) *Visual Perception and World Knowledge: Notes on a Model of Sensory-Cognitive Interaction*, in D.A. Norman & D.E. Rumelhart (eds.): *Explorations in Cognition*, 279–307.

Park, Yong-Woon (1987) *Koryo-sitay-sa (Ha)*, ilcisa, [A History of the Koryo Dynasty, Volume II/the last volume].

Pascal, B. (1951) *Pensées sur la Religion et sur quelques autre sujets*, Éditions du Luxembourg, Paris.

Paul, H. (1920) *Grundzüge der Grammatik*, Halle 1880, 5th ed., 1920.

Peirce, C.S. (1871) *Critical Review of Berkeley's Idealism*, North American Review, Vol. 93:449–472.

Peirce, W.S. (1931 – 1958) *Collected Papers of Charles Sanders Peirce*, edited by C. Hartshorne and P. Weiss, 6 vols. Harvard University Press, Cambridge, Massachusetts.

Peters, S. & Ritchie, R. (1973) "On the Generative Power of Transformational Grammar," *Information and Control* 18:483–501.

Picken, C. (1989) *The Translator Handbook*, 2nd ed., Aslib, London.

Piotrovskij, R.G., K. B. Bektaev, & A.A. Piotrovskaja (1985) *Mathematische Linguistik* (aus d. Russ. übers.), Brockmeyer, Bochum.

Piotrowski, M (1998) *NLP-Supported Full-text Retrieval*, Magisterarbeit, CL lab, Friedrich Alexander Universität Erlangen Nürnberg (CLUE).

Pollard, C. & I. Sag (1987) *Information-Based Syntax and Semantics*, Vol. I, Fundamentals. CSLI Lecture Notes 13. Stanford University.

Pollard, C. & I. Sag (1994) *Head-Driven Phrase Structure Grammar*, CSLI Stanford and The University of Chicago Press, Chicago, Illinois.

Post, E. (1936) "Finite Combinatory Processes — Formulation I," *Journal of Symbolic Logic*, I:103–105.

Putnam, H. (1975a) "The Meaning of 'Meaning'," reprinted in Putnam 1975b, 215–271.

Putnam, H. (1975b) *Mind, Language, and Reality 2*, Cambridge University Press, Cambridge, England.

Quine, W.v.O. (1960) *Word and Object*, M.I.T. Press, Cambridge, Massachusetts.

Rabin, M. O. & D. Scott (1959) "Finite Automata and their Decision Problems," *IBM J. Res.* 3.2:115–125.

Reddy, D.R., L.D. Erman, R.D. Fennell, & R.B. Neely (1973) "The Hearsay Speech Understanding System: An Example of the Recognition Process," *Proceedings of the Third International Joint Conference on Artificial Intelligence*, Stanford, CA.

Reibel, D.A. & S. A. Shane (eds.) (1969) *Modern Studies of English*, Prentice-Hall, Englewood Cliffs, New Jersey.

Rescher, N. (1969) *Many-valued Logic*, McGraw-Hill, New York.

Roget, P.M. (1977) *Roget's International Thesaurus*, Fourth Edition, Revised by Robert L. Chapman, Harper, and Row, New York.

Roukos, S. (1995) "Language Representation," in R. Cole (ed.), 35–41.

Rumelhart, D.E. (1977) *Human Information Processing*, John Wiley and Sons, New York.

Rumelhart, D.E, P. Smolensky, J. McClelland, & G.E. Hinton (1986) "Schemata and Sequential Thought Processes in PDP Models," in Rumelhart, D.E., J.L. McClelland et al. (eds.) *Parallel Distributed Processing*, Vol. II:7–57.

Russell, S.J. & P. Norvig (1995) *Artificial Intelligence, a modern approach*, Prentice Hall, Englewood Cliffs, New Jersey.

Salton, G. & M.J. McGill (1983) *Introduction to Modern Information Retrieval*, McGraw-Hill Book Company, New York.

Salton, G. (1989) *Automatic Text Processing: The Transformation, Analysis, and Retrieval of Information by Computer*, Addison-Wesley, Reading, Massachusetts.

Sapir, E. (1921) *Language, an Introduction to the Study of Speech*, Harvest Books – Harcourt, Brace, and World, New York.

Saussure, F. de (1967) *Grundfragen der Allgemeinen Sprachwissenschaft*, Walter de Gruyter, Berlin.

Saussure, F. de (1972) *Cours de linguistique générale*, Édition critique préparée par Tullio de Mauro, Éditions Payot, Paris.

Schank, R.C. & the Yale A.I. Project (1975) *SAM - A Story Understander*, Research Report No. 43, Yale University, Committee on Computer Science.

Schank, R.C. & R. Abelson (1977) *Scripts, Plans, Goals, and Understanding*, Lawrence Erlbaum, Hillsdale, New Jersey.

Schnelle, H. (1991) *Die Natur der Sprache*, Walter de Gruyter, Berlin.

Schnelle, H. (1996) "Beyond New Structuralism," in R. Sackmann (ed.): *Theoretical Linguistics and Grammatical Description*, John Benjamins, Amsterdam/-Philadelphia.

Schwartz, J.T. (ed.) (1967) *Mathematical Aspects of Computer Science*, Proc. Symposia in Applied Mathematics, 19. Mathematical Society, Providence, R.I.

Schwarz, R. (1996) *Dynamische Aktivierung domänenspezifischer Teillexika*, Magisterarbeit, CL lab, Friedrich Alexander Universität Erlangen Nürnberg (CLUE).

SCG = *Surface Compositional Grammar*, Hausser 1984a.

Scott, D. (1982) "Domains for denotational semantics," *Proceedings of the ICALP '82*, Springer-Verlag, Lecture Notes in Computer Science Vol. 140.

Scott, D. & C. Strachey (1971) "Toward a Mathematical Semantics of Computer Languages," Technical Monograph PRG-6, Oxford University Computing Laboratory, Programming Research Group, 45 Branbury Road, Oxford, England.

Searle, J.R. (1969) *Speech Acts*, Cambridge University Press, Cambridge, England.

Searle, J.R. (1992) *The Rediscovery of the Mind*, M.I.T. Press, Cambridge, Massachusetts.

Searle, J.R. & D. Vanderveken (1985) *Foundations of Illocutionary Logic*, Cambridge University Press, Cambridge, England.

Sells, P. (1985) *Lectures on Contemporary Syntactic Theory*, CSLI Lecture Notes Number 3, Stanford.

Shannon, C.E. & W. Weaver (1949) *The Mathematical Theory of Communication*, University of Illinois Press, Urbana, Illinois.

Sharman, R. (1990) *Hidden Markov model methods for word tagging*, Report 214, IBM UK Scientific Centre, Winchester.

Shieber, S. (1983) "Direct parsing of ID/LP grammars," *Linguistics and Philosophy* 7.2:135–154.

Shieber, S. (1985) "Evidence against the Non-Contextfreeness of Natural Language," *Linguistics and Philosophy* 8:333–343.

Shieber, S., S. Stucky, H. Uszkoreit, & J. Robinson (1983) "Formal Constraints on Metarules," *Proceedings of the 21st Annual Meeting of the Association for Computational Linguistics*, Cambridge, Massachusetts.

Sinclair, J. (1991) *Corpus, Concordance, Collocation*, Oxford Univ. Press, Oxford, England.

Skinnner, B.F. (1957) *Verbal Behavior*, Appleton-Century-Crofts, New York.

Sperling, G. (1960) "The Information available in Brief Visual Processing," *Psychological Monographs*, 11, Whole No. 498.

St.Laurent, S. (1998) *XML. A Primer*, M.I.T. Press, Cambridge, Massachusetts.

Stemberger, P.J. & B. MacWhinney (1986) "Frequency and Lexical Storage of Regularly Inflected Forms," *Memory and Cognition*, 14.1:17–26.

Stoy, J. E. (1977) *Denotational Semantics. The Scott-Strachey Approach to Programming Language Theory.* Cambridge, England.

Stoyan, H. & G. Görz (1984) *LISP, eine Einführung in die Programmierung*, Springer-Verlag, Berlin.

Stubert, B. (1993) *"Einordnung der Familie der C-Sprachen zwischen die kontextfreien und die kontextsensitiven Sprachen,"* CLUE-betreute Studienarbeit der Informatik, Friedrich Alexander Universität Erlangen Nürnberg.

Suppe, F. (ed.) (1977) *The Structure of Scientific Theories*, University of Illinois Press, Urbana, Illinois.

Tarr, M.J. & H.H. Bülthoff (eds.) (1998) *Image-based object recognition in man, monkey and machine*, special issue of *Cognition*, Vol. 67.1&2:1–208.

Tarski, A. (1935) "Der Wahrheitsbegriff in den Formalisierten Sprachen," *Studia Philosophica*, Vol. I:262–405.

Tarski, A. (1944) "The Semantic Concept of Truth," *Philosophy and Phenomenological Research*, 4:341–375.

Taylor, R. (ed.) (1974) *The Empiricists: Locke, Berkeley, Hume*, Anchor Books, Doubleday, Garden City, New York.

Tesnière, L. (1959) *Entwurf einer strukturalen Syntax*, Paris.

Thiel, C. (1995) *Philosophie und Mathematik*, Wissenschaftliche Buchgesellschaft, Darmstadt.

Tomita, M. (1986) *Efficient Parsing for Natural Languages*, Kluwer, Dordrecht.

Turing, A. M. (1950) "Computing Machinery and Intelligence," *Mind*, 59: 433–460.

Uszkoreit, H. & S. Peters (1986) "On Some Formal Properties of Metarules," Report CSLI-85-43, Center for the Study of Language and Information, Stanford University, Stanford.

Valiant, L.G. (1975) "General Context-Free Recognition in less than Cubic Time," *Journal of Computer and System Sciences* 10:2, 308–315.

Wahrig, G. (1986/89) *Deutsches Wörterbuch*, Mosaik Verlag, München.

Wahlster, W. (1993) "Verbmobil, Translation of Face-to-Face Dialogs," *Proceedings of the Fourth Machine Translation Summit*, Kobe, Japan, 127–135.

Wall, L. & R.L. Schwartz (1990) *Programming Perl*, O'Reilly & Associates, Sebastopol, CA.

Webber, B. & N. Nilsson (eds.) (1981) *Readings in Artificial Intelligence*, Morgan Kaufmann, Los Altos, CA.

Wheeler, P. & V. Lawson (1982) "Computing Ahead of the Linguists," *Ambassador International*, 21–22.

Weizenbaum, J. (1965) "ELIZA – A Computer Program for the Study of Natural Language Communications between Man and Machine," *Comm. ACM*, Vol. 9.1 (11).

Wetzel, C. (1996) *Erstellung einer Morphologie für Italienisch in Malaga*, CLUE-betreute Studienarbeit der Informatik, Friedrich Alexander Universität Erlangen Nürnberg.

Wexelblat, A. (ed.) (1993) *Virtual Reality, Applications and Explorations*, Academic Press Professional, Boston.

Weyhrauch, R. (1980) "Prolegomena to a Formal Theory of Mechanical Reasoning," *Artificial Intelligence*, Reprint in Webber & Nilsson (eds.) 1981.

Winograd, T. (1972) *Understanding Natural Language*, Academic Press, Harcourt Brace Jovanovich, San Diego, New York.

Winograd, T. (1983) *Language as a Cognitive Process*, Addison-Wesley, Reading, Massachusetts.

Winston, P.H. & B.K. Horn (1984) *LISP,* 2nd edition, Addison-Wesley, Reading, Massachusetts.

Wittgenstein, L. (1921) "Logisch-philosophische Abhandlung," *Annalen der Naturphilosophie* 14:185–262.

Wittgenstein, L. (1953) *Philosophical Investigations*, Blackwell, Oxford, England.

Wloka, D.W. (1992) *Roboter-Systeme 1*, Springer-Verlag, Berlin.

Younger, D.H. (1967) *"Recognition and Parsing of Context-Free Languages in Time n^3,"* Information and Control 10.2:189–208.

Zadeh, L. (1971) "Quantitative Fuzzy Semantics," *Information Science* 3:159–176.

Zeevat, H., E. Klein, & J. Calder (1987) *"An Introduction to Unification Categorial Grammar,"* in J.N. Haddock, E. Klein, & G. Morris (eds.), Edinburgh Working Papers in Cognitive Science, Vol. 1: Categorial Grammar, Unification Grammar, and Parsing. Center for Cognitive Science, University of Edinburgh.

Zierl, M. (1997) *Ein System zur effizienten Korpusspeicherung und -abfrage*, Magisterarbeit, CL lab, Friedrich Alexander Universität Erlangen Nürnberg (CLUE).

Zipf, G.K. (1932) *Selected Studies of the Principle of Relative Frequency in Language*, Oxford, England.

Zipf, G.K. (1935) *The Psycho-Biology of Language*, Boston, Massachusetts.

Zipf, G. K. (1949) *Human Behavior and the Principle of least Effort*, Cambridge, Massachusetts.

Zue, V., R. Cole, & W. Ward (1995) "Speech Recognition," in R. Cole (eds.) 1998.

Name Index

Subject Index

Printed in Italy by Legoprint S.p.A., Lavis (Trento)